Self-Assembling Systems

Self-Assembling Systems

Theory and Simulation

Edited by

Li-Tang Yan

Department of Chemical Engineering,
Tsinghua University, China

WILEY

This edition first published 2017

Registered Office
John Wiley & Sons, Ltd, The Atrium, Southern Gate, Chichester, West Sussex, PO19 8SQ, United Kingdom

For details of our global editorial offices, for customer services and for information about how to apply for permission to reuse the copyright material in this book please see our website at www.wiley.com.

Library of Congress Cataloging-in-Publication Data

Names: Yan, Li-Tang, 1978–
Title: Self-assembling systems : theory and simulation / Li-Tang Yan.
Description: Chichester, West Sussex : John Wiley & Sons, Inc., 2016. |
Includes bibliographical references and index.
Identifiers: LCCN 2016024926 (print) | LCCN 2016028669 (ebook) | ISBN 9781119113140 (cloth) | ISBN 9781119113157 (pdf) | ISBN 9781119113164 (epub)
Subjects: LCSH: Self-assembly (Chemistry) | Colloidal crystals. | Polymers.
Classification: LCC QD475.Y36 2016 (print) | LCC QD475 (ebook) | DDC 547/.2–dc23
LC record available at https://lccn.loc.gov/2016024926

A catalogue record for this book is available from the British Library.

Set in 10/12pt WarnockPro by SPi Global, Chennai, India

10 9 8 7 6 5 4 3 2 1

Contents

List of Contributors

Saientan Bag
Department of Physics
Indian Institute of Science
Bangalore
India

Anna C Balazs
Department of Chemical and Petroleum Engineering
University of Pittsburgh
USA

Hima Bindu Kolli
Department of Chemistry
University of Oslo
Norway

Jeff Z. Y. Chen
Department of Physics and Astronomy
University of Waterloo
Canada

Giorgio Cinacchi
Departamento de Física Teórica de la Materia Condensada
Instituto de Física de la Materia Condensada (IFIMAC), and Instituto de Ciencias de Materiales "Nicolás Cabreras"
Universidad Autónoma de Madrid
Spain

Alberta Ferrarini
Dipartimento di Scienze Chimiche
Università di Padova
Italy

Elisa Frezza
BMSSI, UMR 5086 CNRS / Institut de Biologie et Chimie des Protéines
Université de Lyon I
France

Achille Giacometti
Dipartimento di Scienze Molecolari e Nanosistemi
Università Ca' Foscari di Venezia
Italy

Zihan Huang
Department of Chemical Engineering
Tsinghua University
Beijing
China

Ying Jiang
School of Chemistry and Environment
Beihang University
Beijing
China

Jehoon Kim
Department of Chemical Engineering
Massachusetts Institute of Technology
Cambridge
USA

Olga Kuksenok
Department of Materials Science and Engineering Clemson University
USA

Yves Lansac
GREMAN, Université François Rabelais
CNRS UMR 7347
Tours
France

Weihua Li
Department of Macromolecular Science
Fudan University
Shanghai
China

Ye Li
State Key Laboratory of Organic–
Inorganic Composites
Beijing University of Chemical
Technology
China

Zhan-Wei Li
Changchun Institute of Applied
Chemistry
Chinese Academy of Sciences
China

Jiaping Lin
School of Materials Science and
Engineering
East China University of Science and
Technology
Shanghai
China

Zhong-Yuan Lu
Institute of Theoretical Chemistry
Jilin University
Changchun
China

Prabal K. Maiti
Deparment of Physics
Indian Institute of Science
Bangalore
India

Ran Ni
School of Chemical and Biomedical
Engineering
Nanyang Technological University
Singapore

Suman Saurabh
Department of Physics
Indian Institute of Science
Bangalore
India

An-Chang Shi
Department of Physics and Astronomy
McMaster University
Hamilton
Canada

Stephen C. Snow
Department of Chemical and Petroleum
Engineering
University of Pittsburgh
USA

Zhao-Yan Sun
Changchun Institute of Applied
Chemistry
Chinese Academy of Sciences
China

Falin Tian
State Key Laboratory of Organic–
Inorganic Composites
Beijing University of Chemical
Technology
China

Jianzhong Wu
Department of Chemical and
Environmental Engineering
University of California
Riverside
USA

Li-Tang Yan
Department of Chemical Engineering
Tsinghua University
Beijing
China

Makoto Yoneya
Advanced Industrial Science and
Technology (AIST)
Higashi
Tsukuba
Japan

Xin Yong
Department of Mechanical Engineering
State University of New York at
Binghamton
USA

Tongtao Yue
State Key Laboratory of Heavy Oil
Processing
China University of Petroleum
(East China)
Qingdao
China

Liangshun Zhang
School of Materials Science and
Engineering
East China University of Science and
Technology
Shanghai
China

Xianren Zhang
State Key Laboratory of Organic–
Inorganic Composites
Beijing University of Chemical
Technology
China

Preface

Self-assembly is one of the most prominent and promising candidates for the development of novel materials with high performance. However, in order to successfully exploit them in technological applications and to ensure efficient scale-up, an in-depth understanding of structure formation, kinetic mechanism and structure–property relationship is required. Theoretical studies and tailored computer simulations offer unique approaches to investigate the evolution and formation of structures as well as to determine structure–property relationships in self-assembled systems. In light of the growing interest in the design and synthesis of building blocks for the self-assembly of complex structures, this book looks at the field of self-assembly from a theoretical perspective, highlighting the importance of computational studies to support the design of new self-assembling materials with useful structural and properties. The aim of this book is to bring together leading scientists working on the issues in theoretical and simulation research of self-assembly, and to offer researchers and graduate students an in-depth review of the most recent developments in this field.

The book begins with discussions on the fundamental principles of self-assembly. In addition to the general aspect and emerging concepts presented in the first chapter, Balazs and coworkers introduce in Chapter 2 developing hybrid modeling methods to simulate self-assembly in polymer nanocomposites, focusing on new computational approaches to model two types of radical polymerization, namely free radical polymerization and atom transfer radical polymerization, in the framework of dissipative particle dynamics. Helical biopolymers and colloidal particles could exhibit liquid crystal phases at high densities. These phases are often tacitly assumed to be the same as those occurring in systems of rod-like particles. To explore the effect of self-assembly of helical polymers and colloids, and in particular to discover whether there is anything special just determined by the helical shape, Giacometti and coworkers have undertaken a comprehensive investigation of the phase behavior of hard helices, interacting through purely steric repulsions, using Monte Carlo simulations and an extension of Onsager theory, a density functional theory that was originally proposed to explain the onset of nematic ordering in a system of hard rods. The details regarding this interesting work are presented in Chapter 3.

The content then moves on to computational tools and techniques for the simulation of self-assembling systems. Two state-of-the-art and important methods or models are emphasized in this section. The first is self-consistent field theory (SCFT) of self-assembling multiblock copolymers. In Chapter 4, Li and Shi introduce the recent applications of different methods of SCFT to the study of the self-assembly of block copolymers. In particular, they focus on a review of recent progress in the study of the self-assembly of multiblock copolymers, alongside some specific details of the numerical techniques. Undoubtedly, this chapter can serve as a very useful reference source for readers who are interested in applying SCFT to multiblock copolymer systems. The second is the simulation model of soft anisotropic particles developed by Li, Sun and Lu. In Chapter 5, they introduce two major kinds of general and effective mesoscale models to describe the aggregation behavior of soft Janus and patchy particles: the soft Janus particle model and soft patchy particle model.

In the last section, the latest research advances of different emerging topics are reviewed and consolidated. The first topic is the application of theory and simulation to self-assembly in biological systems, described in three chapters. In Chapter 6, Kim and Wu analyze the formation of the hepatitis B virus (HBV) core particles, viral maturation mechanisms and essential ingredients of antiviral strategies from a thermodynamic perspective. Importantly, statistical mechanical models are used by them to quantify the thermodynamics of the HBV genome packaging, the dynamic structures for the flexible domains of the viral capsids during maturation, and the stability of the core viral particles before and after RNA encapsidation. In biological systems, hydrogen bonds are known to play a crucial role in the structure formation of a DNA helix, protein alpha-helix and beta-sheet, among others. Compared to this hydrogen-bond-directed self-assembly, coordination-bond-directed self-assembly is less pronounced in biological systems, but has been extensively studied recently in synthetic chemistry. In Chapter 7, Yoneya introduces MD simulation studies of metal-ligand self-assembly, emphasizing the useful applications of their modeling method. This method fills the gap between simulations and real reaction systems. Studying the interaction between biomembrane and nanoparticles is essential for understanding the nature of cellular life and for the safe application of nanoparticles. In Chapter 8, Zhang and coworkers summarize recent developments of computer simulation studies on the nanoparticle–membrane interaction, and particularly stress the results from dissipative particle dynamics simulation. They particularly concentrate on membrane-mediated interaction, internalization pathways and cooperative effect.

The second topic in the last section is the application and development of theoretical approaches to self-assembling systems. In Chapter 9, Jiang and Chen introduce theories for polymer melts consisting of rod-coil polymers. An interesting model based on the wormlike chain model and developed by them is emphasized, which can describe the crossover of the polymer chain from the rod limit to the flexible limit. In the other chapter (Chapter 10), Zhang and Lin describe recent progress in design strategies of structural hierarchy from the view of computational modeling, concentrating on the superstructures of polymer-based systems via the multistep process of self-assembly. The following topics are involved: (1) hierarchical nanostructures self-assembled from block copolymer melts, (2) multicompartment aggregates from block copolymer solutions, and (3) hierarchically ordered nanocomposites formed by organic–inorganic systems. They finally conclude with a brief but very useful outlook on challenges and perspectives.

The third topic in the last section turns to the simulation studies of the self-assemblies in colloidal systems. In Chapter 11, Ni reviews recent progress in nucleation studies in colloidal systems by using computer simulations, which includes the classical description of nucleation, i.e. classic nucleation theory, state-of-the-art simulation methods for studying nucleation, and the new nucleation phenomena observed violating the classic understanding of nucleation as well as the important future directions of nucleation study that have not yet been explored well. The liquid crystalline phase is ubiquitous in nature and is exhibited by a variety of systems like surfactants, nucleic acids, lipid molecules, etc. Simulation methods employed to study liquid crystalline phases range from molecular dynamics, Monte Carlo, Brownian dynamics to dissipative particle dynamics. Maiti and coworkers present a comprehensive view of the various methods used in simulating self-assembly phenomena in liquid crystals in Chapter 12.

Written by specialists in various disciplines, such as polymers, soft matter, nanoparticle self-assembly, biophysics, and so on, this book provides comprehensive knowledge in this emerging and important field, although covering all aspects of the field is impossible. From the editor's point of view, I am very pleased and honored with the principles, methods and outlook conceived in each chapter, and would also like to express my most heartfelt thanks to all contributors who delivered truly excellent topics of research and took the time to write up detailed and pedagogical chapters.

February 2016 *Li-Tang Yan*

1

Theoretical Studies and Tailored Computer Simulations in Self-Assembling Systems: A General Aspect

Zihan Huang and Li-Tang Yan

Key Laboratory of Advanced Materials (MOE), Department of Chemical Engineering, Tsinghua University, Beijing, China

1.1 Introduction

Self-assembly—a governing principle by which materials form—is the autonomous organization of matter into ordered arrangements [1, 2]. It is typically associated with thermodynamic equilibrium, the organized structures being characterized by a minimum in the system's free energy, although this definition is too broad. Self-assembling processes are ubiquitous in nature, ranging, for example, from the opalescent inner surface of the abalone shell to the internal compartments of a living cell [3]. By these processes, nanoparticles or other discrete components spontaneously organize due to direct specific interactions and/or indirectly, through their environment. Self-assembly is one of the few practical strategies for making ensembles of nanostructures. It will therefore be an essential part of nanotechnology. Self-assembly is also common to many dynamic, multicomponent systems, from smart materials and self-healing structures to netted sensors and computer networks. In the world of biology, living cells self-assemble, and understanding life will therefore require understanding self-assembly. The cell also offers countless examples of functional self-assembly that stimulate the design of non-living systems [4, 5].

Self-assembly reflects information coded (as shape, surface properties, charge, polarizability, magnetic dipole, mass, etc.) in individual components; these characters determine the interaction among them. The design of building blocks that organize themselves into desired structure and functions is the key to applications of self-assembly [2]. Much of materials science and soft condensed-matter physics in the past century involved the study of self-assembly of fundamental building blocks (typically atoms, molecules, macromolecules, and colloidal particles) into bulk thermodynamic phases [6]. Today, the extent to which these building blocks can be engineered has undergone a quantum leap. Tailor-made, submicrometer particles will be the building blocks of a new generation of nanostructured materials with unique physical properties [7–11]. These new building blocks will be the "atoms" and "molecules" of tomorrow's materials, self-assembling into novel structures made possible solely by their unique design [2]. For example, patchy particles consisting of various

Self-Assembling Systems: Theory and Simulation, First Edition. Edited by Li-Tang Yan.

compartments of different chemistry or polarity are ideal building blocks of potentially complex shapes with competing interactions that expand the range of self-assembled structures beyond those exhibited by traditional amphiphiles such as surfactants and block copolymers [9]. By controlling the placement of "sticky" patches on the particles, assemblies can be made that mimic atomic bonding in molecules [8]. This greatly expands the range of structures that can be assembled from small components. Extension of the principles to particles of alternative compositions (such as those made from noble metals, semiconductors or oxides) will allow optical, electronic and catalytic materials to be coupled in previously impossible architectures that have potentially new emergent properties. Playing tricks with designer "atoms" also includes the shape of the building blocks [12]. For instance, the local curvature of dumbbell-shaped nanoparticles can be harnessed to control the ionization state of a molecular layer adsorbed on their surfaces and the self-assembly patterns of the particles [13].

Understanding the relation between building blocks and their assemblies is essential for materials design because physical properties depend intimately on structure, which however poses many challenges if considering complex thermodynamic and kinetic behaviors involved in the assembling processes. Indeed, a priori prediction of hierarchically assembled structures from a desired building block requires an in-depth understanding of the delicate balance between entropic and enthalpic interactions [14–17]. Central to this issue is exploring entropy-driven structural organization, because entropy keeps springing non-intuitive findings in the manipulation of the self-assembly of nanoparticles and the structural formation of soft matter systems [14]. On the other hand, directed self-assembly using a template or an external field may also give rise to novel ordered non-equilibrium structures, free from the constraints of entropy maximization, and hence these systems can "reside" in a state of local equilibrium within the global free energy with low entropy states often characterized by complex spatial or coherent spatiotemporal organization [18]. In this case, identifying the possible structures at metastable states is important for controlling the formation of structures or patterns. Not surprisingly, many advances have been made in theoretical models and simulation approaches to predict and analyze structures, dynamics and properties of self-assembling systems; computer simulations offer a unique approach to identify and separate individual contributions to the phenomenon or process of interest [19, 20].

However, the theoretical and computational research of self-assembling systems is far from trivial. These many-body systems cover variations in relevant time and length scales over many orders of magnitude. The assembled structures and macroscopic properties of materials are ultimately to be deduced from the dynamics of the microscopic, molecular level, implicating a lot of demand for new simulation techniques and theoretical approaches. From the computational point of view, the key is to develop methods capable of reaching time and length scales much larger than those accessible by brute force computer simulations on the atomic level [21]. The feat is not a simple one, since it requires a major effort over a wide range of activities, including the development of coarse graining techniques, novel simulation methods and ways to link the different regimes to each other. Even with these challenges, theory and simulations have proven invaluable and indispensable in studies of self-assembling systems, including applications in numerous directions such as development and examination of new principles, predictive science and computer design of complex building blocks, suggesting guidelines of programmable assembly, and exploring entropy interaction in various assembling systems, etc. The purpose of this chapter is therefore to introduce the general aspects of the development and applications of theoretical approaches and computational modeling in self assembling systems, focusing on basic and emerging principles.

1.2 Emerging Self-Assembling Principles

1.2.1 Predictive Science and Rational Design of Complex Building Blocks

Predicting structure from the attributes of a material's building blocks remains a challenge and the central goal for materials science. Here we introduce the rational design and predictive science of two emerging and important building blocks for superstructure construction through self-assembly, that is, polyhedral particles and particles that can self-assemble into helical structures with chirality.

At present, a major focus in material science is to engineer particles with anisotropic shapes and interaction fields that can be self-assembled into complex target structures [1, 2]. Assemblies of anisotropic particles undergo order–disorder transitions involving changes in both translational and rotational degrees of freedom and can lead to phases with partial structural order or "mesophases" [22, 23] such as crystals, plastic crystals and liquid crystals. These ordered assemblies have distinctive electronic, optical and dynamical properties and are highly desirable for fabrication of advanced electronic, photonic and rheological devices [24]. Although numerous theoretical [25, 26] and experimental [27, 28] studies on mesophase behavior of particles with anisotropic shapes have been reported, a roadmap marking out the most probable mesophases that could be formed by constituent particles with particular geometrical features remains incomplete. Exploring such relations will translate into a deeper understanding of the phase behavior of colloidal systems with different particle shapes. The simulation prediction of a dodecagonal quasicrystal with tetrahedra demonstrated the unexpected complexity that could be achieved for particles solely with hard interactions [29]. Escobedo and Agarwal [30] carried out detailed Monte Carlo simulations of six convex space-filling polyhedrons to demonstrate that translational and orientational excluded-volume fields encoded in particles with anisotropic shapes can lead to purely entropy-driven assembly of morphologies with specific order and symmetry. Their simulations reveal the formation of various new liquid-crystalline and plastic-crystalline phases at intermediate volume fractions. They further propose simple guidelines for predicting phase behavior of polyhedral particles: high rotational symmetry is in general conducive to mesophase formation, with low anisotropy favoring plastic-solid behavior and intermediate anisotropy (or high uniaxial anisotropy) favoring liquid-crystalline behavior.

Recently, a more refined structure prediction of polyhedral particles has been attained by Glotzer et al. [31] through investigating 145 convex polyhedra whose assembly arises solely from their anisotropic shape. Their simulations demonstrate that from simple measures of particle shape and local order in the fluid, the assembly of a given shape into a liquid crystal, plastic crystal or crystal can be predicted. Two important shape parameters that were revealed to predict the general category of ordered structure are the coordination number and the isoperimetric quotient (Figure 1.1). Although still unable to predict a specific structure, their results provide an important step toward a predictive

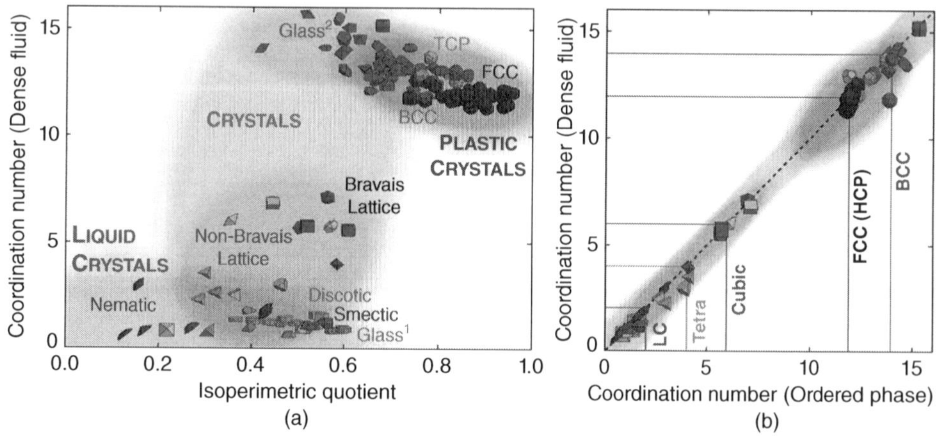

Figure 1.1 (a) The coordination number in the fluid phase, CN_f, is correlated to the isoperimetric quotient (*IQ*) of the polyhedron. Here, *IQ* is a scalar parameter for the sphericity of the shape and coordination number is a measure of the degree of local order. Data points are drawn as small polyhedra, which are grouped according to the assemblies they form. (b) Polyhedra have, in most cases, nearly identical coordination numbers in the ordered phase (CN_o) and the fluid phase (CN_f) close to the ordering transition. Because of this strong correlation, combining CN_f and *IQ* allows for prediction of the assembly category expected for most cases. This figure is reproduced from Ref. [31]. Copyright permission from American Association for the Advancement of Science (2012). *(See insert for color representation of the figure.).*

science of nanoparticle and colloidal assembly, which will be necessary to guide experiments with families of polyhedrally shaped particles that are now becoming available.

Rational design of building blocks for self-assembly can be significantly facilitated if the final structure can be predicted as a function of the building block parameters [30–33]. The interaction fields and anisotropic shapes encoded in the building blocks allow potential approaches for such a prediction. However, considering the complex energy landscape and kinetic pathway, the predictive science of sophisticated supracolloidal structures remains a key challenge. In particular, although numerous studies on the predictive self-assembly of anisotropic particles have been reported [30, 31], a priori prediction of helical supracolloidal structures from rationally designed building blocks still lacks a general roadmap and has yet to be demonstrated. Helical structure represents the principal element responsible for the property of chirality. Control over chirality at nano- and mesoscales is rapidly becoming a goal of great scientific interest because such unique architectures will allow optical, plasmonic and catalysis materials to have distinctively emergent properties [34–36]. This aspect is particularly relevant for photonic applications [37], where the optical properties are significantly influenced by the periodically arranged unit cells. Molecular scaffolds such as DNA origami can enable the high-yield production of superstructures that contain nanoparticles arranged in nanometer-scale helices [38]. However, large-scale fabrication of these scaffolds poses a significant hurdle for many practical applications.

The notion of a directional interaction field encoded by the surface patches suggests that patchy particles can be used to generate supracolloidal helices without fixed templates that offer limited controllability and may penalize the properties of particle assemblies [39, 40]. However, the ability to design and control supracolloidal helices

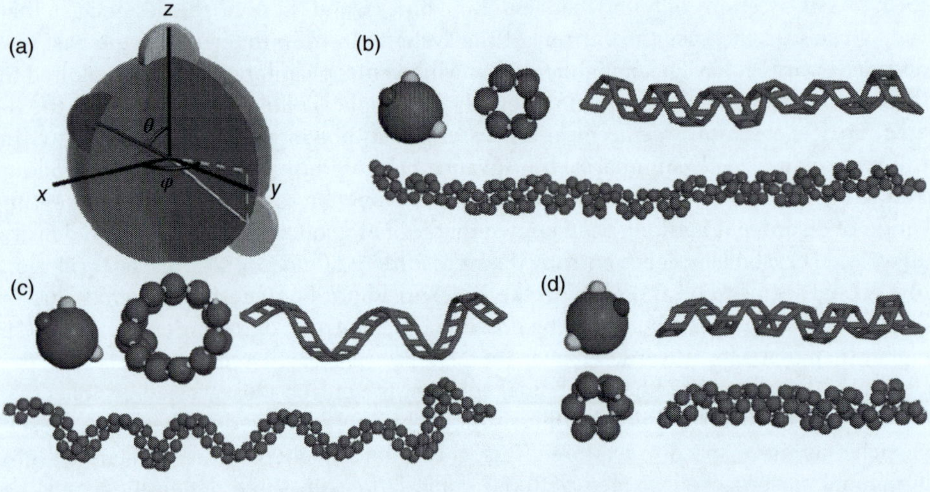

Figure 1.2 Tunable helical supracolloidal structures from a facile particle model. (a) Cartoon of patchy particle model used in the simulations. The top patch is a self-complementary patch while the other two patches are a pair of complementary patches. The relative directions of these patches are determined by angles θ and φ. (b) Right-handed double-stranded helix formed from patchy particles with patch direction of $\theta = 60°$ and $\varphi = 120°$. (c) Right-handed double-stranded helix with larger pitch and radius than those in (b), where the parameters are set as $\theta = 60°$, $\varphi = 150°$. (d) Left-handed double-stranded helix formed from patchy particles with patch direction of $\theta = 60°$ and $\varphi = -120°$. In (b)–(d), the building blocks and the top and side views of the helically supracolloidal structures and their geometrical representation are shown. This figure is reproduced from Ref. 41.Copyright Permission from Nature Publishing Group (2014).

assembled from patchy particles or colloids is limited by the absence of a general prediction principle. Challenges include exploring a facile design rule of patchy particles for helical self-assembly and further establishing a critical prediction principle for such supracolloidal architectures. Recently, inspired by biological helices, Guo et al. [41] showed that the rational design of patchy arrangement and interaction can drive patchy particles to self-assemble into biomolecular mimetic supracolloidal helices. They further derived a facile design rule for encoding the target supracolloidal helices, thus opening the doors to the predictive science of these supracolloidal architectures (Figure 1.2). It is also found that kinetics and reaction pathway during the formation of supracolloidal helices offer a unique way to study supramolecular polymerization, and that well-controlled supracolloidal helices can exhibit tailorable circular dichroism effects at visible wavelengths.

1.2.2 Entropy-Driven Ordering and Self-Assembly

Precise control of self-assembled structures remains a challenge because the structural architectures are governed by an intricate balance of entropic and enthalpic interactions. Central to this issue is exploring entropy-driven structural organization because entropy keeps springing non-intuitive findings in the manipulation of the self-assembly and the structural formation of soft matter systems [14–16, 42]. In fact, understanding entropic contributions to ordering transitions is essential for the design of self-assembling systems with hierarchical structures. Various unexpected structures can form by self-assembly of tailor-made building blocks, and these can be designed so that such structures increase the entropy of the system [43–45]. Indeed, over the past few decades examples have been highlighted in which entropic interactions are exploited to direct self-assemblies, such as self-assembly of complex colloids, shape-entropy mediated particle assembly, hierarchical self-assembly in polymer nanocomposites, etc. In this section, we briefly summarize the advancement of these emerging topics as follow.

Entropy can be a subtle and elusive concept. Although in some sense it is the quantification of disorder, it has long been known that for hard colloidal spheres at high density an ordered crystal has higher entropy than a disordered fluid [46, 47]. Yet the formation of a crystal seems to be at odds with the widespread notion of entropy as a measure of disorder. Actually, this would only be possible if the entropy of the ordered phase were higher than that of the disordered phase at the same density and temperature [14]. In the system of densely packed hard colloidal spheres, the entropy loss caused by collective ordering of the mean atomic positions is more than compensated by the fact that each particle has more space to explore. Thus, the colloidal spheres must crystallize [into hexagonal close-packed or face-centered cubic lattices in three dimensions, and the hexagonal lattice in two dimensions (2D)] to gain accessible volume and thus entropy.

With patchy particles, however, this argument is partially reversed. Granick et al. [48] show that, through decoration of the particle surfaces with a simple pattern of hydrophobic domains, triblock Janus colloidal spheres can be induced to self-assemble into an open kagome structure, contrasting with previously known close-packed periodic arrangements of spheres. In this case, a crucial new factor is the "rotational rattle room," defined by the solid angle that each particle can rotate before any of its sticky patches lose contact with those of its neighbors (Figure 1.3a) [15]. A lost contact would represent a "broken bond," which is effectively forbidden by its relatively high energy penalty. Therefore, the rotational entropy of a patchy particle clearly depends on its environment, in contrast to what occurs for non-patchy hard spheres, which can always undergo complete rotation and for which entropy per particle is always a constant. Recently, Mao et al. [49] showed that under well-motivated limiting assumptions the rotational-entropy contribution is calculable analytically. They constructed an effective description of the interactions in a system of patchy particles with fluctuating lattice positions, and compared it with experimental data on the mode structure of lattice vibrations for triblock Janus spheres. Their calculations determined that in two dimensions the kagome lattice is more stable than the close-packed hexagonal lattice. Both lattices have four bonds per particle and thus the same energy, but at close packing the latter has two additional non-bonded contacts (Figure 1.3b). In a hexagonal crystal of the same density as the kagome lattice (whose maximum density is lower), patchy particles have more room to rattle translationally, but the combined entropy of rotation and vibration is reduced. For densities that are not too high, the latter effect dominates, and the kagome lattice is stabilized.

If the particles are non-spherical, the optimal packing geometry is not always clear. The systematic study of families of idealized colloidal and nanoscale systems by computer simulation has produced overwhelming evidence that shape is implicated in the self-assembly of model systems of particles [50, 51]. In these model systems, the only intrinsic forces between particles are steric, and the entropic effects of shape (which we term "shape entropy") can be isolated. Damasceno et al. [31] have recently shown that arbitrarily shaped hard polyhedra display considerable predictability in their dense

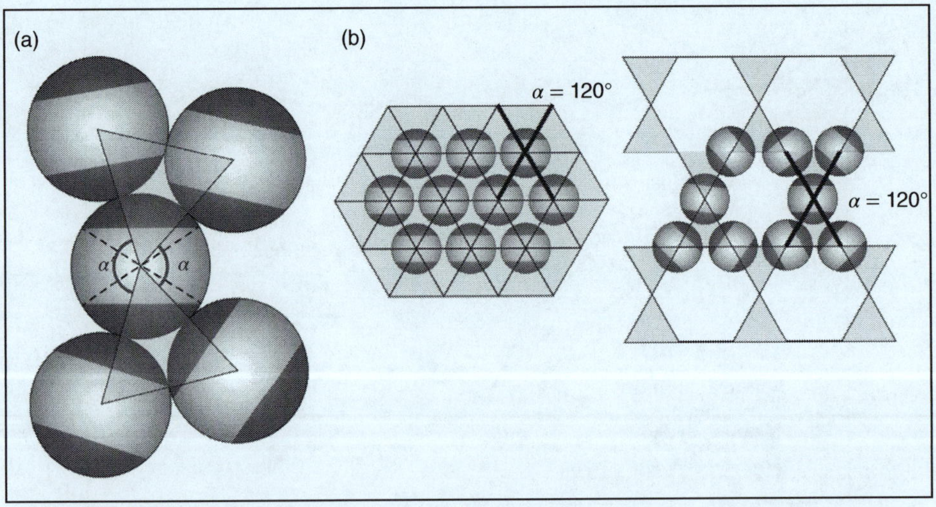

Figure 1.3 The rotational "rattle room" of patchy colloids. (a) Bonded contacts between triblock Janus spheres are maintained as long as the edges of the overlaid triangles cross the attractive dark gray patches. For a fixed particle configuration, the angular excursions that a particle can make without breaking any of its four contacts are independent of the corresponding excursions of neighboring particles. In two dimensions such a rotational entropy can therefore be calculated one particle at a time, and depends only on the bond angles α and α'. (b) The hexagonal (left) and kagome (right) lattices have the same number of bonded contacts and thus the same energy per particle. However, for moderate densities, particles in a kagome lattice have less room to rattle by translation, yet more entropy from rotational and vibrational motions, than particles in a hexagonal lattice at the same density. The latter effect dominates, which favors the formation of the kagome crystal over the hexagonal. This figure is reproduced from Ref. [15]. Copyright permission from Nature Publishing Group (2013).

packing arrangements. Depending on their shape, some polyhedra will form ordered crystals, while others form liquid crystals, "plastic" crystals in which the particles rotate freely, or disordered glasses. The entropic forces promoting the dense phases here heed details of particle shape: in particular, there is often a propensity for particles to sit face to face with facets aligned, creating directional preferences that can lead to ordered self-assembly. However, the origin and strength of these forces are unclear.

Using computer simulations of polyhedral packings, Anders et al. [52] clarify the concept of a shape-dependent directional entropic force, showing that it can be given a rigorous description based on the role of shape and faceting in maximizing the entropy of dense packings. As such, the force is an emergent property of local particle configurations, and it typically manifests itself for polyhedra as a repulsion between corners and an attraction between faces. For these systems the force is typically of the order of a few kT (where k is Boltzmann's constant and T is temperature), making it comparable to van der Waals and depletion forces.

Another common example of entropy-driven self-assembly is polymer nanocomposite in which the mixing of polymers and nanoparticles is opening pathways for engineering flexible composites that exhibit advantageous electrical, optical or mechanical properties [16, 17, 53]. One typical system is diblock copolymer/nanoparticle mixtures where the microphase separation of the copolymer can direct the spatial

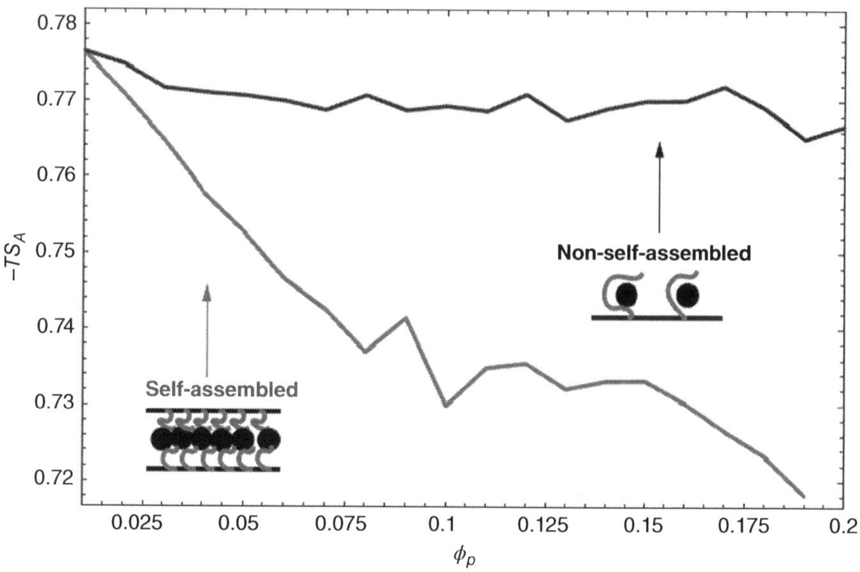

Figure 1.4 *A*-block entropic free energy contribution $-TS_A$ (where S_A is the conformational entropy of the *A* block) per polymer chain, for large particles ($R = 0.3R_0$, gray curve) and small particles ($R = 0.15R_0$, black curve), as a function of ϕ_p. This figure is reproduced from Ref. [56]. Permission from American Association for the Advancement of Science (2001).

distribution of nanoparticles and thereby tailor the properties of the composite [54, 55]. The optical performance of composites, for example, is highly sensitive to the specific location of the particles within the matrix, and the microphase separation of block copolymers can be used to great advantage. However, block copolymers do not simply "template" the arrangement of the nanoparticles. Because the nanoparticles are solids, the polymer chains must stretch around these obstacles, causing a loss in conformational entropy that increases with particle radius. In the absence of specific interactions, larger nanoparticles are expelled from the bulk of the copolymers, whereas smaller particles are not (Figure 1.4). This significantly affects the spatial distribution of nanoparticles within homopolymers and block copolymers and the hierarchical structure of particle-filled systems. Using a combination of self-consistent field- and density-functional theories, Thompson et al. [56] predicted that larger *A*-like particles (i.e., particles that are compatible with the *A* blocks of *AB* copolymers) localize at the center of the *A* microdomains, whereas smaller particles are more uniformly dispersed within a specific microdomain.

The observations above imply that the spatial distribution of nanoparticles in the microphase-separated morphologies can be controlled by tailoring the nanoparticle ligands (i.e., enthalpic effects) and the size of the nanoparticles relative to the radius of gyration of the polymer (i.e., entropic effects). For example, by blending *A*-like nanoparticles of different sizes with symmetric *AB* diblocks, one can fashion gradient materials within the *A* lamellae, where the largest particles are localized in the center, bordered by the next larger particles, which in turn are neighbored by smaller particles.

The entropic penalty associated with chain stretching around particles can also be harnessed to tailor the precisely interfacial organization of Janus nanoparticles in

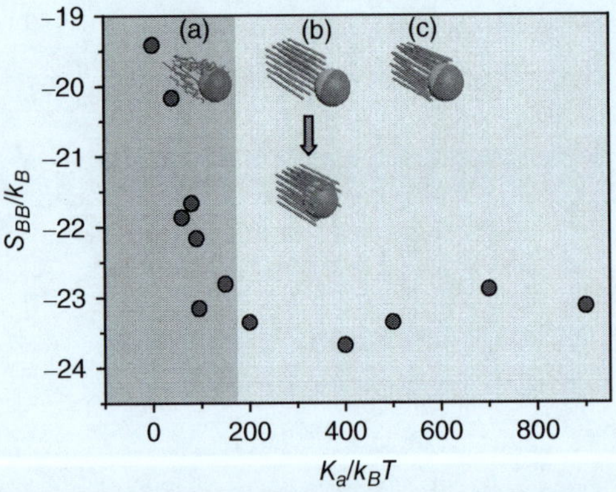

Figure 1.5 The K_a dependence of the excess entropy per bead, S_{BB}, calculated from the pair correlation function between *B*-block beads. The inset diagrams (a)–(c) show the position of Janus nanoparticles and the organization of the stiff *B* blocks around them in response to the increase of the chain stiffness. This figure is reproduced from Ref. [59]. Copyright permission from American Chemical Society (2015).

the scaffold of block copolymers [57–59]. For example, by combining coarse-grained molecular dynamics and a finite difference time domain technique, Dong et al. [59] report entropy-mediated precise interfacial organization of Janus nanoparticles in flexible–semiflexible block copolymers and the resulting optical properties of this heterogeneous material. They find that the stiffness of the semiflexible block can regulate the off-center distribution of symmetric Janus nanoparticles with respect to the phase interface, featured by a roughly 35% deviation from the interface to the utmost extent. The results reveal how entropic and enthalpic effects in this multiphase material contribute to the self-assembled morphologies and, in particular, can lead to entropically driven spatial transition of interfacial nanostructures (Figure 1.5). This might be the very first simulation study to capture the chain stiffness-dependent spatial transition in the interfacial assembly of nanoparticles, and thereby enables applications of chain stiffness in precise control over the interfacial assembly of nanoparticles in polymer scaffolds.

The entropic repulsion between tethered Janus nanoparticles at fluid interfaces can template the mixing of otherwise incompatible nanoparticles, toward perfect mixing of binary nanoparticles at single-nanoparticle level [60, 61]. Indeed, by means of computer simulations, Liu et al. [60] show novel mechanomutable nanocomposites designed by binary mixtures of tethered Janus nanoparticles at the interface of a binary fluid mixture under mechanical pressure. The nanoparticle organization in the systems exhibits a reversible transition between the random state and the long-ranged intercalation state that can be effectively controlled by various structural parameters of the tethered chains and the applied pressure (Figure 1.6). The dynamics mechanism of the transition is explored by examining the detailed trajectories of the interfacial nanoparticles during their 2D diffusion. The origin driving the collective organization of the binary Janus nanoparticles at the fluid–fluid interface is explained by combining computer simulations and theoretical analysis, revealing that both the conformational entropy effect of the tethered chains and sufficient tether disparity dominate the collective nanoparticle organization in the mechanical response behavior. The attractive feature of the procedure proposed by this work is that the interfacial nanostructures are tuned simply by pressing an accessible system containing binary tethered Janus particles in a binary fluid. The motif and findings described in this letter thereby suggest a facile route to create well-defined, responsive and flexible nanopatterns formed by multicomponent

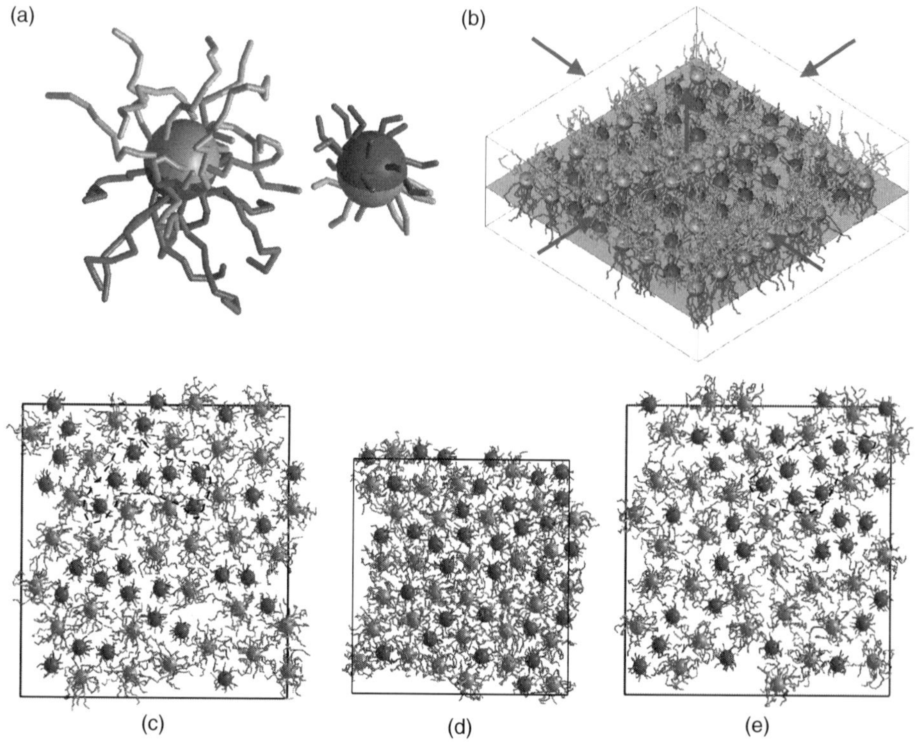

Figure 1.6 (a) Janus nanoparticles tethered with long and short ligand chains. (b) The random arrangement of nanoparticles at the fluid interface for the initial simulations. The gray plane surface represents the fluid interface whereas the fluid beads are not shown for clarity. The arrows indicate the lateral and vertical changes of the simulation box while applying mechanical pressure. (c)–(e) Top-view snapshots of the interfacial nanoparticle patterns where $S = S_0$ (c), $S = 0.64S_0$ (d) and $S = S_0$ (e). The dashed circles in (c) and (e) highlight the phase domains of the same nanoparticle component. This figure is reproduced from Ref. [60]. Copyright permission from American Chemical Society (2014).

nanoparticles in the interface of wide fluid mixtures, and consequently lead to a class of interface-reactive nanocomposites toward technologically important materials and devices.

1.2.3 Programmable Self-Assembly

Programmable self-assembly is an emerging and important concept for future materials fabrication toward precisely controlled structures and optimized properties. For more details of this concept please refer to a recent review by Cademartiri et al. [62]. Here we briefly summarize it and further introduce the theoretical aspect of the relative researches. The programmed assembly of structures from their components requires information [63, 64]—that is, instructions or guidance that direct the reproducible formation of a particular structure from myriad possibilities (Figure 1.7). Such “assembly information” must specify the location and connectivity of the building blocks within the assembled structure and, often, the order and manner in which they are added to it. The information encoded within the building blocks determines the

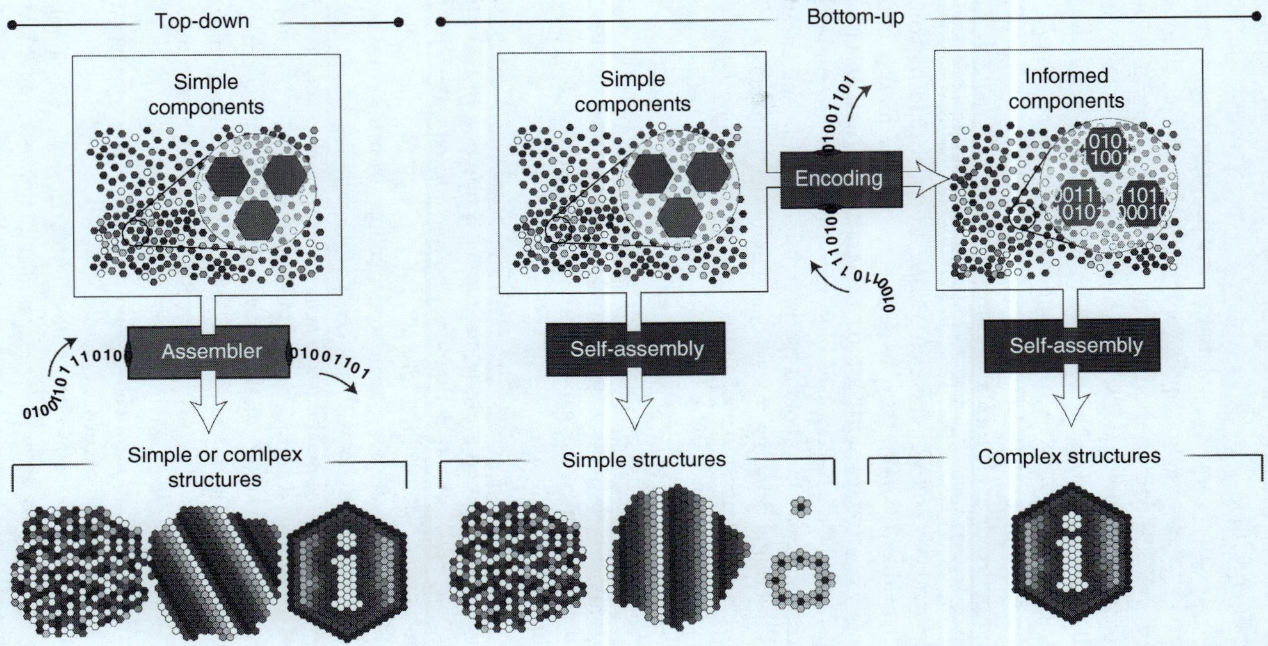

Figure 1.7 The role of information in top-down manufacturing and bottom-up self-assembly. Top-down manufacturing assembles simple building blocks into complex objects by using assemblers. This approach has been the standard strategy for manufacturing, but becomes increasingly difficult at the molecular and nanoscale. By contrast, the bottom-up self-assembly of simple (high-symmetry) components invariably leads to simple structures. The self-assembly of complex structures requires information-rich building blocks (informed components). This figure is reproduced from Ref. [62]. Copyright permission from Nature Publishing Group (2015).

complexity of the final structure [65], which at equilibrium minimizes the system's free energy and is determined by the accessible configurations and their respective energies. In this context, assembly information is encoded into the building blocks in two formats: constraints, which determine the accessible configurations, and interactions, which determine their energy [62].

DNA may be the only practical chemical system that can produce the large numbers of distinct and highly selective interactions preferred by programmable self-assembly [62]. Interactions based on DNA hybridization are well suited for this approach because of the large number of possible interactions, their high specificity and their tunable strength, typically achieved by varying the number of bases in the "sticky ends" of the DNA strands. Early examples of DNA-based interactions between nanoparticles constructed mostly dimers and trimers [66, 67], and have since been extended to achieve the reliable and programmable formation of nanocrystal superlattices [68, 69] and well-defined clusters [70, 71]. In this approach, colloidal building blocks present multiple types of single-stranded DNA on their surface and organize to form structures that maximize DNA hybridization [69]. Particle geometry can also modify both the strength and directionality of attractive surface forces. For example, two cubic particles coated with complementary DNA linkers will bind face to face to maximize DNA hybridization [72]. Recent advances in colloidal synthesis offer trivalent and tetravalent particles that enable directional bonding via specific DNA-based linkers [8]. These particles allow the programmed assembly of finite structures specified by assembly information encoded within selective DNA-based interactions, well-defined surface patches and steric constraints due to particle shape and size.

Theoretical analysis and simulations have played an important role in understanding the process and predicting the formed structures of programmable self-assembly. The results have not only confirmed the experimental results but also stimulate the experiments by providing new predictions.

From the theoretical point of view, the first analytical model was proposed by Tkachenko [73] in 2002, who obtained a phase diagram as a function of two dimensionless parameters. It was suggested that diamond, body-centered cubic (bcc) and other lattices can be obtained by controlling interactions between two particles. Starr et al. [74] have provided theoretical descriptions for equilibrium properties and the dynamics of DNA-linked nanoparticles with concepts from polymer physics, which was considered a successful way to describe novel self-assembly systems. Mirkin et al. [69] have also proposed a rule-based complementary contact model (CCM) to predict the crystal structures of DNA-grafted colloidal nanoparticles. Most recently, they have further constructed a simplified model to study the kinetics of the crystallization process, in which a given particle will transition between free, singly bound and doubly bound states [75]. Based on numerical solving of the thermodynamic equations, the variation of melting temperature and annealing window were predicted, suggesting that the model is useful in the design of future crystals.

From the computer simulation point of view, molecular dynamics (MD) simulation may be the most successful way to describe the dynamics of self-assembly. The first coarse-grained model for MD simulations was introduced by Sciortino et al. [76]. Later, this model was modified by Starr et al. [77]. In 2011, Travesset et al. [78] provided a new modified coarse-grained model to study the crystallization dynamics of DNA-coated nanoparticle systems. A key development of this model is that flanking beads are

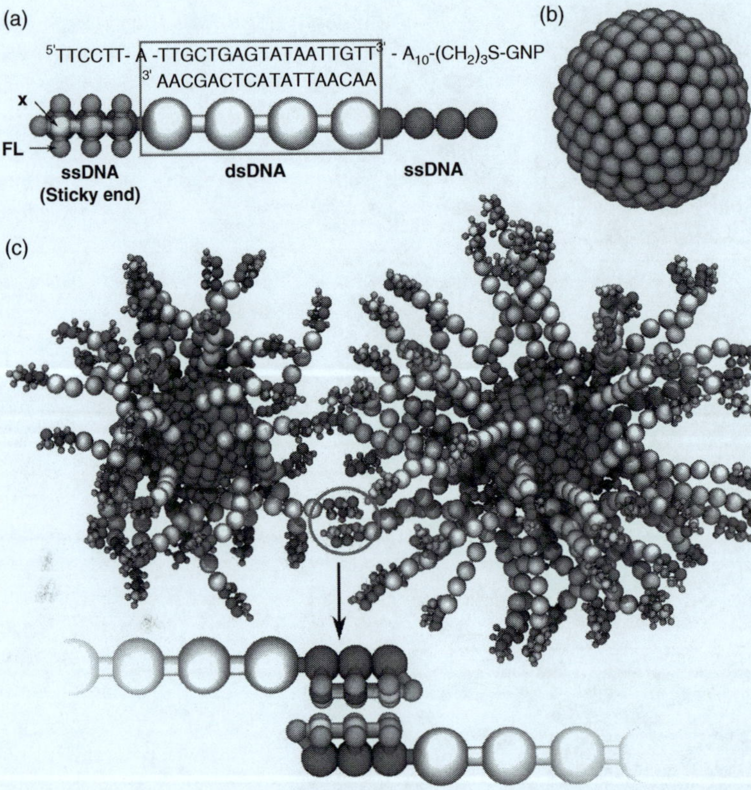

Figure 1.8 Cartoon of the coarse-grained model. (a) A model DNA chain for a sequence used in the experiments. The bead size for the dsDNA portion of the chain is σ, corresponding to a diameter of approximately 2 nm; the bead size for the ssDNA portion is 0.5σ, corresponding to an approximate diameter of 1 nm. (b) A model spherical GNP core. (c) An example of two SNA-GNPs: the particle on the left has an 8#nm gold core and 40 DNA chains; the particle on the right has a 10 nm core and 60 attached DNA chains. For more details, see Ref. [79], from which this figure is reproduced. Copyright permission from American Chemical Society (2012).

utilized to protect any base from binding to more than one complementary base. However, Travesset's model is limited to simulate the formation of many kinds of lattices that have been realized with dsDNA experimentally. To address this issue, de la Cruz et al. [79] proposed a new modified model to better describe these systems. It is to date the most detailed model that is able to capture the crystallization process of many nanoparticles. As seen in Figure 1.8, there are two main modifications in the model. First, the linker DNA chain contains both ssDNA and dsDNA and the dsDNA is modeled with a bead size two times larger than that of ssDNA. Second, the sizes of two nanoparticles, the number and the lengths of grafted DNA chains can be different. Their MD simulation results showed that binary superlattices of bcc, CsCl, AlB_2, Cr_3Si and Cs_6C_{60} can be assembled from binary mixtures of DNA-linked nanoparticles (Figure 1.9). The phase diagrams as a function of the particle size ratio and the DNA coverage ratios for three stoichiometric ratios (1:1, 1:2 and 1:3) were in accordance with experimental results. They also found that the crystallization process is accompanied

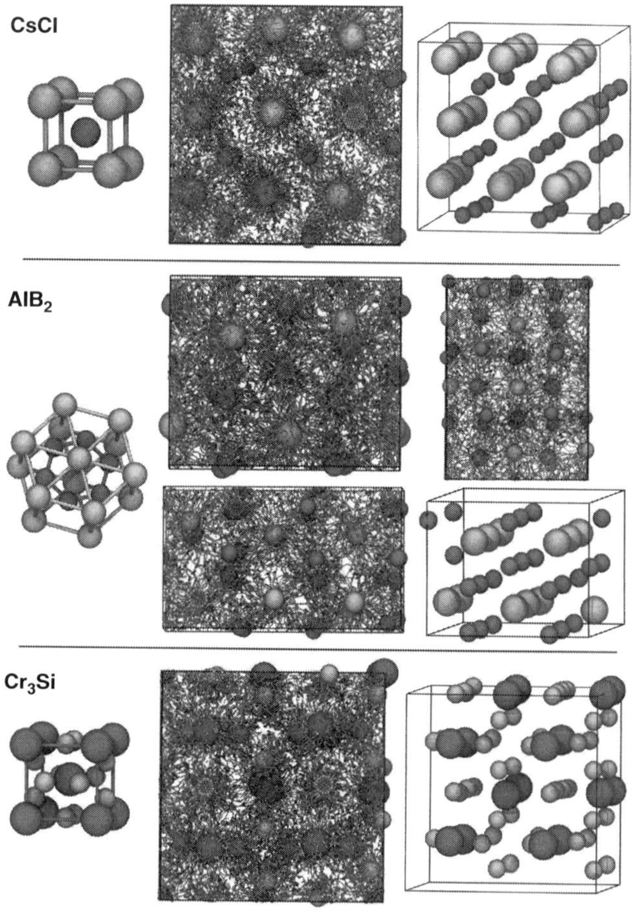

Figure 1.9 Snapshots of the crystal structures obtained from MD simulations. Left to right: The ideal crystal cell, a simulation snapshot after the system fully crystallized and the averaged positions of the SNA-GNPs. For more details, see Ref. [79], from which this figure is reproduced. Copyright permission from American Chemical Society (2012).

by a slight decrease of enthalpy. Furthermore, suitable linker sequences for future nanomaterial designs were proposed.

1.2.4 Self-Assembling Kinetics: Supracolloidal Reaction

The focus of nanoscience is gradually shifting from the synthesis of individual building blocks to the organization of larger nanostructures. The past decade has witnessed great progress in nanoparticle self-assembly, yet the quantitative prediction of the kinetics of their formation remains a challenge. In their pioneering research regarding this important topic, Kumacheva et al. [80, 81] reported on the marked similarity between the self-assembly of metal nanoparticles and reaction-controlled step-growth polymerization. Their results indicate that the kinetics and statistics of step-growth polymerization enable a quantitative prediction of the architecture of linear, branched and cyclic self-assembled nanostructures; their aggregation numbers and size distribution; and the formation of structural isomers. The strong similarity of self-assembly of nanoparticles to polymerization reactions has also been identified in the kinetics of other colloidal systems [8, 83]. Based on this similarity, the molecular concepts of polymer chemistry can be applied to achieve controllable nanoparticle assembly. On the other hand, the ability to visualize nanoparticle assemblies and to exploit characterization tools used in nanoscience offers a unique way to study polymerization reactions. These works bridge the gap between polymerization reactions taking place at a molecular level and nanoparticle self-assembly occurring at a length scale two orders of magnitude larger.

In this context, Guo et al. [41] noted that the construction of supracolloidal helices through systematic computer simulations presents colloidal analogs of the molecular design to some supramolecules with ordered structures [83, 84]. The patchy particles act as multifunctional monomer units that "react" with each other, in a process analogous to supramolecular polymerization [83, 84]. The multi-patchy configuration significantly increases the sophistication of the kinetics and reaction pathway. Figure 1.10a shows the

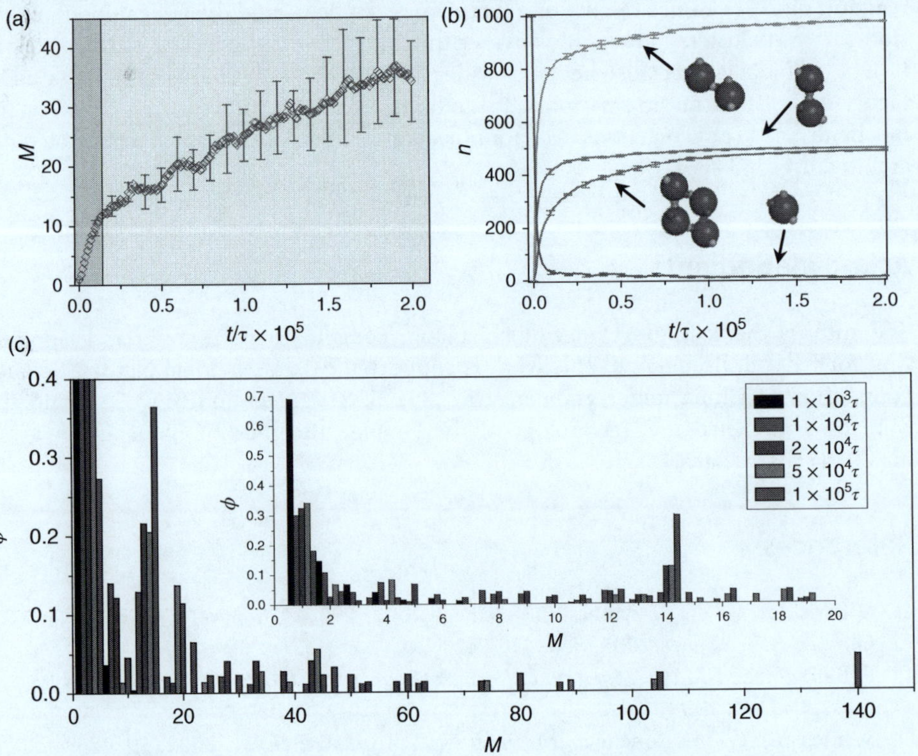

Figure 1.10 Kinetics in the self-assembling process of the helically supracolloidal structure. (a) The number-averaged degree of colloidal clusters as a function of self-assembling time *t*. (b) Variation in the number of patch groups in the course of self-assembly for species shown in inserts. (c) Number-averaged distributions of colloidal clusters at different times. Error bars in (a) and (b) indicate standard deviation. This figure is reproduced from Ref. 41. Copyright Permission from Nature Publishing Group (2014). *(See insert for color representation of the figure.)*.

time dependence of the average number of patchy particles in the supracolloidal chains, M, that resembles the number-averaged degree of polymerization. Interestingly, the $M-t$ plot can be divided into two regimes. In the first regime, M increases linearly with time, being characteristic of reaction-controlled step-growth polymerization. However, the linearity of the $M-t$ plot is relatively suppressed and the plot presents obvious fluctuation in the second regime.

To understand the kinetic mechanisms, they turned to a detailed analysis of the change in the concentrations of different species in the course of self-assembly [41]. Based on patchy interactions directed by their types, four representative species are concerned: individual patchy particle (IP), colloidal cluster linked by self-complementary patches (SP), colloidal cluster linked by a pair of complementary patches (PP) and the elementary colloidal cluster of perfect helix (EC) (Figure 1.10b). Figure 1.10b shows that the concentration of IP is extremely reduced at the initial stage owing to the formation of SP and PP. The complementary patches in SP and PP link each other spontaneously while the formation of EC with long-ranged patchy addition is relatively slow. Thus, the reactions of SP and PP dominate the kinetics in the first regime, resulting in the characteristic of reaction-controlled step-growth polymerization. In the second regime, however, the reaction of EC becomes the major effect, where sophisticated dynamical interconversion between clusters induces the obvious fluctuation in the kinetic plot. The distribution of colloidal clusters also becomes wider in the second regime, as demonstrated by Figure 1.10c. These finding may provide fundamental information for understanding the mechanisms of some reactions in supramolecule chemistry towards helically molecular architectures [83–85].

Acknowledgments

The authors thank Bojun Dong, Junshi Liang, Pengyu Chen, Ye Yang and Guolong Zhu for helpful discussions. This work is supported by the National Natural Science Foundation of China under grant nos. 51273105, 21422403 and 21174080, and the Ministry of Science and Technology of the People's Republic of China under grant no. 2016YFA0202500.

References

1. Whitesides, G. M.; Grzybowski, B. *Science* **2002**, *295*, 2418–2421.
2. Glotzer, S.C. *Science* **2004**, *306*, 419–420.
3. P. Ball, *The Self-Made Tapestry: Pattern Formation in Nature* (Oxford University Press, Oxford, **1999**).
4. Whitesides, G. M.; Boncheva, M. *Proc. Natl. Acad. Sci. U.S.A.* **2002**, *99*, 4769–4774.
5. Philip, D.; Stoddart, J. F. *Angew. Chem. Int. Ed.* **1996**, *35*, 1155–1196.
6. Frenkel, D. *Science* **2002**, *296*, 65–66.
7. Jones, M. R.; Mirkin, C. A. *Nature* **2012**, *491*, 42–43.
8. Wang, Y.; et al. *Nature* **2012**, *491*, 51–55.
9. Zhang, Z.; Glotzer, S. C. *Nano Lett.* **2004**, *4*, 1407–1413.
10. Glotzer, S. C.; Solomon, M. J. *Nat. Mater.* **2007**, *6*, 557–562.

11. Huang, M.; et al. *Science* **2015**, *348*, 424–428.
12. Gang, O.; Zhang, Y. *ACS Nano* **2011**, *5*, 8459–8465.
13. Walker, D. A.; Leitsch, E. K.; Nap, R. J.; Szleifer, I.; Grzybowski, B. *Nat. Nanotech.* **2013**, *8*, 676–681.
14. Frenkel, D. *Nat. Mater.* **2015**, *14*, 9–12.
15. Cates, M. E. *Nat. Mater.* **2013**, *12*, 179–180.
16. Balazs, A. C.; Emrick, T.; Russell, T. P. *Science* **2006**, *314*, 1107–1110.
17. Yan, L. T.; Xie, X. M. *Prog. Polym. Sci.* **2013**, *38*, 369–406.
18. Grzelczak, M.; Vermant, J.; Furst, E. M.; Liz-Marzan, L. M. *ACS Nano* **2010**, *4*, 3591–3605.
19. Schatz, G. C. *Proc. Natl. Acad. Sci. U.S.A.* **2007**, *104*, 85–92.
20. Service, R. F. *Science* **2012**, *335*, 1434–1435.
21. Karttunen, M.; Vattulainen, I.; Lukkarinen, A. *Novel Methods in Soft Matter Simulations* (Springer, Berlin, **2004**).
22. Burda, C.; Chen, X.; Narayanan, R.; El-sayed, M. *Chem. Rev.* **2005**, *105*, 1025–1102.
23. Frenkel, D. *Physica A* **2002**, *313*, 1–31.
24. Gudiksen, M. S.; Lauhon, L. J.; Wang, J.; Smith, D. C.; Lieber, C. M. *Nature* **2002**, *415*, 617–620.
25. Veerman, J. A. C.; Frenkel, D. *Phys. Rev. A* **1992**, *45*, 5632–5648.
26. John, B. S.;Escobedo, F. A. *J. Chem. Phys.* **2008**, *128*, 044909.
27. Yin, J. S.; Wang, Z. L. *Phys. Rev. Lett.* **1997**, *79*, 2570–2573.
28. Maeda, H.; Maeda, Y. *Phys. Rev. Lett.* **2003**, *90*, 018303.
29. Haji-Akbari, A.; et al. *Nature* **2009**, *462*, 773–777.
30. Agarwal, U.; Escobedo, F. A. *Nat. Mater.* **2011**, *10*, 230–235.
31. Damasceno, P. F.; Engel, M.; Glotzer, S. C. *Science* **2012**, *337*, 453–457.
32. Ramano, F.; Sciortino, F. *Nature Comm.* **2012**, *3*, 975–980.
33. Morgan, J. W. R.; Chakrabarti, D.; Dorsaz, N.; Wales, D. J. *ACS Nano* **2013**, *7*, 1246–1256.
34. Harris, A. B.; Kamien, R. D.; Lubensky, T. C. *Rev. Mod. Phys.* **1999**, *71*, 1745–1757.
35. Gansel, J. K. et al. *Science* **2009**, *325*, 1513–1515.
36. Inoue, Y.; Ramamurthy, V. *Chiral Photochemistry* (Marcel Dekker Press, New York, 2004).
37. Soukoulis, C. M.; Wegener, M. *Nature Photon.* **2011**, *5*, 523–530.
38. Kuzyk, A.; et al. *Nature* **2012**, *483*, 311–314.
39. Song, C.; et al. *Nano Lett.* **2013**, *13*, 3256–3261.
40. Fan, Z.; Govorov, A. O. *Nano Lett.* **2010**, *10*, 2580–2587.
41. Guo, R.; Mao, J.; Xie, X. M.; Yan, L. T. *Sci. Rep.* **2014**, *4*, 7021.
42. Szleifer, I. *Nat. Mater.* **2013**, *12*, 693–694.
43. Sheiko, S. S.; Zhou, J.; Arnold, J.; Neugebauer, D.; Matyjaszewski, K.; Tsitsilianis, C.; Tsukruk, V.; Carrillo, J. M. Y.; Dobrynin, A. V.; Rubinstein, M. *Nat. Mater.* **2013**, *12*, 735–740.
44. De Nijs, B.; Dussi, S.; Smallenburg, F.; Meeldijk, J. D.; Groenendijk, D. J.; Filion, L.; Imhof, A.; van Bladderen, A.; Dijkstra, M. *Nat. Mater.* **2015**, *14*, 56–60.
45. Zhang, D.; Carignano, M. A.; Sleifer, I. *Phys. Rev. Lett.* **2006**, *96*, 028701.
46. Alder, B. J.; Wainwright, T. E. *J. Chem. Phys.* **1957**, *27*, 1208–1209.
47. Pusey, P. N.; Van Megen, W. *Nature* **1986**, *320*, 340–342.
48. Chen, Q.; Bae, S.; Granick, S. *Nature* **2011**, *469*, 381–384.

49. Mao, X.; Chen, Q.; Granick, S. *Nature Mater.* **2013**, *12*, 217–222.

50. de Graaf, J.; van Roij, R.; Dijkstra, M. *Phys. Rev. Lett.* **2011**, *107*, 155501.

51. Ni, R.; Gantapara, A. P.; de Graaf, J.; van Roij, R.; Dijkstra, M. *Soft Matter* **2012**, *8*, 8826–8834.

52. van Anders, G.; Klotsa, D.; Ahmed, N. K.; Engel, M.; Glotzer, S. C. *Proc. Natl Acad. Sci. USA* **2014**, *111*, E4812–E4821.

53. Kao, J.; Thorkelsson, K.; Bai, P.; Rancatore, B. J.; Xu, T. *Chem. Soc. Rev.* **2013**, *42*, 2654–2678.

54. Bockstaller, M. R.; Thomas, E. L. *Phys. Rev. Lett.* **2004**, *93*, 166106.

55. Lin Y.; Böker, A.; He, J.; Sill, K.; Xiang, H.; Abetz, C.; Li, X.; Emrick, T.; Long, S.; Wang, Q.; Balazs, A. C.; Russell, T. P. *Nature* **2005**, *434*, 55–59.

56. Thompson, R. B.; Ginzburg, V. V.; Matsen, M. W.; Balazs, A. C. *Science* **2001**, *292*, 2469–2472.

57. Yan, L. T.; Popp, N.; Ghosh, S. K.; Böker, A. *ACS Nano* **2010**, *4*, 913–920.

58. Dong, B.; Guo, R.; Yan, L. T. *Macromolecules* **2014**, *47*, 4369–4379.

59. Dong, B.; Huang, Z.; Chen, H.; Yan, L. T. *Macromolecules* **2015**, *48*, 5385–5393.

60. Liu, Z.; Guo, R.; Xu, G.; Huang, Z.; Yan, L. T. *Nano Lett.* **2014**, *14*, 6910–6916.

61. Huang, Z.; Lu, C.; Dong, B.; Xu, G.; Ji, C.; Zhao, K.; Yan, L. T. *Nanoscale* **2016**, *8*, 1024–1032.

62. Cademartiri, L.; Bishop, K. J. M. *Nat. Mater.* **2015**, *14*, 2–9.

63. Lehn, J. M. *Angew. Chem. Int. Ed.* **2013**, *52*, 2836–2850.

64. Cartwright, J. H. E.; Mackay, A. L. *Phil. Trans. R. Soc. A* **2012**, *370*, 2807–2822.

65. Soloveichik, D.; Winfree, E. *SIAM J. Comput.* **2007**, *36*, 1544–1569.

66. Mirkin, C. A.; Letsinger, R. L.; Mucic, R. C.; Storhoff, J. J. *Nature* **1996**, *382*, 607–609.

67. Alivisatos, A.; et al. *Nature* **1996**, *382*, 609–611.

68. Biancaniello, P.; Kim, A.; Crocker, J. *Phys. Rev. Lett.* **2005**, *94*, 58302.

69. Macfarlane, R. J.; et al. *Science* **2011**, *334*, 204–208.

70. Feng, L.; Dreyfus, R.; Sha, R. J.; Seeman, N. C.; Chaikin, P. M. *Adv. Mater.* **2013**, *25*, 2779–2783.

71. Yan, W.; et al. *J. Am. Chem. Soc.* **2012**, *134*, 15114–15121.

72. Jones, M. R.; et al. *Nat. Mater.* **2010**, *9*, 913–917.

73. Tkachenko, A. V. *Phys. Rev. Lett.* **2002**, *89*, 148303.

74. Hsu, C. W.; Sciortino, F.; Starr, F. W. *Phys. Rev. Lett.* **2010**, *105*, 55502.

75. Macfarlane, R. J.; Thaner, R. V.; Brown, K. A.; Zhang, J.; Lee, B.; Nguyen, S. T.; Mirkin, C. A. *Proc. Natl. Acad. Sci. USA* **2014**, *111*, 14995–15000.

76. Starr, F. W.; Sciortino, F. *J. Phys.: Condens. Matter* **2006**, *18*, L347–L353.

77. Vargas Lara, F.; Starr, F. W. *Soft Matter* **2011**, *7*, 2085–2096.

78. Knorowski, C.; Burleigh, S.; Travesset, A. *Phys. Rev. Lett.* **2011**, *106*, 21550.

79. Li, T.; Sknepnek, R.; Macfarlane, R. J.; Mirkin, C. A.; de la Cruz, M. O. *Nano Lett.* **2012**, *12*, 2509–2514.

80. Liu, K.; et al. *Science* **2010**, *329*, 197–200.

81. Klinkovaa, A.; Thérien-Aubina, H.; Choueiria, R. M.; Rubinsteinb, M.; Kumacheva, E. *Proc. Natl. Acad. Sci. USA* **2013**, *110*, 18775–18779.

82. Chen, Q.; Whitmer, J. K.; Jiang, S.; Bae, S. C.; Luijten, E.; Granick, S. *Science* **2011**, *331*, 199–202.

83. van Gorp, J. J.; Vekemans, J.; Meijer, E. W. *J. Am. Chem. Soc.* **2002**, *124*, 14759–14769.

84. Kim, H. J.; Zin, W. C.; Lee, M. *J. Am. Chem. Soc.* **2004**, *126*, 7009–7014.

85. Ho, R. M.; Chiang, Y. W.; Lin, S. C.; Chen, C. K. *Prog. Polym. Sci.* **2011**, *36*, 376–453.

2

Developing Hybrid Modeling Methods to Simulate Self-Assembly in Polymer Nanocomposites

Xin Yong[1], *Stephen C. Snow*[2], *Olga Kuksenok*[3] *and Anna C. Balazs*[2]

[1] *Department of Mechanical Engineering, State University of New York at Binghamton, USA*
[2] *Department of Chemical and Petroleum Engineering, University of Pittsburgh, USA*
[3] *Department of Materials Science and Engineering, Clemson University, USA*

2.1 Introduction

One of the grand challenges in nanotechnology is designing nanocomposites with well-defined properties and functionalities. A particularly vital functionality for the next generation of nanocomposites is the ability to undergo self-repair or even self-regeneration; these attributes allow the materials to exhibit significantly greater sustainability and lifetimes of use. The creation of effective self-healing nanocomposites is challenging because the systems require multiple components that must interact in a spatially and temporally well-coordinated manner. In effect, the materials should mimic salient features of the complex biological behavior that enables wound healing. Polymeric nanocomposites constitute ideal materials for creating biomimetic systems since the soft matrix allows the material to be flexible and compliant, much as a biological muscle. Moreover, the polymer "arms" that interlink the hard components in the composites enable the materials to reorganize or reconfigure, aiding in the self-healing process. Finally, the hard components within the nanocomposites provide an effective "skeleton" that greatly improves the mechanical behavior of the material, as well as imparts desired optical, magnetic, or electrical properties.

A vast range of nanocomposites can be fabricated by radical polymerization, which constitutes one of the most versatile and widely applied polymerization techniques [1]. Generally speaking, radical polymerization can be categorized into two types: free radical polymerization (FRP) [2, 3] and controlled/living radical polymerization [4, 5]. In both processes, a free radical is formed at the active chain end of a growing polymer chain, which reacts with a monomer and thereby successively adds this monomer to the chain. The relevant reactions in radical polymerization are initiation, propagation, termination, and chain transfer. In atom transfer radical polymerization (ATRP) [6–9], which is a living radical polymerization technique, the termination of growing polymer chains is absent and the initiation is fast relative to the propagation. Due to these two features, chains polymerized by ATRP are relatively monodisperse as compared to chains polymerized by FRP.

Self-Assembling Systems: Theory and Simulation, First Edition. Edited by Li-Tang Yan.

To design and synthesize well-tailored nanocomposites that can be utilized in a range of applications, including self-healing, we must establish a better fundamental understanding of the coupling between radical polymerization reactions and the dynamics of the host solution, which often contains a variety of nanoinclusions. Computer simulations provide a powerful means of probing these interactions at mesoscopic length scales, and thus can yield guidelines for optimizing the macroscopic properties and functionalities of the nanocomposites.

Recently, we developed a dissipative particle dynamics (DPD) framework to model the formation of polymer gels by both polymerization processes, FRP and ATRP [10, 11]. DPD is an effective mesoscale particle-based method to simulate the behavior of complex fluids and multi-component mixtures. Hence, by adapting this approach to simulate radical polymerization reactions, one can capture effects arising from the coupling of the reaction kinetics and dynamics of the host solution.

In Section 2.2, we describe our approach for modeling both FRP and ATRP processes within the DPD framework. In Section 2.3, we first focus on modeling bulk polymerization using both FRP and ATRP approaches. We validate our respective approaches by comparing our results to available experimental and/or simulation data. For example, we show that our DPD approach successfully captures living copolymerization of monomers and bifunctional cross-linkers to form a polymer gel.

We then utilize both simulation techniques to model two distinct types of polymer nanocomposites. In the first example, we rely on ATRP and exploit the properties of interfacially active nanorods to design composites that undergo self-regeneration. In the second example, we utilize FRP to model the fabrication of polymer–clay composite gels. The latter materials constitute ideal systems for creating adaptive, biomimetic composites. In particular, in the presence of external stimuli, the "muscle-like" gel [12, 13] can drive the actuation of the composite, and thereby yield a strong, reconfigurable material.

2.2 Methodology

2.2.1 Dissipative Particle Dynamics

Dissipative particle dynamics (DPD) is a coarse-grained, particle-based method for simulating complex fluids and multi-component mixtures. In this approach, a volume of fluid is modeled by individual beads, where each bead represents a cluster of molecules. The temporal evolution of this many-body system is dictated by the motion of the beads. In particular, the movement of each bead is governed by Newton's equation of motion, $m\, d\mathbf{v}_i/dt = \mathbf{f}_i$. Each bead i experiences a force $\mathbf{f}_i(t)$ that is the sum of three pairwise additive forces: $\mathbf{f}_i(t) = \sum(\mathbf{F}^{\mathrm{C}}_{ij} + \mathbf{F}^{\mathrm{D}}_{ij} + \mathbf{F}^{\mathrm{R}}_{ij})$, where $\mathbf{F}^{\mathrm{C}}_{ij}$, $\mathbf{F}^{\mathrm{D}}_{ij}$, and $\mathbf{F}^{\mathrm{R}}_{ij}$ are the conservative force, drag or dissipative force, and random force, respectively. All pairwise forces are truncated at a certain cutoff radius r_{c}, i.e., the sum only applies on all beads j within a distance of r_{c} from bead i.

The conservative force is a repulsive force given by $\mathbf{F}^{\mathrm{C}}_{ij} = a_{ij}(1 - r_{ij})\hat{\mathbf{r}}_{ij}$, where $r_{ij} = |\mathbf{r}_i - \mathbf{r}_j|/r_{\mathrm{c}}$ and $\hat{\mathbf{r}}_{ij} = (\mathbf{r}_i - \mathbf{r}_j)/|\mathbf{r}_i - \mathbf{r}_j|$. In contrast to hard-core potentials (e.g., the Lennard-Jones potential) in which the repulsion diverges when the distance between beads approaches zero, this soft-core force permits a degree of overlap between beads.

As a result, we can apply larger time steps than those typically used in molecular dynamics (MD) simulations, which commonly involve hard-core potentials. Combined with the coarse-grained representation of multi-component systems, the use of soft potentials allows DPD to capture physical phenomena occurring on relatively larger length and time scales than those normally captured by MD, and makes DPD an ideal computational simulation tool to model self-assembly in polymer nanocomposites.

The interaction parameter a_{ij} determines the maximum repulsion between beads i and j, which is given in terms of $k_B T/r_c$. Here, k_B is the Boltzmann constant and T is the temperature of the system. We set r_c as the characteristic length scale, room temperature as the reference temperature, and the corresponding energy scale $k_B T_0$ at $T_0 = 25\,°C$ as the characteristic energy scale. Thus, we set $k_B T = 1$ at $T = 25\,°C$. The reduced temperature is then introduced as $T^* = T/T_0$ (e.g., $T^* = 1.01$ for $T = 28\,°C$). All the interaction parameters are presented as functions of the reduced temperature T^*.

The drag force and random force work together as a thermostat for producing a canonical ensemble. These forces are applied on pairs of neighboring beads such that the momentum is conserved locally, and hydrodynamic behavior emerges in relatively small systems [14–16]. In particular, the drag force is $\mathbf{F}_{ij}^{D} = -\gamma\omega_D(r_{ij})(\hat{\mathbf{r}}_{ij} \cdot \mathbf{v}_{ij})\hat{\mathbf{r}}_{ij}$, where γ is a simulation parameter related to viscosity arising from the interactions between beads. ω_D is a weight function that goes to zero at the cutoff radius r_c, and the relative velocity is $\mathbf{v}_{ij} = \mathbf{v}_i - \mathbf{v}_j$. The random force is given by $\mathbf{F}_{ij}^{R} = \sigma\,\omega_R(r_{ij})\xi_{ij}\hat{\mathbf{r}}_{ij}$, where ξ_{ij} is a zero-mean Gaussian random variable of unit variance and σ is the amplitude of the noise. The fluctuation-dissipation theorem relates σ to γ as $\sigma^2 = 2k_B T\gamma$ [15]. Finally, the weight functions take the following form: $\omega_D(r_{ij}) = \omega_R(r_{ij})^2 = (1 - r_{ij})^2$ for $r_{ij} < r_c$.

The velocity Verlet algorithm is applied to integrate the equation of motion in time. Knowing the characteristic length, mass, and energy, the characteristic time scale for the DPD simulation is then defined as $\tau = \sqrt{mr_c^2/k_B T_0}$. We use time step $\Delta t = 0.02\,\tau$ for all simulations. By setting $\gamma = 4.5$, we obtain a relatively rapid equilibration of the temperature in the system and ensure the numerical stability of the simulations for the specified time step [16]. The remaining simulation parameter is the average bead number density of the entire system, ρ_{sys}, which is set to 3 unless stated otherwise. All simulations are performed by modifying and extending the particle dynamics software code LAMMPS [17].

2.2.2 Polymer Chains, Gels, and Nanoparticles

Polymer nanocomposites incorporate a variety of nanoparticles dispersed in the polymer matrix, which consists of physically entangled or chemically cross-linked polymer chains. As a particle-based approach, DPD can readily capture the molecular architecture of the chains and the detailed topology of the networks. Moreover, it allows us to incorporate nanoparticle inclusions of different types, shapes, and sizes into the polymer matrix and to explicitly specify the interactions between these inclusions and other components in the system. Below, we describe how we construct hydrogels filled with nanorods or nanoclay particles in the framework of DPD.

The semi-flexible polymer chains are modeled as a sequence of DPD beads that are connected by harmonic bonds. The interaction potential between neighboring beads consists of a bond potential and an angle potential. The functional form of the potential is given by $E = \frac{1}{2}K_{bond}(r - r_0)^2 + K_{angle}[1 + \cos(\theta)]$. The first term is the elastic energy of a harmonic bond, where K_{bond} is the elastic constant and r_0 is the equilibrium bond distance. The second term represents the bending energy with the rigidity parameter K_{angle} [18, 19]. Here, θ is the bond angle between two adjacent bonds.

The chemically cross-linked gels can be constructed from the polymer chains in two ways. In the simulations of nanorod-filled gel, we introduce cross-linking beads to connect different polymer strands and specify the topology of the gel. Namely, we take the initial configuration of the gel to be a finite-sized tetra-functional network with a diamond-like topology [20]. Each strand consists of 30 beads. This finite-sized network contains cross-links, which are beads with a connectivity of four, and dangling ends, which are located on the surface of the lattice and have a connectivity of less than four. The bond and angle potential parameters of the gel are set respectively as $K_{bond} = 128$, $r_0 = 0.5$, and $K_{angle} = 4$ to prevent bond crossing, and produce a polymer concentration comparable to the experimental results for hydrogels [21–23]. The gel layer shown in Figure 2.1a is periodic in the lateral (x and y) directions and contains 16 488 beads in total (540 strands, 252 cross-links and 36 dangling ends).

In contrast, the gels with nanoclay particles modeled in the simulations do not have a specified topology and do not contain chemical cross-linking agents. The polymer network emerges in a polymer-grafted nanoclay solution due to covalent bonds formed during free radical combination reactions, which we detail further below. In particular, the connectivity is established via the combination of growing chains from different clay particles (as sketched in Figure 2.1b), which forms the inter-particle cross-linking chains (X-chains). Each clay particle now serves as the multi-functional cross-linkers that maintain the integrity of the network. Meanwhile, we note that the combination reaction also generates intra-particle looped chains (L-chains), in which both ends are attached to one particle; these L-chains do not contribute to the formation of the gel network.

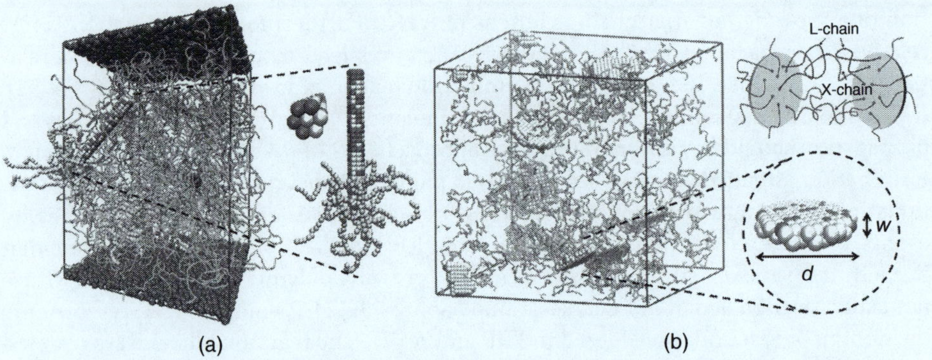

Figure 2.1 (a) Equilibrium state of a nanorod-filled polymer gel in a good solvent (not displayed). Inset: enlarged view of initiator modified nanorod with one end functionalized with polymer chains and a cross-section of the hexagonal nanorod. (b) Snapshot of nanoclay-embedded polymer network at monomer conversion 95%. Inset top: schematic representation of the network. Inset bottom: detailed structure of the nanoclay particle. Reproduced from Ref. [10] with permission from the American Chemical Society. *(See insert for color representation of the figure.)*.

The nanoparticles considered in our simulations are nanorods that are modified to be interfacially active by grafting polymer chains, and exfoliated nanoclay particles that have discoid shapes as shown in Figure 2.1. These nanoparticles are modeled as clusters of DPD beads and the constituent beads of each particle move collectively as a rigid body. Here, the shaft of each nanorod is formed from 210 beads, with the cross-section having hexagonal structure (see inset in Figure 2.1a). The lattice constant of the hexagonal structure is 0.5 and the distance between layers of hexagonal cross-sections is also 0.5. The rod has a total length of 15 dimensionless units.

We functionalize one end of the nanorod with 12 polymer chains. The length covered by the chains is approximately one fifth of the total length. Each chain is formed from 20 connected beads with the same bond and angle potentials as described above for the gel. The grafting is achieved by bonding an end bead of the chain to a surface bead of the rod, whose position is randomly picked in the chain-covered region. The remaining surface of the rod is modified with initiators for the polymerization reaction; i.e., a certain fraction of the surface beads are replaced with initiator beads. We define the initiator density σ_i as the number of initiator beads on each rod divided by 144, which is the total number of beads on the remaining rod's surface (excluding the portion of the surface functionalized by the grafted chains). In this manner, the polymer chains will grow from the surface of the rod during polymerization.

The clay particle is modeled as a nanodisk with two layers of DPD beads arranged in a face-centered cubic lattice structure (see inset in Figure 2.1b). The lattice constant of the FCC structure is 0.7 and each clay particle has 232 beads in total. The size of the clay particle is 6 dimensionless units in diameter and 1 dimensionless unit in thickness. A certain fraction of the particle beads are replaced with the initiator beads in the same way as we modified the nanorod surface.

2.2.3 Radical Polymerization Model

We develop a DPD framework to model the formation of polymer gels by FRP and ATRP. The major difference between these two polymerization processes is that the termination of a growing polymer chain is absent in ATRP. Furthermore, the initiation is fast relative to propagation in ATRP. These two features lead to chain length being relatively monodisperse. In contrast, the termination reactions are important in FRP [2–5], which should be taken into account in the simulation. Also, the rates of initiation and propagation should be comparable in FRP approaches [2–5]. Two mechanisms of chain termination are introduced in our model, namely radical combination and disproportionation. In order to form polymer networks, the chain transfer reactions should be rare events, and we exclude these reactions in both FRP and ATRP simulations. We note that the activation and deactivation processes of the growing polymer chain end in ATRP are not explicitly modeled in our coarse-grained approach [24, 25]. Figure 2.2 presents only the dominant reactions considered in FRP and ATRP. The reaction kinetics are coupled to the dynamics of the system, which are evolved by the DPD simulations. This reactive DPD model is particularly useful for capturing both hydrodynamics and chemical reactions in the presence of solid inclusions with complex geometries.

In our simulations, the reactive components include the initiator, monomer, and cross-linker, and all three are modeled as coarse-grained DPD beads. We use different bead types to represent the states of the reactive species. For example, an unreacted

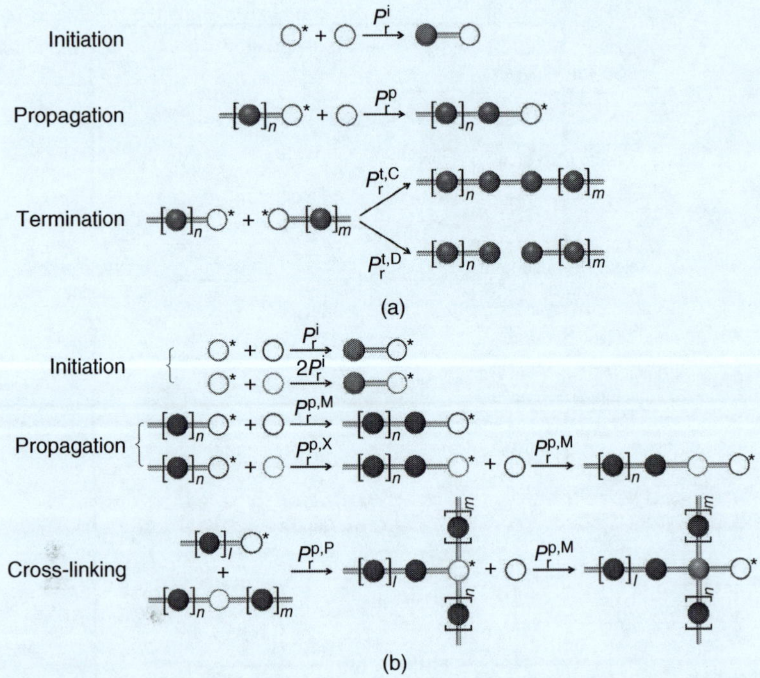

Figure 2.2 Schematic of the elemental reactions in (a) standard FRP and (b) living radical copolymerization of monomer and cross-linker. An asterisk indicates an active radical. Open beads show unreacted or partially reacted species and filled beads show fully reacted species. P_r^i, $P_r^{p,M}$, $P_r^{p,X}$, $P_r^{p,P}$, $P_r^{t,C}$, and $P_r^{t,D}$ are probabilities of the respective reactions (initiation, propagation with monomer, propagation with unreacted bifunctional cross-linker, cross-linking with partially reacted cross-linker with pendent functional group, termination through combination, and termination through disproportionation). Reproduced from Ref. [11] with permission from Elsevier.

monomer is specified as type 1, but this monomer changes to type 2 when it is added to a growing chain and becomes the new active end. Once the monomer has fully reacted, it is labeled as type 3. Our cross-linker consists of two reactive cross-linking units, which simulate a divinyl cross-linker [25] that forms a bridge to connect two chains. Due to the existence of two reactive units, the cross-linker has six different states corresponding to different effective reactivity (i.e., the extent to which it has reacted). In addition to the dominant reactions shown in Figure 2.2, we account for such relatively rare events as a reaction between an initiator and partially reacted cross-linker, reactions between two cross-linker beads in different states, and a reaction between a monomer and a cross-linker with two active radicals [25]. The result of the specific chemical reaction is modeled by updating the bead types after the reaction has occurred [25–27] (see the reaction scheme in Figure 2.3). Below, we describe the reaction scheme in detail.

A list of beads with free radicals (denoted R-beads), which can be activated by initiators or active chain ends, is constructed at the beginning of every reaction step. In each reaction step, we select an R-bead from the list as the reaction center and search all reactive beads within an interaction radius $r_i = 0.7$ from the given R-bead. If there

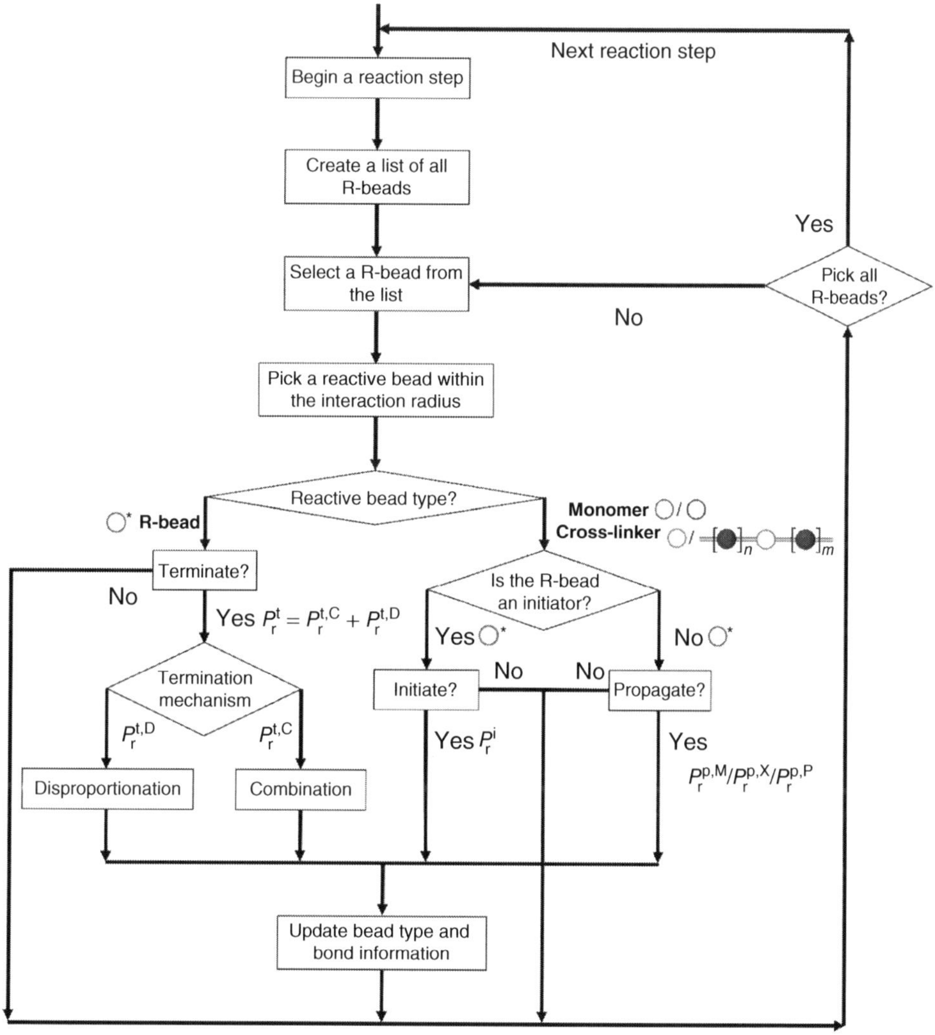

Figure 2.3 Flowchart of the reaction scheme in the computational model of radical polymerization. Details of the individual steps are described in the text. Reproduced from Ref. [11] with permission from Elsevier.

are reactive beads within this interaction range, one is picked at random and a reacting pair is formed. Based on the types of the R-bead and the selected reactive bead, the corresponding elemental reaction is determined.

We then use reaction probabilities P_r^x to characterize the reaction kinetics [24, 28–30], where the superscript x stands for the type of reaction. A successful reaction changes the type of the reacting beads accordingly. Depending on the reaction, the reacting pair of beads can result in an irreversible bond formation, with the energy of the bond given by $E_{\mathrm{bond}} = \frac{1}{2}K_{\mathrm{bond}}(r - r_0)^2$. Here, $K_{\mathrm{bond}} = 128$ and $r_0 = 0.5$ are the same as the parameters for the predefined polymer chains. Notably, the chains created during the polymerization reactions do not contain an angle potential between three successive beads (see above); hence, the stiffness of the chains in the newly generated gel is lower than that in the predefined gel. Within every reaction time step, the above steps are conducted for the entire list of R-beads.

Every R-bead within the list is only allowed to go through the reaction procedure once per reaction step regardless of whether the reaction occurs or not (as dictated by the corresponding reaction probabilities). The reaction steps are separated by the reaction interval $\tau_r = 0.2\,\tau$, and thus the reactions are performed every 10 time steps [24]. By choosing different probabilities for the reactions, P_r^i, $P_r^{p,M}$, $P_r^{p,X}$, $P_r^{p,P}$, $P_r^{t,C}$, and $P_r^{t,D}$, we can effectively adjust the rate constants of the respective reactions: initiation, propagation with monomer, propagation with unreacted bifunctional cross-linker, cross-linking with partially reacted cross-linker with pendent functional group, termination through combination, and termination through disproportionation [28].

2.3 Results and Discussions

2.3.1 Modeling Bulk Polymerization Using FRP and ATRP

We first model bulk polymerization in a three-dimensional system to obtain a polymer solution using FRP. The simulation box is $20 \times 20 \times 20$ units in size, with periodic boundary conditions being imposed in all three directions. The initial configuration of the system is generated by randomly distributing initiator, monomer, and solvent beads in the simulation box, modeling a solution with the dissolved components. The ratio of the initial concentrations of initiator and monomer is $[\mathrm{Ini}]_0/[\mathrm{M}]_0 = 1/75$. The solvent concentration is $[\mathrm{S}]_0 = 50\%$, which results in a polymer volume fraction equal to 0.5 in the solution when the polymerization reaches full conversion. The solution with the dissolved components was equilibrated for 1×10^5 simulation time steps before the polymerization starts. The polymerization time is 1×10^6 time steps. We consider the initiator and monomer to be chemically compatible in the solvent. Therefore, we model the athermal solvent condition at the reference temperature with the interaction parameters between any two beads, a_{ij}, set to $a_{ij} = 25$.

We choose the following reaction probabilities for our reference case: the initiation probability $P_r^i = 0.005$, the monomer addition (propagation) probability $P_r^{p,M} = 0.005$, the total termination probability $P_r^t = P_r^{t,C} + P_r^{t,D} = 0.001$. We assume that chain combination is the dominant one of the two termination mechanisms in the considered polymerization (e.g., one involving vinyl radicals) [31]. Thus, we set the ratio between the probabilities of the two termination processes to $P_r^{t,C}/P_r^{t,D} \approx 4.7$ [31].

We characterize the detailed polymerization kinetics by monomer conversion, molecular weights, and the polydispersity index (PDI), as shown in Figure 2.4. To obtain the molecular weight of polymer chains from the degree of polymerization, we assume that the molecular weight of each monomer bead is 1. Figure 2.4a shows the evolution of the monomer conversion and $\ln([\mathrm{M}]_0/[\mathrm{M}])$ as a function of the simulation time steps. The curve is concave downward, which indicates the decline of polymerization speed as time elapses. This behavior is attributed to the termination reactions that diminish the number of active chain ends [2]. Correspondingly, the number average and weight average molecular weights M_n and M_w both increase significantly at the late stage of polymerization due to the combination of growing polymer chains, which produces

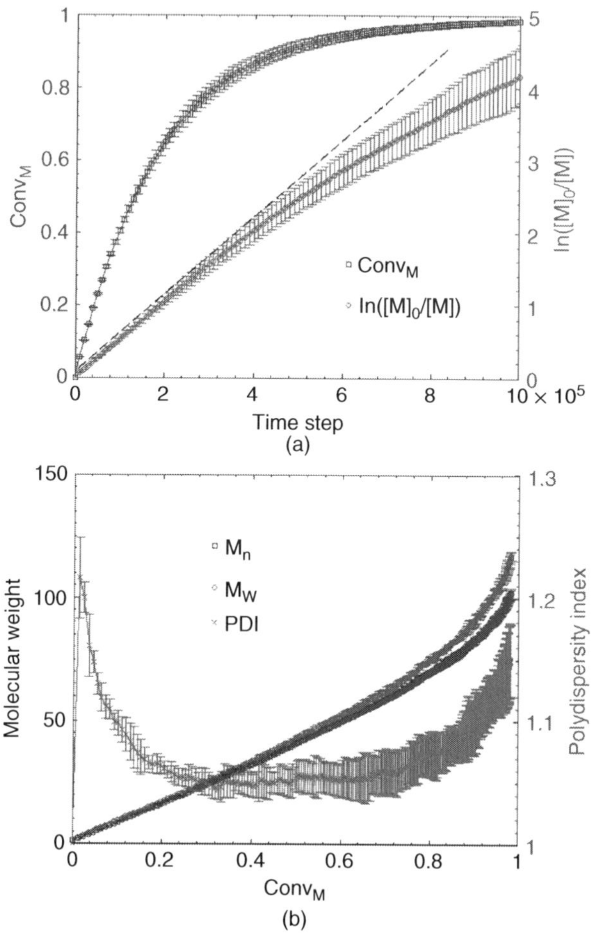

Figure 2.4 (a) Monomer conversion (squares) and $\ln([M]_0/[M])$ (circles) as a function of the number of simulation time steps. (b) Number average molecular weight M_n (squares), weight average molecular weight M_w (open circles), and polydispersity index PDI $= M_w/M_n$ (crosses) as functions of the monomer conversion. Error bars arise from averaging over four independent runs. Reproduced from Ref. [11] with permission from Elsevier.

long chains in short times. The PDI shown in Figure 2.4b provides a quantitative characterization of the molecular weight distribution of the growing chains. The PDI shows a non-monotonic behavior as monomer conversion increases. It first increases rapidly before reaching a peak value and then decreases to a plateau. The large peak at the beginning of the polymerization indicates that the growth of all the polymer chains does not occur simultaneously; this primarily results from comparable rates of initiation and propagation. At the late stage of the polymerization, the PDI increases again due to the chain combination, which also produces chains with a wide range of lengths [28, 32].

Our reaction model allows us to probe the chain propagation and termination and the state of growing chains in greater detail. In particular, a disproportionation reaction reduces the number of polymer chains with active ends by two and increases the number of non-reactive polymer chains by two. Whereas, a combination reaction reduces the number of polymer chains with active ends by two, but only increases the number of non-reactive polymer chains by one, because the total number of polymers decreases by one due to combination. As shown in Figure 2.5a, the increase in the monomer conversion leads to a decrease in the total number of polymers because of the chain combination reactions. Both disproportionation and combination yield an increase in the number of non-reactive polymers and a decrease in the number of polymer chains with active ends. Figure 2.5b confirms that the ratio between the number of combination reactions versus the number of disproportionation reactions is consistent with the ratio between corresponding reaction probabilities stated above.

With the well-characterized reference case, we can now explore how three reaction probabilities, namely P_r^i, $P_r^{p,M}$, and P_r^t, affect the polymerization kinetics. We first vary the initiation probability P_r^i while keeping all other parameters the same as those in the above case. When we decrease P_r^i to 0.0005 (i.e., one tenth of the propagation probability), there is no noticeable difference in the time evolution of $\ln([M]_0/[M])$ and the PDI of the final polymer solution (see Figures 2.6a and 2.6b). This behavior is in good

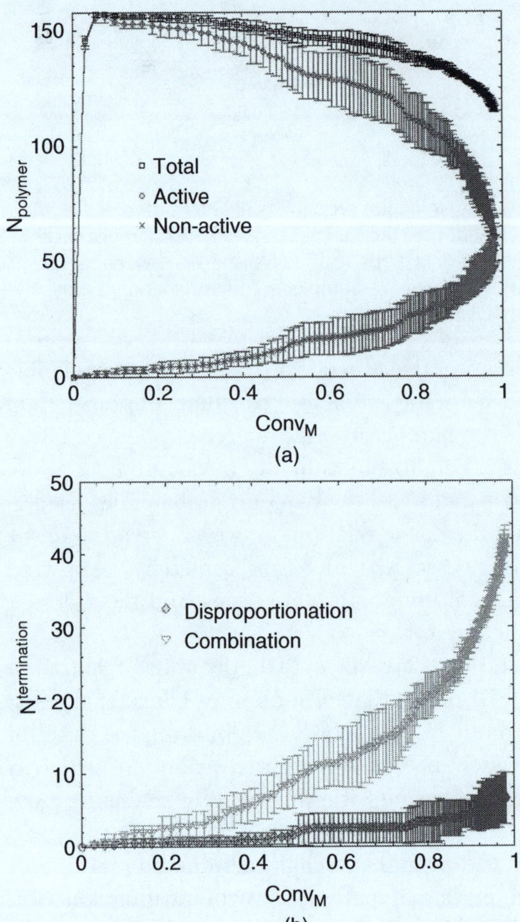

Figure 2.5 (a) Numbers of total polymer chains (squares), chains with active ends (circles), and non-reactive (crosses) polymer chains as functions of the monomer conversion. (b) Numbers of termination reactions, combination (downward triangles), and disproportionation (rhombuses) as functions of monomer conversion. Error bars arise from averaging over four independent runs. Reproduced from Ref. [11] with permission from Elsevier.

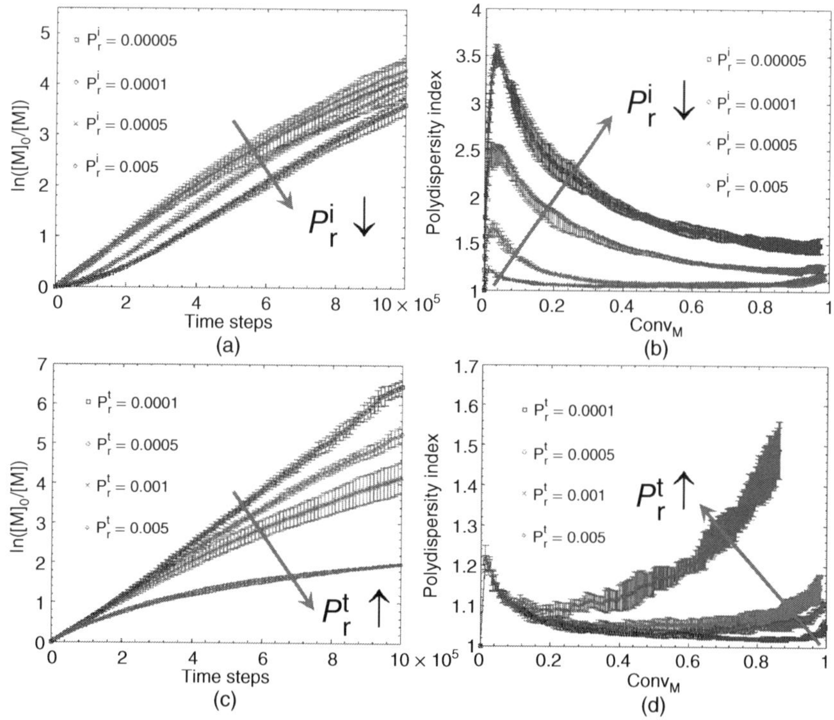

Figure 2.6 Polymerization kinetics with varying (a, b) initiation probability P_r^i and (c, d) total termination probability P_r^t. Other reaction probabilities are the same as those in the reference case. (a, c) Dependence of ln($[M]_0/[M]$) on the simulation time steps. (b, d) Polydispersity index as a function of the monomer conversion. Error bars arise from averaging over four independent runs. Reproduced from Ref. [11] with permission from Elsevier.

agreement with the experimental observation that the effect of initiation is negligible unless $k_p/k_i > DP_n/4$ [33], where k_p and k_i are the actual rate constants of propagation and initiation, respectively. Here, DP_n is the number average degree of polymerization at the end of the polymerization process. Due to the difficulty in obtaining accurate conversions between the physical rate constants and the reaction probabilities used in the simulation, we cannot directly verify the above relation. However, we observe the effect of slow initiation when P_r^i is further decreased, which is presented by the upward curvature of the ln($[M]_0/[M]$) curve at small monomer conversion and the different PDI curves at the late stage of the polymerization.

Unlike in the ln($[M]_0/[M]$) curve and the late stage PDI, the rate of initiation significantly influences the rise of the PDI peak at small monomer conversion, even when $P_r^{p,M}/P_r^i = 10$ (see Figure 2.6b). The higher peak of the PDI during the onset of the polymerization process means a wider molecular weight distribution, which is attributed to the fact that slower initiation permits the propagating chains to grow longer before a new chain is initiated.

The characteristic feature of FRP is the prominent chain termination, and thus we examine the effect of termination probability on the polymerization kinetics. Here, we vary the total termination probability P_r^t, but keep the ratio between the probabilities of the two termination processes, combination and disproportionation, constant at $P_r^{t,C}/P_r^{t,D} \approx 4.7$ [31]. We plot $\ln([M]_0/[M])$ and the PDI in Figures 2.6c and 2.6d, respectively. At the lowest value of P_r^t considered here (which corresponds to the weakest termination reaction), the behavior of $\ln([M]_0/[M])$ is nearly linear and shows the first-order reaction kinetics with constant concentration of the active chain ends [2]. The termination effect is clearly reflected in the downward curvature of the $\ln([M]_0/[M])$ curves and their clear deviation from the linear behavior as P_r^t increases. As shown in Figure 2.6c, the overlap of the PDI curves at small monomer conversion suggests that the termination has no effect at the early stage of the polymerization, when the effect of initiation is dominant due to the small concentration of the active chain ends. In contrast, the termination effect prevails at the late stage. The late stage PDI increases as P_r^t increases, as demonstrated in Figure 2.6d, indicating that longer chains with more heterogeneous chain lengths are formed.

Finally, we observe that the propagation probability $P_r^{p,M}$ only dictates the monomer conversion reached at the end of the polymerization. In particular, higher probability leads to higher monomer conversion. In contrast to tuning P_r^i and P_r^t as discussed above, varying $P_r^{p,M}$ has no consequence on the shape of the PDI curve.

After we explored the kinetics of three-dimensional bulk polymerization via FRP, we simulate the living copolymerization of monovinyl monomers and divinyl cross-linkers to form polymer gels. As we noted in the methodology section, we neglect chain termination reactions by setting $P_r^t = 0$ and model fast initiation compared with propagation due to the nature of ATRP [8]. Hence, we only list the initiation, propagation, and cross-linking reactions in Figure 2.2b. For ATRP, the values of $P_r^{p,M}$ should be sufficiently small to achieve controlled polymerization growth in the kinetically controlled reaction regime [29]. Here we take $P_r^i = P_r^{p,M} = 0.0025$ and $P_r^{p,X} = 2P_r^{p,M}$. The reaction probability $P_r^{p,P}$ controlling the cross-linking process is chosen as $P_r^{p,P} = 3P_r^{p,M}$ to account for the fact that the bifunctional cross-linker is modeled by a single bead [25, 26]. Similar to the FRP simulation, the initial configuration of beads is a random distribution of initiator, monomer, cross-linker, and solvent beads in the simulation box. The initiators, monomers, and cross-linkers are chemically compatible and dissolve in an athermal solvent. Therefore, the interaction parameters between any two beads, a_{ij}, is set to $a_{ij} = 25$. Using these values, we investigate the effect of varying the volume fraction of cross-linker on the gel point.

Our purpose is to validate the first-order kinetics and to compare the characteristic features of gelation with those obtained in previous experiments [25, 34]. We probe the polymerization in the solution as we did in the bulk polymerization through FRP and measure the gel point, which is defined as the monomer conversion at which the reduced degree of polymerization (RDP) reaches a peak value [26]. The RDP is the weight-average degree of polymerization of all macromolecules, except the longest chain. The simulation results and previous experimental data are summarized in Table 2.1 Figure 2.7a shows that the copolymerization with initial concentration ratios of monomer, cross-linker, and initiator equal to $[M]_0/[X]_0/[Ini]_0 = 78/20/2$ clearly exhibits first-order kinetics for both monomer (solid line) and cross-linker (dashed line), which validates our ATRP model. Notably, gelation occurs at 34% monomer conversion and 57% cross-linker conversion with these parameters, as can be seen from the black curve in Figure 2.7b.

Table 2.1 Comparison among the gel points in ATRP experiments and DPD simulations

			ATRP exp[a]		DPD sim[b]	
$[Ini]_0$(%)	$[X]_0$(%)	$[M]_0$(%)	$conv_M$	$conv_X$	$conv_M$	$conv_X$
2	1	97	no gelation		no gelation	
2	1.2	96.8	no gelation		no gelation	
2	1.5	96.5	no gelation		no gelation	
2	2	96	no gelation		no gelation	
2	2.2	95.8	0.97	1.00	0.99	1.00
2	3	95	0.78	1.00	0.89	0.99
2	6	92	0.61	0.81	0.62	0.86
2	10	88	0.42	0.66	0.49	0.74
2	20	78	0.32	0.43	0.34	0.57

a) Experimental values are taken from Refs. [21, 30].
b) Simulated gel points by using DPD ATRP scheme with total bead number of 81 000.

We further investigate the effect of the ratio of cross-linker to initiator ($[X]_0/[Ini]_0$) on gelation. We vary the cross-linker amount $[X]_0$ while fixing the initial concentration of initiator as $[Ini]_0 = 2\%$. We plot the RDP as a function of monomer conversion in Figure 2.7b, in which the sharp peaks in the curves correspond to the gel points. We observe that the gel point shifts toward lower monomer conversion as the volume fraction of the cross-linker increases. In other words, gelation occurs earlier for an increasing amount of cross-linker. These results are in quantitative agreement with previous experiments [25, 34].

Figure 2.7c confirms that our DPD approach successfully captures living copolymerization of monomer and bifunctional cross-linker to form a polymer gel. We plot the simulated gel point as a function of the initial concentration ratios of cross-linker to initiator, $[X]_0/[Ini]_0$, showing that the gel points obtained with our DPD technique are in good agreement with previous experimental data [25, 34]. This behavior indicates that our DPD method can accurately take into account the spatial distribution and long-range diffusion of pendant functional groups that are subject to the topological constraints in formed macromolecules [25].

2.3.2 Modeling Regeneration of Severed Polymer Gels with Interfacially Active Nanorods

Designing synthetic systems that mimic the remarkable ability of amphibians to regrow limbs is considered as one of the holy grails of materials science. While self-healing materials that can mend local defects have been developed [35–38], there are virtually no examples of systems that can regenerate bulk sections of a severed material. The advent of such regenerative materials can dramatically extend the useful lifetime of manufactured products by giving them the ability to regrow when damaged.

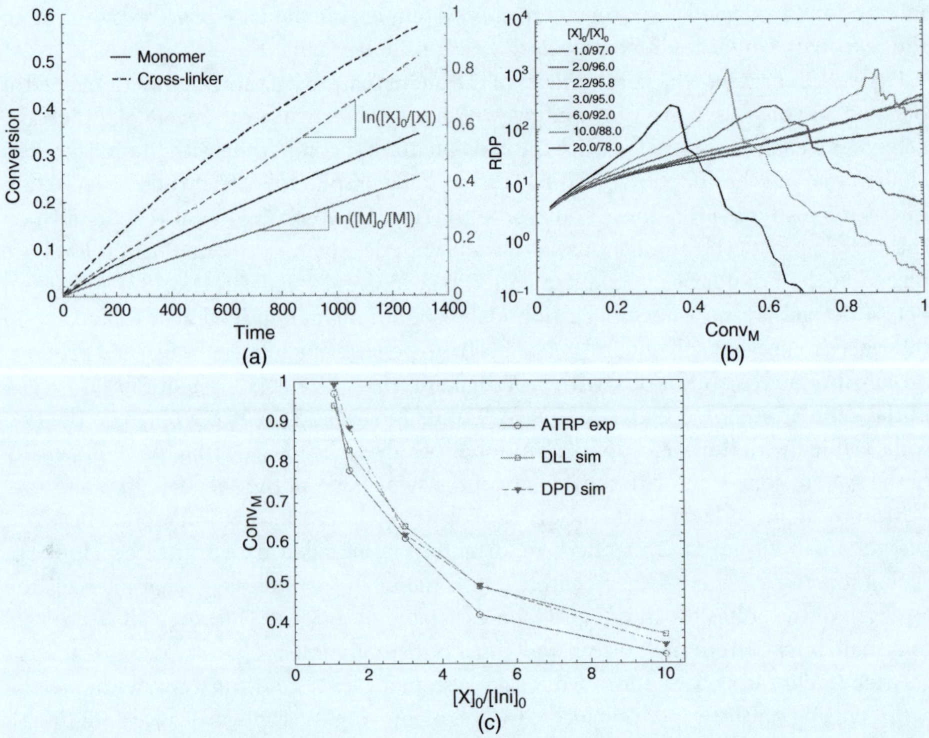

Figure 2.7 (a) Conversions and $\ln([M]_0/[M])$ of monomer (solid lines) and $\ln([X]_0/[X])$ of cross-linker (dashed lines) on reaction times during copolymerization with initial concentration ratios $[M]_0/[X]_0/[Ini]_0 = 78/20/2$. (b) Evolution of the reduced degree of polymerization (RDP) with monomer conversion for a systematic variation of the initial concentration ratios of cross-linker to initiator $[X]_0/[Ini]_0$, while fixing $[Ini]_0 = 2\%$. (c) Comparison of DPD simulated gel points with ATRP experimental values. Reproduced from Ref. [10] with permission from the American Chemical Society.

Our efforts are inspired by the biological processes in tissue regeneration [39], which is guided by an internalized "instruction set": initiation, propagation, and termination [40], signaling molecules mediate the dynamic cascade of the biological events in the regrowth [40]. Analogous to biological systems, fully synthetic self-regenerating materials should incorporate the following elements: first, the system must contain a component that not only senses the removal of material but also initiates the regrowth; second, the system must involve a means of continuing or propagating the desired growth; finally, when the material reaches a certain size, the ongoing reactions must terminate.

Using the DPD gel model and ATRP reaction scheme, we design a nanorod-filled gel that undergoes these processes to regenerate the gel matrix when a layer of the material is sliced off. With this layer removed, the interfacially active nanorods diffuse to the newly formed interface and extend into the outer solution, which contains monomers and a small fraction of cross-linkers. Polymerization initiated from the rods' surfaces grows chains that become cross-linked to form a new gel that resembles the severed layer. After the initial cut, the regeneration requires no external intervention; synergistic interactions among all components in this system enable the vital processes leading to the regrowth, which could be repeated with subsequent cuts.

Figure 2.1a presents the components of the uncut material: nanorods functionalized at one end and dispersed in a polymer gel, which is swollen in a good solvent at 28 °C. The solvent is weakly incompatible with the rods' shafts, but compatible with the end-grafted chains, making the functionalized rods essentially amphiphilic. The rods also contain initiator sites for ATRP. These rods serve to trigger the ensuing, vital reactions when the sample is cut. The amorphous solid walls with density $\rho_{wall} \sim 3$ and height $h = 1$ are placed at the boundaries of the simulation box in the normal direction (z) above and below the gel layer. Bounce-back boundary conditions are applied at the fluid–solid interfaces to prevent solvent and gel beads from penetrating into the walls, and produce no-slip boundary conditions with minimal interfacial density oscillations [41]. The initial simulation box is $24.6 \times 24.6 \times 34.9$ units in size and the space between the two walls is filled with the corresponding number of solvent beads, keeping the total density of the system consistent with the density of the swollen gel at the specified temperature.

Moreover, the gel layer is attached to the substrate via an adhesive interaction between the gel and wall beads. This effective attraction is modeled by a truncated Hookean spring force [42] $F_{adh} = -K_{adh} r_{ij}$ with a cutoff radius of r_c, where K_{adh} characterizes the strength of the adhesion and K_{adh} is 6 unless stated otherwise. This force allows the gel to remain anchored to the bottom wall in all our simulations.

After equilibrating the nanorod-filled gel system for 1.5×10^6 time steps, we model the act of cutting off the upper portion of the gel by moving the top wall upward (in the $+z$ direction) and adding outer solution beads to this newly created space (see Figure 2.8a).

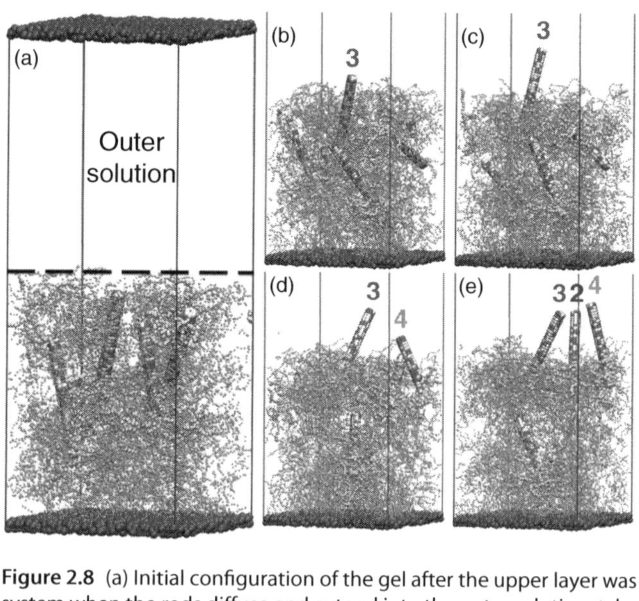

Figure 2.8 (a) Initial configuration of the gel after the upper layer was cut off. (b–e) Snapshots of the system when the rods diffuse and extend into the outer solution, taken at the following times: (b) $t = 2500$, (c) $t = 10\,000$, (d) $t = 23\,700$, and (e) $t = 34\,300$. Numbers identify the rods that extend into the outer solution. Reproduced from Ref. [10] with permission from the American Chemical Society.

The simulation box changes from 24.6 × 24.6 × 34.9 to 24.6 × 24.6 × 64.9, with the total density being $\rho_{sys} = 3$. The outer solution is partially immiscible with the inner solution.

To properly model the thermo-responsive behavior of a realistic gel (e.g., poly(*N*-isopropylacrylamide), PNIPAAm), we relate the repulsive parameter between a polymer and solvent bead, a_{ps}, to the Flory–Huggins parameter characterizing the polymer–solvent interaction, χ_{ps}, as follows: $a_{ps} = a + k_B T \chi_{ps}/0.306$ [16]. In studies of PNIPAAm, it is typically assumed that χ_{ps} depends on temperature and polymer concentration [23, 42–44]. Hence, we assume that $\chi_{ps}(T, \phi) = \chi_1(T) + \chi_2\phi_p$, where ϕ_p is the polymer volume fraction in the gel and $\chi_1(T) = (\delta h - T\delta s)/k_B T$, with δh and δs being the respective changes in enthalpy and entropy [23, 45]. Note that ϕ_p is calculated as $\phi_p = \overline{\rho}_p/\overline{\rho}_{gel}$, where ρ_p is the time-averaged polymer number density and $\overline{\rho}_{gel}$ is the time-averaged total number density of the gel including the polymer and solvent beads. Here, we set $a = 25\ T^*$ and take $\delta h = -14.331 \times 10^{-14}$ erg, $\delta s = -5.452 \times 10^{-16}$ erg·K^{-1}, and $\chi_2 = 0.596$ to reproduce the experimentally observed lower critical solution temperature (LCST) volume phase transitions of the PNIPAAm gels [23, 46]. Notably, this is the first DPD simulation that captures such temperature-induced volume phase transitions in gels.

Our choices of the interaction parameters between the different components, a_{ij}, in the system are listed in Table 2.2. The components can be separated into hydrophilic and hydrophobic moieties: the inner solution and end-grafted chains on the rods are hydrophilic; the outer solution (including the monomers and cross-linkers) and the nanorods are hydrophobic. The gel is slightly hydrophobic, as characterized by positive experimental values of the χ_{ps} parameter [23]. We set the interaction parameter between the gel and end-grafted chains on the rods to be 20 (i.e., we assume some relatively weak attractive interactions between these moieties). Finally, we set the rod–solvent interaction to 35 only upon introducing the outer solution. During the initial equilibration of the nanorod-filled gel system, we set this interaction parameter to 25 (i.e., the rods are hydrophilic) to prevent the clustering of the rods in the original sample and to ensure their uniform distribution in the gel matrix. Experimentally, such a change in the properties of the rod–solvent interactions can be achieved, for example, if rods are additionally functionalized with photosensitive moieties that are hydrophilic in the dark and become hydrophobic in light [47].

We first probe the cascade of cooperative events that occur when the upper half of the nanorod-filled gel is removed (see Figure 2.8). With this dramatic cut, the system is now exposed to the outer fluid, which is both weakly incompatible with the solvent in the gel

Table 2.2 List of DPD interaction parameters between different components in units of reduced temperature T^*

	Inner solution	Outer solution	Gel	Rod	Rod-grafted chain
Inner solution	25	35	$25 + \chi_{ps}/0.306$	35	25
Outer solution	—	25	35	25	35
Gel	—	—	25	35	20
Rod	—	—	—	25	35
Rod-grafted chain	—	—	—	—	25

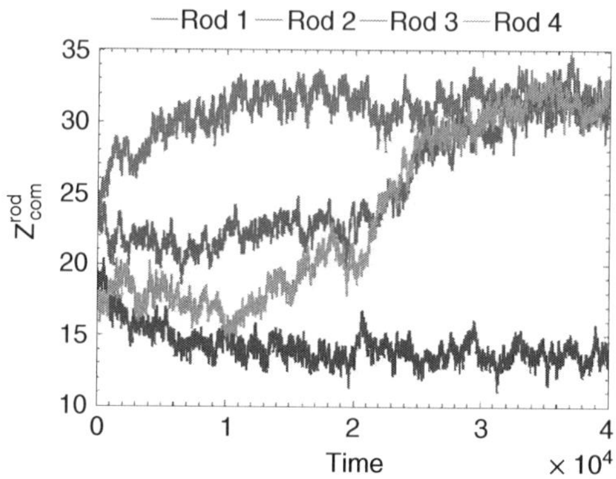

Figure 2.9 Time evolution of the center of mass of the four rods in the *z*-direction. The rod numbers correspond to the respective rods in Figure 2.8. Reproduced from Ref. [10] with permission from the American Chemical Society.

and compatible with the rods. The outer solution also contains dispersed monomer and a small concentration of cross-linking molecules to provide materials for the regrowth of the polymer matrix. The amphiphilic rods act as crucial sensors, indicating a change in the state of the system. Specifically, rods near the new interface encounter the more compatible outer solvent and hence diffuse toward this boundary to maximize their interaction with this fluid. Figures 2.8b–e illustrate the rods' progression to the cut and the final orientation at the cut, and Figure 2.9 shows the rods' center of mass as a function of time. Because the chains on the rods are attracted to the gel and incompatible with the outer solution, they prevent these particles from diffusing out of the gel matrix. Hence, the grafted chains anchor the nanorods at the interface where the cut gel meets the outer solvent.

We note that, since the rods diffuse toward this newly formed interface only upon exposure to the outer fluid, rods buried deep within the network remain within the gel (see Figure 2.8e and Figure 2.9). Moreover, steric hindrance from rods already at the interface can inhibit particles deeper in the matrix from reaching the cut. We assert that the extension of the rods into the outer solution also depends on their orientation. Namely, when a rod is oriented downward (i.e., the unfunctionalized end points to the bottom of the gel), the grafted chains are closer to the interface. Due to the unfavorable enthalpic interactions between the grafted polymer chains and the outer solvent, the rod is likely to diffuse away from the cut. Furthermore, topological constraints imposed by the gel matrix hinder the reorientation of these downward-pointing rods.

To explore the effect of the initial proximity of the rods to the cut on the number of rods that remain buried in the matrix, we conduct two separate sets of simulations involving the vertically aligned, upward-pointing rods. In the first set, the rods were placed a distance of $3r_c$ closer to the interface than in the second set. (Recall $r_c = 1$ in the simulations.) The runs from the first set have 25% fewer buried rods than the runs from the second set.

Figures 2.10a and 2.10b demonstrate the effect of the rods' orientation. We plot the trajectories of rods in eight independent runs for samples containing rods that are initially randomly distributed and ones with rods initially vertically aligned and pointing upward, respectively. The results show that the latter samples yield a higher fraction of rods localized at the interface at the end of simulation. (Notably, 31% fewer rods remain buried in the gel for the case in Figure 2.10b than for the case in

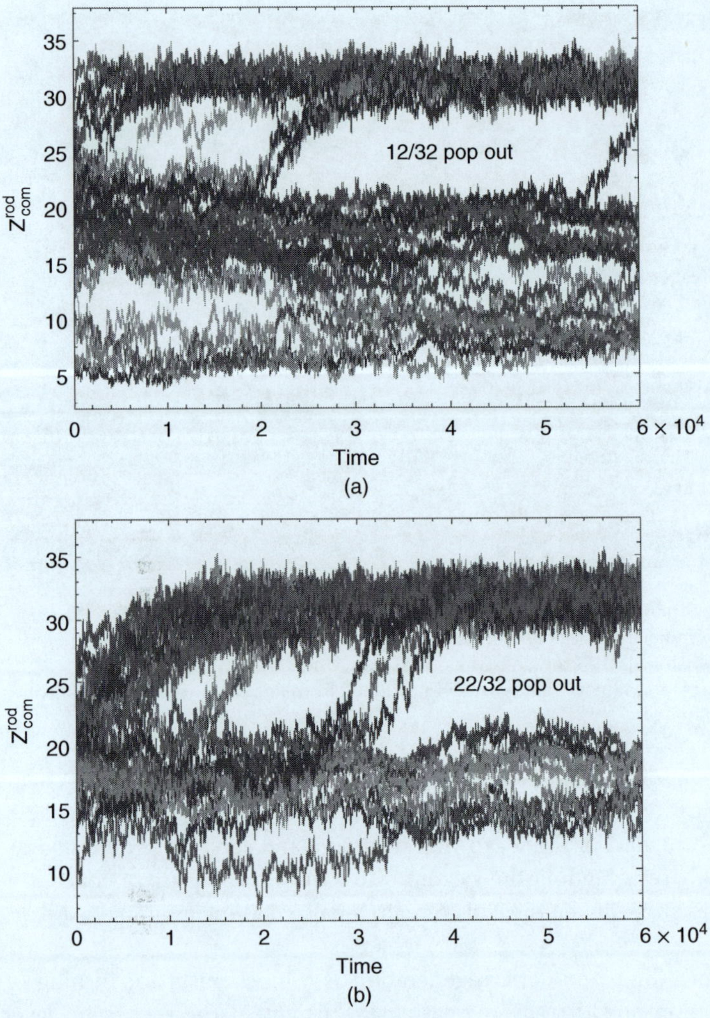

Figure 2.10 Time evolution of the z-coordinate of the center of mass of the rods for eight independent runs (total 32 rods) with: (a) random initial distribution (position and orientation) of rods and (b) rods that were initially aligned vertically and facing upward and have $\Delta z_{rod-cut} = 4.4$, which is the distance between the tip of the rods and the interface. Reproduced from Ref. [10] with permission from the American Chemical Society.

Figure 2.10a.) Figure 2.10a also reveals that even when the rods are initially randomly distributed in the gel, 38% of them become localized at the new interface.

With rods at the interface, the initiator sites now become exposed to the monomer-laden outer solution. Hence, the system undergoes one of the critical events for regeneration: the initiation process. Importantly, chain growth in ATRP is significantly slower than the diffusion rate of rods to the interface [48, 49]. Thus, the polymerization primarily occurs once the rods are at the cut (not within the gel's bulk). Also, the high surface area provided by the rods facilitates changes in initiator concentration, which plays a significant role in the structure of the new layer as discussed further below.

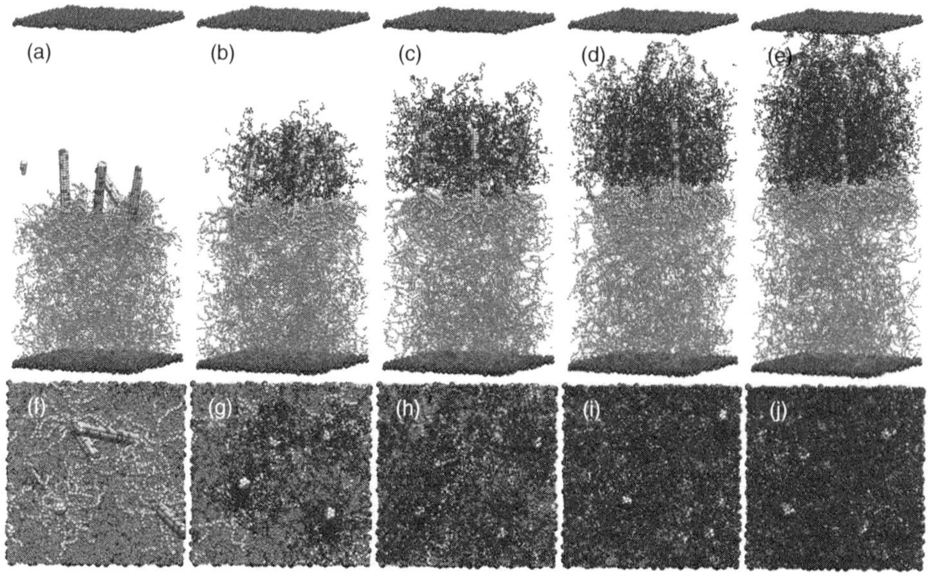

Figure 2.11 (a–e) Regrowth of the top layer at the following monomer conversions: (a) 0%, (b) 25%, (c) 50%, (d) 75%, (e) 96%. Initial monomer and cross-linker concentrations are $[M]_0 = 30\%$ and $[X]_0 = 0.53\%$. Initiator densityσ_i is 0.25, which corresponds to a concentration $[Ini]_0 = 0.27\%$. (f–j) Top-down views corresponding to frames a–e. Reproduced from Ref. [10] with permission from the American Chemical Society.

Figure 2.11 shows snapshots of the critical propagation step, as monomers form chains from the surface of the rods via the ATRP. As shown in the reaction diagram (Figure 2.2b) the cross-linkers become part of these growing polymers and can bridge two separate neighboring chains, thereby binding the system into a network. For these concentrations of monomer and cross-linker, the rods remain localized at the interface as the gel continues to grow.

The final component for biomimetic regeneration is a mechanism for turning off the propagating reaction and thereby regulating the height of the regrowing layer. The concentration of monomer and cross-linker are the critical variables in the termination process; once these species have completely reacted, the growth halts. Figure 2.12a shows that with initial concentrations of 30% and 0.5% for the monomer and cross-linker respectively, both species become almost fully converted into the gel within the time scale of the simulation. The density profiles in Figure 2.12b demonstrate the vital role of the percent conversion of the monomers in the density of the regrown gel, for a fixed initiator concentration; at 38% conversion, the densities of the two gels are quite similar; at 96% conversion, the new gel exhibits a higher density than the original material.

More importantly, the properties of the new gel can be tailored to resemble those of the original material by tuning the initiator density on the rod, σ_i, and initial monomer concentration $[M]_0$. Increasing $[M]_0$ leads to increases in the density and height of the new layer (Figure 2.13a). The densities of the two layers can be reasonably matched by the appropriate choice of the initiator concentration, σ_i (Figure 2.13b). The density of

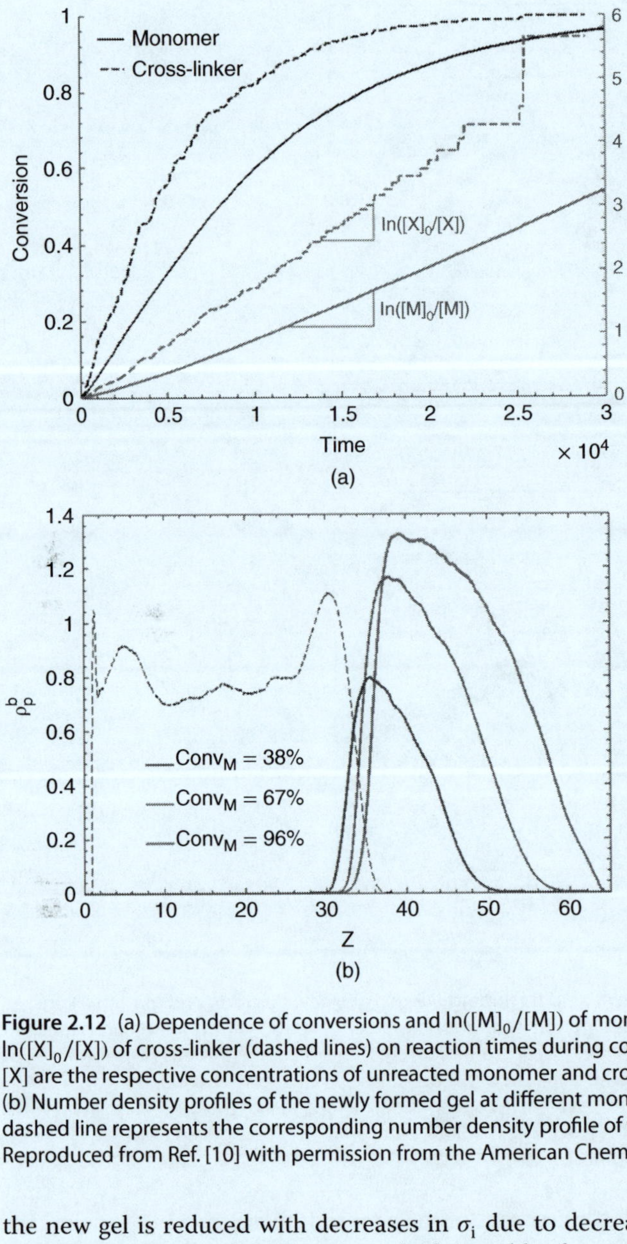

Figure 2.12 (a) Dependence of conversions and $\ln([M]_0/[M])$ of monomer (solid lines) and $\ln([X]_0/[X])$ of cross-linker (dashed lines) on reaction times during copolymerization, where [M] and [X] are the respective concentrations of unreacted monomer and cross-linker at a given time. (b) Number density profiles of the newly formed gel at different monomer conversions. The black dashed line represents the corresponding number density profile of the original gel at $t = 3 \times 10^4$. Reproduced from Ref. [10] with permission from the American Chemical Society.

the new gel is reduced with decreases in σ_i due to decreases in the number of growing polymer chains per unit volume [24]. Notably, the structure of the regenerated gel layer is relatively heterogeneous, exhibiting a dense network with a high fraction of cross-links in the proximity of the rods (see Figure 2.14). While the fraction of inter-rod cross-links is relatively insensitive to σ_i (increasing only slightly at $[M]_0 = 50\%$ with decreasing σ_i), it increases dramatically with increases in $[M]_0$for any given value of σ_i (Figure 2.15).

Although we showed that the density of the new gel can be tuned to match that of the uncut gel, the resulting composite is not ideal due to the lack of embedded nanorods in

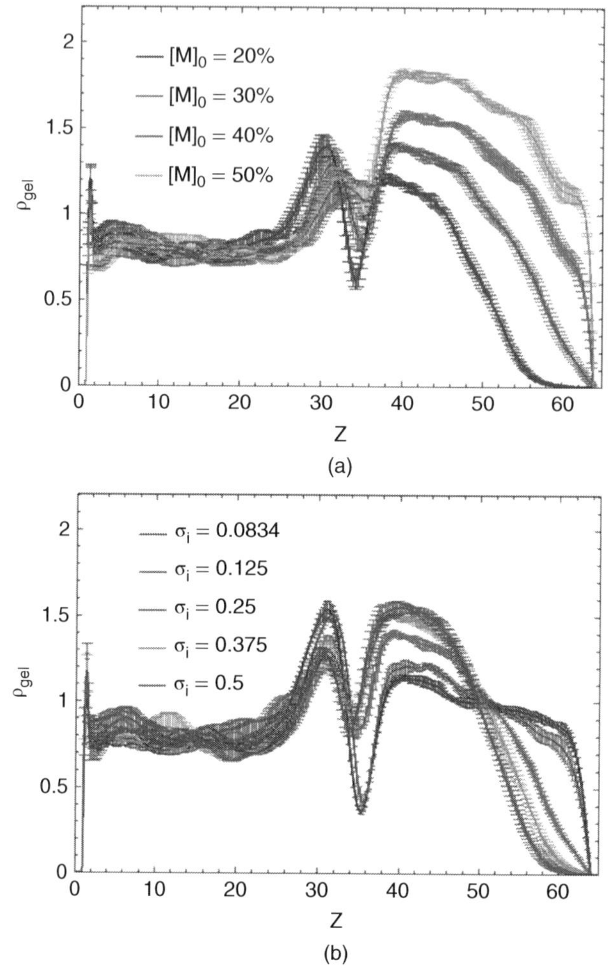

Figure 2.13 Density profiles of the system (including original gel, new gel, rods, and grafted chains) at monomer conversion 95% with: (a) different initial monomer concentrations $[M]_0$ for initiator density on the rod surface $\sigma_i = 0.25([\text{Ini}]_0 = 0.27\%)$ and for initial cross-linker concentrations $[X]_0$ varying with $[M]_0$ by fixing $[M]_0/[X]_0 \approx 56$; (b) with different initial densities σ_i ($[\text{Ini}]_0$ varies from 0.09% to 0.54%) for $[M]_0 = 30\%$ and $[X]_0 = 0.53\%$. Insets are the corresponding snapshots of the gel system in (a) with $[M]_0 = 20\%$ and $[M]_0 = 50\%$, and in (b) with $\sigma_i = 0.00834$ and $\sigma_i = 0.5$. Reproduced from Ref. [10] with permission from the American Chemical Society.

the regenerated layer. Thus, it is important for the nanorods to disperse in the new layer so that the characteristics of the uncut material are replicated as closely as possible. Notably, the presence of nanorods in the regenerated layer also impart additional strength to the new gel [50–53]. However, when the nanorods, initially anchored at the interface, escape into the regrowing gel as the amount of new gel increases during the polymerization (see Figure 2.16), their anchoring effect diminishes. Consequently, the structural integrity between the cut gel and the new gel is compromised due to the repulsive interaction between the respective moieties that leads to the development of a gap at the interface.

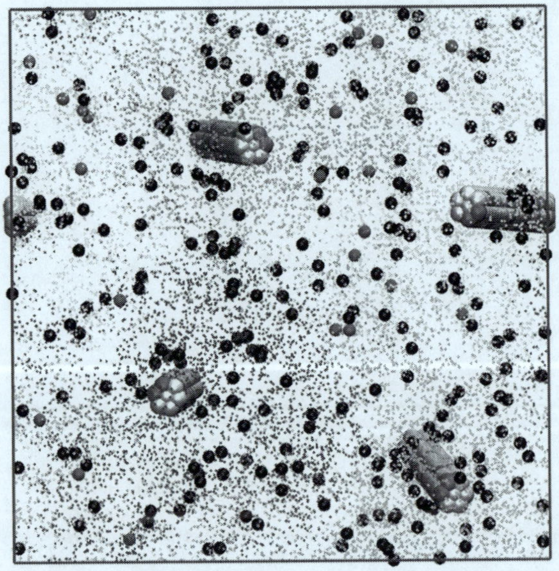

Figure 2.14 Top-down view of the spatial distribution of the cross-links at monomer conversion 96% for the regenerated gel system with $\sigma_i = 0.25$ and $[M]_0 = 30\%$ (corresponding to the simulation in Figures 2.11e and 2.11j). Reproduced from Ref. [10] with permission from the American Chemical Society.

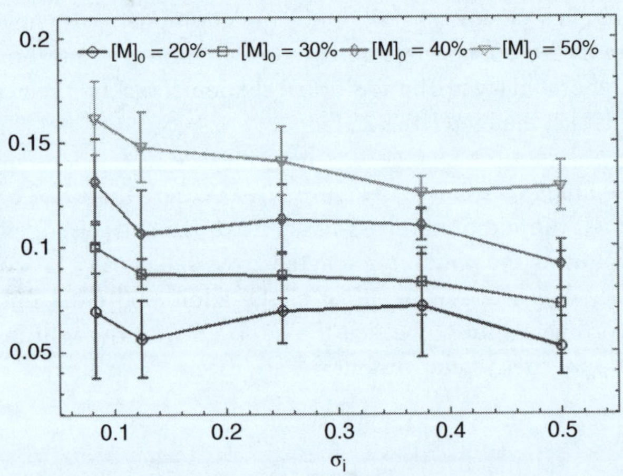

Figure 2.15 Fraction of the number of inter-rod cross-links ($X_{inter\text{-}rod}$) with respect to the total number of cross-links formed (X_{total}) as a function of initiator density and initial monomer concentration. Data are taken at cross-linker conversion of at least 96%. Error bars indicate the variations among four independent runs. Reproduced from Ref. [10] with permission from the American Chemical Society.

In order to produce a true nanocomposite polymer matrix that resembles the original composite while preserving a coherent interface, we propose to introduce a new type of cross-linker to the system (below, we refer to it as the additional cross-linker), which binds the original gel and the regrown gel. Namely, additional cross-linkers constitute a fraction of the original gel chain beads. The modified uncut gel is now essentially a copolymer network with one monomer containing functional groups that remain inert during the polymerization reaction until they are activated by external stimuli,

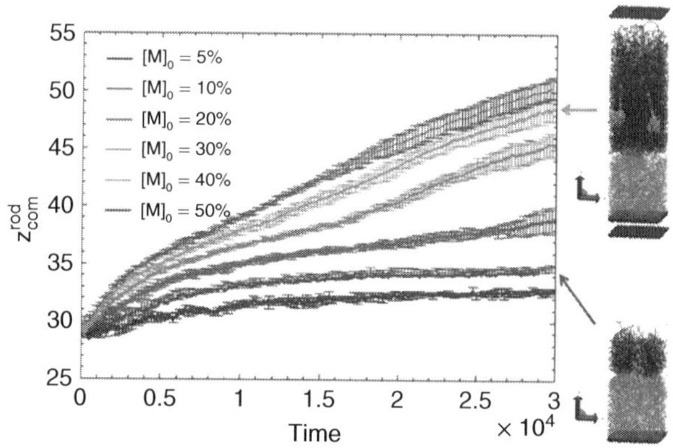

Figure 2.16 Position of center of mass of nanorods. Left: the z position of the center of mass of the four nanorods is measured over time for different initial monomer concentrations in the solvent. Right: system snapshots for $[M]_0 = 40\%$ (top) and $[M]_0 = 10\%$ (bottom) at the end of the simulation. The simulation box changes from 24.6 × 24.6 × 34.9 to 24.6 × 24.6 × 94.9 after the cut to incorporate more materials for regeneration. Reproduced from Ref. [10] with permission from the American Chemical Society.

resembling the agents used in click chemistry [54]. Once the additional cross-linkers are activated, they participate in the polymerization reaction to form covalent bonds that bridge the cut gel and regenerated layer. The additional elemental reaction relevant to the additional cross-linker is sketched in Figure 2.17.

Our results show that the structure change at the interface is evident after incorporating additional cross-linkers, as shown in Figure 2.18a. Where there had once been a separation at the interface, the additional cross-linker holds the two gels together through covalent bonds. By counting the number of reacted cross-linkers as a function of z position, we quantitatively characterize the effect of the additional cross-linker. Importantly, the cross-linker profile exhibits a peak at the interface with the additional cross-linker in contrast to a dip for the system without the additional cross-linker.

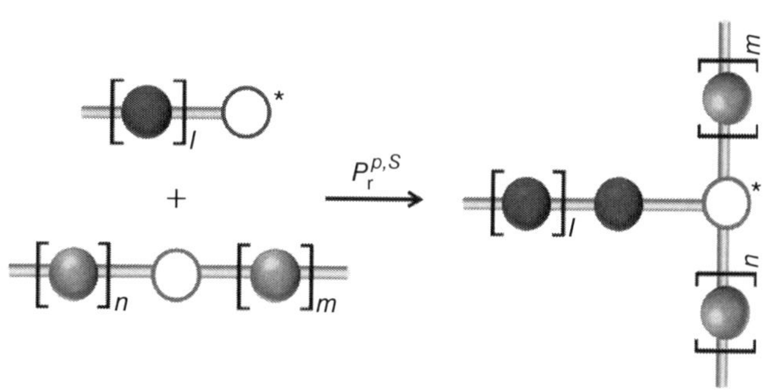

Figure 2.17 Formation of an inter-gel cross-link. The radical from an active end (indicated by an asterisk) is first transferred to the additional cross-linker. At this point, the additional cross-linker with the radical may now form another bond. An open circle indicates that the particle has not yet fully reacted. $P_r^{p,S} = 0.075$ is the reaction probability between a monomer with the radical and an additional cross-linker.

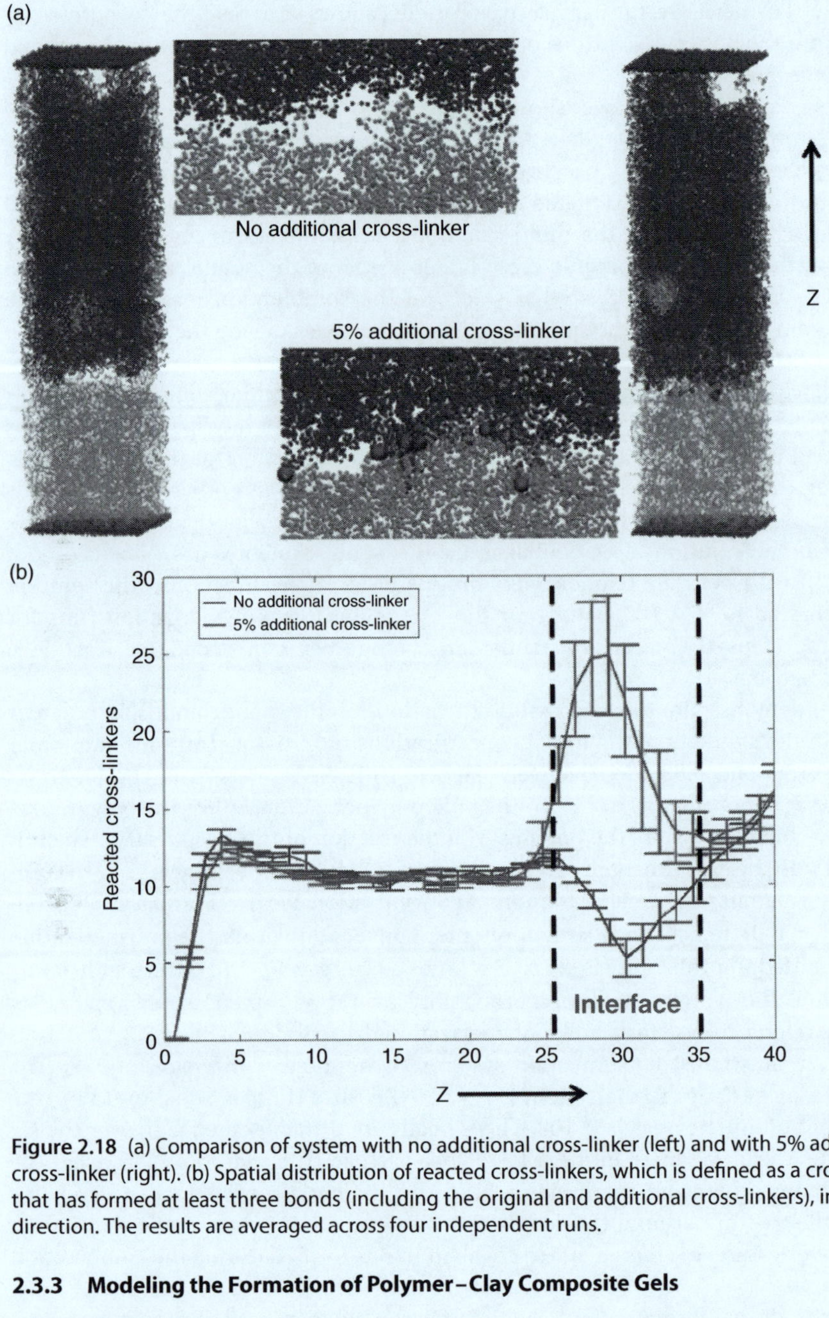

Figure 2.18 (a) Comparison of system with no additional cross-linker (left) and with 5% additional cross-linker (right). (b) Spatial distribution of reacted cross-linkers, which is defined as a cross-linker that has formed at least three bonds (including the original and additional cross-linkers), in the z direction. The results are averaged across four independent runs.

2.3.3 Modeling the Formation of Polymer–Clay Composite Gels

The synthesis of polymer nanocomposites often involves *in situ* polymerization that occurs within complex, multi-component fluids [50–53]. A better understanding of the coupling of the polymerization reactions to the dynamics of the host solution is vital for precise control of the polymerization, and hence the properties of the final products. Here, we study the FRP of hydrogels with embedded clay particles using our

DPD model. Furthermore, the particle-based model allows us to readily incorporate clay particles into the polymer solutions or gels and to explicitly specify clay–polymer and clay–solvent interactions.

Previous experiments have shown that nanocomposite gels consisting of poly(*N*-isopropylacrylamide), PNIPAAm, and inorganic clay prepared by FRP exhibit excellent mechanical properties compared to chemically cross-linked gels [55–57]. In these gels, the exfoliated clay particles in an aqueous solution connected by polymer chains serve as the multi-functional cross-linkers to form the polymer network. In the absence of organic cross-linker agents in the solution, the formation of polymer–clay composite gels relies solely on the combination reactions of active chain ends emanating from the surface of the clay particles during the polymerization [55–57].

To model the preparation of such composite gels, the initial solution comprises solvent beads, initiator beads, monomer beads, and initiator-modified clay particles (as described in Section 2.2.2). We consider an athermal solution for the clay particles by choosing the interaction parameters between clay beads and other solution beads (including solvent, initiator, and monomer) all as 25. The clay particles are randomly dispersed in the solution after equilibrium. The simulation box has dimensions of $30 \times 30 \times 30$ with periodic boundary conditions in all three directions. The solution was equilibrated for 5×10^5 simulation time steps and the polymerization runs for 3×10^6 time steps to reach sufficiently large monomer conversion (at least 90% monomer conversion).

We first demonstrate the successful formation of the nanocomposite polymer network. In the simulation, the initial concentrations of initiator $[\text{Ini}]_0$ and monomer $[\text{M}]_0$ are 0.13% and 9.83%, respectively. Hence, $[\text{Ini}]_0/[\text{M}]_0 \approx 1/74$ is chosen to be similar to that in our reference case for the bulk polymerization. When the polymerization reaches full conversion, the polymer volume fraction of the gel ϕ_p is 0.1, which is consistent with the experimentally prepared polymer–clay composite gels [57]. To compensate for a significantly smaller monomer concentration, we use a propagation probability $P_\text{r}^{\text{p,M}} = 0.01$ to achieve a large monomer conversion within the polymerization time. The initiation probability is set to ten times as large as $P_\text{r}^{\text{p,M}}$ to ensure sufficiently fast initiation. The overall termination probability is set to $P_\text{r}^\text{t} = 0.002$, so that $P_\text{r}^{\text{p,M}}/P_\text{r}^\text{t} = 5$ (as in the reference case for a bulk polymerization discussed in Section 2.3.1).

Figure 2.2b illustrates the simulation systems at monomer conversion of 95%. The number of clay particles in the system is $N_\text{C} = 6$. Because $[\text{Ini}]_0$ is equal to 0.13%, the total number of initiator beads is 108. These beads are distributed equally over the six clay particles, resulting in 18 initiator beads per particle. The snapshots show that the polymer chains grow from the initiator beads on the surface. The emanating chains form a shell structure around the center clay particle. At large monomer conversion, the gel network starts to emerge due to covalent bonds formed during the combination reactions.

To quantify the gel formation, we characterize the polymerization kinetics by investigating the dependence of $\ln([\text{M}]_0/[\text{M}])$ on time. A deviation from linear first-order kinetic behavior is evident at large monomer conversion (Figure 2.19a), indicating the effects of chain termination. We observe a monotonic decrease in the number of polymers with active ends with time. Notably, we find that active chain ends are consumed faster at the early stage of the polymerization. This behavior is different from that in the

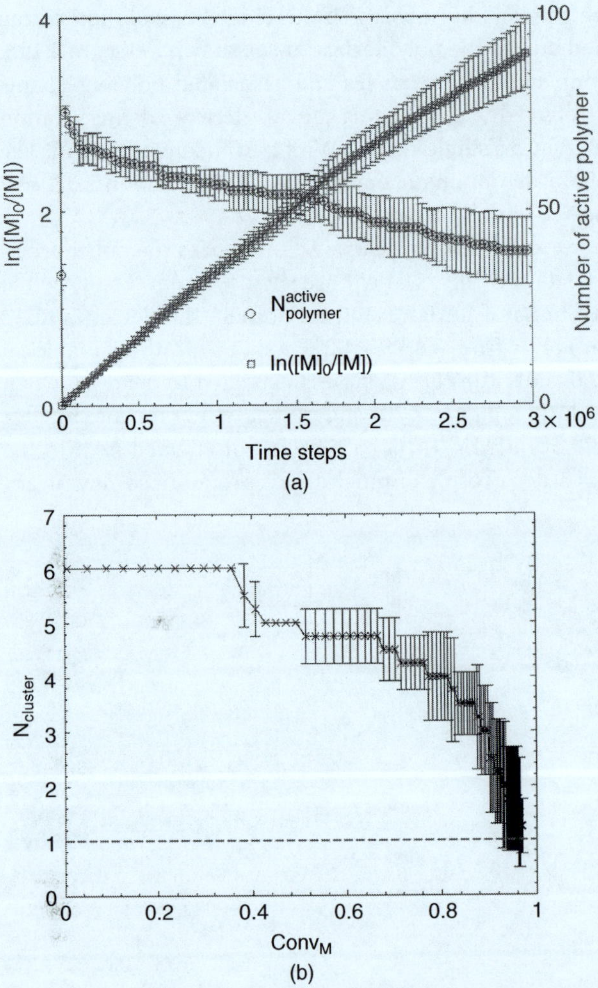

Figure 2.19 (a) $\ln([M]_0/[M])$ (squares) and the number of active polymers (circles) as functions of reaction time steps during polymerization. The number of clay particles in the solution is $N_C = 6$. (b) Number of clusters in the simulation system as a function of monomer conversion for the system with $N_C = 6$. Error bars arise from averaging over four independent runs.

bulk polymerization discussed in Section 2.3.1, in which the rapid decrease in the number of polymers with active ends occurs at the late stage (see Figure 2.5a). To understand this behavior, recall that at the beginning of polymerization, the distribution of the active chain ends in the solution are highly heterogeneous in space because the initiator is concentrated near clay platelets [55–57]. Hence, the probability of chain ends meeting each other is highest near the clay surface when polymerization just starts. As chains grow, the volume that active chain ends can explore increases, which leads to more homogeneous distribution. Thus, we attribute this observed fast consumption of active chain ends at the early stage to the high local concentration of propagating chains around particles.

Furthermore, we characterize the gelation process by tracing the evolution of the number of clusters $N_{cluster}$ formed during the polymerization, as shown in Figure 2.19b. Each cluster consists of interconnected clay particles and associated polymer chains that connect different particles (referred to as X-chains above). Here, we define gelation when all particles are incorporated into a single cluster. With four independent runs, we observe that the gelation only occurs at monomer conversions ~93%. Gelation is defined when the averaged $N_{cluster}$ with error bars is close to 1.

Finally, we explore how the number of clay particles, N_C, influences the polymerization kinetics and gelation. To isolate the effects of N_C, we vary N_C from 3 to 12 while maintaining the initial concentrations of initiator and monomer. The corresponding volume fraction of clay particles varies from 1.1% to 4.2%. As a result, the initial local concentration of active chain ends near the clay particle is expected to decrease when there are more particles in the solution. Figure 2.20a shows that except for the smallest value of $N_C = 3$, the dependence of $\ln([M]_0/[M])$ on time is not affected (within the magnitudes of error bars). $N_C = 3$ does result in much more pronounced downward

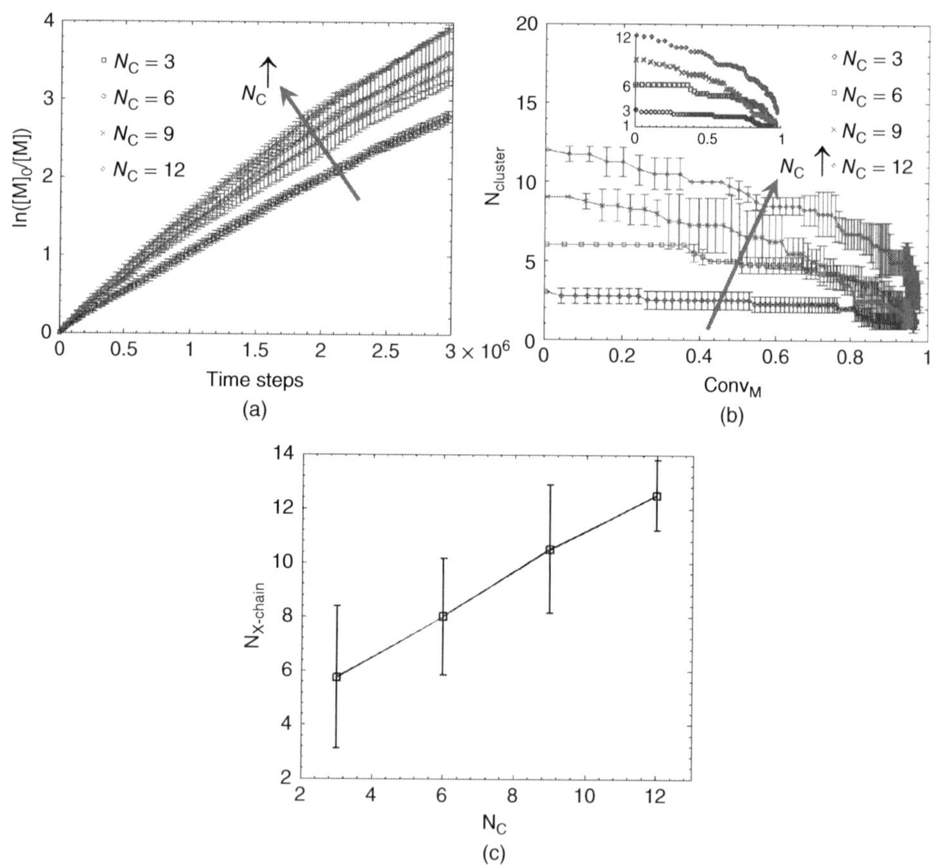

Figure 2.20 Effect of N_C on the evolution of (a) $\ln([M]_0/[M])$ and (b) the number of clusters. (c) Number of X-chains formed due to the combination of growing polymer chains as a function of the number of clay particles N_C. Error bars arise from averaging over four independent runs.

curvature associated with frequent termination reactions, which are due to the large initial concentration of active chain ends around the particles discussed above. Notably, this heterogeneous distribution of initiator is most pronounced for the smallest number of particles; an increase in N_C leads to a more homogeneous distribution of initiator and, correspondingly, a more homogeneous distribution of active chain ends at early times.

The number of clusters exhibits complicated behavior when we vary N_C. Figure 2.20b shows that the number of clusters decreases as the monomer conversion increases, indicating that the system is undergoing gelation. Successful gelation occurs for $N_C = 3$, 6, and 9, but none of the four runs reaches gelation when $N_C = 12$. On the one hand, as noted above, a larger number of particles results in more homogeneous distribution of active chain ends. In principle, this could allow more active chain ends to be preserved until the growing chains from different particles are long enough to meet and form the X-chains. As shown in Figure 2.20c, the number of X-chains increases monotonically as N_C increases. This behavior is consistent with experimental observations [57]. It has been shown that more X-chains would lead to better mechanical properties [57]. Therefore, a large number of X-chains is desired in the composite gels. A large number of particles, however, increases the difficulty of satisfying our chosen criterion for gelation ($N_{cluster} = 1$). Based on this criterion, there is a tradeoff between an increase in the number of X-chains and decrease in the chance to form a single cluster. In our simulations, 3.2% volume fraction of clay particles ($N_C = 9$) appears to yield an optimal composite gel with the largest number of X-chains.

2.4 Conclusions

We develop a reactive DPD approach to explore the self-assembly behavior in polymer nanocomposites in the course of polymerization. Our model allows us to simulate FRP and ATRP in the same computational framework. Chemical reactions change the bead types of the components involved and can also lead to the formation of a covalent bond between the reacting pair of beads, where experimentally relevant reaction probabilities are employed to govern the reaction kinetics. The reactions are coupled to the dynamics of the DPD system, which is evolved according to Newton's equations of motion. The characteristic features of FRP and ATRP are successfully reproduced in the simulations. More importantly, this coupling captures the interplay between hydrodynamics and chemical reactions in multi-component mixtures.

In the bulk polymerization via FRP, we systematically varied the reaction probabilities relative to the set of reference parameters in order to understand the effects of initiation, propagation, and termination on the polymerization kinetics. The slow initiation effect is exhibited in the simulations as the upward curvature of the $\ln([M]_0/[M])$ curve and the increase of the PDI peak at small monomer conversion. The termination probability has a significant influence on the polymerization. In particular, the polydispersity index and maximum degree of polymerization significantly increase with an increase in the termination probability. By modeling a living copolymerization of monomers and cross-linkers, we observed that the network-forming gelation is in quantitative agreement with the corresponding experiments. An increasing amount of cross-linkers facilitates the gelation, which provides evidence that the effects of the polymer geometry and the spatial distribution of the pendant functional group are captured in the model.

We then modeled a polymer nanocomposite that enables self-regeneration of the gel when a significant portion of the material is severed. The cut instigates the dynamic cascade of cooperative events leading to regrowth. The information for the regeneration is inherent to the system; the "popping up" of the rods at the interface provides the signal to initiate the polymerization that leads to the formation of the new gel. The living copolymerization is halted when monomer and cross-linker are consumed, yielding a layer with a specified height, cross-link density, and monomer density. As some of the rods remain buried in the bulk of the gel, subsequent cuts below the newly regenerated region can reposition the previously buried nanorods to the interface, and hence to repeated regrowth. These subsequent events do, however, require the addition of new monomers and cross-linkers to the host solution. In this system, the interfacially active rods play a crucial role in the initiation and propagation steps of the regrowth by sensing the damage, migrating to the cut, and thereby delivering controllable quantities of the initiators to the critical location. Notably, the chains grafted to the ends of the rods also serve an important function: they provide a means of tailoring the effective adhesion between the newly regenerated and old gels. Furthermore, by allowing the nanorods to diffuse into the regenerated layer and introducing an additional cross-linker into the original gel that bridges the cut gel and regenerated layer, we demonstrated the regeneration of a true nanocomposite polymer matrix that resembles the original composite while preserving a coherence interface.

Finally, we investigated the preparation of polymer–clay composite gels by FRP with exfoliated inorganic clay in an aqueous solution. We successfully demonstrated gelation of the polymer–clay composites caused by the combination reactions of active chain ends emanating from different clay particles, where each particle acts as a multi-functional cross-linker. We showed that the number of clay particles strongly affects the gelation process, and we observed that a 3.2% volume fraction of clay particles produces a composite gel with the largest number of X-chains.

To summarize, our reactive DPD approach provides an effective means to model simultaneous self-assembly and polymerization processes involving multi-component mixtures. Our findings provide insight into the polymerization kinetics of FRP and ATRP, as well as potential guidelines for synthesizing polymer nanocomposites that exhibit the desired properties. The regrowth of a severed gel layer outlined here helps pave the way for creating materials that are both self-regulating and self-replenishing [58], leading to systems with improved functionality over significantly extended lifetimes.

Acknowledgments

The authors gratefully acknowledge support from the U.S. Department of Energy, Office of Basic Energy Sciences, Division of Materials Sciences and Engineering under Award DEFG-02-02ER45998 for the development of the computer simulations, and from the Army Research Office under grant number W911NF-13-1-0489 for applications of the model toward 4D printing.

References

1. G. Odian, *Principles of Polymerization* (John Wiley & Sons, Inc., Hoboken, 2004), 4th edn.
2. K. Matyjaszewski, T. P. Davis, eds., *Handbook of Radical Polymerization* (John Wiley & Sons, Inc., Hoboken, 2002).
3. G. Moad, D. H. Solomon, *The Chemistry of Radical Polymerization* (Elsevier Science Ltd, Amsterdam, 2006), 2nd edn.
4. M. Krzysztof, *Controlled Radical Polymerization.* ACS Symposium Series (American Chemical Society, 1998), vol. 685, p. 500.
5. W. A. Braunecker, K. Matyjaszewski, Controlled/living radical polymerization: Features, developments, and perspectives. *Progress in Polymer Science* **32**, 93–146 (2007).
6. K. Matyjaszewski, N. V. Tsarevsky, Macromolecular engineering by atom transfer radical polymerization. *Journal of the American Chemical Society* **136**, 6513–6533 (2014).
7. K. Matyjaszewski, Atom transfer radical polymerization (ATRP): Current status and future perspectives. *Macromolecules* **45**, 4015–4039 (2012).
8. K. Matyjaszewski, J. Xia, Atom transfer radical polymerization. *Chemical Reviews* **101**, 2921–2990 (2001).
9. J.-S. Wang, K. Matyjaszewski, Controlled/"living" radical polymerization. Atom transfer radical polymerization in the presence of transition-metal complexes. *Journal of the American Chemical Society* **117**, 5614–5615 (1995).
10. X. Yong, O. Kuksenok, K. Matyjaszewski, A. C. Balazs, Harnessing interfacially active nanorods to regenerate severed polymer gels. *Nano Letters* **13**, 6269–6274 (2013).
11. X. Yong, O. Kuksenok, A. C. Balazs, Modeling free radical polymerization using dissipative particle dynamics. *Polymer* **72**, 217–225 (2015).
12. X. He et al., Synthetic homeostatic materials with chemo-mechano-chemical self-regulation. *Nature* **487**, 214–218 (2012).
13. A. Shastri et al., An aptamer-functionalized chemomechanically modulated biomolecule catch-and-release system. *Nature Chemistry* **7**, 447–454 (2015).
14. P. J. Hoogerbrugge, J. M. V. A. Koelman, Simulating microscopic hydrodynamic phenomena with dissipative particle dynamics. *Europhysics Letters* **19**, 155–160 (1992).
15. P. Espanol, P. Warren, Statistical mechanics of dissipative particle dynamics. *Europhysics Letters* **30**, 191–196 (1995).
16. R. D. Groot, P. B. Warren, Dissipative particle dynamics: Bridging the gap between atomistic and mesoscopic simulation. *Journal of Chemical Physics* **107**, 4423–4435 (1997).
17. S. Plimpton, Fast parallel algorithms for short-range molecular dynamics. *Journal of Computational Physics* **117**, 1–19 (1995).
18. H. Masoud, A. Alexeev, Controlled release of nanoparticles and macromolecules from responsive microgel capsules. *Acs Nano* **6**, 212–219 (2012).
19. W. Pan, D. A. Fedosov, G. E. Karniadakis, B. Caswell, Hydrodynamic interactions for single dissipative-particle-dynamics particles and their clusters and filaments. *Physical Review E* **78**, 046706 (2008).

20. P. K. Jha, J. W. Zwanikken, F. A. Detcheverry, J. J. de Pablo, M. Olvera de la Cruz, Study of volume phase transitions in polymeric nanogels by theoretically informed coarse-grained simulations. *Soft Matter* 7, 5965 (2011).
21. P. Nikunen, I. Vattulainen, M. Karttunen, Reptational dynamics in dissipative particle dynamics simulations of polymer melts. *Physical Review E* **75**, 036713 (2007).
22. A. Karatrantos, N. Clarke, R. J. Composto, K. I. Winey, Topological entanglement length in polymer melts and nanocomposites by a DPD polymer model. *Soft Matter* **9**, 3877 (2013).
23. S. Hirotsu, Softening of bulk modulus and negative Poisson ratio near the volume phase-transition of polymer gels. *Journal of Chemical Physics* **94**, 3949–3957 (1991).
24. H. Liu, M. Li, Z.-Y. Lu, Z.-G. Zhang, C.-C. Sun, Influence of surface-initiated polymerization rate and initiator density on the properties of polymer brushes. *Macromolecules* **42**, 2863–2872 (2009).
25. H. Gao, P. Polanowski, K. Matyjaszewski, Gelation in living copolymerization of monomer and divinyl cross-linker: Comparison of ATRP experiments with Monte Carlo simulations. *Macromolecules* **42**, 5925–5932 (2009).
26. P. Polanowski, J. K. Jeszka, K. Matyjaszewski, Modeling of branching and gelation in living copolymerization of monomer and divinyl cross-linker using dynamic lattice liquid model (DLL) and Flory–Stockmayer model. *Polymer* **51**, 6084–6092 (2010).
27. P. Polanowski, J. K. Jeszka, W. Li, K. Matyjaszewski, Effect of dilution on branching and gelation in living copolymerization of monomer and divinyl cross-linker: Modeling using dynamic lattice liquid model (DLL) and Flory–Stockmayer (FS) model. *Polymer* **52**, 5092–5101 (2011).
28. J. Genzer, In silico polymerization: Computer simulation of controlled radical polymerization in bulk and on flat surfaces. *Macromolecules* **39**, 7157–7169 (2006).
29. S. Turgman-Cohen, J. Genzer, Computer simulation of controlled radical polymerization effect of chain confinement due to initiator grafting density and solvent quality in "grafting from" method. *Macromolecules* **43**, 9567–9577 (2010).
30. S. Turgman-Cohen, J. Genzer, Simultaneous bulk- and surface-initiated controlled radical polymerization from planar substrates. *Journal of the American Chemical Society* **133**, 17567–17569 (2011).
31. A. Fahr, A. H. Laufer, Direct rate measurements of the combination and disproportionation of vinyl radicals. *Journal of Physical Chemistry* **94**, 726–729 (1990).
32. T. Zhao, Y. Zheng, J. Poly, W. Wang, Controlled multi-vinyl monomer homopolymerization through vinyl oligomer combination as a universal approach to hyperbranched architectures. *Nature Communications* **4**, 1873 (2013).
33. K. Matyjaszewski, A. H. E. Müller, Naming of controlled, living and "living" polymerizations. *Am. Chem. Soc., Polym. Prepr.* **38**, 6–9 (1997).
34. H. Gao, W. Li, K. Matyjaszewski, Synthesis of polyacrylate networks by ATRP: Parameters influencing experimental gel points. *Macromolecules* **41**, 2335–2340 (2008).
35. Y. Yang, M. W. Urban, Self-healing polymeric materials. *Chemical Society Reviews* **42**, 7446–7467 (2013).
36. B. J. Blaiszik et al., Self-healing polymers and composites. *Annual Review of Materials Research* **40**, 179–211 (2010).

37. M. D. Hager, P. Greil, C. Leyens, S. van der Zwaag, U. S. Schubert, Self-healing materials. *Advanced Materials* **22**, 5424–5430 (2010).

38. D. Y. Wu, S. Meure, D. Solomon, Self-healing polymeric materials: A review of recent developments. *Progress in Polymer Science* **33**, 479–522 (2008).

39. M. J. Webber, J. A. Kessler, S. I. Stupp, Emerging peptide nanomedicine to regenerate tissues and organs. *Journal of Internal Medicine* **267**, 71–88 (2010).

40. J.-D. Benazet, R. Zeller, Vertebrate limb development: Moving from classical morphogen gradients to an integrated 4-dimensional patterning system. *Cold Spring Harbor Perspectives in Biology* **1**, a001339 (2009).

41. D. A. Fedosov, I. V. Pivkin, G. E. Karniadakis, Velocity limit in DPD simulations of wall-bounded flows. *Journal of Computational Physics* **227**, 2540–2559 (2008).

42. M. Quesada-Pérez, J. A. Maroto-Centeno, J. Forcada, R. Hidalgo-Alvarez, Gel swelling theories: The classical formalism and recent approaches. *Soft Matter* **7**, 10536 (2011).

43. S. Hirotsu, Phase transition of a polymer gel in pure and mixed solvent media. *Journal of the Physical Society of Japan* **56**, 233–242 (1987).

44. B. Erman, P. J. Flory, Critical phenomena and transitions in swollen polymer networks and in linear macromolecules. *Macromolecules* **19**, 2342–2353 (1986).

45. P. J. Flory, *Principles of Polymer Chemistry* (Cornell University Press, Ithaca, NY, 1953).

46. H. G. Schild, Poly(N-isopropylacrylamide): Experiment, theory and application. *Progress in Polymer Science* **17**, 163–249 (1992).

47. A. Manna et al., Optimized photoisomerization on gold nanoparticles capped by unsymmetrical azobenzene disulfides. *Chemistry of Materials* **15**, 20–28 (2003).

48. M. Shibayama, Y. Isaka, Y. Shiwa, Dynamics of probe particles in polymer solutions and gels. *Macromolecules* **32**, 7086–7092 (1999).

49. M. Zhong, K. Matyjaszewski, How fast can a CRP be conducted with preserved chain end functionality? *Macromolecules* **44**, 2668–2677 (2011).

50. S. S. Ray, M. Okamoto, Polymer/layered silicate nanocomposites: A review from preparation to processing. *Progress in Polymer Science* **28**, 1539–1641 (2003).

51. F. Hussain, M. Hojjati, M. Okamoto, R. E. Gorga, Review article: Polymer-matrix nanocomposites, processing, manufacturing, and application: An overview. *Journal of Composite Materials* **40**, 1511–1575 (2006).

52. D. R. Paul, L. M. Robeson, Polymer nanotechnology: Nanocomposites. *Polymer* **49**, 3187–3204 (2008).

53. K. Haraguchi, Synthesis and properties of soft nanocomposite materials with novel organic/inorganic network structures. *Polymer Journal* **43**, 223–241 (2011).

54. J. Xu, J. Ye, S. Liu, Synthesis of well-defined cyclic poly(*N*-isopropylacrylamide) via click chemistry and its unique thermal phase transition behavior. *Macromolecules* **40**, 9103–9110 (2007).

55. K. Haraguchi, T. Takehisa, S. Fan, Effects of clay content on the properties of nanocomposite hydrogels composed of poly(*N*-isopropylacrylamide) and clay. *Macromolecules* **35**, 10162–10171 (2002).

56. K. Haraguchi, T. Takehisa, Nanocomposite hydrogels: A unique organic–inorganic network structure with extraordinary mechanical, optical, and swelling/de-swelling properties. *Advanced Materials* **14**, 1120–1124 (2002).

57. K. Haraguchi, T. Takada, Synthesis and characteristics of nanocomposite gels prepared by in situ photopolymerization in an aqueous system. *Macromolecules* **43**, 4294–4299 (2010).

58. A. C. C. Esteves, Y. Luo, M. W. P. van de Put, C. C. M. Carcouet, G. de With, Self- replenishing dual structured superhydrophobic coatings prepared by drop-casting of an all-in-one dispersion. *Advanced Functional Materials* **24**, 986–992 (2014).

3

Theory and Simulation Studies of Self-Assembly of Helical Particles

Giorgio Cinacchi [1], Alberta Ferrarini [2], Elisa Frezza [2,3], Achille Giacometti [4] and Hima Bindu Kolli [5]

[1] Departamento de Física Teórica de la Materia Condensada, Instituto de Física de la Materia Condensada (IFIMAC) and Instituto de Ciencias de Materiales "Nicolás Cabreras", Universidad Autónoma de Madrid, Spain
[2] Dipartimento di Scienze Chimiche, Università di Padova, Italy
[3] BMSS, IUMR 5086 CNRS/Univ. Lyon I, Institut de Biologie et Chimie des Protéines, France
[4] Dipartimento di Scienze Molecolari e Nanosistemi, Università Ca' Foscari di Venezia, Campus Scientifico, Edificio Alfa, Italy
[5] Department of Chemistry, University of Oslo, Norway

3.1 Introduction: Why Hard Helices?

Crystallization as an ordering phenomenon is neither restricted to molecular systems, nor to attractive interactions [1]. Even in the absence of explicit attractive interactions, systems might be spontaneously forming ordered aggregates driven by entropic gain, a well-know example being the formation of a crystal structure in a hard colloidal sphere systems [2]. This is one of several counterintuitive examples where there is a spontaneous ordering driven by entropy [3]. While in this case the number of possible crystal structures was still limited by the aspecificity of the steric interactions, the new developments in the chemical synthesis of colloids paved the way to constructing crystals with prescribed crystal structures by changing the shape of the colloidal building blocks [4].

Together with the shape, also the chirality of the building blocks may play an important, though not yet fully explored, role in self-assembly processes [5]. Leaving aside the issue of the emergence of homochirality in biological systems from equally probable molecular chiral moieties, a crucial problem in the current understanding of the origin of life, the use of chiral particles finds important applications in photonic metamaterials. Helically nanostructured materials have been attracting increasing attention, also because of their unique electrical and mechanical properties [6–8]. Control of their three-dimensional organization is also a crucial step in pushing the chiral properties to a mesoscale range.

While it is possible to control the enantioselective process by using depletion interactions and faceting of the building blocks [5], helices are among the natural

Self-Assembling Systems: Theory and Simulation, First Edition. Edited by Li-Tang Yan.

objects to focus on when dealing with chirality. New functional materials [9, 10] can be produced by exploiting the intrinsic chirality of the helical structures, which are useful in catalysis and demixing of enantiomers [11, 12]. The importance of the helix in nature is unquestionable: proteins, polysaccharides, DNA and RNA, the so-called molecules of life, have a helical structure. The helical shape is exhibited in nature also by microorganisms, like filamentous viruses, and cell organelles, like bacterial flagella. Filamentous viruses are formed by a DNA of RNA core, wrapped by a coating of helically arranged proteins. Well-known examples are the Tobacco Mosaic Virus (TMV), the first discovered virus [13], and viruses related to the filamentous phage *fd*, whose mutants are present in nature (M13, fd), while others can be obtained by genetic engineering. They have been widely investigated as models of highly anisotropic, colloidal systems, with the advantages of being essentially monodisperse and that their length, of the order of a micrometer, makes them suitable for imaging techniques, such as optical microscopy. Bacterial flagella are helical macromolecular structures assembled from a single protein (flagellin). Their helical shape can be tuned with high precision by regulating external parameters such as temperature or pH, and their large size, of the order of microns, makes them very handy for optical observations.

Because of their shape anisotropy, helical biopolymers and colloidal particles may exhibit liquid crystal phases at high densities [14]. These phases are often tacitly assumed to be the same as those occurring in systems of rod-like particles. However, it cannot be taken for granted that at such high densities the intrinsic helicity of the shape can be neglected. To explore the effect of self-assembly of helical polymers and colloids, and in particular to discover whether there is anything special just determined by the helical shape, we have undertaken a comprehensive investigation of the phase behavior of hard helices [15–19], interacting through purely steric repulsions, using Monte Carlo simulations and an extension of Onsager theory [20], a density functional theory (DFT) that was originally proposed to explain the onset of nematic ordering in a system of hard rods. These studies have revealed an unexpectedly rich phase behavior, the most interesting result being the existence of a special phases characterized by screw-like ordering. Such an organization had been proposed for DNA, based on theoretical considerations [21], and had been observed in dense suspensions of flagellar filaments [22]. Hard helices are an athermal system: phase transitions are controlled by density and are driven by the entropy gain on moving from the less to the more ordered phase. This is a minimalist model, possibly insufficient to account entirely for the complexity of real systems. It is nonetheless useful to obtain a general picture, lacking in the previous literature, that will represent a useful reference for interpreting the behavior of systems that may be more complicated.

The remainder of this chapter is organized as follows. In Section 3.2, we briefly recall the main features of liquid crystal phases, with special attention to their chiral versions. Then, we will describe our model systems and the methods used to investigate them. In Section 3.5, we give an outline of Onsager theory, whereas in Section 3.6 we extend Onsager theory to helically modulated nematic phases. In Section 3.7, we review the definitions of the order parameters and correlation functions necessary to identify the various phases. In Section 3.8, we discuss the physical origin of the chiral nematic phases formed by helical particles, and the next Section 3.9 gives an overview of the phase behavior of hard helices. In Section 3.10, theoretical predictions are compared to experimental data, and finally Section 3.11 presents the conclusions and possible perspectives.

3.2 Liquid Crystal Phases

Anisometric molecules and particles can form liquid crystal phases, which are fluid states characterized by long-range orientational order, while long-range translational order is absent or only partial. The phase transitions can be induced by changes of temperature or density, and in the two cases one speaks of thermotropic and lyotropic systems, respectively. To the former class, mostly represented by organic low molar mass molecules and polymers, belong the materials used in electro-optical applications. The latter class includes various kinds of systems, like surfactants, lipids, anisotropic colloids and stiff or semi-flexible polymers. Important examples of lyotropic liquid crystals can be found in nature, e.g. biomembranes and DNA.

Liquid crystals comprise a variety of phases, differing from each other in the kind and symmetry of order [23, 24]. Probably the most common is the *nematic* (N), in which the centers of mass of molecules or particles are randomly distributed in space, but their long axes are preferentially aligned with each other. The average alignment axis is denoted as the *director* ($\hat{\mathbf{n}}$). This is the phase generally found in the proximity of the isotropic phase, at lower temperature in thermotropic or higher density in lyotropic systems. With further decreasing temperature or increasing density, smectic (Sm) phases may be found, which exhibit an additional one-dimensional order: molecules or particles are preferentially located in layers, and the director may be either parallel (e.g smectic A, smectic B) or tilted (e.g. smectic C) with respect to the layer normal (z axis). In the smectic A and C phases the positions of molecules or particles are randomly distributed within layers, whereas in the smectic B phase there is hexatic short-range order within the layers. Other, less common, phases can be found for specific systems. One of them is the *biaxial* N phase (N_b), which differs form the conventional ("uniaxial") N phase because the orientational distribution in the plane perpendicular to the director $\hat{\mathbf{n}}$ is anisotropic: thus, two other directors, $\hat{\mathbf{b}}$ and $\hat{\mathbf{c}}$, can be defined, with $\hat{\mathbf{n}} = \hat{\mathbf{c}} \times \hat{\mathbf{b}}$ [25, 26]. Biaxial N phases (N_b) were detected in solutions of surfactants that self-assemble into biaxial micelles [27] and in colloidal suspensions of board-like particles [28], whereas their existence in thermotropic systems is more controversial. In this context, bent mesogens were proposed as suitable candidates, but the results are less straightforward.

Chiral molecules or particles can impart the liquid crystal phase a chiral character [29]. The chiral analog of the nematic is the *cholesteric* or *twisted nematic* (N^*) phase, where the director $\hat{\mathbf{n}}$ rotates in helical way around a perpendicular axis, rather than being uniform. Handedness and pitch ($\mathcal{P}$) of the cholesteric helix are determined by the structure at the molecular level, but the connection is not straightforward [30, 31]. The general features are that pitches are orders of magnitude longer than the molecular size (from hundreds of nanometers to millimeters) and cholesteric phases formed by enantiomers have identical pitch and opposite handedness. Some chiral systems also exhibit, between the isotropic and the cholesteric phase, one or more blue phases [32], which can be described as fluid lattices of defects, with cubic symmetry and lattice periods of the order of the wavelength of visible light. They are locally chiral, since directors are locally arranged in double-twist cylinders. On the other boundary of the N^* phase, the smectic phases may also be chiral: in the so-called smectic C chiral (SmC^*) phase, the director, tilted with respect to the layer normal, rotates in a helical way from layer to layer, again with typical pitches longer than 100 nm. In the case of short-pitch cholesterics, between the cholesteric and the smectic, *twist grain boundary* (TGB) phases may be found. These are frustrated structures, first predicted by de Gennes [33] and Lubensky [34], and observed soon after [35], composed of smectic slabs, rotated with respect to one another and separated by defect walls. Such a helical superstructure results from the competition between the cholesteric organization and smectic layering, which cannot be simultaneously realized without the formation of defects.

3.3 Hard Helices: A Minimal Model

Our model helices are formed by chains of N_b partially fused hard spherical beads, each of diameter D, rigidly arranged in a right-handed helical fashion as shown in Figure 3.1. All lengths will eventually be expressed in units of D that will set the length scale. The morphology of the helix will be defined by providing the radius r and the pitch p, along with an additional parameter that can be chosen either to be the contour length L or the Euclidean length Λ. Upon fixing the contour length L and changing r and p independently, the aspect ratio will be changing. Conversely, upon fixing the Euclidean length Λ and changing r and p independently, the contour length L will be modified. We can easily find the relation between Λ and L. The centers of the beads can be identified by the following helix parametric equations:

$$\begin{aligned} x_i &= r\cos(2\pi t_i) \\ y_i &= r\sin(2\pi t_i) \qquad 1 \le i \le N_b \\ z_i &= p t_i \end{aligned} \tag{3.1}$$

The centers of the beads lie on an inner cylinder of radius r, whereas the diameter of the outer cylinder $(2r + D)$ is the width of the helix r_{max}. The long axis of the helix $\hat{\mathbf{u}}$ passes through the center of the helix. Given r, p and L, the increment $\Delta t = t_{i+1} - t_i$ can be computed as

$$\frac{L}{14} = 2\pi\Delta t\sqrt{r^2 + \left(\frac{p}{2\pi}\right)^2}. \tag{3.2}$$

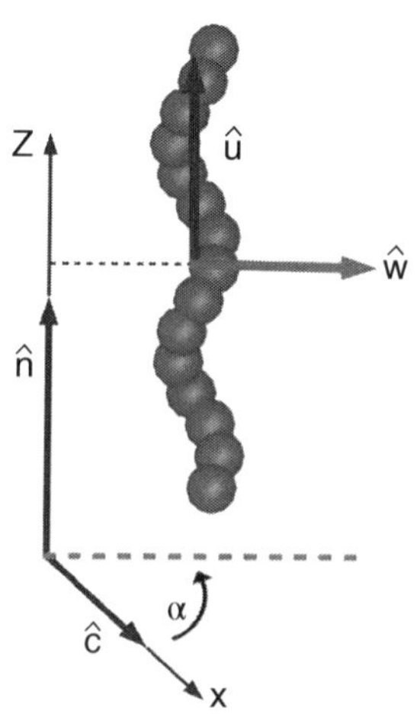

Figure 3.1 Model helix made up of a chain of partially overlapping hard spheres. The orientation of the helix in space is univocally defined by the direction of its main axis $\hat{\mathbf{u}}$ and its two-fold symmetry axis $\hat{\mathbf{w}}$. Reproduced from Ref. [18] with permission from the Royal Society of Chemistry.

The parametric equations could alternatively be written using the number of turns as the fixed value instead of fixed L [15]. The Euclidean length Λ is measured as the component parallel to the long axis of the helix of the distance between first and last bead. Different helix shapes—from a slender rod to a highly coiled helix—can be achieved by varying r and p independently, as illustrated in Figure 3.2. The limit case of $r = 0$ corresponds to a rod-like shape that can be contrasted with known results from the phase diagram of hard spherocylinders [36].

In our studies we have generally focused on helices of $N_b = 15$ beads and constant contour length $L = 10$ (see Figure 3.2).

In a system of our model helices, the only interactions are hard-core repulsions; that is, beads belonging to different helices interact as follows:

$$u(r_{ij}) = \infty, \quad r_{ij} < D, \tag{3.3}$$

$$u(r_{ij}) = 0, \quad r_{ij} > D, \tag{3.4}$$

where r_{ij} is the distance between a pair of beads belonging to different helices.

3.4 Numerical Simulations

The problem of calculating the properties of a condensed matter system by solving the equations of motion appears a dreadful task in view of the large number of particles ($\approx 10^{23}$) involved. Statistical mechanics helps to relate the equilibrium thermodynamic properties of the system to the average of particle properties by treating them in a probabilistic way, thus bridging microscopic properties with thermodynamical quantities. In this framework, Monte Carlo (MC) methods have been evolved in the past decades as one of the most efficient and reliable tools for assessing, and calculating thermodynamical properties and drawing the corresponding phase diagrams.

One of the crucial issues for generating proper microstates is ergodicity. Macroscopic properties measured in experiments are time-averaged quantities. The time scale of the observation is very large when compared to the switching time scale from one microstate to another. Because of this, we can assume that the system visits all accessible microstates during the observation time. Hence, a macroscopic quantity obtained by taking an average over a suitable ensemble is equivalent to the time-averaged quantity. In order to ensure that the system is in equilibrium, the sampling of microstates should obey "detailed balance," which requires the forward and backward rates of any transition between states to be equal.

Different ensembles, corresponding to the different independent thermodynamical variables, can be used by performing appropriate Legendre transformations. In our studies we have employed the Monte Carlo technique under isothermal–isochoric (NVT) and isothermal–isobaric conditions (NPT) to study the phase behavior of systems of hard helices. Here, N, V, P and T denote the number of particles, the total volume, the pressure and the temperature of the system, respectively.

3.4.1 Monte Carlo in Various Ensembles

The first MC simulation was implemented on hard disks by Metropolis in 1953 [37]. Since then, MC has evolved into a reliable methodology for studying virtually all soft

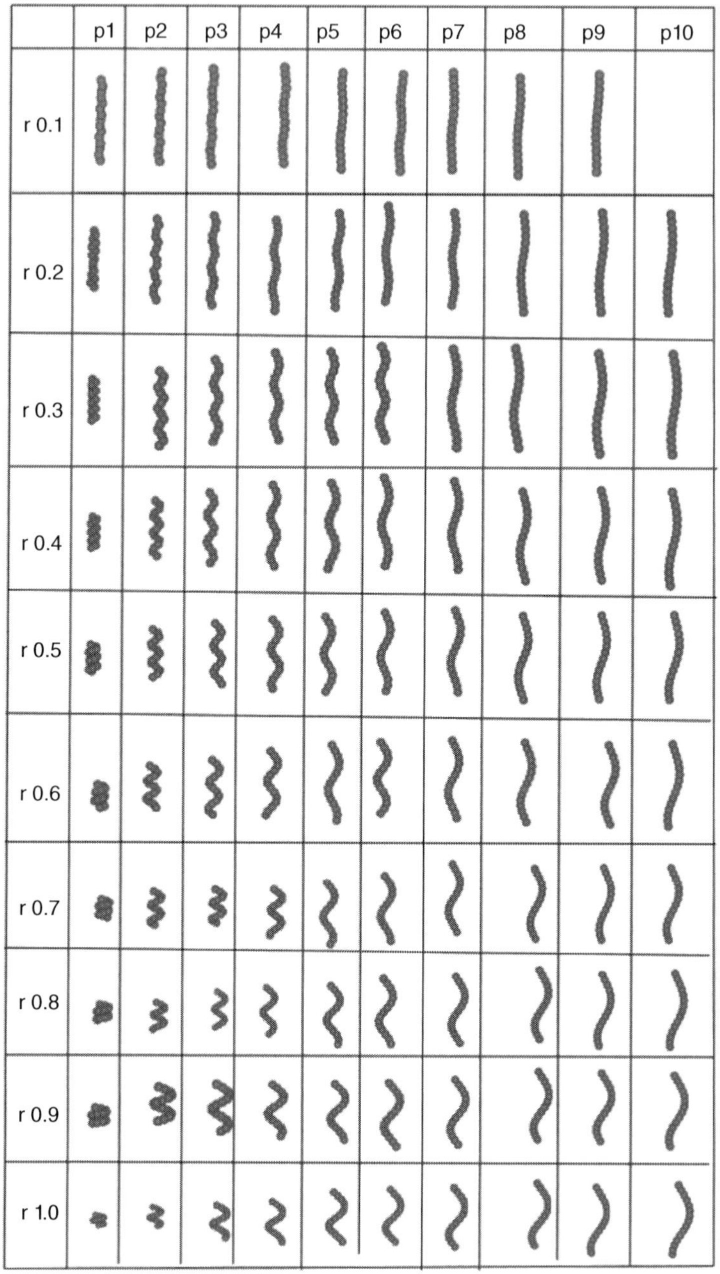

Figure 3.2 Helical shapes studied, with radius *r* ranging from 0.1 to 1.0, pitch *p* ranging from 1 to 10, and constant contour length $L = 10$.

matter systems, ranging from simple and molecular fluids, to polymers, surfactants and proteins. This list includes mesogens forming liquid crystal phases.

In general, a MC approach consists of the three steps: (i) Determining the microstate probability distribution for the ensemble of interest. (ii) Determining the set of MC moves accomplishing changes in all fluctuating quantities. (iii) Imposing the detailed balance condition to find the acceptance criterion.

3.4.1.1 Canonical Monte Carlo Simulations (NVT–MC)

The most common and convenient ensemble for fluid systems is usually the canonical (NVT) ensemble, where N particles are inserted into a fixed computational box of volume V at a given temperature T. In this case, the pressure is computed via the virial theorem [38],

$$P = \rho k_B T + \frac{1}{3V}\left\langle \sum_{i<j} f_{ij} \cdot (\mathbf{R}_i - \mathbf{R}_j) \right\rangle, \tag{3.5}$$

where ρ is the number density, and f_{ij} is the force on particle i due to particle j, located at $\mathbf{R}_i$ and $\mathbf{R}_j$, respectively. For isotropic potentials, depending only on the distance $R_{ij} = |\mathbf{R}_i - \mathbf{R}_j|$ between particles, this can be easily translated in terms of an integral involving the pair correlation function $g(R)$. However, the same procedure is not as straightforward for non-spherical hard-core discontinuous potentials.

3.4.1.2 Isothermal–Isobaric Monte Carlo Simulations (NPT–MC)

Isothermal–isobaric (NPT) Monte Carlo simulations were first employed by W. W. Wood [39] in the study of a fluid of hard disks. The isothermal–isobaric ensemble is widely used because most of the experiments are done at controlled pressure and temperature. In NPT–MC simulations the system is assumed to be in thermal contact with a large heat bath, and mechanically coupled with a barostat. The mechanical coupling allows the system to change its volume in order to keep its pressure constant. The problem of finding the correct probability distribution (point (i) above) is a standard textbook topic [38], so we will focus here on the problem of finding the correct moves for generic biaxial particles.

3.4.2 Details on the MC Simulation of Hard Helices

The position of a helix is defined by the coordinates of its center of mass, while its orientation is specified by the unit vectors $\hat{\mathbf{u}}$, parallel to its axis, and $\hat{\mathbf{w}}$, parallel to its two-fold symmetry axis (see Figure 3.1). The combination of the two unit vectors provides the equivalent information to the three Euler angles $\omega = (\theta, \phi, \psi)$.

Unlike spherical particles, uniaxial objects, such as spherocylinders, require rotational moves of their C_∞ axis ($\hat{\mathbf{u}}$) in addition to the translation of their center of mass. Biaxial particles require an additional rotation of the $\hat{\mathbf{w}}$ transversal axis. There are several ways of accomplishing this task. One possibility, due to Barker and Watts [38], consists of selecting with equal probability one of the three axes of the computational box, and performing a random rotation of $\hat{\mathbf{u}}$ around that axis, supplemented by an additional random rotation of $\hat{\mathbf{w}}$ around $\hat{\mathbf{u}}$ in the case of biaxial particles. A convenient (and equivalent) alternative hinges on the use of quaternions, thus avoiding the repeated use of trigonometric equations that are typically time consuming. A quaternion can be defined as the unit vector in four-dimensional space $Q \equiv (q_0, q_1, q_2, q_3)$ with $q_0^2 + q_1^2 + q_2^2 + q_3^2 = 1$. The quaternion offers an efficient way to generate uniform random vectors on the four-dimensional unit sphere [40]. The one-to-one correspondence between the Euler angle and the quaternion representation is given by

$$
\begin{aligned}
q_0 &= \cos\left(\frac{\theta}{2}\right) \cos\left(\frac{\phi + \psi}{2}\right) \\
q_1 &= \sin\left(\frac{\theta}{2}\right) \cos\left(\frac{\phi - \psi}{2}\right) \\
q_2 &= \sin\left(\frac{\theta}{2}\right) \sin\left(\frac{\phi - \psi}{2}\right) \\
q_3 &= \cos\left(\frac{\theta}{2}\right) \sin\left(\frac{\phi + \psi}{2}\right),
\end{aligned}
$$

and the rotation matrix defining the rotation is given by

$$
\begin{pmatrix}
q_0^2 + q_1^2 - q_2^2 - q_3^2 & 2(q_1q_2 - q_0q_3) & 2(q_1q_3 + q_0q_2) \\
2(q_1q_2 + q_0q_3) & q_0^2 - q_1^2 + q_2^2 - q_3^2 & 2(q_2q_3 - q_0q_1) \\
2(q_1q_3 - q_0q_2) & 2(q_2q_3 + q_0q_1) & q_0^2 - q_1^2 - q_2^2 + q_3^2
\end{pmatrix}.
$$

Finally, in an NPT calculation, a volume move is required, where one or more lengths of the box are randomly changed so volume is increased or decreased. This is a global move, involving the whole system, and hence should be performed with a relative frequency of $1/N$ with respect to the local translational and rotational moves of a single helix. On average, we then attempted $N/2$ translations and $N/2$ rotations with equal probability, and one volume move within each MC step. Typical equilibration times were of the order of 10^9 MC steps, followed by production runs for collecting statistics of the order of 3×10^9 steps.

Simulations with hard particles consume more computational time compared to those with soft particles. This is because of the overlap check for each pair of particles. In our model systems (see Figure 3.1), the distance between two particles cannot be less than the diameter D of a bead. The overlap check in such non-convex particles needs some tricky algorithms to reduce the computer time. In our calculations we first inserted a helix into a suitable spherocylinder, for which a fast test for overlapping is available, and then tested for possible overlaps at the level of single spheres forming the helix only in the event of overlapping of the embedding spherocylinders. In this way, the computational time to simulate a fluid of hard helices formed by 12 partially fused spheres, and having a fixed contour length of $L/D = 12$, is roughly a factor 8 larger compared with the corresponding fluid of spherocylinders.

Most of our simulations were carried out with periodic boundary conditions (PBC) on a floppy (i.e. shape adaptable) triclinic box, but no significant differences were found when using a cuboidal box. Such conditions are fully compatible with the existence of helical order with small periodicity, comparable with the particle length. However, they do not allow in general the development of a cholesteric organization because systems of prohibitively large sizes would be required when the periodicity is much longer than the scale of particle chirality. For this reason, using PBC in our simulations we could find nematic phases, with uniform rather than twisted $\hat{\mathbf{n}}$ directors (i.e. $\mathcal{P} \rightarrow \infty$).

In our simulations we used a number of particles varying from 900 to 2000, always finding rather stable results. The value 900 was then used for most of the simulations.

3.5 Onsager (Density Functional) Theory

This section describes the Onsager theory for the isotropic-to-nematic phase transition in a system of hard, slender, rod-like particles [20]. Beside serving as a basis for all the specific calculations described in the following sections, this theory is of general importance for the whole of (soft) condensed matter science. Onsager formulated it in the forties [20]. That is, in the same period when Kirkwood remarkably conjectured that a hard-sphere system, despite the absence of any attractive interactions, will ultimately exhibit, on increasing density, a disorder–order transition between a fluid phase and a crystal phase [41]. Since no other interactions are operative but hard steep repulsions, this *ordering* phase transition is purely driven by the sole *entropy*. That is, at a sufficiently high density, the crystal is thermodynamically more stable than the fluid because its entropy is larger. Such a statement was rather counterintuitive and not easily accepted at that time. Yet, the Kirkwood conjecture was later confirmed in the earliest applications of the numerical simulation techniques [39, 42]. The Onsager theory proceeded, in essence, along the same lines yet, exploiting the peculiarity of a system of hard, long and thin rods, it proved analytically, rather than numerically, the entropic origin of the disorder–order transition between an isotropic liquid phase and a nematic liquid crystal phase. Thanks to these basic works, the concept of *ordering entropy* progressively gained acceptance and is nowadays well established.

Let us suppose we have a system of hard rod-like particles. For example, the one shown in Figure 3.3, formed by prolate ellipsoids that interact only through hard repulsive interactions. Looking at this figure, the particle centroids appear uniformly distributed in space while the particle main (C_∞) axes roughly align along a common direction, the *director* $\hat{\mathbf{n}}$: this is a snapshot of a system in a nematic liquid crystal phase.

Hard-particle systems of this sort mimic rather closely suspensions of rod-like colloidal particles interacting through short-range repulsive interactions. In fact,

Figure 3.3 Image of a system of hard prolate ellipsoids in the nematic liquid crystal phase. The image was created with the program QMGA [43].

Onsager set out to develop his theory to explain earlier experimental results on suspensions of this kind, either of inorganic (mineral) origin, such as the suspensions made of V_2O_5 colloids [44], or organic (biological) origin, such as those formed by the tobacco mosaic virus particles [13]. In these experiments, a transition between two phases was observed at a low concentration of colloids. Onsager showed that the transition, involving an isotropic liquid phase and a nematic liquid crystal phase, was due to the highly anisometric shape of the colloidal particles.

One can view a system of N hard, equally sized and rigid rods, uniformly distributed in a container of volume V and free to rotate, as a (infinite-component) mixture, each component being identified by the orientation, ω. With the symbol ω one intends a set of variables required to define the orientation of a particle in the laboratory frame of reference. For example, in the case of cylindrically symmetric particles, the components of the unit vector $\hat{\mathbf{U}}$ along the particle's C_∞ axis. One can then define a distribution function $f(\omega)$ such that $Nf(\omega)d\omega$ gives the number of particles whose orientation is comprised within the interval $[\omega, \omega + d\omega]$. The distribution function $f(\omega)$ is normalized such that $\int d\omega f(\omega) = 1$.

In full analogy with the expression of the entropy of an ideal multi-component mixture, one can write the expression of the *ideal* entropy of the hard-rod system as

$$S_{\text{id}} = -Nk_{\text{B}} \left[\log \left(\frac{\rho \Lambda_{\text{tr}}^3 \Theta_{\text{rot}}}{T} \right) - 1 + \int d\omega f(\omega) \log f(\omega) \right], \tag{3.6}$$

with k_{B} the Boltzmann constant, Λ_{tr} is the thermal wavelength and Θ_{rot} is the rotational temperature [45]. The integral in Eq. (3.6) plays the role of a mixing entropy.

In the aforementioned experiments, significant deviations from the ideal behavior were, however, observed even at low concentration. This is due to the hard interactions between the particles, whose effect is appreciable even at low concentration provided the particles are sufficiently anisometric, and therefore that effect significantly depends on particle orientation.

In a dilute and uniform system of particles interacting with short-range interactions, the pressure P can be expressed as a *virial* series in the number density $\rho = N/V$:

$$P = \rho k_{\text{B}} T \; [1 + B_2(T)\rho + B_3(T)\rho^2 + \cdots], \tag{3.7}$$

with T the absolute temperature and the coefficients B_ℓ, $\ell = 1, 2, \ldots$ are termed virial coefficients. The first coefficient $B_1 = 1$, while the successive coefficients, starting with the second virial coefficient B_2, depend on particle interactions and T.

Originally devised as an empirical way to account for deviations from the ideality, the virial series was given a statistical-mechanical foundation by the remarkable work of Mayer and Mayer [46]. They showed that the virial coefficients can be written as integrals, of rapidly increasing complexity as the order ℓ increases, whose integrands depend on the Mayer function M. The latter is defined as:

$$M(\mathbf{R}_i, \omega_i, \mathbf{R}_j, \omega_j) = e^{-\frac{u(\mathbf{R}_j - \mathbf{R}_i, \omega_i, \omega_j)}{k_{\text{B}} T}} - 1, \tag{3.8}$$

with $\mathbf{R}_i$ the vector giving the position of particle i's centroid in the laboratory frame of reference and $u(\mathbf{R}_j - \mathbf{R}_i, \omega_i, \omega_j)$ the potential energy of interaction between particles i

and j. In particular, the expression for the leading second virial coefficient is

$$B_2(\omega_i, \omega_j | T) = -\frac{1}{2} \int d\omega_i f(\omega_i) \int d\omega_j f(\omega_j) \int d\mathbf{R} M(\mathbf{R}, \omega_i, \omega_j). \tag{3.9}$$

Observe that for hard interactions, the Mayer function takes on two values only: -1 if particles overlap and 0 if they do not. The dependence on temperature drops off: a hard-particle system is athermal. The second virial coefficient can thus be rewritten as:

$$B_2(\omega_i, \omega_j) = \frac{1}{2} \int d\omega_i f(\omega_i) \int d\omega_j f(\omega_j) v(\omega_i, \omega_j), \tag{3.10}$$

with $v(\omega_i, \omega_j)$ the *volume* that particle i *excludes* to particle j. The expressions for the next virial coefficients are increasingly more complicated, to such an extent that the computational burden to evaluate them rapidly becomes prohibitively heavy even for hard spheres, the simplest hard-particle model. Thus, exact summation of a virial series is in general precluded and one has to resort to approximations whose reliability should always be tested. The system of hard slender rods is, however, an exception in this respect.

In fact, Onsager made the crucial observation that, as long as the particles are needle-like, the terms of order ℓ higher than two can be neglected in Equation (3.7). In this case, the system's *excess* free energy F_{ex}, obtained from the non-ideal part of the equation of state by standard thermodynamic integration, results:

$$\mathsf{F}^{ex} = N k_B T \frac{\rho}{2} \int d\omega f(\omega) \int d\omega' f(\omega') v(\omega, \omega'). \tag{3.11}$$

Summing Equations (3.6) and (3.11) provides the system's total free energy F:

$$\frac{\mathsf{F}}{N k_B T} = \log\left(\frac{\rho \Lambda_{tr}^3 \Theta_{rot}}{T}\right) - 1 + \int d\omega f(\omega) \log f(\omega) + \frac{\rho}{2} \int d\omega f(\omega) \int d\omega' f(\omega') v(\omega, \omega'). \tag{3.12}$$

Observe that the free energy is a *functional* of the distribution function $f(\omega)$: the Onsager theory is a precursor of modern density functional theories. The thermodynamic stable state of the system at a given ρ corresponds to that specific distribution function $f(\omega)$ that minimizes the free energy. If $f(\omega)$ is constant, that is, the particle orientations are uniformly distributed, the orientational entropy term is maximized. Conversely, the density-dependent excluded-volume term is minimal when particles align. It is from the competition between these two terms that the isotropic-to-nematic phase transition arises at sufficiently high density.

The rigorous minimization of the free energy functional under the constraint of distribution function normalization leads to the following integral equation:

$$\log K f(\omega) = -\rho \int d\omega' f(\omega') v(\omega, \omega'), \tag{3.13}$$

with K a constant whose value is determined by ensuring the normalization condition. Its analytic solution is not available. Onsager bypassed this difficulty by resorting to an approximate trial function $f(\omega)$ whose form can be regulated by the value of a single parameter α and minimized directly the free energy, Equation (3.12), with respect to α. This procedure can be extended to functional forms containing an arbitrarily large number of variational parameters. A common choice is that of using a truncated expansion of $f(\omega)$ on a set of orthonormal basis functions. The expansion coefficients are the orientational order parameters. The integral Equation (3.13) can nonetheless be solved by an iterative algorithm that, starting from a guessed orientational distribution function, ends whenever self-consistency is satisfied within a given tolerance.

One can thus see that, beside the isotropic-liquid solution of Equation (3.13), corresponding to a constant $f(\omega)$ and always present for any value of ρ, an additional anisotropic-liquid solution emerges provided the value of ρ is sufficiently high [47, 48]. Inserting the isotropic and anisotropic solutions into the expression for the pressure, the equations of state for the isotropic and nematic phases can respectively be obtained. From standard thermodynamics, the chemical potential expressions for the two branches can also be obtained. Equating pressure and chemical potential, the value of ρ at which the isotropic phase and nematic phase are in equilibrium can thus be determined. The Onsager theory predicts the isotropic-to-nematic phase transition to be first order, in agreement with experimental evidence. Provided the experimental rod-like colloids have a sufficiently large aspect ratio, like the tobacco mosaic virus particles, the agreement is essentially quantitative [49]. Definite confirmation of the validity of the Onsager theory came from numerical experiments on model hard-particle systems such as that shown in Figure 3.3.

Deviations from Onsager predictions increase with decreasing aspect ratio, because of the progressively higher contribution of virial coefficients beyond second order [50]. In earlier calculations approximate values of B_3 were obtained [51, 52], and recently the eighth virial level has been reached for hard uniaxial ellipsoids [53]. Surprisingly good agreement between theory and simulations (in systems of hard rod-like particles) could be obtained using the Parsons–Lee expression [54], where many-body effects are introduced in an approximate way, by multiplying the Carnahan–Starling form of the excess free energy of hard spheres [55] by the second virial coefficient specific to the particles under investigation [56]. So, the excess free energy reads:

$$\mathsf{F}^{\mathrm{ex}} = Nk_{\mathrm{B}}TG(\eta)\frac{\rho}{2}\int d\omega f(\omega)\int d\omega' f(\omega')v(\omega,\omega'), \tag{3.14}$$

where $G(\eta) = (4-3\eta)/4(1-\eta)^2$. Here, $\eta = \rho v_0$ is the volume fraction, with v_0 being the volume of a particle. In Ref. [15] we used the Parsons–Lee approximation to investigate the isotropic-nematic transition of hard helical particles and compared with the results of MC simulations (see Figure 3.5).

The Onsager theory forms the basis of all successive developments in the field. The case of hard helices is no exception in this respect.

3.6 Onsager-Like Theory for the Cholesteric and Screw-Nematic Phases

We have considered two kinds of helically modulated nematic phases: cholesteric and screw-nematic. In the former, the nematic director $\hat{\mathbf{n}}$, i.e. the average alignment axis of the helical axes ($\hat{\mathbf{u}}$), spirals around a perpendicular axis (the y axis of the laboratory frame in Figure 3.1). In the screw-nematic (Ns^*) phase it is the secondary polar director $\hat{\mathbf{c}}$, i.e. the average alignment axis of the $\hat{\mathbf{w}}$ axes of helices, that spirals around $\hat{\mathbf{n}}$ (the z

axis of the laboratory frame in Figure 3.1). For such phases the Onsager theory outlined in the previous section has to be generalized to include explicit dependence upon positional variables. In both cases a single positional variable is needed, i.e. the position along the helical axis, because there is translational homogeneity perpendicular to such an axis. In the case of a helically modulated phase with the helical axis along the direction h and pitch $\mathcal{P}$, we can introduce the orientational distribution function at the h position, $f(\omega|h)$, which has periodicity $\mathcal{P}$ too, and is normalized such that $\int d\omega f(\omega|h) = 1$, irrespective of h. By retaining only the second and third terms in the virial expansion, the excess Helmholtz free energy of such a system is given by:

$$\frac{\mathrm{F}^{\mathrm{ex}}}{Vk_{\mathrm{B}}T} = \frac{\rho^2}{2\mathcal{P}}\int_0^{\mathcal{P}} dh \int d\omega f(\omega|h) \int dh' \int d\omega' f(\omega'|h') \Bigg[a_{\mathrm{excl}}(h,\omega,h',\omega') + \frac{\rho}{3}\int dh'' \int d\omega'' f(\omega''|h'') a_3(h,\omega,h',\omega',h'',\omega'')\Bigg], \quad (3.15)$$

where the functions $a_{\mathrm{excl}}(h,\omega,h',\omega')$ and $a_3(h,\omega,h',\omega',h'',\omega'')$ have been introduced. The first is given by:

$$a_{\mathrm{excl}}(h,\omega,h',\omega') = -\int d\mathbf{R}'_{\perp} M(0,h,\omega,\mathbf{R}'_{\perp},h',\omega'), \quad (3.16)$$

where M is the Mayer function, Equation (3.8), and $\mathbf{R}_{\perp}$ is the vector position in the plane perpendicular to h. a_{excl} is interpreted as the area of the surface obtained by cutting with a plane perpendicular to the helical axis at position h' the volume excluded to a particle with orientation ω' by a particle at position h with orientation ω [57]. The second function in Equation (3.15) is given by:

$$a_3(h,\omega,h',\omega',h'',\omega'') = -\int d\mathbf{R}'_{\perp} \int d\mathbf{R}''_{\perp} M(0,h,\omega,\mathbf{R}'_{\perp},h',\omega') M(0,h,\omega,\mathbf{R}''_{\perp},h'',\omega'') M(\mathbf{R}'_{\perp},h',\omega',\mathbf{R}''_{\perp},h'',\omega''). \quad (3.17)$$

Since a_{excl} and a_3 actually depend on the differences $\zeta' = h' - h$ and $\zeta'' = h'' - h$, Equation (3.15) above can be rewritten as:

$$\frac{\mathrm{F}^{\mathrm{ex}}}{k_{\mathrm{B}}TV} = \frac{\rho^2}{2}\int d\omega f(\omega|h=0) \int d\zeta' \int d\omega' f(\omega'|\zeta') \Bigg[a_{\mathrm{excl}}(\omega,\zeta',\omega') + \frac{\rho}{3}\int d\zeta'' \int d\omega'' f(\omega''|\zeta'') a_3(\zeta',\omega,\omega',\zeta'',\omega'')\Bigg]. \quad (3.18)$$

If the Parsons–Lee approximation is used, the following expression is obtained:

$$\frac{\mathrm{F}^{\mathrm{ex}}}{k_{\mathrm{B}}TV} = \frac{\rho^2}{2}G(\eta)\int d\omega f(\omega|h=0) \int d\zeta' \int d\omega' f(\omega'|\zeta') a_{\mathrm{excl}}(\omega,\zeta',\omega'). \quad (3.19)$$

For a uniform nematic phase $f(\omega|h=0) = f(\omega)$, and if the expansion is truncated at the second virial term, Equation (3.14) is recovered. The equations above are equally valid for the cholesteric and screw-like nematic phase. In the former case, $h = Y$ and $\mathcal{P}$ is the cholesteric pitch; in the latter case, $h = Z$ and $\mathcal{P}$ coincides with p, the pitch of a single helix.

The free energy of the cholesteric and screw-nematic phases bears a dependence upon the helical pitch $\mathcal{P}$, introduced by the orientational distribution function $f(\omega|\zeta)$.

Thus, determination of the equilibrium state at a given density requires minimization of the free energy with respect to this parameter, in addition to the other quantities that define the orientational distribution function. The numerical handling of the free energy of helically modulated nematic phases is significantly more demanding than for the uniform case, not only because of the positional dependence which explicitly appears in Equations (3.16), but also because of the inherent biaxiality of such modulated phase. Similar approach has been recently implemented for the twist-bend nematic phase [58]. When dealing with the cholesteric phase, some simplifying assumptions can be used, taking advantage of the long scale of the twist deformation, compared to the scale of interparticle interactions. A first consequence of the long cholesteric pitch is that the biaxiality is very small, so as a first approximation it can be neglected. This was done in Refs. [59, 60], where a Parson-Lee treatment was proposed based on expressions of the free energy in terms of uniaxial order parameters and the wavenumber q. The Straley approach can also be adopted [61], where further assumptions are made, i.e. that $f(\omega)$ in the N* phase is identical to that in the undeformed N phase at the same density, but with respect to a director that rotates in a helical way, and that the q dependence of the excess free energy can be accounted for by the following truncated Taylor expansion:

$$\mathsf{F}^{\mathrm{ex}} = \mathsf{F}^{\mathrm{ex}}|_{q=0} + \left.\frac{d\mathsf{F}^{\mathrm{ex}}}{dq}\right|_{q=0} q + \frac{1}{2}\left.\frac{d^2\mathsf{F}^{\mathrm{ex}}}{dq^2}\right|_{q=0} q^2. \tag{3.20}$$

Here, the first term on the right-hand side is the free energy density of the undeformed nematic phase. So, the difference between the free energy of the twisted and the uniform nematic phase, $\mathsf{F}^{\mathrm{ex}} - \mathsf{F}^{\mathrm{ex}}|_{q=0}$, can be compared with the elastic energy of the cholesteric phase, which according to the continuum elastic theory reads [23]:

$$\frac{\mathsf{F}_{\mathrm{el}}}{V} = k_2\, q + \frac{1}{2}K_{22}\, q^2, \tag{3.21}$$

where K_{22} is the *twist elastic constant* and k_2 is the *chiral strength*. The former is a positive quantity accounting for the energetic cost of twist deformations, and has typical values of the order of piconewtons. The chiral strength k_2 is a pseudoscalar, which has opposite values for systems which are the mirror image of each other and vanishes in the achiral nematic phase. The equilibrium pitch of a cholesteric phase is obtained by minimization of the elastic energy and is given by:

$$\mathcal{P} = -2\pi K_{22}/k_2; \tag{3.22}$$

it becomes infinitely long in achiral nematic liquid crystals.

Comparing Equations (3.20) and (3.21), we can write the microscopic expression for the chiral strength and the twist elastic constants:

$$k_2 = \left.\frac{d\mathsf{F}^{\mathrm{ex}}}{dq}\right|_{q=0}, \tag{3.23}$$

$$K_{22} = \left.\frac{d^2\mathsf{F}^{\mathrm{ex}}}{dq^2}\right|_{q=0}. \tag{3.24}$$

These two quantities behave differently under inversion: the former is invariant (scalar), whereas the latter changes its sign (pseudoscalar). Thus, for chiral particles that are the

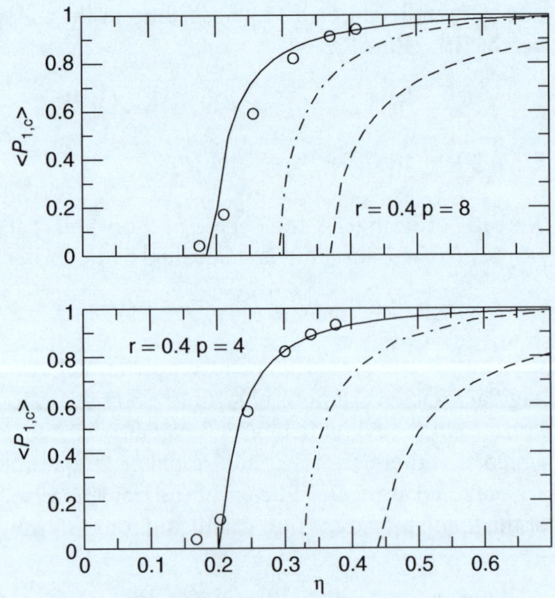

Figure 3.4 Comparison between numerical simulations (circles) and Onsager-like theory for perfectly parallel helices, in the plane screw-nematic order parameter–volume fraction. Lines are theory results using the II virial approximation (dashed), the II virial approximation with Parson–Lee correction (dash-dotted) and the III virial approximation (solid).

mirror image of each other (enantiomers), the free energy of the undeformed nematic phase and the twist elastic constant are identical, whereas k_2 takes opposite values.

In our calculation for the cholesteric phase of hard helices [17] we followed the Straley approach, along the lines described in Ref. [62]. In the case of the screw-nematic phase, we adopted the simplifying assumption of perfectly parallel helical axes; thus, helices had only translational and azimuthal freedom [18]. This could be justified by the high degree of orientational order that characterizes the screw-nematic phase. However, the high value of density at which the $\mathrm{N}-\mathrm{N}_S^*$ transition occurs called for considering also the explicit contribution of the third virial terms. Indeed, it was seen that their inclusion is needed to achieve a nearly quantitative agreement with corresponding numerical simulation data, as illustrated in Figure 3.4. The generalization of these results to systems of freely rotating hard helices is planned for the near future. While conceptually clear [Equations (3.15)–(3.18)], it is expected to require a considerable numerical effort not only because of the complexity of a helically modulated phase but also because of the incorporation of the third virial term contribution. This task will nonetheless be rewarding in that it will allow the treatment of both cholesteric and screw-like nematic phases at the same level of theory and the accurate study of their mutual phase transition and the transition between these and the isotropic phase.

3.7 Order Parameters and Correlation Functions

Different phases are characterized by appropriate order parameters capturing their distinctive symmetry and the corresponding degrees of ordering, and by suitable correlation functions. In the following we will present the order parameters and correlation functions that have been used to characterize the different phases formed by helical particles. Some of the latter are conveniently resolved along the $\hat{\mathbf{n}}$ director (||)

and perpendicular to it ($\perp$), according to the following decoupling of the interparticle vector $\mathbf{R}_{ij} = \mathbf{R}_i - \mathbf{R}_j$:

$$\mathbf{R}_{ij} = \mathbf{R}_{ij}^{\parallel} + \mathbf{R}_{ij}^{\perp} = (\mathbf{R}_{ij} \cdot \hat{\mathbf{n}})\hat{\mathbf{n}} + |\mathbf{R}_{ij} \times \hat{\mathbf{n}}|\hat{\mathbf{R}}_{ij}^{\perp}. \tag{3.25}$$

3.7.1 Nematic Order Parameter $\langle P_2 \rangle$

Nematic ordering, i.e. the degree of (non-polar) alignment of the $\hat{\mathbf{u}}$ axes of helices with respect to the $\hat{\mathbf{n}}$ director, is quantified by the order parameter:

$$\langle P_2 \rangle = \left\langle \frac{3}{2}\cos^2\theta - \frac{1}{2} \right\rangle, \tag{3.26}$$

where P_2 is the second Legendre polynomial and $\cos\theta = \hat{\mathbf{u}} \cdot \hat{\mathbf{n}}$. Here and henceforth, angular brackets denote ensemble averages. $\langle P_2 \rangle$ is zero in the isotropic phase and nonzero in the nematic phase. It can also be nonzero in other more ordered phases, in which case there are other additional order parameters. In a simulation, average values are obtained as mean values over the trajectory; so, $\langle P_2 \rangle$ would be obtained as the mean of the following quantities, calculated for each trajectory frame:

$$(P_2)_{\text{frame}} = \frac{1}{N}\sum_{i=1}^{N} \frac{3}{2}(\hat{\mathbf{n}} \cdot \hat{\mathbf{u}}_i)^2 - 1, \tag{3.27}$$

where the sum is over all particles. However, one does not know at the outset the orientation of the director $\hat{\mathbf{n}}$, as particles can align in principle along any direction in space. A convenient approach is to use the second rank Veilliard–Baron tensor [63]

$$Q_{\alpha\beta} = \frac{1}{N}\sum_{i=1}^{N}\left(\frac{3}{2}u_i^{\alpha}u_i^{\beta} - \frac{\delta_{\alpha\beta}}{2}\right), \tag{3.28}$$

where $\alpha, \beta = x, y, z$ are indices referring to the laboratory frame, and u_i^{α} is the component of the unit vector $\hat{\mathbf{u}}$ of particle i along α. The $\mathbf{Q}$ matrix represents a symmetric, traceless second rank tensor. The three eigenvalues $\lambda_1, \lambda_2, \lambda_3$ of $\mathbf{Q}$ are a measure of nematic order in the three orthogonal directions defined by the corresponding eigenvectors. In case of uniaxial nematics, $-2\lambda_1 = -2\lambda_2 = \lambda_3$. The largest (in absolute value) eigenvalue, λ_3, is the nematic order parameter $\langle P_2 \rangle$, and the corresponding eigenvector is the nematic director $\hat{\mathbf{n}}$.

Figure 3.5 shows the order parameter $\langle P_2 \rangle$ as a function of the volume fraction η, for helices with radius $r = 0.2$ and pitch $p = 4$. We can see the typical behavior, with $\langle P_2 \rangle$ close to zero at low density (isotropic phase) and then jumping to a high value beyond a given density. This jump is a signature of the first-order isotropic-to-nematic transition. The fact that $\langle P_2 \rangle$ does not exactly vanish in the isotropic phase is a finite-size effect. Likewise, we find $\lambda_1 \approx \lambda_2$ within a 5% discrepancy, which, however, can be taken as compatible with the uniaxial character of the conventional N phase.

3.7.2 Screw-Like Nematic Order Parameter

The phase diagram of helices presents phases that, unlike the conventional nematic, are characterized by screw-like order. This is quantified by the order parameter,

$$\langle P_{1,c} \rangle = \langle \hat{\mathbf{w}} \cdot \hat{\mathbf{c}} \rangle, \tag{3.29}$$

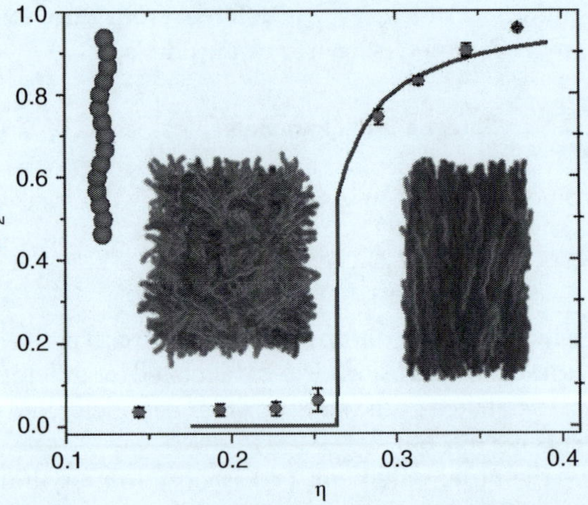

Figure 3.5 The nematic order parameter, $\langle P_2 \rangle$, as a function of the volume fraction η for helices with radius $r = 0.2$ and pitch $p = 4$, whose morphology is displayed in the inset. The plot shows the results of NPT–MC simulations (filled circles) and those of Onsager theory with Parsons–Lee correction, see Equation (3.14) (line). Representative snapshots from simulations in the isotropic (left) and the nematic phases (right) are shown.

which measures the average alignment along a common direction ($\hat{\mathbf{c}}$) of the two-fold symmetry axes ($\hat{\mathbf{w}}$) of helices. In phases with screw-like ordering the $\hat{\mathbf{c}}$ director rotates in a helical way around the $\hat{\mathbf{n}}$ director, with periodicity $\mathcal{P}$ equal to the pitch p of the helical particles. To calculate $\langle P_{1,c} \rangle$ from simulations it is expedient to perform a preliminary rotation of $-2\pi(\mathbf{R}_i \cdot \hat{\mathbf{n}})/p$ around $\hat{\mathbf{n}}$ of the particle coordinates. The $\hat{\mathbf{c}}$ director and the $\langle P_{1,c} \rangle$ order parameter are then determined for the untwisted structure, following the same procedure used for the calculation of the nematic order parameter.

In smectic phases there is also the possibility of an in-layer transversal polar order without correlation between layers. Even in this case an order parameter defined according to Equation (3.29) can be used, but the average has to be meant over all particles belonging to individual layers. To stress this point and distinguish this case from screw-like order, we will use for it the symbol $\langle P_{1,c}^{\text{layer}} \rangle$.

Figure 3.6 shows the polar order parameter, $\langle P_{1,c} \rangle$, as a function of the volume fraction, η, for helices with radius $r = 0.2$ and pitch $p = 4$. Comparing with the plot of the nematic order parameter, $\langle P_2 \rangle$, calculated for the same helices (see Figure 3.5), we can see that $\langle P_{1,c} \rangle$ is negligibly small in both the isotropic and nematic phases, but at higher density it raises, reaching values close to 1. The different gray scales in Figure 3.6 refer to different

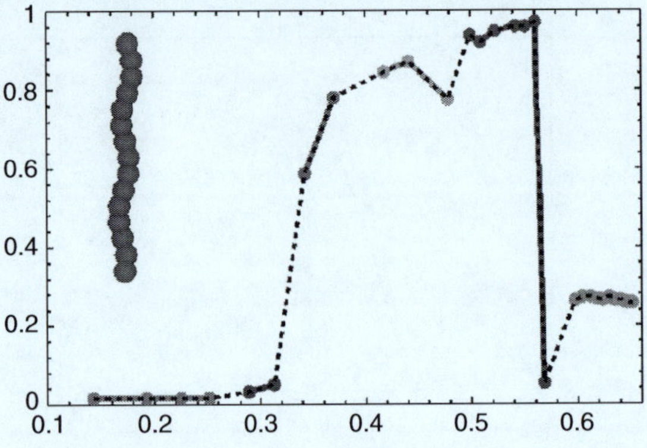

Figure 3.6 Polar order parameter, $\langle P_{1,c} \rangle$, as a function of the volume fraction η for helices with radius $r = 0.2$ and pitch $p = 4$ (see inset). The symbols show the results of NPT–MC simulations; the lines are meant as a guide to eye. I(isotropic), N(nematic), N_s^*(screw-nematic), $Sm_{A,s}^*$(screw smectic A), $Sm_{B,s}^*$(screw smectic B), C(compact). Phase labeling is reported in Table 3.1; C is a high density compact phase.

phases in which $\langle P_{1,c} \rangle$ is different from zero; to distinguish between them, additional order parameters have to be introduced.

3.7.3 Smectic Order Parameter

Smectic order can be characterized by the translational order parameter

$$\langle \tau_1 \rangle = \left| \left\langle \exp\left(2\pi i \frac{z}{d}\right) \right\rangle \right|, \tag{3.30}$$

where z is the position along the director $\hat{\mathbf{n}}$ and d is the layer spacing. Since such spacing is not known a priori, $\langle \tau_1 \rangle$ is calculated for different values of d. The maximum $\langle \tau_1 \rangle$ value is then taken as the smectic order parameter and the corresponding d value as the layer spacing [64, 65].

The order parameter $\langle \tau_1 \rangle$ is close to unity in the case of perfect layering and close to zero for non-layered phases, like I, N and N_s^*. A representative example is reported in Figure 3.7, where $\langle \tau_1 \rangle$ is plotted as a function of the layer spacing d calculated at different η values for helices with $r = 0.2$ and $p = 4$. The black line in the figure, which is always close to zero, corresponds to the N_s^* phase at $\eta = 0.370$. The lines for $\eta = 0.417$, 0.441 and 0.478 respectively, showing peaks at $d \approx 11.5$, correspond to smectic phases and are in order of increasing volume fraction η. The maximum values of peaks are their corresponding smectic order parameters. With increase in η, the order parameter increases and the corresponding layer spacing shows a small tendency to decrease.

3.7.4 Hexatic Order Parameter

The $\langle \tau_1 \rangle$ order parameter captures the onset of smectic ordering but is unable to distinguish between the smectic B phase, where there is six-fold bond-orientational (hexatic) ordering, and other phases, like the smectic A, where there is no in-plane ordering. This discrimination can, however, be accomplished by calculating the hexatic order parameter $\langle \psi_6 \rangle$ [64, 66]:

$$\langle \psi_6 \rangle = \left\langle \frac{1}{N} \sum_{i=1}^{N} \left| \frac{1}{n(i)} \sum_{j=1}^{n(i)} \exp(6i\theta_{ij}) \right| \right\rangle. \tag{3.31}$$

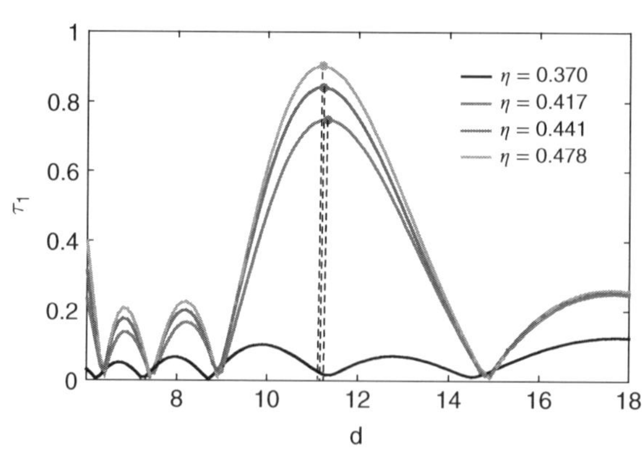

Figure 3.7 The dependence of translational order parameter $\tau_1 \equiv \langle \tau_1 \rangle$ on layer spacing for helices having $r = 0.2$, $p = 4$ on either side of the N_s^*–$Sm_{A,s}^*$ transition. The maximum value of τ_1 is taken as the smectic order parameter.

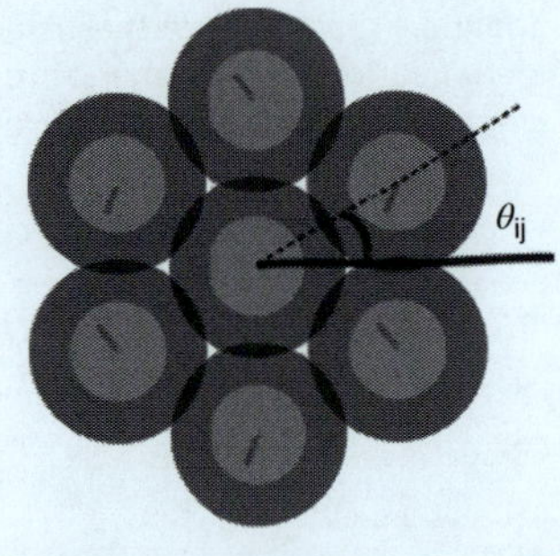

Figure 3.8 Cartoon showing the quantities used to define the hexatic order parameter $\langle \psi_6 \rangle$, Equation (3.31). The gray circles are transversal sections of cylinders enclosing the helices and the thick short line inside each circle indicates the orientation of the helix $\widehat{\mathbf{w}}$ vector. The bond angle θ_{ij} is the angle between a reference axis (thick solid line) in a plane perpendicular to the director $\widehat{\mathbf{n}}$, and the projection of the interparticle vector $\mathbf{R}_{ij}$ into this plane (dashed line). Reproduced from Ref. [19] with permission from the Royal Society of Chemistry.

Here, the inner sum is over all nearest neighbors of particle i within a single layer, whose number is $n(i)$, and θ_{ij} is the angle between the unit vector $\widehat{R}_{ij}^{\perp}$ defined in Equation (3.25) and a randomly fixed reference axis in a plane perpendicular to $\widehat{\mathbf{n}}$. (see Figure 3.8)

$\langle \psi_6 \rangle$ is equal to 1 in the case of perfect hexatic ordering and otherwise vanishes. The piece of information stemming from $\langle \psi_6 \rangle$ can be supported by calculation of the average number of nearest neighbors $\langle n \rangle$ within each layer, which would be equal to 6 for perfect hexagonal order. $\langle n \rangle$ is computed by taking for any helix the number of its nearest neighbors in a plane and averaging this number over all helices and all configurations. Both ψ_6 and $\langle n \rangle$ are very sensitive to the definition of the nearest neighbor distance. We used the value corresponding to the first minimum of the radial distribution function.

We note that helices display smectic phases with or without screw-like ordering. We can then identify $\langle P_{1,c} \rangle$ to check whether the phase is screw-smectic or simply smectic. It turns out that the latter does not occur for all cases studied in our simulations, while we find both screw-smectic A ($\mathrm{Sm}^*_{\mathrm{A,s}}$, no in-plane hexagonal order) and screw-smectic B ($\mathrm{Sm}^*_{\mathrm{B,s}}$, with in-plane hexagonal order). Conversely, we do not have any evidence of a simple Sm_{A} phase in our results for appreciable curliness ($r > 0.1$). The reason is likely to be ascribed to the fact that before encountering a smectic phase, the system is in a screw-nematic phase, and there is no reason for losing this organization when initially entering into a smectic phase. The screw organization may or may not be lost at higher densities in the presence of short range in-plane hexagonal order. In the former case, the helices secondary axes $\widehat{\mathbf{w}}$ tend to be correlated within each layer but uncorrelated from one layer to the neighboring ones. We call this organization smectic B polar $\mathrm{Sm}_{\mathrm{B,p}}$. In the latter case, the $\widehat{\mathbf{w}}$ are further correlated from one layer to the neighboring ones, thus maintaining the original screw-like organization. We call this configuration screw-smectic B ($\mathrm{Sm}^*_{\mathrm{B,s}}$).

Table 3.1 summarizes the different phases exhibited by hard helices and the relative order parameters.

3.7.5 Parallel and Perpendicular Pair Correlation Functions

In addition to global order parameters, it proves essential to introduce pair correlation functions that provide detailed insights into the structural features of the phase. In particular, it is useful to consider correlation functions resolved along the $\widehat{\mathbf{n}}$ director and perpendicular to it, according to the decoupling of the interparticle vector shown in Equation (3.25). Thus, two relevant correlation functions are $g_{\parallel}(R_{\parallel})$ and $g_{\perp}(R_{\perp})$,

Table 3.1 Summary of the different phases identified in the MC simulations of hard helices, along with the corresponding order parameters.

Phase	Code	Order parameter
Nematic	N	$\langle P_2 \rangle$
Screw-nematic	N_s^*	$\langle P_2 \rangle$, $\langle P_{1,c} \rangle$
Smectic A	Sm_A	$\langle \tau_1 \rangle$
Screw-smectic A	$Sm_{A,s}^*$	$\langle \tau_1 \rangle$, $\langle P_{1,c} \rangle$
Smectic B	Sm_B	$\langle \tau_1 \rangle$, $\langle \psi_6 \rangle$
Polar smectic B	$Sm_{B,p}$	$\langle \tau_1 \rangle$, $\langle \psi_6 \rangle$, $\langle P_{1,c}^{\text{layer}} \rangle$
Screw-smectic B	$Sm_{B,s}^*$	$\langle \tau_1 \rangle$, $\langle \psi_6 \rangle$, $\langle P_{1,c} \rangle$

which are sensitive to interparticle correlations along the director and perpendicular to it, respectively.

The perpendicular positional correlation function $g_\perp(R_\perp)$ is related to the probability that, given a particle, another one is found at a distance that, when projected onto a plane perpendicular to the director, is equal to $R_\perp$, and is defined as [66]:

$$g_\perp(R_\perp) = \frac{1}{2\pi R_\perp N} \left\langle \frac{1}{\rho L_z} \sum_{i=1}^{N} \sum_{j\neq i}^{N} \delta(R_\perp - |\mathbf{R}_{ij} \times \hat{\mathbf{n}}|) \right\rangle, \tag{3.32}$$

where $\delta(\cdots)$ is the Dirac *delta* function and L_z is the length of the simulation box along the director $\hat{\mathbf{n}}$. Note that the number density is taken inside the average as the volume V of the box may change in the course of simulations. The volume is calculated as $V = L_x L_y L_z$, where L_x and L_y are the lengths of the simulation box perpendicular to director $\hat{\mathbf{n}}$.

The parallel positional correlation function, $g_\parallel(R_\parallel)$, is related to the probability that, given a particle, another one is found on a plane perpendicular to the director $\hat{\mathbf{n}}$, at the distance $R_\parallel$, and is defined as:

$$g_\parallel(R_\parallel) = \frac{1}{N} \left\langle \frac{1}{\rho L_x L_y} \sum_{i=1}^{N} \sum_{j\neq i}^{N} \delta(R_\parallel - \mathbf{R}_{ij} \cdot \hat{\mathbf{n}}) \right\rangle. \tag{3.33}$$

One important correlation function, designed to identify the onset of screw-like ordering, is:

$$g_{1,\parallel}^{\hat{\mathbf{w}}}(R_\parallel) = \left\langle \frac{\sum_{i=1}^{N} \sum_{j\neq i}^{N} \delta(R_\parallel - \mathbf{R}_{ij} \cdot \hat{\mathbf{n}})(\hat{\mathbf{w}}_i \cdot \hat{\mathbf{w}}_j)}{\sum_{i=1}^{N} \sum_{j\neq i}^{N} \delta(R_\parallel - \mathbf{R}_{ij} \cdot \hat{\mathbf{n}})} \right\rangle. \tag{3.34}$$

The function $g_{1,\parallel}^{\hat{\mathbf{w}}}(R_\parallel)$ detects correlations between the C_2 symmetry axes of helices with the value of its maximum coinciding with $\langle P_{1,c} \rangle$. It also allows a direct measure of the pitch of the phase, as seen in Figure 3.9 for helices with $r = 0.2$ and $p = 3$ and $p = 6$. The plots show pronounced oscillations, whose amplitude increases with increasing density, with periodicity equal to the pitch p of the helical particles.

The set of correlation functions help to distinguish different phases: In the isotropic and nematic phases, both $g_\parallel(R_\parallel)$ and $g_\perp(R_\perp)$ exhibit the characteristic behavior of a

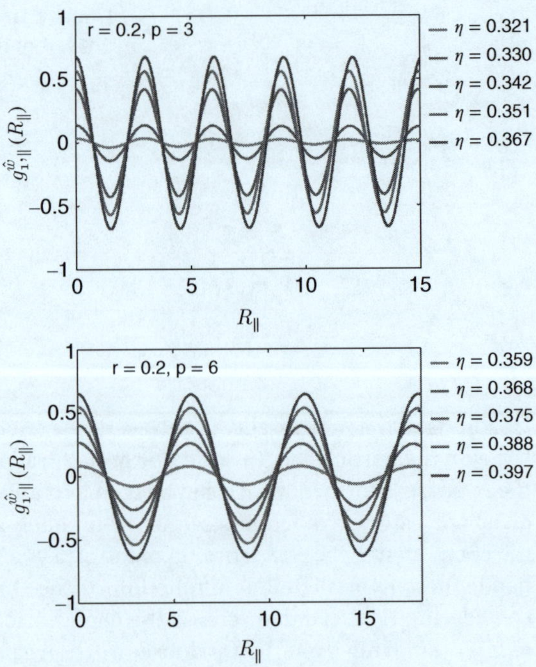

Figure 3.9 Correlation function $g^{\hat{\mathbf{w}}}_{1,\parallel}(R_\parallel)$ as a function of the projection of the interparticle distance along the director $\hat{\mathbf{n}}$, from MC simulations of hard helices with radius $r = 0.2$ and pitch $p = 3$ and $p = 6$, at different values of the volume fraction η.

liquid; in the smectic A phase $g_\perp(R_\perp)$ is still liquid-like, whereas $g_\parallel(R_\parallel)$ displays the characteristic peaks of one-dimensional ordering; in the smectic B phase, $g_\perp(R_\perp)$ will additionally acquire the structure characteristic of hexatic ordering; finally, $g^{\hat{\mathbf{w}}}_{1,\parallel}(R_\parallel)$ exhibits oscillations with periodicity identical to the particle pitch p in the presence of screw-like ordering, both in the nematic and smectic phases.

3.8 The Physical Origin of Cholesteric and Screw-Like Order

Elongated particles such as rigid spherocylinders interacting only via excluded volume form a nematic phase upon compression because of the gain in translational entropy accompanying the onset of orientational order. Rigid hard helical particles will display a similar tendency to align their main axes along a common director $\hat{\mathbf{n}}$, but with two important differences.

The first difference stems from the fact that a helix is a chiral object. Again, formation of the nematic phase is driven by the gain in translational entropy accompanying the orientational order, for helices with sufficiently high aspect ratio. However, due to the particle chirality, pair configurations with right- and left-handed twists of the helix axes ($\hat{\mathbf{u}}$) are not equivalent. So, they give different contributions to the average excluded volume [17, 59, 60]. Depending on the morphology and the state point, right- or left-handed contributions may prevail for helices with a given handedness. Anyway, the net effect is very small, since it derives from the imbalance of oppositely signed and very similar contributions. The macroscopic result is that the director is twisted; a right-/left-handed cholesteric phase is formed if pair configurations with a right-/left-handed twist have on average a smaller excluded volume. Due to the

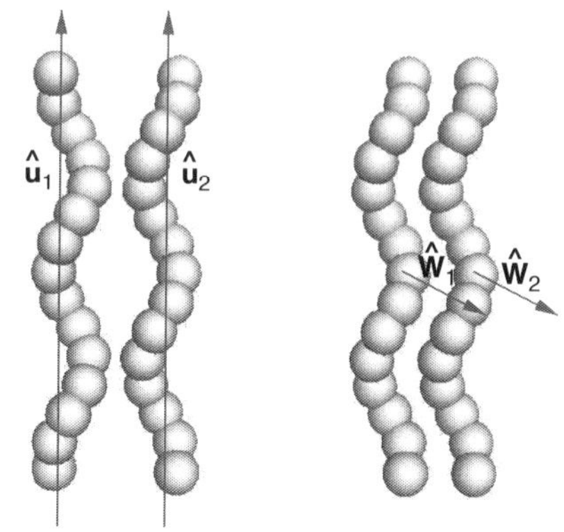

Figure 3.10 Pairs of helices in antiphase (left) and in phase (right).

smallness of the net chirality of the average excluded volume, the cholesteric pitch $\mathcal{P}$ is large on the particle length scale (for non-chiral particles the net chirality vanishes and the N phase is formed, which can be seen as cholesteric with infinite pitch). Moreover, a given cholesteric handedness cannot be uniquely associated with a given handedness of the constituent helices. According to the Straley model [61], inversion of the cholesteric handedness is predicted as a function of the thread pitch p for hard threaded rods: considering right-handed screws, the cholesteric handedness would switch from right to left on moving from tight to looser pitch. Analogous results were obtained in recent studies of hard helical particles, which additionally showed that a given system may exhibit N* handedness inversion upon increasing density [17, 59, 60].

The second important difference from rod-like particles originates from the particular shape of the helix that triggers a tendency for neighboring parallel helices in a nematic organization to slide one over another in a screw-like fashion. This can be rationalized as follows. Imagine having two parallel helices with no azimuthal correlation between their respective orientations in the plane perpendicular to $\hat{\mathbf{n}}$ (see Figure 3.10, left). Under these conditions, the system looks locally nematic. At higher densities, neighboring parallel helices are expected to have a significant intrusion between grooves, thus providing an effective azimuthal correlation stemming from having helices in phase (see Figure 3.10, right), with a corresponding significant loss of rotational entropy about their own axes. This can only be compensated by a translational motion of the helix along the nematic direction $\hat{\mathbf{n}}$ in a screw-like fashion. We call this the screw-nematic phase, and denote it as N_s^*. A similar mechanism also occurs in smectic phases, and will be termed screw-smectic. Screw-like order is characterized by a phase periodicity $\mathcal{P}$ and handedness identical to the pitch p and handedness of the constituent helices, at any density. Another distinctive consequence is the emergence of a local polar order ($\hat{\mathbf{c}}$ is a polar director), which is missing in both N and N* phases.

3.9 The Phase Diagram of Hard Helices

We have performed NPT−MC simulations for helices of different morphology (see Figure 3.2), and using the order parameters and correlation functions defined in Section 3.7 we have identified the sequence of phases formed by them. In the following we will present some representative results. Scaled quantities will be used throughout, with the bead diameter D taken as the unit of length and $k_B T$ as the unit of energy. In all calculations right-handed helices were considered. For left-handed helices the same phase diagrams would be obtained, but in the case of helical phases the handedness would be reversed.

A special comment has to be reserved for cholesteric phases. For the reasons mentioned in Section 3.4, no cholesteric phase could be found in our NPT–MC simulations due to the periodic boundary conditions. However, this does not compromise the significance of the phase diagrams, since the existence of a large-scale director twist does not essentially modify the thermodynamic and ordering properties of the N phase. On the other hand, using an Onsager-like theory (Section 3.6) we calculated the cholesteric pitch in the state points corresponding to nematic phases, and we obtained $\mathcal{P}$ values of the order of 100–1000 D [19]. So, what throughout the paper, based on the NPT–MC results, we have denoted as nematic phases, should rather be considered as actual cholesteric phases.

3.9.1 The Equation of State

In the volume fraction-pressure plane, the pressure displays a jump in the case of a first-order transition or a change in slope for a continuous transition. However, the differences between liquid crystal phases can be small, and even a first-order transition may be very weak. As an example, Figure 3.11 shows the equation of state for helices with $r = 0.2$ and pitches $p = 4$ and $p = 8$. In both cases we can see the sequence I–N–N_s^*–Sm with increasing density. This is a common behavior. The isotropic–nematic and nematic–smectic transitions are found to be always first order,

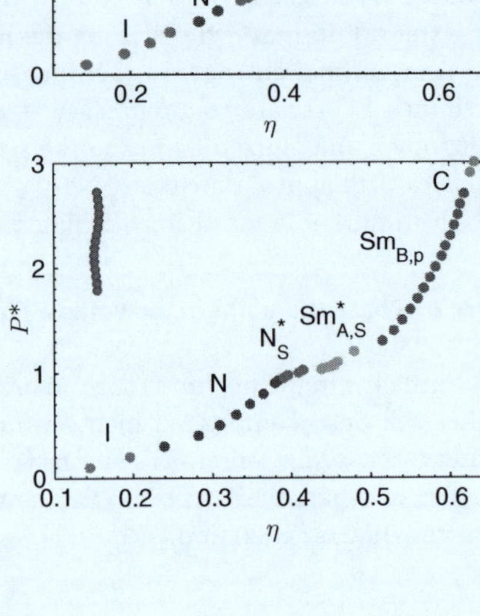

Figure 3.11 Reduced pressure $P^* = PD^3/k_B T$ versus volume fraction η for helices with $r = 0.2$ and pitch $p = 4$ (top) and $p = 8$ (bottom). The corresponding shape of the helix is displayed in the insets. Different phases are identified by different grays and labeled accordingly. *(See insert for color representation of the figure.).*

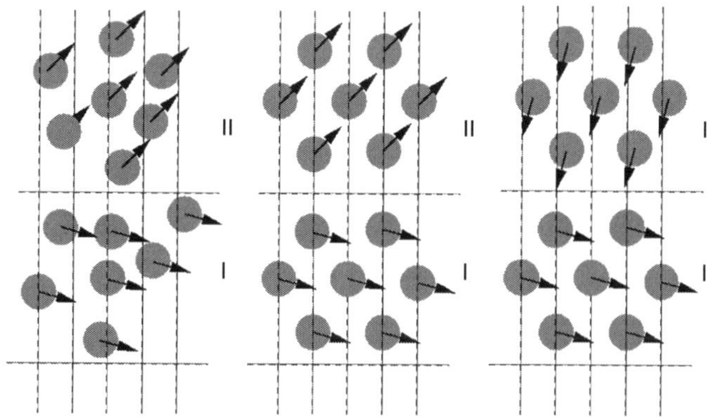

Figure 3.12 Cartoon of a top view of two neighboring smectic layers (I) and (II) with different in-plane organizations. $Sm^*_{A,s}$ (left), $Sm^*_{B,s}$ (center) and $Sm_{B,p}$ (right).

as indicated by the jump in the nematic $\langle P_2 \rangle$ and smectic $\langle \tau_1 \rangle$ order parameters. Conversely, the transition between nematic N and screw-nematic N^*_s phase appears to be continuous, on the basis of the behavior of $\langle P_{1,c} \rangle$.

As we can see in Figure 3.11, the relative amplitudes of the different phases depend on the helix morphology. The N^*_s phase occurs at a higher volume fraction than the N phase, which can be rationalized considering that high packing is required for the screw-like mechanism to be advantageous from the entropic view point. For curly helices the N^*_s phase tends to widen and, in some cases, for very curly ones, we have found direct transition from the isotropic to the screw-nematic phase. At even higher density, N^* is superseded by a smectic phase which preserves the screw-like order. In this phase, denoted as $Sm^*_{A,s}$, the centers of mass of the helices are homogeneously distributed within each layer, with their orientation in phase, that is, the $\hat{\mathbf{w}}$ vectors tend to be parallel, and different layers are correlated in a screw-like fashion. This is sketched in Figure 3.12 (left). Upon increasing the density, we observe two alternative scenarios.

In the $Sm^*_{B,s}$ phase, Figure 3.12 (center), the system maintains the same global organization as in the $Sm^*_{A,s}$ case, but with additional in-plane hexatic order. Essentially, helices stack on top of each other with the appropriate azimuthal orientation to form a system of infinite helices along the $\hat{\mathbf{n}}$ direction, in this way having a significant gain in translational entropy. In the $Sm_{B,p}$ phase, Figure 3.12 (right), in-plane hexatic and azimuthal correlations are preserved, but different layers are fully uncorrelated. Here, entropy is maximized by losing the screw order in favor of an optimal packing of the helices with appropriate offsets.

The different kinds of smectic phases are shown in Figure 3.13.

3.9.2 Phase Diagrams in the Volume Fraction–Pitch Plane

We are now in the position to summarize the phase behavior of different helical shapes. This will be accomplished in the volume fraction–pitch plane for the two different radii discussed in previous sections ($r = 0.2$ and $r = 0.4$). The case $r = 0.1$ will also be shown as a paradigmatic example of a very slender shape not showing any tendency to a screw-like organization.

Figure 3.13 Snapshots from NPT–MC simulations of helices with $r = 0.2$, $p = 4$ (top) and $r = 0.2$, $p = 8$ (bottom). Density increases on moving from left to right, displaying the sequence of phases. Different grays are given according to the projection of the local tangent to helices onto a plane perpendicular to the director $\hat{\mathbf{n}}$ [43]. Adapted from Ref. [19] with permission from the Royal Society of Chemistry.

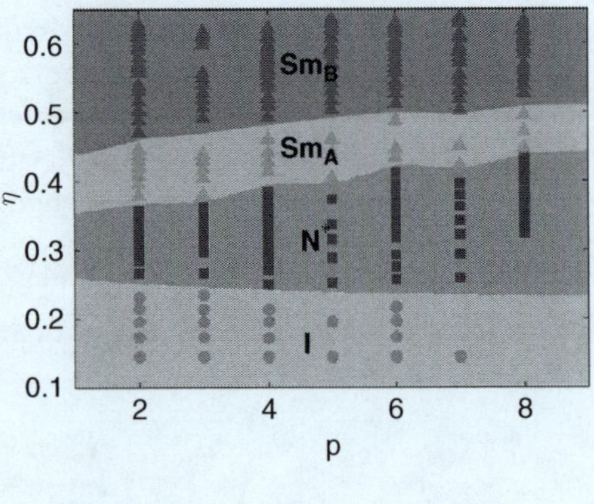

Figure 3.14 Phase diagram in the volume fraction η versus pitch p plane for helices having $r = 0.1$. Symbols correspond to calculated points. Different phases are labeled as indicated in Table 3.1. The symbol N* is used for state points that in NPT–MC simulations were found in the nematic phase, and then their cholesteric pitch was determined by an Onsager-like theory. Reproduced from Ref. [19] with permission from the Royal Society of Chemistry.

3.9.2.1 Phase Diagram for $r = 0.1$

As can be seen in Figure 3.2, helices with $r = 0.1$ are almost rod-like particles with high aspect ratio. Therefore we expect a phase behavior not very different from that of their spherocylinder counterparts. Figure 3.14 shows this to be the case: as in spherocylinders, we find isotropic to nematic N and nematic to smectic A and B transitions upon increasing η, the difference between Sm_A and Sm_B being the hexatic order of the latter, as discussed.

The only qualitative difference from the phase diagram of hard spherocylinders is in the nematic phase, which here is cholesteric rather than uniform nematic, as shown

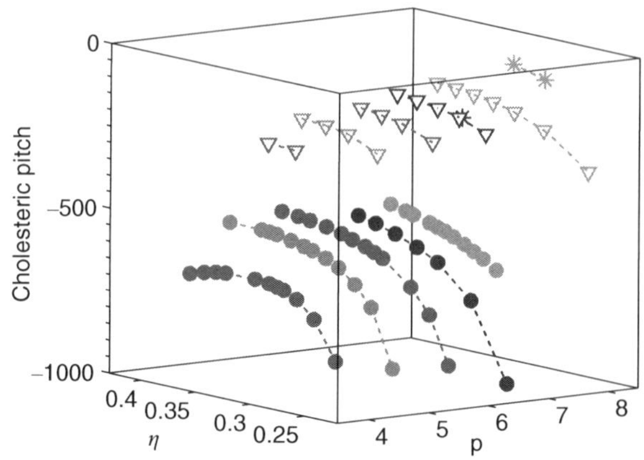

Figure 3.15 Cholesteric pitch $\mathcal{P}$, calculated in state points indicated as N* in the phase diagrams in Figure 3.14 ($r = 0.1$, full circles), Figure 3.16 ($r = 0.2$, open triangles) and Figure 3.17 ($r = 0.4$, asterics). η is the volume fraction and p is the pitch of the helical particles (only p values greater than 3 are shown).

by the values of $\mathcal{P}$ predicted by an Onsager-like theory and shown in Figure 3.15. Left-handed N* phases are predicted ($\mathcal{P} < 0$), with pitches longer than 500 D, which tend to decrease with increasing density.

The absence of screw-like order in this case is clearly due to the weak curliness of the helices, which is not sufficient to trigger the screw-like behavior. The aspect ratios are not significantly different from each other, ranging between 8.78 and 9.14. As the pitch p increases, the aspect ratio slightly increases and stabilizes the nematic phase at lower η, consistent with the spherocylinder counterpart.

3.9.2.2 Phase Diagram for $r = 0.2$

Figure 3.16 shows the phase diagram of helices with $r = 0.2$. Comparing with the case $r = 0.1$ we can appreciate significant differences, with the presence of several new phases, most of which are chiral. As in the case $r = 0.1$, at a given density these systems exhibit a transition from the isotropic to the nematic phase. And again, calculations based on an Onsager-like theory predict this to be a twisted nematic phase. The pitch, shown in Figure 3.15, is much smaller than for $r = 0.1$, the sign of a stronger chirality of interactions between the curlier helices. Then at higher density the screw-nematic phase appears, which is followed by smectic phases specific to helices. The increase of helix radius from $r = 0.1$ to $r = 0.2$ brings about a significant curliness in helical shape to promote a screw-like ordering at higher densities.

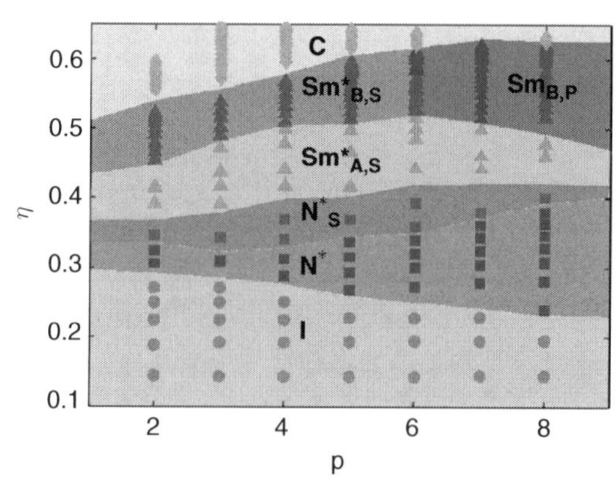

Figure 3.16 Phase diagram in the volume fraction η versus pitch p plane for helices having $r = 0.2$. Symbols correspond to calculated points. Different phases are labeled as indicated in Table 3.1; C is a high density compact phase. The symbol N* is used for state points that in NPT–MC simulations were found in the nematic phase, and then their cholesteric pitch was determined by an Onsager-like theory. Reproduced from Ref. [19] with permission from the Royal Society of Chemistry.

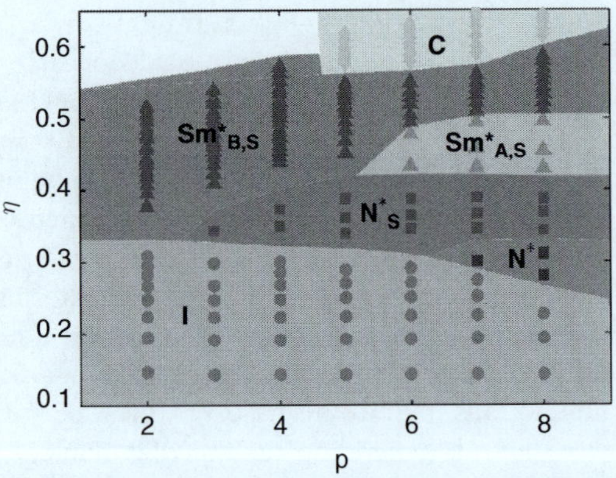

Figure 3.17 Phase diagram in the volume fraction η versus pitch p plane for helices having $r = 0.4$. Symbols correspond to calculated points. Different phases are labeled as indicated in Table 3.1; C is a high density compact phase. The symbol N* is used for state points that in NPT–MC simulations were found in the nematic phase, and then their cholesteric pitch was determined by an Onsager-like theory. Reproduced from Ref. [19] with permission from the Royal Society of Chemistry.

An important difference occurs for p values below and above $p = 6$. For $p < 6$ the $Sm^*_{A,S}$ develops into a $Sm^*_{B,S}$ ordering upon increasing η. Here, the screw-like coupling between different layers is then favored as hexatic ordering is gradually setting in. Conversely, for $p > 6$ this is lost in favor of a rearrangement of neighboring layers to achieve an optimal packing. A glance back to Figure 3.2 reveals the reason for this being again related to the fact that for $p > 6$ helices are slender enough to make unfavorable the screw-like organization. As we will see below, this will not be the case for $r = 0.4$.

3.9.2.3 Phase Diagram for $r = 0.4$

Figure 3.17 shows the phase diagram of helices with $r = 0.4$. A notable difference with respect to the case $r = 0.2$ hinges on the increased stability of the N^*_s phase with respect to the N^* counterpart, and the disappearance of the $Sm_{B,p}$ phase in favor of a wider $Sm^*_{B,s}$ phase that extends over all p values above a certain η. This was anticipated above, and can be ascribed to the curliness of all helices with $r = 0.4$, as can be inferred again from Figure 3.2. Another difference with the $r = 0.2$ case is given by the presence of several triple points, a signature of a rich polymorphism. At higher densities, one finds rather compact structures, denoted as "C," with the exception of helices with low p, for which the situation is less clear. Here one could expect the presence of rotator phases in analogy with what occurs in the spherocylinder counterpart. The N^* phase appears only in a narrow region, for helices with a long pitch. According to an Onsager-like theory, in this phase the director is spontaneously twisted, with a pitch $\mathcal{P}$ smaller than in the systems previously examined.

3.10 Helical (Bio)Polymers and Colloidal Particles

It is interesting to see how results obtained for hard helices compare with the behavior of systems made of helical particles, several examples of which can be found in natural or synthetic polymers and colloids. Cholesteric phases have been found in solutions of stiff or semi-flexible, covalent or supramolecular, helical polymers, either in water or in organic solvent. In such systems a variety of interactions is generally present and this, in addition to the intrinsic flexibility of polymers, may affect the features of the cholesteric phase [31], as well as the relative stability of phases [66]. The presence of linear self-assembly, which introduces a state-dependent length dispersion of polymers, makes the situation even less straightforward in the case of supramolecular polymers. The clear picture of the phase behavior that we have reached represents nonetheless a useful, general reference, previously lacking, even for systems which may be much more complex. Indeed, the complexity of real systems calls for clear guidelines, as can be obtained from simple models that can be fully understood.

Aqueous solutions of double stranded B-DNA (> 100 base pairs), which is right-handed, form a left-handed cholesteric phase with a pitch $\mathcal{P}$ of the order of some micrometers [67, 68]. Relatively short DNA duplexes (< 20 base pairs), whose aspect ratio would be too low to induce liquid crystal order, also exhibit N* phases upon self-assembling into long helical aggregates, promoted by end-to-end stacking interactions [69]. In this case, however, both right- and left-handed cholesteric phases were found, depending on the oligonucleotide sequence and length. Based on the Straley model [61], a right-handed N* phase would be predicted for the B-DNA helix morphology [70], a prediction confirmed by calculations based on Onsager theory for a coarse-grained B-DNA model [62, 71]. This is also in agreement with the results obtained for our model of hard helices, which indicate the formation of a right-handed N* phase in the case of right-handed helices with small pitch p [17, 59, 60]. The variety of behaviors exhibited by B-DNA oligomers, with a prevalence of left-handed cholesterics, was ascribed to the competing effect of other interactions, specifically electrostatic ones, superimposed on steric repulsions, which would promote an opposite twist of particles [71, 72].

Other helical polymers that in organic solvents form cholesteric phases are polypeptides, and particularly well studied is the case of poly(γ-benzyl-L-glutamate) (PBLG). This has a right-handed α-helical structure and may form either a right- or a left-handed N* phase, depending on the solvent, with pitches of the order of hundreds of nanometers. In this uncharged system a special role of dispersion interactions was invoked, whose effect would depend on the relative dielectric constant of polymer and solvent [73], and again on the competition between dispersion interactions and the underlying steric repulsions [74, 75].

Cholesteric phases have also been observed in helical colloidal systems: a well-known example is that of filamentous viruses, which are formed by a DNA core wrapped by a coating of helically arranged proteins. *fd* and M13 viruses were found to form a left-handed N* phase with pitch $\mathcal{P}$ ranging from tens to hundreds of micrometers [76]; however, the *fd* Y21M mutant, which differs from *fd* only by having a methionine in place of a tyrosine as the 21rst amino acid in the coat protein, forms a right-handed N* phase with a pitch almost an order of magnitude larger [77]. Either the competition between steric and electrostatic interactions [78] or a key role of helical shape fluctuations [76] were proposed for these systems, but no comprehensive understanding has been reached, yet.

Experimental evidence of screw-nematic order has been observed in concentrated suspensions of helical flagella isolated from *Salmonella typhimurium* [22]. While flagella with a rod-like shape exhibit a nematic phase, filaments with a pronounced helical character were found to undergo a direct transition from the isotropic to a modulated nematic phase with pitch $\mathcal{P}$ in the micrometer scale, which in the original paper was denoted as conical. It can be easily verified that the experimental results are fully compatible with what we have called the screw-nematic phase, with the helix axes ($\hat{\mathbf{u}}$) preferentially aligned along the same direction (the main director $\hat{\mathbf{n}}$) and the two-fold symmetry axes of helices $\widehat{\mathbf{w}}$ spiraling around $\widehat{\mathbf{n}}$. This was discussed in detail in Refs. [16, 18, 19]. Direct transition from the isotropic to the N_s^* phase can be seen in the phase diagram that we have calculated for helices having $r = 0.4$ and various values of the pitch (Figure 3.17). In such a phase diagram the N_s^* phase is superseded by smectic phases at higher density, at variance with the experimental system, which did not exhibit any smectic phase. This difference can be ascribed to the length polydispersity of the helical flagella, which inhibits the formation of layers, but is fully compatible with the existence of screw-like order.

It is also worth noting that in the experiments on helical flagella, no cholesteric phase has been observed. It would be interesting to investigate in detail what is the role of length polydispersity in the stability of a cholesteric phase in systems of helical particles.

To our knowledge, there has been no evidence of screw-like order in polymeric systems. A possible reason is that polymers, owing to their shape and flexibility, do not meet the requirements for the stabilization of phases with screw organization. This is what occurs for the slightly helical particles whose phase diagram is shown in Figure 3.14. Another possible reason is the difficulty in experimental detection of space modulations with pitch $\mathcal{P}$ in the nanometer range. We hope that new, targeted experiments can shed light on this issue, possibly involving the design and synthesis of polymers with a more pronounced helical shape.

3.11 Conclusions and Perspectives

In this chapter we have tried to account for some recent results obtained by considering a fluid of rigid helices interacting only sterically, that is, via excluded volume only. To this end, we have used a combination of Monte Carlo simulations and density functional theory.

Notwithstanding the fundamental role of helical motifs in nature, a detailed study of this fluid was surprisingly missing, likely because the phase behavior of helical particles was assumed to be a minor variation of that of hard straight rods. Conversely, we found a rich and unconventional polymorphism of entropically driven liquid crystal phases that, as well as conventional nematic and smectic (A and B) phases, also appearing in a fluid of rod-like particles, display specific phases, whose origin we have discussed.

A rigid helical particle is a chiral object, in that left-handed helices cannot be superimposed on their right-handed counterparts. Because of this chirality, neighboring parallel helices in a nematic phase experience a twist that propagates in the direction perpendicular to the original plane following a helical path with a pitch typically much larger than the particle itself. Therefore the phase looks nematic on a short scale but is cholesteric on a much larger scale. We found this phase to be favored at moderate volume fractions and moderate curliness.

At larger densities and for more pronounced curliness, when there is sufficient interlocking between neighboring helices to promote azimuthal coupling, a different mechanism sets in, where there is a tendency for the parallel helices to slide up and down along the nematic director to increase their translational entropy, compensating for the corresponding loss of rotational entropy due to the azimuthal coupling. This is the driving mechanism for the formation of a screw-nematic phase whose origin is rooted in the helical shape of the particles.

We have further discussed how a similar mechanism favors the formation of screw-smectic phases (both A and B), in addition to another smectic phase (called "B polar") where this organization is replaced by a different one with horizontal sliding and rotation of neighboring smectic layers which differs from the conventional smectic B by the presence of polar transversal order.

These findings suggest significant sensitivity of the liquid crystal phases to the shape of the helices that is to be accounted for in all those experimental systems for which hard helices can be reckoned as a good minimal model. This includes biological systems, such as helical flagella, but also helically nanostructured materials exhibiting special optical properties that are of interest for photonic metamaterials.

The above model could be made more realistic by adding charges, localized interactions, by allowing mixing right- and left-handedness, as well as by including flexibility in the helices. These model variants will be useful in describing different systems.

Yet, there are a number of possible further studies even maintaining the helices as hard and rigid. One interesting issue hinges on the possible presence of columnar phases in a system of monosized hard helices, which are exhibited for instance by *fd* viruses [80]. Equally deserving a dedicated study is a detailed analysis of the high-densities ordered phases, as they might provide some surprising features in view of the non-convex nature of the helices [5]. Other important, still unexplored, issues concern the behavior of mixtures of helices, differing in length and morphology, mixtures of enantiomers, as well as the dynamics of the onset of screw-like phases.

All these issues will be pursued in the near future.

Acknowledgments

G.C. thanks the Government of Spain for the award of a Ramón y Cajal research fellowship and for financial support under the grant FIS2013-47350-C5-1-R, and the grant MDM-2014-0377. A.F. and A.G. acknowledge financial support from MIUR PRIN-COFIN2010-2011 (contract 2010LKE4CC). The use of the SCSCF multiprocessor cluster at the Università Ca' Foscari Venezia is gratefully acknowledged.

References

1. Whitesides G.M, M. Boncheva, *Proc. Natl. Acad. Sci.* **99**, 4769 (2002); Whitesides G. M., B. Grzybowski, *Science* **295**, 2418 (2002).
2. Pusey P.N., W. Van Megen, *Nature* **320**, 340 (1986).
3. Frenkel D., *Nature Mat.* **14**, 9 (2014).
4. Glotzer S. C., M. J. Solomon, *Nature Mat.* **6**, 557 (2007).
5. Damasceno P. F., A. S. Karas, B. A. Schultz, M. Engel, S. C. Glotzer, *Phys. Rev. Lett.* **115**, 158303 (2015).
6. Yang M., N. A. Kotov, *J. Mater. Chem.* **21**, 6775 (2011).
7. Gibbs J., A. G. Mark, T.-C. Lee, S. Eslami, D. Schamel, P. Fischer, *Nanoscale* **6**, 9457 (2014).
8. Liu L., L. Zhang, S. M. Kim, S. Park, *Nanoscale* **6**, 9355 (2014).

9. Douglas S. M., H. Dietz, T. Liendl, B. Hogberg, F. Graf, W. H. Shih, *Nature* **459**, 414 (2009).
10. Seeman N. C., *Nature* **421**, 427 (2003).
11. Nakano T., Y. Okamoto, *Chemical Review* **101**, 4013 (2001).
12. Yashima E., K. Maeda, H. Idia, Y. Furusho, K. Nagai, *Chemical Review* **109**, 6102 (2009).
13. Bawden F. C. et al., *Nature* **138**, 1051 (1936).
14. Rey A. D., *Soft Matter* **6**, 3402 (2010).
15. Frezza E., A. Ferrarini, H. B. Kolli, A. Giacometti, G. Cinacchi, *J. Chem. Phys.* **138**, 164906 (2013).
16. Kolli H. B., E. Frezza, G. Cinacchi, A. Ferrarini, A. Giacometti, T. S. Hudson, *J. Chem. Phys.* **140**, 081101 (2014).
17. Frezza, E., A. Ferrarini, H. B. Kolli, A. Giacometti, G. Cinacchi, *Phys. Chem. Chem. Phys.* **16**, 16225 (2014).
18. Kolli H. B., E. Frezza, G. Cinacchi, A. Ferrarini, A. Giacometti, C. de Michele, F. Sciortino, *Soft Matter* **10**, 8171 (2014).
19. Kolli H. B., G. Cinacchi, A. Ferrarini, A. Giacometti, *Faraday Disc.* **186**, 171 (2016).
20. Onsager L., *Ann. N. Y. Acad. Sci.*, **51**, 627 (1949).
21. Manna F., V. Lorman, R. Podgornik, B. Zeks, *Phys. Rev. E* **75**, 030901(R) (2007).
22. Barry E., Z. Hensel, Z. Dogic, M. Shribak, R. Oldenbourg, *Phys. Rev. Lett.* **96**, 018305 (2006).
23. de Gennes P. G., J. Prost, *The Physics of Liquid Crystals* (Clarendon Press, Oxford, 1993), 2nd edn.
24. Demus D., J. Goodby, G. W. Gray, H. W. Spiess, V. Vill, *Handbook of Liquid Crystals* (Wiley-VCH Verlag GmbH, Weinheim, 1998).
25. Luckhurst G. R., T. J. Sluckin, eds. *Biaxial Nematic Liquid Crystals: Theory, Simulation, and Experiment* (John Wiley and Sons, Ltd, Chichester, 2015).
26. Berardi R., L. Muccioli, S. Orlandi, M. Ricci, C. Zannoni, *J. Phys.: Condens. Matter* **20**, 463101 (2008).
27. Yu L. J., A. Saupe, *Phys. Rev. Lett.* **45**, 1000 (1980).
28. van den Pol E., A. V. Petukhov, D. M. E. Thies-Weesie, D. V. Byelov, G. J. Vroege, *Phys. Rev. Lett.* **103**, 258301 (2009).
29. Kitzerow H., C. Bahr, eds. *Chirality in Liquid Crystals* (Springer, Heidelberg, 2001).
30. Pieraccini S., S. Masiero, A. Ferrarini, G. P. Spada, *Chem. Soc. Rev.* **40**, 258 (2011).
31. Katsonis N., E. Lacaze, A. Ferrarini, *J. Mater. Chem.* **22**, 7088 (2012).
32. Wright D. C., N. D. Mermin, *Rev. Mod. Phys.* **61**, 385 (1989).
33. de Gennes P. G., *Sol. State Commun.* **10**, 753 (1972).
34. Renn S. R., T. C. Lubensky, *Phys. Rev. A* **38**, 2132 (1988).
35. Goodby J. W., M. A. Waugh, S. M. Stein, E. Chin, R. Pindak, J. S. Patel, *Nature* **337**, 449 (1989).
36. Bolhuis P., D. Frenkel, *J. Chem. Phys.* **106**, 666 (1997).
37. Metropolis N., A. W. Rosenbluth, A. H. Teller, E. Teller, *J. Chem. Phys.* **21**, 1087 (1953).
38. Allen M. P., D. J. Tildesley, *Computer Simulation of Liquids* (Clarendon Press, Oxford, 1989).
39. Wood W. W., J. D. Jacobson, *J. Chem. Phys.* **27**, 1207 (1957).
40. Vesely F. J., *J. Comput. Phys.* **47**, 291 (1982).

41. Kirkwood J. G., *J. Chem. Phys.* 7, 919 (1939).
42. Alder B. J., T. E. Wainwright, *J. Chem. Phys.* **27**, 1208 (1957).
43. Gabriel A., T. Meyer, G. Germano, *J. Chem. Theory. Comput.* **4**, 468 (2008).
44. Zocher H., *Z. Anorg. Chem.* **147**, 91 (1925).
45. McQuarrie D. A., *Statistical Mechanics* (University Science Books, Sausalito, CA, 2000).
46. Mayer J. E., *Equilibrium Statistical Mechanics* (Pergamon Press, New York, 1968).
47. Kayser R. F., H. J. Raveché, *Phys. Rev. A* **17**, 2067 (1978).
48. Herzfeld J., A. E. Berger, J. W. Wingate, *Macromolecules* **17**, 1718 (1984).
49. Oldenbourg R., X. Wen, R. B. Meyer, *Phys. Rev. Lett.* **61**, 1851 (1988); Fraden S., G. Maret, D. L. D. Casper, R. B. Meyer, *Phys. Rev. Lett.* **63**, 2068 (1989).
50. Frenkel D., *J. Phys. Chem.* **91**, 4912 (1987).
51. Straley J. P., *Molec. Crys. Liq. Cryst.* **24**, 7 (1973).
52. Mulder B. M., D. Frenkel, *Mol. Phys.* **55**, 1193 (1985); Tjipto-Margo B., G. T. Evans, *J. Chem. Phys.* **93**, 4254 (1990).
53. Denniston M., A. J. Masters, *Phys. Rev. E* **84**, 021709 (2011).
54. Parsons J. D., *Phys. Rev. A* **19**, 1225 (1979); Lee S.-D., *J. Chem. Phys.* **87**, 4972 (1987); Lee S.-D., *J. Chem. Phys.* **89**, 7036 (1989).
55. Carnahan N. F., K. E. Starling, *J. Chem. Phys.* **51**, 635 (1969).
56. Camp P. J., C. P. Mason, M. P. Allen, A. A. Khare, D. A. Kofke, *J. Chem. Phys.* **105**, 2837 (1996).
57. Cinacchi G., L. Mederos, E. Velasco, *J. Chem. Phys.* **121**, 3854 (2004).
58. Greco C., A. Ferrarini, *Phys. Rev. Lett.* **115**, 147801 (2015).
59. Belli S., S. Dussi, M. Dijkstra, R. van Roij, *Phys. Rev. E* **90**, 0250503(R) (2014).
60. Dussi S., S. Belli, R. van Roij, M. Dijkstra, *J. Chem. Phys.* **142**, 074905 (2015).
61. Straley J. P., *Phys. Rev. A* **14**, 1835 (1976).
62. Tombolato F., A. Ferrarini, *J. Chem. Phys.* **122**, 054908 (2005).
63. Veilliard-Baron J., *Molec. Phys.* **28**, 809 (1974).
64. Memmer R., *Liq. Cryst.* **29**, 483 (2002).
65. Cifelli M., G. Cinacchi, L. de Gaetani, *J. Chem. Phys.* **125**, 164912 (2006).
66. Cinacchi G., L. De Gaetani, *Phys. Rev. E* **77**, 051705 (2008).
67. Rill R. R., *Proc. Natl. Acad. Sci.* **83**, 342 (1986).
68. Livolant F., A. Leforestier, *Progress in Polymer Science* **21**, 1115 (1996).
69. Zanchetta G., F. Giavazzi, M. Nakata, M. Buscaglia, R. Cerbino, N. A. Clark, T. Bellini, *Proc. Natl. Acad. Sci.* **107**, 17497 (2010).
70. Proni G., G. Gottarelli, P. Mariani, G. P. Spada, *Chem. Eur. J.* **6**, 3249 (2000).
71. Frezza E., F. Tombolato, A. Ferrarini, *Soft Matter* **7**, 9291 (2011).
72. Wensink H. H., *Europhys. Lett.* **107**, 36001 (2014).
73. Samulski T. V., E. T. Samulski, *J. Chem. Phys.* **67**, 824 (1977).
74. Osipov M. A., *Nuovo Cimento D* **10**, 1249 (1988).
75. Emelyanenko A. V., *Phys. Rev. E* **67**, 031704: (2003).
76. Grelet E., S. Fraden, *Phys. Rev. Lett.* **90**, 198302: (2003).
77. Tombolato F., A. Ferrarini, E. Grelet, *Phys. Rev. Lett.* **96**, 258302 (2006).
78. Barry E., D. Beller, Z. Dogic, *Soft Matter* **5**, 2563 (2009).
79. Zhang Z., E. Grelet., *Soft Matter* **9**, 1015 (2013).
80. Grelet E., *Phys. Rev. X* **4**, 021053 (2015).

4

Self-Consistent Field Theory of Self-Assembling Multiblock Copolymers

Weihua Li[1] *and An-Chang Shi*[2]

[1] *State Key Laboratory of Molecular Engineering of Polymers, Collaborative Innovation Center of Polymers and Polymer Composite Materials, Department of Macromolecular Science, Fudan University, Shanghai, China*
[2] *Department of Physics and Astronomy, McMaster University, Hamilton, Canada*

4.1 Introduction

Block copolymers are heterogeneous polymers composed of chemically distinct subchains or blocks tethered together. This combination of chemical diversity and chain connectivity makes block copolymers a fascinating class of soft materials with unique structural and mechanical properties [1, 2]. In particular, the self-assembly of block copolymers has attracted tremendous attention due to their ability to form various ordered structures with domain sizes at the nanometer scale. These nanoscopic ordered morphologies possess potentials for advanced applications in a wide range of areas. From a thermodynamic point of view, the self-assembly of block copolymers is governed by a delicate balance of two opposing trends, i.e. the repulsive interactions between the chemically distinct blocks and the connectivity of the polymer chains. The repulsive interaction between chemically different blocks tends to drive the system to phase separate, whereas the connectivity of the blocks prevents a phase separation at the macroscopic scale. From this perspective block copolymers represent a classical example of frustrated physical systems. To alleviate this frustration, phase separation occurs in block copolymer systems via the formation of nanoscopic domains composed of different blocks. The packing of these domains can further assume different symmetries, leading to the formation of various periodic nanostructures. For the simplest block copolymer composed of A and B blocks, which are linked at their ends to form an AB diblock copolymer, its self-assembly behavior is mainly controlled by two characteristic parameters, viz the volume fraction of the A block, f, and the interaction parameter, χN, which is the product of the Flory–Huggins parameter (χ) and the total polymerization index N. By varying these two variables, AB diblock copolymers could self-assemble into ordered structures including lamellae (L), hexagonally packed cylinders (HEX), spherical domains arranged on a hexagonally close-packed (HCP) or body-centered cubic (BCC) lattice, and bicontinuous phases composed of networked cylinders with different symmetries—gyroid or *Fddd* (O^{70})

Self-Assembling Systems: Theory and Simulation, First Edition. Edited by Li-Tang Yan.

[3, 4]. Furthermore, conformational asymmetry of the A and B blocks could lead to the formation of complex spherical phases such as the A15 phase and the Frank–Kasper σ-phase [5].

Due to the availability of different monomers and chain architectures, the type of block copolymers is unlimited. In recent years, the development of modern synthetic techniques makes complex block copolymers consisting of more than two blocks (i.e. multiblock) and more chemically distinct components accessible, as demonstrated in numerous recent experiments [6–8]. The vast library of chain architectures, including linear, star, comb, and loop topologies, provides unlimited possibilities for the fabrication of diverse structures. As a simple extension of the AB diblock copolymers, a complex block copolymer is the ABC triblock copolymers composed of three different blocks with different topologies (e.g. linear, ring, or loop). It has been well established that chain topology could have a large effect on the self-assembly behavior of ABC triblock copolymers. Even for a given topology, an ABC triblock copolymer could exhibit complicated self-assembly behavior, and thus be capable of forming more ordered structures than the AB diblock copolymer. This richer phase behavior of ABC triblock copolymers is partly due to the very large phase space with at least five independent parameters including three interaction parameters and two independent composition parameters [9]. For linear ABC triblock copolymers, experimental and theoretical studies have revealed a rich array of intricate and novel structures, such as a square array of cylinders [10, 11], helical supercylinders [12, 13], and knitting pattern [14, 15]. Furthermore, for AB-type block copolymers, new structures can be obtained by utilizing designed chain architectures. For example, the perforated lamellar phase, which is metastable in simple AB diblock or ABA triblock copolymers, could become a stable phase in AB_m miktoarm and comb block copolymers [4]. More recently, it has been predicted that AB_m ($m = 3$) miktoarm block copolymers could exhibit a phase transition sequence from disorder, HCP, BCC, and σ to A15 phases [5]. From these previous studies it can be concluded that the self-assembly of multiblock copolymers provides a robust platform for the fabrication of rich ordered structures.

In general, the study of the phase behavior of block copolymers could be formulated as a forward problem or an inverse problem. In the forward problem, the possible phases that could be formed from a given block copolymer are explored, both experimentally and theoretically. The resulting phase diagram can then be used as library to obtain desirable structures. In the inverse problem, a desired structure or ordered phase is specified first, then possible block copolymer architectures and compositions are explored to obtain this ordered structure as an equilibrium phase of the system. In principle, the forward problem has a unique solution, whereas the inverse problem could have multiple solutions. In order to solve the inverse problem, or to design block copolymers for the engineering of given ordered structures, efficient methods and strategies to explore possible phases of a given copolymer are needed. More importantly, an understanding of the principles of structural formation is needed so that the desired structures could be designed following these principles. As an example of the inverse problem, Xie et al. have demonstrated a guiding principle allowing the design of multiblock terpolymers for the formation of a large number of targeted structures in the form of binary crystalline phases [16]. This basic idea can be extended to tailored multiblock copolymers for various desired structures if the underlying self-assembly mechanism is understood.

In order to realize the idea of designing multiblock copolymers for desired structures, it is essential to have a robust and efficient theoretical framework that is capable of predicting the phase behavior and morphologies of block copolymers with any specific molecular architecture. Ideally, the theory should take the molecular parameters as input and be able to predict thermodynamically stable phases and the phase transition boundaries between them. In the past, a number of theoretical methods have been developed to study the phases and phase transitions of block copolymer systems. One of the most successful theoretical frameworks for inhomogeneous polymeric systems such as block copolymers is the self-consistent field theory (SCFT). The SCFT has its origin in the work by Edwards in the 1960s [17]. This theoretical framework was explicitly adapted to treat block copolymers by Helfand in 1975 [18]. From the mathematical point of view, block copolymer theory presents a complex problem, and it is difficult to obtain exact solutions. Different approaches have been developed to solve the SCFT equations for block copolymer systems. In 1980 Leibler proposed an approximate analytical theory by assuming that the free energy functional could be expanded around the homogeneous state [19]. Therefore, Leibler's theory is valid in the weak-segregation regime. A similar theory was introduced by Ohta and Kawasaki for the study of AB diblock copolymers in 1986 [20]. The Ohta–Kawasaki theory has been extended to more complex block copolymer systems, such as to AB/C blends [21] and very recently to ABC linear triblock copolymers [22]. Another approximate analytical theory, valid in the strong-segregation regime, was introduced by Semenov in 1985 [23]. These approximate theories have been applied to a variety of block copolymer systems, leading to valuable insights into the phases and phase transitions of block copolymers. Besides the efforts of developing analytic methods, an alternative approach to block copolymer theory is to solve the SCFT equations using numerical techniques. The earliest attempts to obtain numerical solutions for diblock copolymers were made by Helfand and coworkers [24]. Later, Shull as well as Vavasour and Whitmore constructed phase diagrams of block copolymer melts and solutions using approximate numerical techniques [25, 26]. The first accurate three-dimensional solutions of the different ordered phases for diblock copolymers were obtained by Matsen and Schick in 1994 using the spectral method, or the reciprocal-space method, to solve the SCFT equations [27]. The spectral method utilizes the space-group symmetry of the ordered phases and is able to provide exact numerical solutions to SCFT equations. This technique has been applied to a large number of block copolymer systems [3, 4, 11]. In the early applications of the spectral method, the symmetry of the ordered phases is assumed to be known, leading to an efficient numerical method to obtain accurate SCFT solutions. Replacing the symmetry-restricted basis functions by a generic Fourier series, the spectral method could be extended for discovering new phases, as demonstrated in the case of ABC linear triblock copolymers by Guo et al. [28]. Complementary to the reciprocal-space method, an alternative method is to solve the SCFT equations in real space. In particular, Drolet et al. proposed that the real-space SCFT could be used to discover new ordered structures starting from random initial conditions [29]. In 2002, a new numerical method, which solves SCFT equations by taking advantage of both real-space and reciprocal-space methods, was developed by Tzeremes and coworkers [30]. The pseudo-spectral method is highly accurate and numerically efficient because it takes advantage of the high efficiency of the fast Fourier transform (FFT) algorithm. Moreover, it was further developed to be able to calculate the free energy of phases with a higher order of accuracy [31]. These previous studies have demonstrated that the SCFT based on the Gaussian model of the flexible polymer chains provides a powerful theoretical framework for the study of block copolymers. In this chapter we provide a review of various approaches to solving the SCFT equations, emphasizing applications to the study of the self-assembly of multiblock copolymers.

The theory of block copolymers is a very active research area in polymer science. Significant progress is continuously being made. Therefore it is virtually impossible to include all the current theoretical developments in a short review chapter. Instead, we focus on the recent applications of different methods of SCFT to the study of the self-assembly of block copolymers. In particular, we focus on a review of the recent progress in the study of the self-assembly of multiblock copolymers, alongside some specific details of the numerical techniques. We hope that this chapter will serve as a reference source for readers who are interested in applying the SCFT to multiblock copolymer systems. There is a large body of literature on block copolymer theories, including a number of valuable reviews and books [32–34]. Systematic derivations of the formulation of SCFT for block copolymers are provided in these references, and are not repeated in this chapter. Rather, the presentation of the material in this chapter relies heavily on these review articles and on our own work. Furthermore, we focus on the SCFT of block copolymers in this chapter, and thus a number of different theoretical approaches are not mentioned. We apologize in advance for our failure to cite all of the relevant and important references.

4.2 Theoretical Framework: Self-Consistent Field Theory of Block Copolymers

The formulation of the self-consistent field theory of block copolymers starts with a molecular model of the polymers. For block copolymers, one of the most successful molecular models, or the "standard model," is a coarse-grained model in which the polymer chains are modeled as flexible Gaussian chains and the interactions between the distinct monomers are described by short-range contact potentials [32]. Furthermore, the hard-core repulsive interactions are treated using the incompressibility condition. The advantage of this simple model is that the thermodynamic behavior of the system can be formulated in terms of a field theory, the self-consistent field theory (SCFT), which allows systematic studies using a variety of techniques. In particular, the mean-field approximation of the theory results in a powerful theoretical framework for the study of block copolymer phase behavior. The solutions of the SCFT equations have provided a quantitative phase diagram for diblock copolymers, which is in good agreement with experiments. For more complex block copolymers, SCFT solutions also provide useful guides to experiments. Further development of the SCFT on studying the effect of anisotropic Gaussian fluctuations on ordered phases provides an understanding of the nature of these fluctuations, as well as a technique to study the stability, kinetic pathways, and scattering function of ordered structures [33]. Recently, SCFT has been combined with the string method to study the minimal free energy paths (MFEPs) that dictate the transition pathways connecting different phases in bulk [35] or the defective structures in external fields [36, 37]. In what follows, we briefly formulate the SCFT equations. For more details about the derivation and underlying physics we refer readers to the literature [32–34].

We use AB diblock copolymer melts as an example to formulate the SCFT. Extension to more complex block copolymers as well as block copolymer blends and solutions is straightforward. We use a canonical ensemble approach and consider n copolymer chains in a volume V at a given temperature T. Each copolymer chain consists of N segments of species $K = A, B$. The blocks A and B contain $N_A = fN$ and $N_B = (1-f)N$ segments, respectively. Each block has an associate segment length (i.e. Kuhn length) $b_K = \varepsilon_K b$, where b is a reference Kuhn length. The segments are assumed to have the same segment density, ρ_0, which is defined as segments per unit volume; thus the volume occupied by one segment is ρ_0^{-1}. It is convenient to choose the spatial length unit as the Gaussian radius of gyration of the chain, $R_g = b\sqrt{N/6}$. The chain contour length is scaled by the segment number N. When each copolymer chain is treated as a spatial curve, the free energy per chain within the mean field approximation can be expressed as [32–34]

$$\frac{F}{nk_BT} = -\ln Q + \frac{1}{V}\int d\mathbf{r}\{\chi N\phi_A(\mathbf{r})\phi_B(\mathbf{r}) - w_A(\mathbf{r})\phi_A(\mathbf{r}) - w_B(\mathbf{r})\phi_B(\mathbf{r}) - \xi(\mathbf{r})[1-\phi_A(\mathbf{r})-\phi_B(\mathbf{r})]\}, \tag{4.1}$$

where $\phi_K(\mathbf{r})$ $(K = A, B)$ are the local concentration or volume fraction of the segment K. The quantity Q is the partition function of a single copolymer chain interacting with the mean fields of $w_K(\mathbf{r})$, which are produced by the surrounding chains. The field function $\xi(\mathbf{r})$ is a Lagrange multiplier used to enforce the incompressibility condition, $\phi_A(\mathbf{r}) + \phi_B(\mathbf{r}) = 1$. Minimization of the free energy functional with respect to the compositional and field functions leads to the following SCFT equations:

$$w_A(\mathbf{r}) = \chi N\phi_B(\mathbf{r}) + \xi(\mathbf{r}), \tag{4.2}$$

$$w_B(\mathbf{r}) = \chi N\phi_A(\mathbf{r}) + \xi(\mathbf{r}), \tag{4.3}$$

$$\phi_A(\mathbf{r}) = \frac{1}{Q}\int_0^f ds\, q(\mathbf{r}, s)q^\dagger(\mathbf{r}, s), \tag{4.4}$$

$$\phi_B(\mathbf{r}) = \frac{1}{Q}\int_f^1 ds\, q(\mathbf{r}, s)q^\dagger(\mathbf{r}, s), \tag{4.5}$$

$$Q = \frac{1}{V}\int d\mathbf{r} q(\mathbf{r}, s)q^\dagger(\mathbf{r}, s), \tag{4.6}$$

$$\phi_A(\mathbf{r}) + \phi_B(\mathbf{r}) = 1. \tag{4.7}$$

In the above equations, the quantities $q(\mathbf{r}, s)$ and $q^\dagger(\mathbf{r}, s)$ are the end-segment distribution functions measuring the probability of finding the segment s at the position $\mathbf{r}$ when starting from the end $s = 0$ or $s = 1$, respectively. The distribution functions satisfy the modified diffusion equations,

$$\frac{\partial q(\mathbf{r}, s)}{\partial s} = \varepsilon(s)\nabla^2 q(\mathbf{r}, s) - w(\mathbf{r}, s)q(\mathbf{r}, s), \tag{4.8}$$

$$-\frac{\partial q^\dagger(\mathbf{r}, s)}{\partial s} = \varepsilon(s)\nabla^2 q^\dagger(\mathbf{r}, s) - w(\mathbf{r}, s)q^\dagger(\mathbf{r}, s), \tag{4.9}$$

where $w(\mathbf{r}, s) = w_A(\mathbf{r})$, $\varepsilon(s) = \varepsilon_A$ for $s < f_A$, otherwise $w(\mathbf{r}, s) = w_B(\mathbf{r})$, $\varepsilon(s) = \varepsilon_B$. The initial conditions are $q(\mathbf{r}, 0) = 1$ and $q^\dagger(\mathbf{r}, 1) = 1$.

4.3 Numerical Methods of SCFT

A number of numerical methods, formulated in real-space or in reciprocal-space, have been developed to solve the SCFT equations since SCFT was formulated in the 1970s. Generically, all the numerical approaches share a common technique, i.e. a self-consistent iteration process (also referred to as a Picard-type algorithm) [27, 29]. This iteration process consists of the following steps: (i) initializing the mean fields $w_K(\mathbf{r})$, (ii) calculating the propagators $q(\mathbf{r}, s)$ and $q^{\dagger}(\mathbf{r}, s)$ as well as the single-molecular partition function Q via solving the modified diffusion equation, (iii) calculating the density distributions $\phi_K(\mathbf{r})$, and (iv) updating the mean fields with the new density profiles. Convergence of the iteration process leads to the solution of SCFT equations with converged density profile and corresponding free energy that specifies the morphology and determines the stability of the structure, respectively. The details of implementing the SCFT in real-space and in reciprocal-space are different, and will be discussed in the following subsections.

4.3.1 Reciprocal-Space Method

The reciprocal-space method, or the spectral method, is based on the observation that, for an ordered phase of block copolymers, all spatial functions of interest, i.e. the density distributions, the mean fields, and the propagators, are periodic functions. In principle, these periodic spatial functions can be expanded in terms of a set of basis functions composed of the plane waves $e^{i\mathbf{G}\cdot\mathbf{r}}$, where $\mathbf{G}$ are the reciprocal lattice vectors of the given ordered phase. Therefore, an ordered phase is specified by the set of expansion coefficients of these basis functions. For a given ordered phase, its structure is specified by its space-group symmetry, which in turn determines the permissible reciprocal lattice vectors. For example, the ordered phases from the self-assembly of diblock copolymers are the one-dimensional lamellar phase, the two-dimensional (2D) hexagonal phase with the plane group $p6m$, the BCC spherical phase with the space group $Im3m$, and the double gyroid phase with the space group $Ia\bar{3}d$. The corresponding reciprocal lattice vectors are determined using these real-space structures.

Besides the translational symmetry, the point-group symmetry of a given phase ensures that some of the reciprocal lattice vectors are symmetrically equivalent, thus having equal magnitudes. These symmetrically equivalent reciprocal lattice vectors could be grouped together, forming a star of reciprocal lattice vectors. The basis function of a star could be constructed as a linear combination of the plane waves of the corresponding vectors. The idea of reconstructing new basis functions is critical in the reciprocal-space method (also named as the spectral method) because it dramatically reduces the number of basis functions and thereby enhances the computational efficiency.

Specifically, each of these new basis functions is a linear combination of the form [32, 33]

$$f_n(\mathbf{r}) = \frac{1}{\sqrt{N_n}} \sum_{i \in n} S_i^n e^{i\mathbf{G}_i^n \cdot \mathbf{r}}, \tag{4.10}$$

where the wave vectors $\mathbf{G}_i^n$ are related by the point group symmetry and satisfy the relation $|\mathbf{G}_i^n|^2 = \lambda_n$, the factor $S_i^n = \pm 1$ according to the space group, and N_n is the number of reciprocal lattice vectors belonging to the nth star. The values of S_i^n and N_n can be found from the International Table of Crystallography. The new basis functions are orthonormal [i.e., $V^{-1}\int f_i(\mathbf{r}f_j)(\mathbf{r})d\mathbf{r} = \delta_{ij}$], and are eigenfunctions of the Laplacian operator, $\nabla^2 f_n(\mathbf{r}) = -\lambda_n f_n(\mathbf{r})$. Usually, the set of basis functions starting with $f_1(\mathbf{r}) = 1$ are ordered in a nondecreasing series of λ_n. As a consequence, the spatial functions of the ordered phase under consideration can be expanded in terms of the set of the basis functions:

$$\phi_K(\mathbf{r}) = \sum_n \phi_{K,n} f_n(\mathbf{r}), \tag{4.11}$$

$$w_K(\mathbf{r}) = \sum_n w_{K,n} f_n(\mathbf{r}), \tag{4.12}$$

$$q(\mathbf{r}, s) = \sum_n q_n(s) f_n(\mathbf{r}). \tag{4.13}$$

With the expansion, the partial differential diffusion equation of $q(\mathbf{r}, s)$ [or $q^\dagger(\mathbf{r}, s)$] becomes the normal differential equations of their coefficient functions $q_n(s)$ [or $q_n^\dagger(s)$],

$$\begin{aligned}
\frac{dq_n(s)}{ds} &= \begin{cases} \sum_m A_{nm} q_m(s), & \text{if } s < f, \\ \sum_m B_{nm} q_m(s), & \text{if } s > f, \end{cases} \\
A_{nm} &= -\lambda_n \varepsilon_A \delta_{nm} - \sum_l w_{A,l} \Gamma_{nml}, \\
B_{nm} &= -\lambda_n \varepsilon_B \delta_{nm} - \sum_l w_{B,l} \Gamma_{nml},
\end{aligned} \tag{4.14}$$

where $\Gamma_{nm} = V^{-1}\int f_n(\mathbf{r}) f_m(\mathbf{r}) f_l(\mathbf{r}) d\mathbf{r}$. The initial condition is $q_n(0) = \delta_{n1}$. The differential equations of the coefficient functions $q_n^\dagger(s)$ are similar. The set of linear differential equations is readily solved by diagonalizing the matrix A or B, leading to the solution of the amplitudes of the propagator functions. Accordingly, the SCFT equations in Equations (4.2)–(4.7) in terms of the amplitudes of all spatial functions become

$$\begin{aligned}
w_{A,n} &= \chi N \phi_{B,n} + \xi_n, \\
w_{B,n} &= \chi N \phi_{A,n} + \xi_n, \\
\phi_{A,n} &= \frac{1}{q_1(1)} \int_0^f ds \sum_{ml} q_m(s) q_l^\dagger(s) \Gamma_{nml}, \\
\phi_{B,n} &= \frac{1}{q_1(1)} \int_f^1 ds \sum_{ml} q_m(s) q_l^\dagger(s) \Gamma_{nml}, \\
Q &= V q_1(1), \\
\phi_{A,n} + \phi_{B,n} &= \delta_{n,1}.
\end{aligned} \tag{4.15}$$

The free energy can also be written in terms of these amplitudes as

$$\frac{F}{nk_\mathrm{B}T} = -\ln[q_1(1)] - \chi N \sum_n \phi_{A,n} \phi_{B,n}. \tag{4.16}$$

Besides the two controlling parameters, f and χN, the free energy of a given ordered phase is an implicit function of the wavelength or the period, D, of the structure because the wave vector **G** is inversely proportional to D. The equilibrium value of the period D is obtained by minimizing the free energy with respect to D. The phase diagram is constructed by comparing the free energy of different phases. Importantly, the accuracy of the free energy is determined by the number of basis functions in the spectral method, which, in turn, dictates the computational cost. With the development of high-performance computing technologies, larger numbers of basis functions can be included in the calculations using the spectral method, which guarantees a reliable accuracy for the free energy of each known ordered phase. In other words, as long as the candidate ordered phases are known for a given block copolymer, its phase diagram can be constructed accurately with the spectral method. For example, since the relatively accurate phase diagram of the simplest AB diblock copolymer, which consists of lamellar, cylindrical, double gyroid, and BCC spherical phases, was first constructed by the spectral method in 1994 [27], two new phases of HCP sphere and bicontinuous *Fddd* have been included in the phase diagram. The accuracy of the phase boundaries surrounding the gyroid phase in the strong segregation region has been improved, resulting in an up-to-date phase diagram of AB diblock copolymers (Figure 4.1) [3, 4].

Although the above formulation of the SCFT is for AB diblock copolymers, it is straightforward to extend the spectral method to polymeric systems with different molecular architectures as well as multi-component block copolymer systems. For example, the spectral method has been applied to determine the phase diagrams of diverse block copolymer systems including copolymer/homopolymer blends [38] and multiblock copolymers [4]. For example, Matsen has calculated the phase diagrams of AB-type multiblock copolymers with various architectures including ABA triblock, AB_m miktoarm, $(AB)_m$ star, ABAB $\cdots$ linear multiblock, and comb consisting of an A backbone and regularly spaced teeth of B blocks. His results suggest that the topology of the phase diagrams is relatively unaffected by differences in architecture, but the phase boundaries shift significantly in the $f - \chi N$ plane. In particular, the perforated-lamellar (PL) phase, which is a metastable phase in the AB and ABA triblock copolymers, could become stable in the comb multiblock copolymers (Figure 4.2). This prediction implies that the design of molecular architectures of multiblock copolymers could provide an

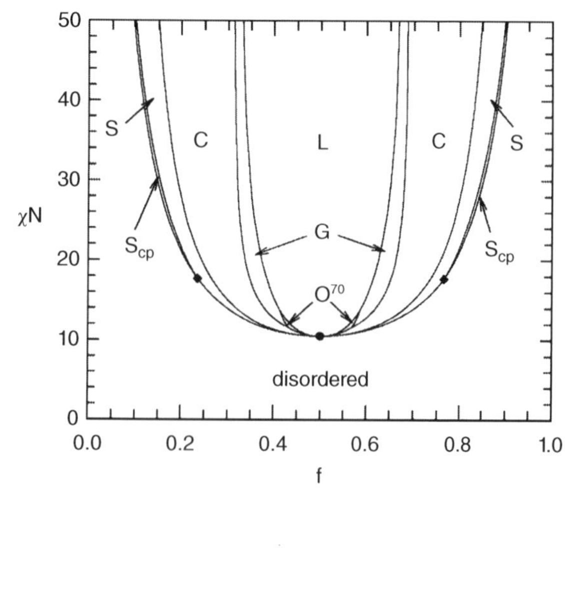

Figure 4.1 Phase diagram of AB diblock copolymer melts, consisting of the stability regions of lamellar (L), cylindrical (C), BCC spherical (S), HCP spherical (S_{cp}), gyroid (G), and *Fddd* (O^{70}) morphologies. Reproduced with permission from Ref. [4].

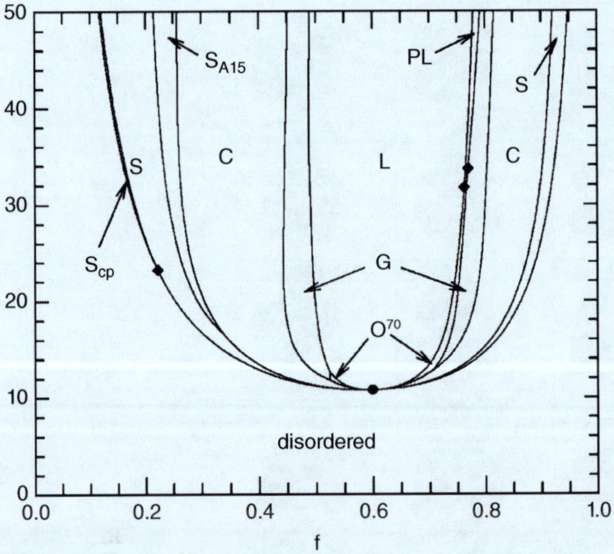

Figure 4.2 Phase diagram of a melt of an AB comb copolymer which is composed of a B-block backbone and regularly spaced teeth of A-blocks. Here, N is the number of segments of each tooth. Reproduced with permission from Ref. [4].

opportunity to obtain new ordered morphologies. The spectral method of SCFT with its accuracy and efficiency provides a powerful tool to determine the phase diagram of new designed multiblock copolymers.

Initially, the spectral method would require the symmetry of ordered phases as an input, thus limiting its application to exploring unknown structures. However, this limitation could be overcome by replacing the restricted basis functions with the full set of plane waves. Because any function can be expanded in terms of a set of the plane waves, this approach provides a generic description of different ordered phases. The utilization of generic Fourier series leads to an increase in the number of basis functions, and thereby increases the computational cost significantly. With a reasonable number of basis functions, Guo et al. demonstrated that the *generic* spectral method is capable of discovering new structures in ABC linear triblock copolymers (Figure 4.3). In practice, the combination of the generic reciprocal-space method and the spectral method enables us to study the phase behavior of complex block copolymer systems. The generic reciprocal-space method is used to construct possible candidate structures in the parameter space of interest, while the spectral method is used to compute accurate free energies of these structures, which are used to construct phase diagrams.

4.3.2 Real-Space Method

A natural method to solve the SCFT equations is to discretize the system in real space. While the real-space approach has been used in a number of previous studies, a three-dimensional computational method to solve the SCFT equations in real space was proposed by Drolet and Fredrickson [29]. In this approach, the system of SCFT

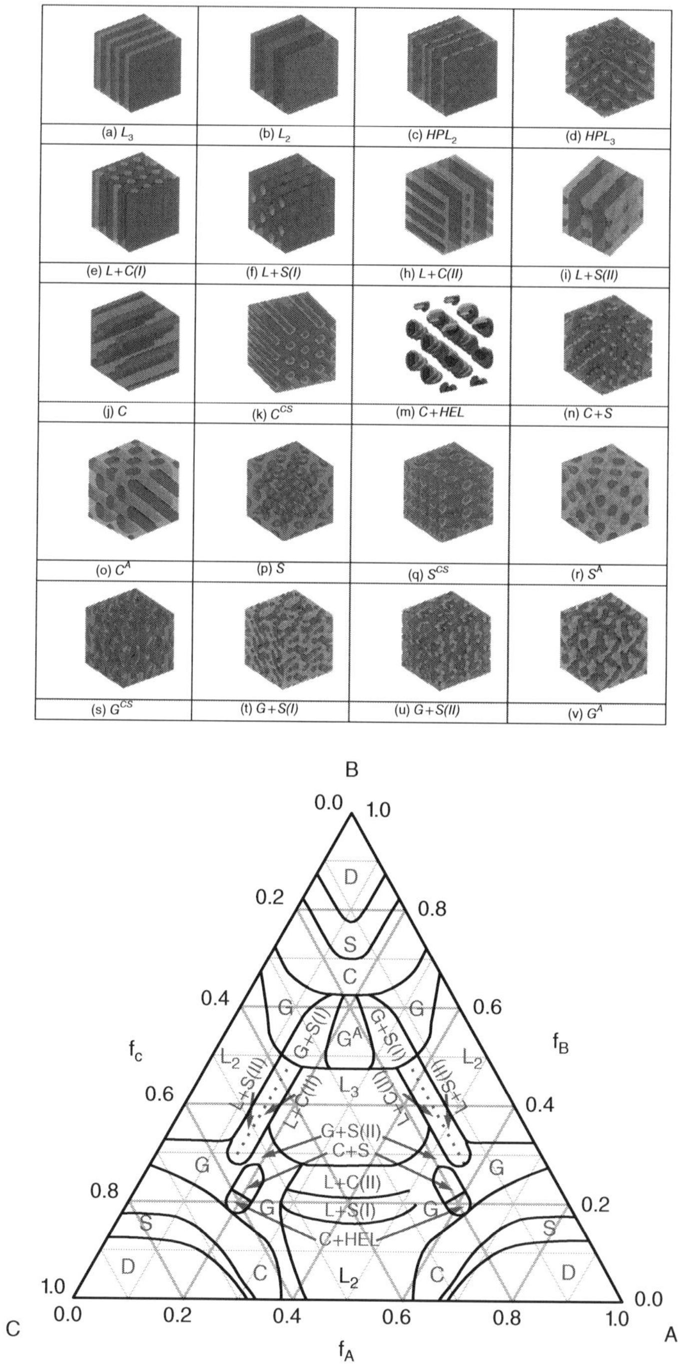

Figure 4.3 (Top) Morphologies self-assembled in the ABC linear triblock copolymers with $\chi_{AB}N = \chi_{BC}N = 35$ and $\chi_{AC}N = 15$, discovered by the general spectral method of SCFT. (Bottom) Phase diagram of these morphologies. Reproduced with permission from Ref. [28]. *(See insert for color representation of the figure.).*

equations is placed on a discrete grid and all spatial functions [e.g. $w_K(\mathbf{r})$, $\phi_K(\mathbf{r})$, $q(\mathbf{r}, s)$, and $q^\dagger(\mathbf{r}, s)$] are expressed as a set of values on the lattices. The steps of the iteration algorithm in real space are implemented as follows. First, an initial guess for the value of $w_K(\mathbf{r})$ on each lattice point is constructed with a random number generator. Alternatively, the fields $w_K(\mathbf{r})$ can be obtained from Equations (4.2) and (4.3) with $\xi(\mathbf{r}) = 0$ and the density functions $\phi_K(\mathbf{r})$ are randomly initialized with the reasonable range of values $0 < \phi_K(\mathbf{r}) < 1$. Second, the partial differential equations of $q(\mathbf{r}, s)$ and $q^\dagger(\mathbf{r}, s)$ are solved using a Crank–Nicholson scheme that is an alternating direction implicit (ADI) method with second-order accuracy of the integration step Δs and is unconditionally stable. Third, new density profiles are obtained by evaluating Equations (4.4)–(4.6). Finally, the mean fields are updated with the new density profiles and the Lagrange function or the pressure function $\xi(\mathbf{r})$ that is constructed from the incompressibility condition. In order to maintain the stability of the solution, a linear mix of new and old fields is usually used for the next iteration. These steps are repeated until the mean fields and free energy converge and the incompressibility condition is satisfied.

The most prominent feature of the real-space method proposed by Drolet and Fredrickson is the utilization of random initial conditions for the mean fields or the density profiles. This method offers one possibility of searching new morphologies self-assembled by complex block copolymers. One strategy for using the real-space method is to perform a large number of parallel runs in the parameter space of interest, hopefully to obtain many possible candidate structures that may be either stable or metastable. As a demonstration, Drolet and Fredrickson applied this strategy to ABCA tetrablock copolymers, and obtained a few novel 2D ordered phases for this system (Figure 4.4). In general, although the real-space method with random initial configurations could provide a number of candidate structures, it is possible that using random initial conditions alone could miss many of the candidate phases as demonstrated by Xu et al. [39]. The best practice is to use a combination of different initialization schemes to obtain candidate structures [39].

4.3.3 Pseudo-Spectral Method

The main computational task in the real-space method is to obtain solutions of the modified diffusion equations. This is usually done by using the Crank–Nicholson scheme in which the temporal integration is carried out by solving a triangular matrix for different directions in sequence. It is known that there are some severe constraints on the spatial and temporal resolutions of this method. To improve the stability and accuracy, Tzeremes et al. proposed a new scheme, or the pseudo-spectral method, to solve the modified diffusion equations [30]. The pseudo-spectral method utilizes both the real-space and reciprocal-space descriptions of the functions. Computationally, the interchange between these two descriptions could be carried out efficiently by taking advantage of the split-step Fourier transform. Due to the availability of a number of fast Fourier transform (FFT) packages, the pseudo-spectral method can be highly efficient and accurate.

The pseudo-spectral method starts with the observation that the solution of the modified diffusion equation Equation (4.8) can be expressed as a time-evolution operation,

$$q(\mathbf{r}, s + ds) = \exp[ds\ (\varepsilon(s)\nabla^2 - w(\mathbf{r}, s))]\ q(\mathbf{r}, s). \tag{4.17}$$

Figure 4.4 Two-dimensional morphologies obtained by solving the SCFT equations for ABCA tetrablock copolymer melts in real space. Regions rich in A, B, or C monomers are shown in gray, black, and white, respectively. The interaction parameters are fixed as $\chi_{AB}N = \chi_{AC}N = \chi_{BC}N = 35$, and the compositions of the four panels are: (a) $(f_A, f_B, f_C) = (0.26, 0.37, 0.37)$, (b) $(f_A, f_B, f_C) = (0.34, 0.33, 0.33)$, (c) $(f_A, f_B, f_C) = (0.48, 0.26, 0.26)$, and (d) $(f_A, f_B, f_C) = (0.24, 0.56, 0.20)$, respectively. Reproduced with permission from Ref. [29].

This solution is exact. However, a direct evaluation of the time-evolution operation is not straightforward due to the fact that the Laplacian operator ∇^2 does not commute with the position operator. One effective method to overcome this difficulty is to find an approximate solution of the time-evolution operation by splitting the two operators in the exponential. Within an error of order ds^3, the operator-splitting leads to the following approximate solution:

$$q(\mathbf{r}, s + ds) \approx \exp\left[-\frac{ds}{2}w(\mathbf{r}, s)\right] \exp[ds\,\varepsilon(s)\nabla^2] \exp\left[-\frac{ds}{2}w(\mathbf{r}, s)\right] q(\mathbf{r}, s). \tag{4.18}$$

The separation of the operators in Equation (4.18) can be used to develop an efficient numerical method for solving the modified diffusion equations. Specifically, the operator $\exp[-dsw(\mathbf{r}, s)/2]$ is simply implemented in real space, whereas the operator $\exp[ds\varepsilon(s)\nabla^2]$ is conveniently implemented in the Fourier space. Using this strategy leads to a numerical implementation for solving the modified diffusion equation via

Equation (4.18):

$$q(\mathbf{r}, s + ds) \approx \exp\left[-\frac{ds}{2}w(\mathbf{r}, s)\right] \mathrm{FFT}^{-1}\{\exp[-ds\varepsilon(s)k^2] \times \mathrm{FFT}\left(\exp\left[-\frac{ds}{2}w(\mathbf{r}, s)\right] q(\mathbf{r}, s)\right)\}, \tag{4.19}$$

where FFT and FFT^{-1} represent the forward and backward (inverse) fast Fourier transform (FFT), respectively. The wave vector **k** has a standard definition in a discrete space [34]. Highly efficient FFT algorithms enable calculations with a large grid size, e.g. 64^3, for a three-dimensional (3D) system on a single processor. Larger grid sizes are accessible by using parallel programming techniques, especially with the availability of well-developed FFT software packages such as FFTW [40]. A large grid size enables one to choose small grid spacing, thus ensuring accurate computation of the free energy of the system. In contrast to the spectral method in which the integration along the contour length is done analytically, the discretization degree of the contour length ds becomes the main factor influencing the accuracy of the free energy calculation. This is especially true for the case of strong segregation. Fortunately, the effect of ds becomes negligible when $ds \leq 0.01$ for weak or intermediate segregation degree. Because of these advantages, the pseudo-spectral method has become the method of choice for the study of the self-assembly behaviors of many block copolymer systems, including geometrically confined systems [41, 42] and multiblock copolymer systems [16, 39, 43].

The pseudo-spectral algorithm to solve the modified diffusion equation provides a fast and accurate method for finding solutions of the SCFT equations. Because the mean fields and density profiles are represented in real space and reciprocal space, the initialization of the iterative procedure could be performed in real and reciprocal space. This flexibility of initialization makes it possible to search for new structures as proposed by Drolet and Fredrickson in the case of the real-space method. On the other hand, as pointed out by Xu et al. [39], the strategy of starting with random initial configurations for the search for new structures may miss many local minima. The best practice is to use a combination of different methods to construct initial configurations, which could lead to a large number of possible candidate structures [39]. Furthermore, for a given block copolymer system with fixed parameters and fixed system sizes, random initial conditions could lead to the formation of many defective morphologies. This problem becomes more pronounced for the complex ordered phases such as continuous networked phases (gyroid or *Fddd*) and some hierarchical phases of multiblock terpolymers. Relaxation of these defective structures to reach a local or global minimum is usually a formidable task. In order to overcome this limitation of random initialization, a number of approaches to initialization procedures have been developed, as summarized by Xu et al. [39]. One specialized initialization procedure proposed by Li and coworkers utilizes specially tailored initial mean fields to obtain a given structure as the solution of the SCFT equations. For example, for the hexagonal cylinder phase, the density profile at the strong segregation limit can be readily generated. This density profile is then inserted into Equations (4.2) and (4.3) to produce the initial mean fields. The calculation is further accelerated by using a rectangular computational box with a fixed side ratio of $L_y/L_x = \sqrt{3}$ and an appropriate estimated value of L_x, which consists of two cylindrical domains. Using this initial mean field with proper symmetry and length scale, the SCFT calculation converges quickly to the solution of the hexagonal phase. It is then straightforward to minimize the free energy with respect to L_x leading to the equilibrium morphology and domain period. This initialization strategy is readily extended to more complex structures. In the following section, the application of the pseudo-spectral method with the special initialization scheme to the study of the phase behavior of multiblock copolymers is illustrated using three examples.

4.3.4 Fourth-Order Pseudo-Spectral Method

It is known that the pseudo-spectral method is accurate to the second order of the discrete step ds. While the spectral method is adequate for block copolymer systems in the weak to intermediate segregation regimes, a direct application of this method to the strong-segregation regime could lead to inaccurate results. For strongly segregated systems with very sharp interfaces, e.g. for AB diblock copolymers with $\chi N > 60$, the error induced by a finite step ds could become considerable. A naive approach to rectify this limitation is to simply reduce the step size ds, which requires longer computational time. On the other hand, it has been shown that the influence of the computational error on the location of the phase boundaries, which are determined by the relative values of the free energies, is less sensitive to the accuracy of the free energy itself [44]. Therefore, the second-order algorithm still provides accurate results for the phase boundaries.

Alternatively, the accuracy of the computation could be significantly increased by using a fourth-order pseudo-spectral method to solve the modified diffusion equations [31]. In the fourth-order algorithm, the Laplacian operator is treated implicitly with a fourth-order backward differentiation formula (BDF4), whereas the source term is discretized explicitly using the fourth-order accurate Adams–Bashford formula, leading to the following discretization scheme:

$$\begin{aligned}\frac{25}{12}q_{n+1} - 4q_n + 3q_{n+1} - \frac{4}{3}q_{n-2} + \frac{1}{4}q_{n-3} \\ = ds\nabla^2 q_{n+1} - dsw(q_{n+1} + 4q_n - 6q_{n-1} + 4q_{n-2} - q_{n-3}),\end{aligned} \tag{4.20}$$

where q_{n+i} denotes $q(\mathbf{r}, s + ids)$. The initial values required to apply this formula are obtained using backward Euler and Richardson's extrapolations. This fourth-order scheme is also unconditionally stable. This highly accurate algorithm has been applied to identify the phase diagram of AB diblock copolymers extending to very strong segregation degree, i.e. $\chi N \leq 100$, aiming to focus on the stability region of the double gyroid phase (Figure 4.5). The phase diagram indicates that the contraction of the phase region of the gyroid slows down as χN increases, and hence the gyroid phase remains a stable phase within a narrow region at large values of $\chi N = 100$.

4.4 Application of SCFT to Multiblock Copolymers

The development of numerical methods to solve the SCFT equations has led to great progress in the study of the phases and phase transitions of block copolymer systems. There have been numerous applications of SCFT to the study of the self-assembly of block copolymer melts, solutions, and block copolymers under confinement. In this

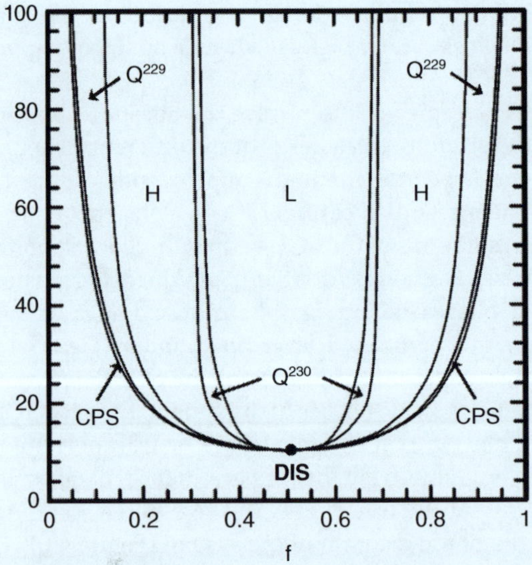

Figure 4.5 Phase diagrams of AB diblock copolymer identified by the fourth-order pseudo-spectral method, extending to a strong segregation region of $\chi N \leq 100$. Reproduced with permission from Ref. [31].

section we use three examples to illustrate the application of SCFT to the discovery and understanding of complex ordered phases formed from multiblock copolymer melts.

The first example concerns the formation of complex structures from ABC triblock copolymers. The self-assembled structures from block copolymers become increasingly complicated as the complexity of the block copolymer architecture increases. For example, addition of one more type of block changes an AB diblock copolymer to an ABC triblock copolymer. This seemingly trivial change leads to a tremendous increase in the dimension of the phase space for the system. Instead of two parameters, f and χN, for AB diblock copolymers, the phase behavior of an ABC triblock copolymer melt depends on at least five parameters, i.e. two independent volume fractions of the blocks and three independent interaction parameters. Furthermore, the topology of the triblock copolymers could be linear or star-like, resulting in richer phase behavior. For multiblock copolymers with more types of blocks, the dimensionality of the phase space is even larger. Exploring the phase behavior in a phase space with such a large dimensionality, e.g. five dimensions for ABC triblock copolymers, is a formidable task. Instead of attempting to construct a complete phase diagram of the system, it is useful to focus on the understanding of the formation mechanisms of particular ordered phases of interest. One particular interesting ordered structure is the helical morphologies formed in various soft matter systems. Helical structures are ubiquitous in nature, especially in biological systems. In the case of block copolymers, a hexagonally arranged helical super-cylindrical phase has been observed in experimental studies of linear PS-*b*-PB-*b*-PMMA triblock copolymers. In this complex morphology the PMMA-core cylinder is decorated by a substructure consisting of PB helices. Understanding the formation mechanism of the helical superstructures from the self-assembly of achiral block copolymers presents an interesting challenge in soft matter theory. Another interesting question about these helical structures concerns the number of helices preferred by the super-cylinder. Initially, the number of helices was speculated to be four from transmission electron microscopy (TEM) images [12]. Later on it was determined to be double helices by the 3D transmission electron microtomography

(TEMT) technique [45]. However, it is still not fully understood why double helices are formed, or whether other numbers of helices cannot be formed, such as single/triple or even quadruple.

The SCFT provides an ideal theoretical framework to resolve the question about the optimum number of helices because the accurate free energy of the different structures could be obtained by using the pseudo-spectral method coupled with the special initialization algorithm. Specifically, using initial configurations constructed from helical structures it is possible to obtain solutions of the SCFT equations corresponding to various helical super-cylindrical phases (e.g. single/double/triple/quadruple) as well as other competing super-cylindrical phases including sphere-on-cylinder, straight cylinder-on-cylinder, ring-on-cylinder, and perforated layer-on-cylinder (Figure 4.6). All these intricate morphologies are self-consistent solutions of the SCFT equations, thus these structures represent metastable or stable phases of the system. Detailed SCFT calculations by Li et al. led to the conclusion that the quadruple helical morphology is an unstable structure, whereas single/double/triple helices are metastable or stable phases [13]. Furthermore, the availability of the free energy of the different candidate structures allows the construction of the phase diagram of the system (Figure 4.6). The phase diagram indicates that the double and triple helical phases are stable, whereas the single helical phase is metastable. Besides the total free energy of the ordered phases, SCFT calculations provide more detailed information about the structures, thus enabling us to probe the self-assembly mechanism of the helical super-cylindrical phases. The competition between the interfacial energy and the stretching energy leads to preferred diameters of the helical domain and the core cylinder for a given set of parameters, resulting in a specific length ratio of the helices to the core cylinder. When the length ratio deviates from an integer value, the B-block domain tends to curve into helices around the C-core cylinder such that the variable pitch angle of the helices

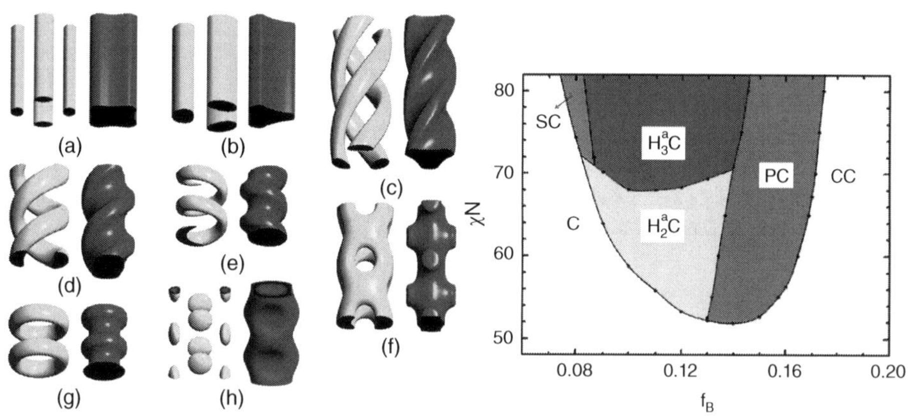

Figure 4.6 (Left) Candidate super-cylindrical morphologies produced by the pseudo-spectral method coupled with a special initialization scheme. (Right) Two-dimensional cross-section of phase diagram of ABC linear triblock copolymers with fixed $\chi_{AC}N = 20$ and $f_A = 0.70$, showing the stability regions of super-cylindrical phases including sphere-on-cylinder (SC), double helices-on-cylinder (H^a_2C), triple helices-on-cylinder (H^a_3C), and perforated lamella-on-cylinder (PC). The superscript "a" of H^a_2C or H^a_3C denotes that the helical super-cylinders have an anti-chirality arrangement and possess half left-handed and half right-handed helices. Reproduced with permission from Ref. [13].

regulates the length ratio. The scenario of the varying length ratio of the helical domain to the core cylinder and the pitch angle of the helices rationalizes the stability of double and triple helical super-cylindrical phases. The prediction of the stable triple helical phase was confirmed in a later experiment [46].

The second example explores the possibility of forming binary crystals of spherical domains from multiblock terpolymers. A class of ordered structures self-assembled from block copolymers is the spherical phase that resembles the space symmetry of atomic crystals. These ordered phases could be referred to as "mesocrystals" in which the spherical domain could be regarded as an artificial macromolecular "atom" (AMA). For the simple AB diblock copolymers, the dominant spherical phase or mesocrystal is a body-centered cubic (BCC) crystal in which the spherical A-domains, or the AMAs, are located on a BCC lattice and the B-blocks form a matrix. For linear ABC triblock copolymers with equal-sized A and C spherical domains, the stable phase is that the A and C spheres occupy the two sub-lattices of the BCC structure, forming a mesocrystal with the symmetry of a CsCl crystal. Compared with their atomic counterparts, the mesocrystals formed from block copolymers have a much larger size and tunable period. These characteristic lengths could be regulated by the molecular weight of the block copolymers. In contrast to the rich array of binary ionic crystals, much smaller numbers of ordered spherical phases have been observed in the self-assembly of sphere-forming block copolymers. In particular, the commonly observed stable binary spherical phase formed by ABC linear triblock copolymers is the CsCl crystal, as compared with the large number of binary ionic crystals or metallic alloy crystals. Designing block copolymers such that they spontaneously form spherical phases corresponding to the different binary atomic crystals presents an open and challenging question to the inverse problem of structures by design using block copolymers.

As mentioned in the introduction, the inverse problem could have multiple solutions. Furthermore, because the types of multiblock copolymers are virtually unlimited due to the possibility of different blocks, architectures, and topologies, from this perspective, it is impracticable and impossible to search for a desired ordered phase by simply exploring all possible types of block copolymers. It is essential to establish guiding principles for a given class of morphologies, so that the discovery of ordered phases does not become an impractical problem [47]. For the case of binary mesocrystals, Xie et al. have proposed a useful guiding principle for the molecular design of multiblock terpolymers for the formation of binary crystal phases [16]. The basic idea is to use the length of bridging block chains to regulate the nearest neighbor distance, thus controlling the number of nearest neighbors, or the coordination number (CN). According to this design principle, they proposed to use $B_1AB_2CB_3$ pentablock or AB_2CB_3 tetrablock terpolymers, in which the A and C blocks self-assemble into A and C spherical domains, to solve the inverse problem of stabilizing different binary mesocrystals. The length of the middle B_2 block of these multiblock terpolymers could be used to tune the effective bond length and the average magnitude of the CNs (Figure 4.7). Moreover, the asymmetry of the CNs between the A and C spherical domains could be regulated by the relative lengths of the two tail blocks B_1 and B_3. Once the architecture of the multiblock copolymers is specified, their phase diagram is explored to establish stable regions of the desired ordered phases. In order to obtain solutions of the SCFT equations corresponding to a given binary mesocrystal, initial configurations of the system, which specify the positions of the A and C spheres, are constructed according to the space symmetry of

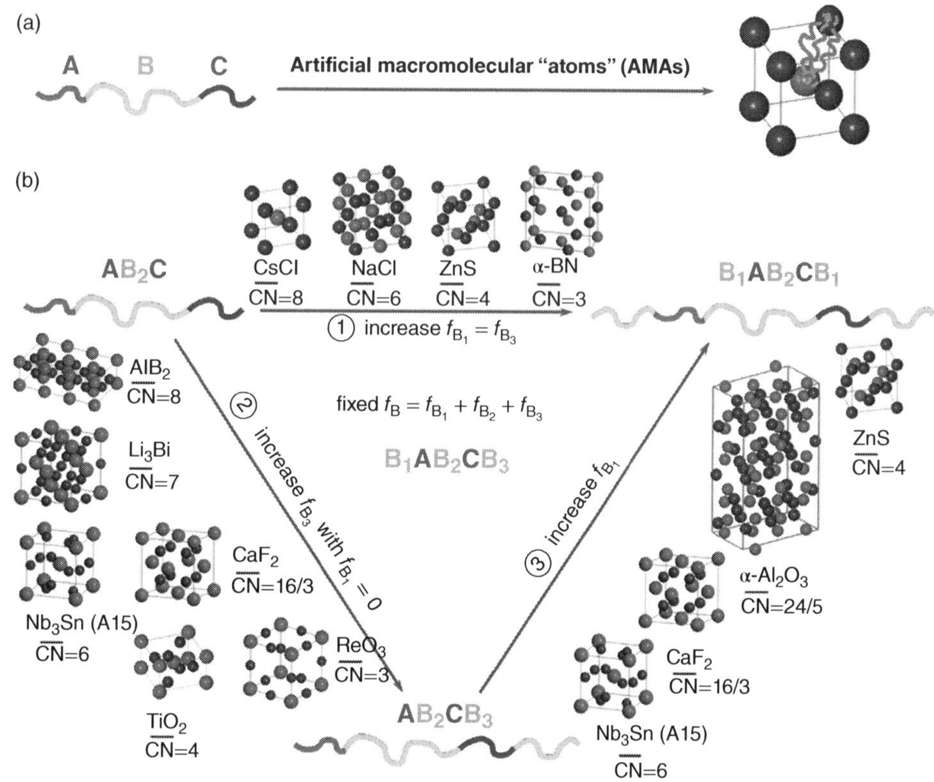

Figure 4.7 Design principle of multiblock terpolymers for binary soft mesocrystals. (a) Artificial macromolecular "atoms" (AMAs) self-assembled by linear ABC triblock copolymers pack into the CsCl crystal lattice. (b) AMAs formed by multiblock terpolymers can be programmed to assemble into a variety of crystallographic arrangements by tailoring the polymer architectures. Three possible paths are demonstrated to design series of mesocrystals with varying magnitudes and asymmetries of coordination numbers (CNs) by tuning the relative lengths among the B blocks while keeping the total B component fixed. Reproduced with permission from Ref. [16]. *(See insert for color representation of the figure.)*.

the desired crystal structure. Once the positions of the A and C spheres are given, the density profiles of the system could be constructed assuming phase separation is complete. The initial mean fields can then be generated according to the SCFT equations. With the properly constructed initial mean fields, it is possible to obtain solutions of the SCFT equations using the pseudo-spectral method or the spectral method. These solutions usually lead to the ordered phases corresponding to the desired ordered binary mesocrystals. Using this strategy of designed initial configurations, Xie et al. obtained a large number of candidate structures for the system, each of these ordered morphologies corresponding to a local minimum of the free energy landscape. The phase diagram of the multiblock copolymers is then constructed by comparing the free energy of these candidate phases, as shown in Figure 4.8. These phase diagrams validate the concept of regulating the CN of the structure using the length of the middle block. Specifically, decreasing the length of the middle B_2 block results in the phase transition

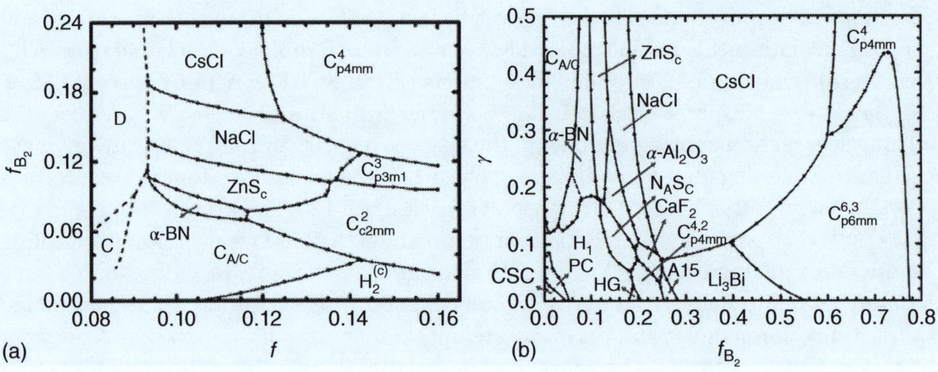

Figure 4.8 (a) Phase diagram of symmetric $B_1AB_2CB_1$ pentablock copolymers in the $f = f_A = f_C - f_{B_2}$ plane for fixed $\chi_{AB}N = \chi_{AC}N = \chi_{BC}N = \chi N = 80$. (b) Phase diagram of asymmetric $B_1AB_2CB_3$ pentablock copolymers in the $f_{B_2} - \gamma$ plane, where $\gamma = f_{B_1}/f_{B_3}$ quantifies the asymmetric degree of the molecule. Reproduced with permission from Ref. [16].

sequences from CsCl, NaCl, ZnS, to α-BN shown in Figure 4.8a and from NaCl, Al_2O_3, CaF_2, to Nb_3Sn (A15) shown in Figure 4.8b. These phase transition sequences reveal that the magnitude and asymmetry of the CNs of the A and C spheres could be controlled by varying the relative length of the middle B_2 block and the asymmetry of the two tail B blocks. In addition, this design principle is further validated in the case of two-dimensional binary crystal phases with *p6mm*, *p4mm*, and *p4gm* plane symmetries.

The third example studies the formation of complex spherical packing phases from AB-type block copolymers. For linear AB diblock copolymers, the commonly observed spherical phase is the BCC phase in which spherical domains are located at the lattice sites of a body-centered-cubic lattice. On the other hand, there are two well-known examples of packing problems of spherical domains, i.e. hard spheres and equal-sized bubbles. It is well known that hard spheres form close-packed structures such as the face-centered-cubic (FCC) or hexagonally close-packed (HCP) phases. The packing of equal-sized bubbles presents a more interesting problem, known as the Kelvin problem, in which a space-filling arrangement of the bubbles requires a minimal surface area. Initially it was conjectured that the solution to the Kelvin problem is the BCC lattice [48]. It took more than a century for researchers to find the correct solution, the A15 lattice or its variations, to the Kelvin problem. It is well known that the packing of spherical domains of block copolymers is dictated by two competing factors, the entropic and enthalpic contributions to the free energy [32]. This competition could lead to different spherical phases. It has been established that the spherical phase of AB diblock copolymers is mainly the BCC phase except for a very narrow region of the HCP phase at the vicinity of the order–disorder transition (ODT). Besides these two commonly considered spherical phases, theoretical studies based on SCFT have predicted that the A15 morphology could become an equilibrium phase for nonlinear AB-type block copolymers, e.g. AB_m miktoarm copolymers [4, 49]. More interestingly, experimental studies have revealed that a new and complex spherical phase, the Frank–Kasper σ-phase with a unit cell containing 30 spheres, could be formed in PI-*b*-PLA diblock copolymers and PS-*b*-PI-*b*-PS-*b*-PEO tetrablock copolymers [8]. Inspired by these previous experimental and theoretical studies, Xie et al. examined the mechanisms for the stabilization of the Frank–Kasper σ-phase. Specifically, these authors carried out SCFT calculations for conformationally asymmetric AB diblock copolymers and AB_m miktoarm copolymers and obtained solutions of the SCFT equations corresponding to the Frank–Kasper σ-phase, using a special initialization scheme based on the strong segregation approximation and the space symmetry of the crystal lattice. Phase diagrams including the Frank–Kasper σ-phase have been obtained for these conformationally asymmetric AB-type block copolymers (Figure 4.9). Their results predicted that the Frank–Kasper σ-phase could appear before the A15 phase for decreasing annealing temperature or increasing χN. The A15 phase becomes stable in the region of large composition and high χN for AB_m miktoarm copolymers with $m \geq 3$. The predicted phase transition sequence for decreasing temperature is from BCC to the Frank–Kasper σ-phase, and then to the A15 phase. It is interesting to note that this phase transition sequence has been observed in a number of different soft matter systems, such as the packing of micelles formed by amphiphilic superbranching liquid crystals [50].

In summary, it could be concluded that efficient and accurate methods to solve the SCFT equations have been developed. In particular, the pseudo-spectral method coupled with a designed initialization scheme provides a robust platform to explore and discover ordered phases of block copolymers. It has been demonstrated that this combination can be used to obtain a large number of candidate structures [39]. The availability of the candidate phases makes it possible to construct phase diagrams for multiblock copolymers. This strategy could be used to solve the forward problem of the self-assembly of block copolymers, i.e., constructing phase diagrams for given block copolymers. Furthermore, the availability of efficient and accurate SCFT algorithms coupled with the capability of finding solutions for given structures makes it possible to explore the reverse problem of structures by design using multiblock copolymers. The validity of this approach has been verified by two examples, the formation of various binary mesocrystals and the stabilization of the Frank–Kasper σ-phase. The most important ingredient in the design of multiblock copolymers for desired structures is the understanding of the formation mechanism of the given ordered phase. This understanding is then used to establish design principles for a given morphology. Besides changing the architecture of the block copolymers, there are various routes to regulate the phase behavior. For example, it is well established that blending homopolymers or diblock copolymers into more complex block copolymers could provide an efficient route to regulate the phase behavior of the system due to selective space-filling by the added polymers [51].

4.5 Conclusions and Discussions

In this chapter the self-consistent field theory (SCFT) of inhomogeneous polymeric systems and its numerical implementation have been reviewed, focusing on the study of block copolymer self-assembly. Special attention has been paid to the introduction of various numerical approaches to solving the SCFT equations. The spectral method, or reciprocal-space method, developed by Matsen and Schick, which takes advantage of the symmetry of the ordered phases, is highly efficient and accurate for the calculation of the free energy of given ordered structures. Accordingly, the spectral method is often used to identify phase diagrams of relatively simple block copolymer systems, leading

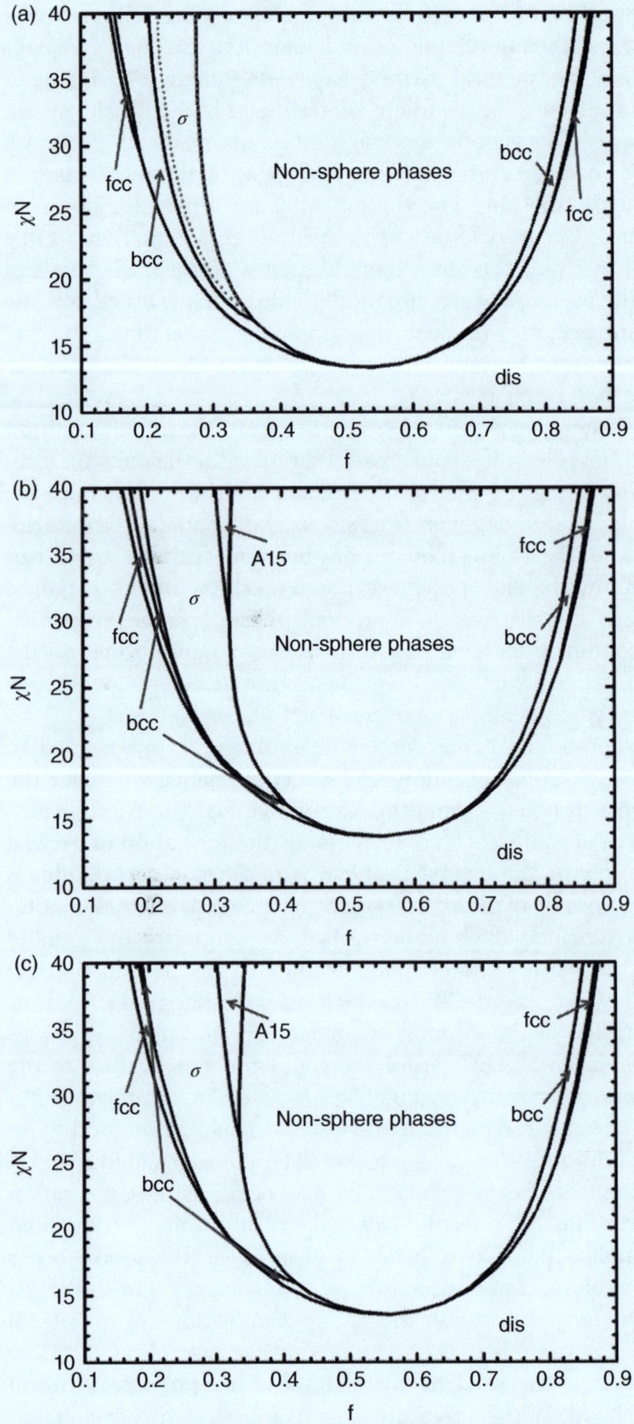

Figure 4.9 Phase diagrams of miktoarm AB_m block copolymers with $m = 2$, 3, and 4 in (a), (b), and (c), respectively, from top to bottom. In figure (a) of AB_2, the boundaries between the A15 phase and the BCC phase as well as the cylindrical phase without considering the σ-phase are plotted in dashed lines. Reproduced with permission from Ref. [5].

to a comprehensive understanding of the self-assembly mechanism of simple block copolymer systems that serves as the foundation for understanding the phase behavior of complex block copolymers. The spectral method has been extended to a generic reciprocal-space method by replacing the symmetry-determined basis functions with the complete set of plane waves. This generic spectral method is capable of the search and discovery of new ordered phases. Alternatively, the real-space method with random initial conditions proposed by Drolet and Fredrickson provides a useful addition to the methods for the discovery of new structures. The third numerical method in this chapter, the pseudo-spectral method, combines the advantages of the real-space and reciprocal-space methods. The pseudo-spectral method is able to achieve a reasonable balance between efficiency and accuracy because of its implementation of the spectral collocation. Thus the pseudo-spectral method has become the method of choice for the numerical solution of the SCFT equations.

Because the overall self-consistent algorithm is iterative in nature, the initial condition of the iteration plays an essential role in the converged solution of the calculation. Previous studies have demonstrated that, coupled with special initialization schemes, the pseudo-spectral method could be used to generate a rich array of candidate structures. This strategy of using diverse initialization schemes has been successfully applied to studying the self-assembly of multiblock copolymers. Based on these previous studies, a two-step strategy to explore and discover new ordered phases emerges, in which the fast SCFT algorithm is combined with various initialization schemes to produce a large number of possible candidate structures, then the accurate free energy of these candidate structures is computed and compared to construct phase diagrams.

Besides solving the forward problem by constructing phase diagrams for given block copolymers, the SCFT coupled with initialization schemes could be used to solve the inverse problem of structures by design using multiblock copolymers. The question is to design molecular architectures of multiblock copolymers for the formation of desired structures. A crucial step to solving the inverse problem is to develop useful guiding principles based on the understanding of the self-assembly mechanisms of block copolymers. The validity of the design principles could be verified via computing the stability regions of the desired structures using the coupling scheme of the pseudo-spectral method and the special initialization scheme. The feasibility of this strategy of structures by design has been demonstrated in the examples of binary crystal spherical phases from specifically designed multiblock ABC terpolymers and the stabilization of the Frank–Kasper σ-phase in conformationally asymmetric AB-type block copolymers.

In contrast to the accurate spectral method, the pseudo-spectral method is of second-order accuracy in the discrete step ds along the chain contour length. Therefore, pronounced errors in the free energy calculations may occur as the segregation strength is increased. The error due to the second-order algorithm could be eliminated by extending this second-order scheme to the fourth-order one. The fourth-order scheme has been applied to diblock copolymers at large values of χN and the result establishes the stability of the gyroid phase in the strong segregation regime of AB diblock copolymers.

Besides the numerical schemes discussed in this chapter, other potentially useful approaches have been developed in the literature. For example, a highly efficient pseudo-spectral method has been developed by Ceniceros and Fredrickson, which adopts a semi-implicit algorithm [52]. The most prominent feature of this method is that it employs asymptotic small scale information about the nonlocal density operators from the random phase approximation. For AB-type block copolymers, the iteration of SCFT equations with the implementation of this algorithm converges in only tens of iteration steps. It is plausible that a combination of these newly developed algorithms with specially designed initialization schemes could provide a powerful platform for the study of the self-assembly of multiblock copolymers.

Acknowledgments

We would like to thank Dr Li-Tang Yan for the invitation to contribute this chapter. The study on the self-assembly of multiblock copolymers was carried out in fruitful collaborations with Drs Nan Xie, Meijiao Liu, Yuci Xu, Weiquan Xu, Kai Jiang, Mr Hanlin Deng, and especially Profs. Feng Qiu and Pingwen Zhang. W. H. Li acknowledges the support of the National Natural Science Foundation of China (Grant nos. 21322407 and 21574026); A.-C. Shi acknowledges the support of the Natural Science and Engineering Research Council (NSERC) of Canada.

References

1. F. S. Bates and G. H. Fredrickson, *Ann. Rev. Phys. Chem.* **41**, 525 (1990).
2. I. W. Hamley, *The Physics of Block Copolymers* (Oxford University Press, Oxford 1998).
3. C. A. Tyler and D. C. Morse, *Phys. Rev. Lett.* **94**, 208302 (2005).
4. M. W. Matsen, *Macromolecules* **45**, 2161 (2012).
5. N. Xie, W. H. Li, F. Qiu, and A. C. Shi, *ACS Macro Lett.* **3**, 906 (2014).
6. J. Masuda, A. Takano, Y. Nagata, A. Noro, and Y. Matsushita, *Phys. Rev. Lett.* **97**, 098301 (2006).
7. G. Fleury and F. S. Bates, *Macromolecules* **42**, 1691 (2009).
8. S. Lee, M. J. Bluemle, and F. S. Bates, *Science* **330**, 349 (2010).
9. F. S. Bates and G. H. Fredrickson, *Physics Today* **52**, 32 (1999).
10. Y. Mogi, et alet al., *Macromolecules* **25**, 5408 (1992).
11. M. W. Matsen, *J. Chem. Phys.* **108**, 785 (1998).
12. U. Krappe, R. Stadler, and I. Voigt-Martin, *Macromolecules* **28**, 4558 (1995).
13. W. H. Li, F. Qiu, and A. C. Shi, *Macromolecules* **45**, 503 (2012).
14. H. Ott, V. Abetz, and V. Altstadt, *Macromolecules* **34**, 2121 (2001).
15. M. J. Liu, W. H. Li, F. Qiu, and A. C. Shi, *Macromolecules* **45**, 9522 (2012).
16. N. Xie, M. J. Liu, H. L. Deng, W. H. Li, F. Qiu, and A. C. Shi, *J. Am. Chem. Soc.* **136**, 2974 (2014).
17. S. F. Edwards, *Proc. Phys. Soc.* **85**, 613 (1965).
18. E. Helfand, *J. Chem. Phys.* **62**, 999 (1975); *Macromolecules* **8**, 552 (1975).
19. L. Leibler, *Macromolecules* **13**, 1602 (1980).
20. T. Ohta and K. Kawasaki, *Macromolecules* **19**, 2621 (1986).
21. T. Ohta and A. Ito, *Phys. Rev. E* **52**, 5250 (1995).
22. P. C. Millett, *Phys. Rev. E* **92**, 022602 (2015).
23. A. N. Semenov, *Sov. Phys. JETP* **61**, 733 (1985).

24. E. Helfand and Z. R. Wasserman, *Macromolecules* **9**, 879 (1976).
25. K. R. Shull, *Macromolecules* **25**, 2122 (1992).
26. J. D. Vavasour and M. W. Whitmore, *Macromolecules* **25**, 5477 (1992).
27. M. W. Matsen and M. Schick, *Phys. Rev. Lett.* **72**, 2660 (1994).
28. Z. J. Guo, G. J. Zhang, F. Qiu, H. D. Zhang, Y. L. Yang, and A. C. Shi, *Phys. Rev. Lett.* **101**, 028301 (2008).
29. F. Drolet and G. H. Fredrickson, *Phys. Rev. Lett.* **83**, 4317 (1999).
30. G. Tzeremes, K. Ø. Rasmussen, T. Lookman, and A. Saxena, *Phys. Rev. E* **65**, 041806 (2002).
31. E. W. Cochran, C. J. Garcia-Cervera, and G. H. Fredrickson, *Macromolecules* **39**, 2449 (2006).
32. M. W. Matsen, *J. Phys.: Condens. Matter* **14**, R21 (2002).
33. A. C. Shi, in *Developments in Block Copolymer Science and Technology*, Edited by I. W. Hamley (Wiley, New York 2004).
34. G. H. Fredrickson, *The Equilibrium Theory of Inhomogeneous Polymers* (Oxford University Press, Oxford 2006).
35. X. Y. Cheng, L. Lin, W. N.E, P. W. Zhang, and A. C. Shi, *Phys. Rev. Lett.* **104**, 148301 (2010).
36. H. Takahashi, N. Laachi, K. T. Delaney, S. M. Hur, C. J. Weinheimer, D. Shykind, and G. H. Fredrickson, *Macromolecules* **45**, 6253 (2012).
37. W. H. Li, P. F. Nealey, J. J. de Pablo, and M. Müller, *Phys. Rev. Lett.* **113**, 168301 (2014).
38. M. W. Matsen, *Macromolecules* **36**, 9647 (2003).
39. W. Xu, K. Jiang, P. W. Zhang, and A. C. Shi, *J. Phys. Chem. B* **117**, 5296 (2013).
40. M. Frigo and S. G. Johnson, *Proc. ICASSP* **1998**, 1381 (1998).
41. W. H. Li, R. A. Wickham, and R. A. Garbary, *Macromolecules* **39**, 806 (2006).
42. W. H. Li and R. A. Wickham, *Macromolecules* **39**, 8492 (2006).
43. W. H. Li and A. C. Shi, *Macromolecules* **42**, 811 (2009).
44. Y. C. Xu, W. H. Li, F. Qiu, H. D. Zhang, Y. L. Yang, and A. C. Shi, *J. Polym. Sci. Part B: Polym. Phys.* **48**, 1101 (2010).
45. H. Jinnai, T. Kaneko, K. Matsunaga, C. Abetz, and V. Abetz, *Soft Matter* **5**, 2042 (2009).
46. T. Higuchi, et alet al., *Macromolecules* **46**, 6991 (2013).
47. F. S. Bates, M. A. Hillmyer, T. P. Lodge, C. M. Bates, K. T. Delaney, and G. H. Fredrickson, *Science* **336**, 434 (2012).
48. W. Thomson, *Philos. Mag.* **24**, 503 (1887).
49. G. M. Grason, B. A. DiDonna, and R. D. Kamien, *Phys. Rev. Lett.* **91**, 058304 (2003).
50. X. Zeng, G. Ungar, Y. Liu, V. Percec, A. E. Dulcey, and J. K. Hobbs, *Nature* **428**, 157 (2004).
51. M. J. Liu, B. K. Xia, W. H. Li, F. Qiu, and A. C. Shi, *Macromolecules* **48**, 3386 (2015).
52. H. D. Ceniceros and G. H. Fredrickson, *Multiscale Modeling and Simulation* **2**, 452 (2004).

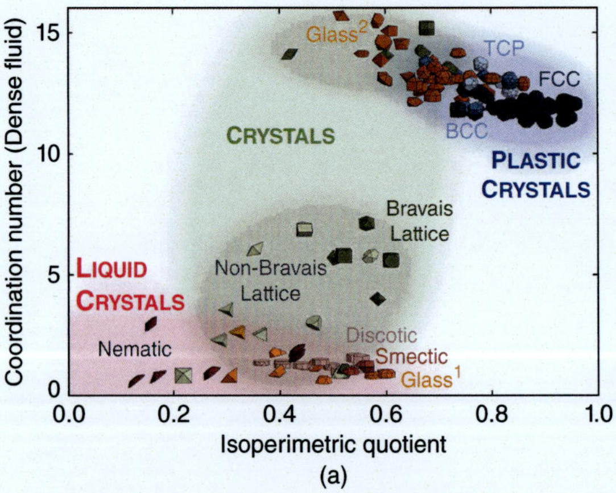

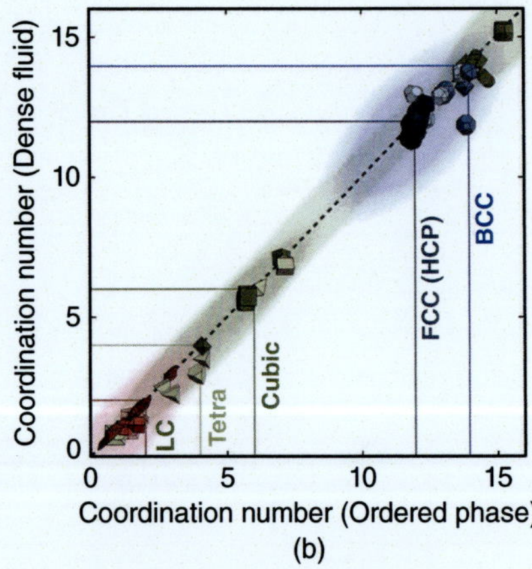

Figure 1.1 (a) The coordination number in the fluid phase, CN_f, is correlated to the isoperimetric quotient (IQ) of the polyhedron. Here, IQ is a scalar parameter for the sphericity of the shape and coordination number is a measure of the degree of local order. Data points are drawn as small polyhedra, which are colored and grouped according to the assemblies they form. (b) Polyhedra have, in most cases, nearly identical coordination numbers in the ordered phase (CN_o) and the fluid phase (CN_f) close to the ordering transition. Because of this strong correlation, combining CN_f and IQ allows for prediction of the assembly category expected for most cases. This figure is reproduced from Ref. [31]. Copyright permission from American Association for the Advancement of Science (2012).

Self-Assembling Systems: Theory and Simulation, First Edition. Edited by Li-Tang Yan.

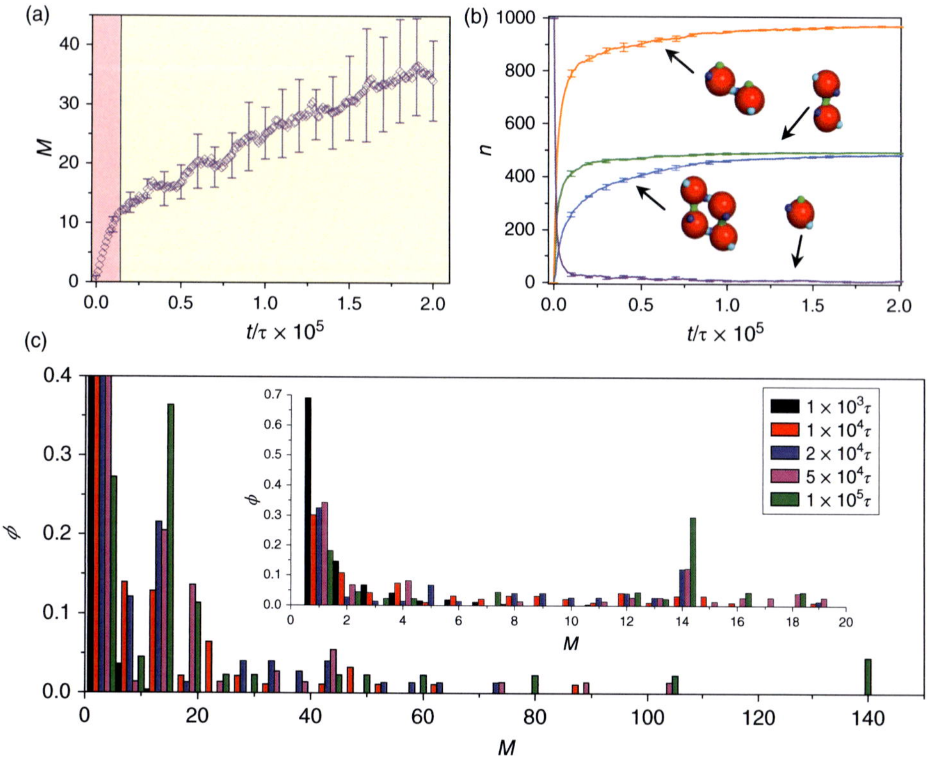

Figure 1.10 Kinetics in the self-assembling process of the helically supracolloidal structure. (a) The number-averaged degree of colloidal clusters as a function of self-assembling time t. (b) Variation in the number of patch groups in the course of self-assembly for species shown in inserts.
(c) Number-averaged distributions of colloidal clusters at different times. Error bars in (a) and (b) indicate standard deviation. This figure is reproduced from Ref. 41. Copyright Permission from Nature Publishing Group (2014).

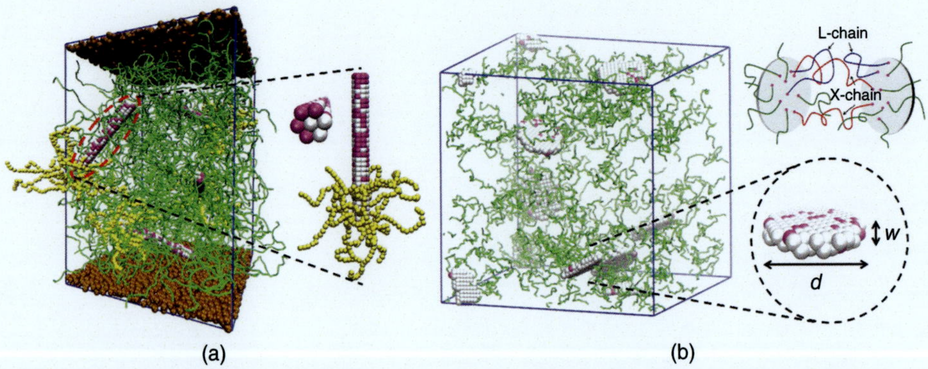

Figure 2.1 (a) Equilibrium state of a nanorod-filled polymer gel in a good solvent (not displayed). Green chains represent polymer strands and brown beads form amorphous solid walls. Inset: enlarged view of initiator (magenta beads) modified nanorod (white beads) with one end functionalized with polymer chains (yellow beads) and a cross-section of the hexagonal nanorod. (b) Snapshot of nanoclay-embedded polymer network at monomer conversion 95%. Inset top: schematic representation of the network with chain color representing grafted chain (green), cross-linking chains (red), and looped chain (blue). Note that all types of the chains are represented by the green chains in the snapshot. Inset bottom: detailed structure of the nanoclay particle. Reproduced from Ref. [10] with permission from the American Chemical Society.

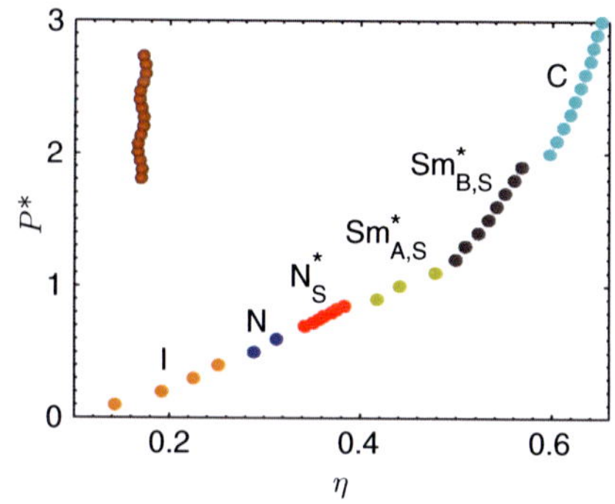

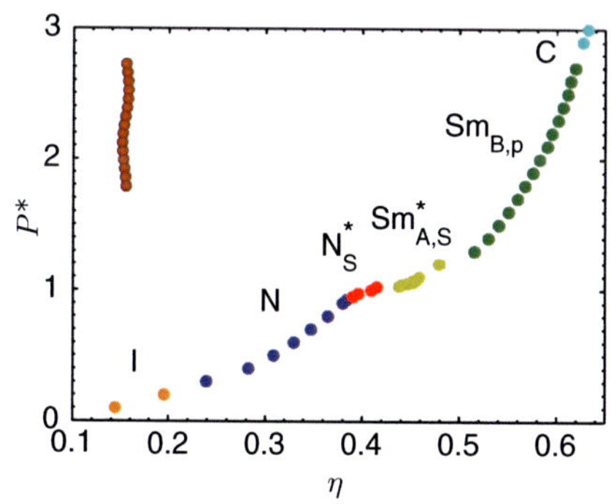

Figure 3.11 Reduced pressure $P^* = PD^3/k_BT$ versus volume fraction η for helices with $r = 0.2$ and pitch $p = 4$ (top) and $p = 8$ (bottom). The corresponding shape of the helix is displayed in the insets. Different phases are identified by different colors and labeled accordingly.

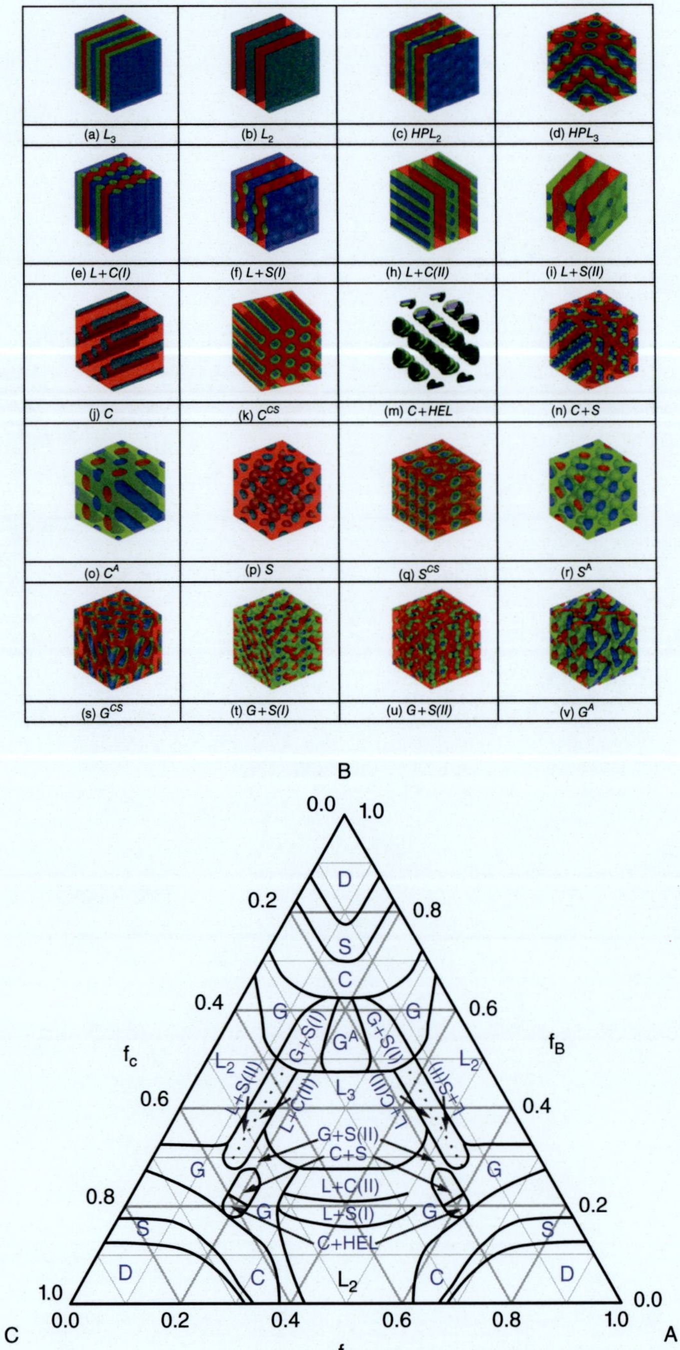

Figure 4.3 (Top) Morphologies self-assembled in the ABC linear triblock copolymers with $\chi_{AB}N = \chi_{BC}N = 35$ and $\chi_{AC}N = 15$, discovered by the general spectral method of SCFT. (Bottom) Phase diagram of these morphologies. Reproduced with permission from Ref. [28].

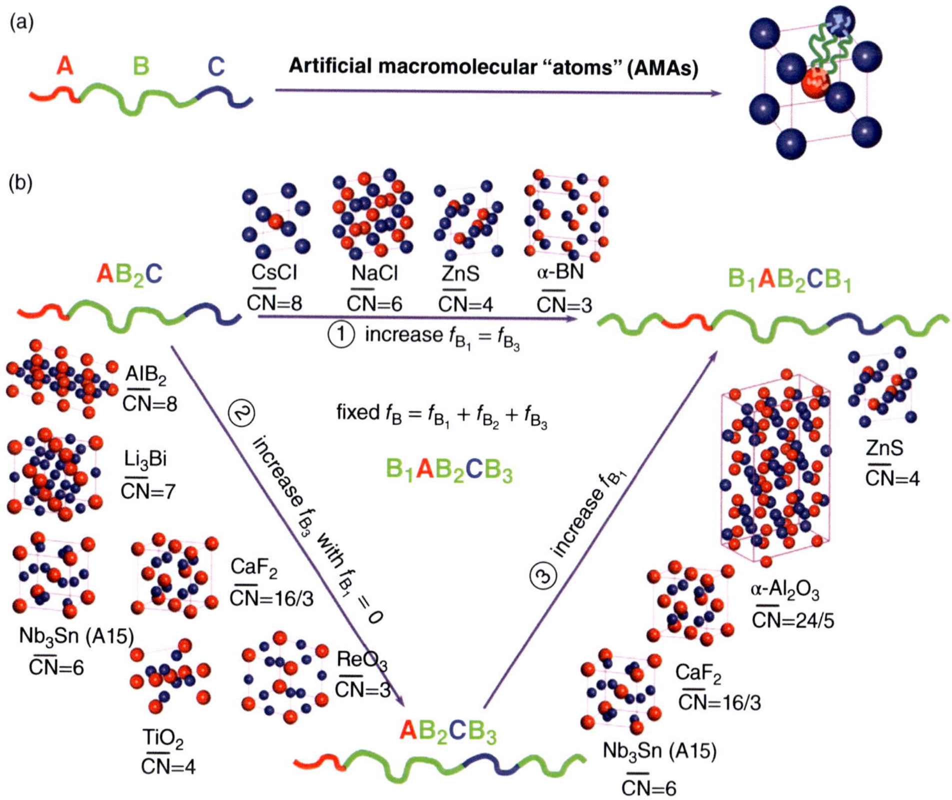

Figure 4.7 Design principle of multiblock terpolymers for binary soft mesocrystals. (a) Artificial macromolecular "atoms" (AMAs) self-assembled by linear ABC triblock copolymers pack into the CsCl crystal lattice. (b) AMAs formed by multiblock terpolymers can be programmed to assemble into a variety of crystallographic arrangements by tailoring the polymer architectures. Three possible paths are demonstrated to design series of mesocrystals with varying magnitudes and asymmetries of coordination numbers (CNs) by tuning the relative lengths among the B blocks while keeping the total B component fixed. Reproduced with permission from Ref. [16].

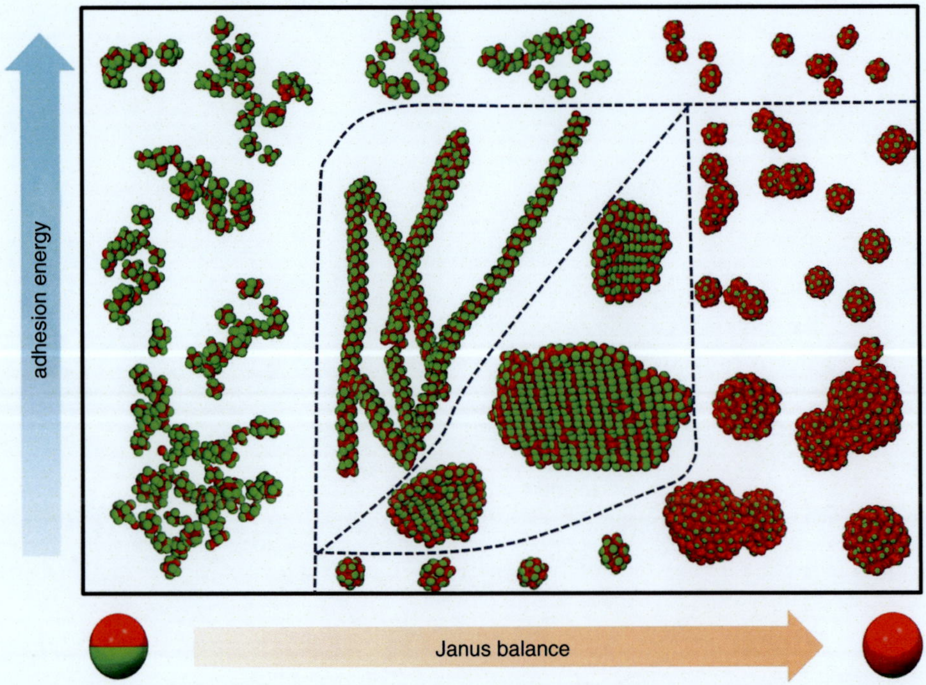

Figure 5.4 Representative equilibrium superstructures self-assembled from soft one-patch Janus particles by properly tuning the Janus balance β and the strength of attraction between attractive patches α_{ij}^{A} while keeping $\alpha_{ij}^{\mathrm{R}} = 396$, $\phi = 5\%$, and $\nu = 1/2$.

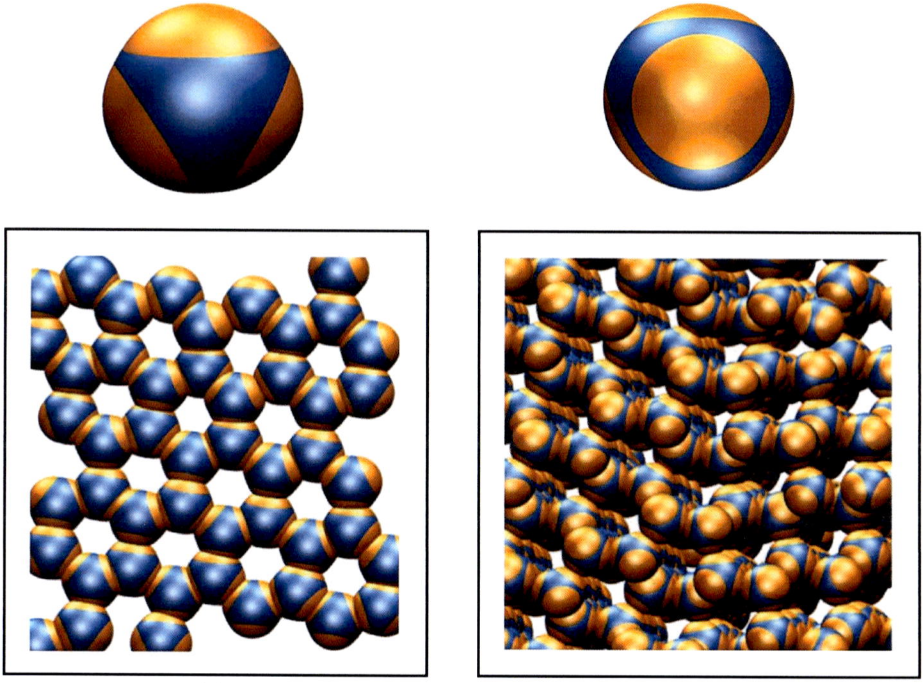

(a) (b)

Figure 5.8 (a) Graphene-like two-dimensional structure with a hexagonal honeycomb lattice self-assembled from soft three-patch particles with regular triangular arrangement of the patches. (b) Diamond lattice structure self-assembled from soft four-patch particles with regular tetrahedral arrangement of the patches. In the simulations, $\alpha_{ij}^{R} = 396$, $\alpha_{ij}^{A} = 220$ ($G \approx 10.00\ k_B T$), $\nu = 0.5$, and $\theta_m^{\kappa} = 45°$.

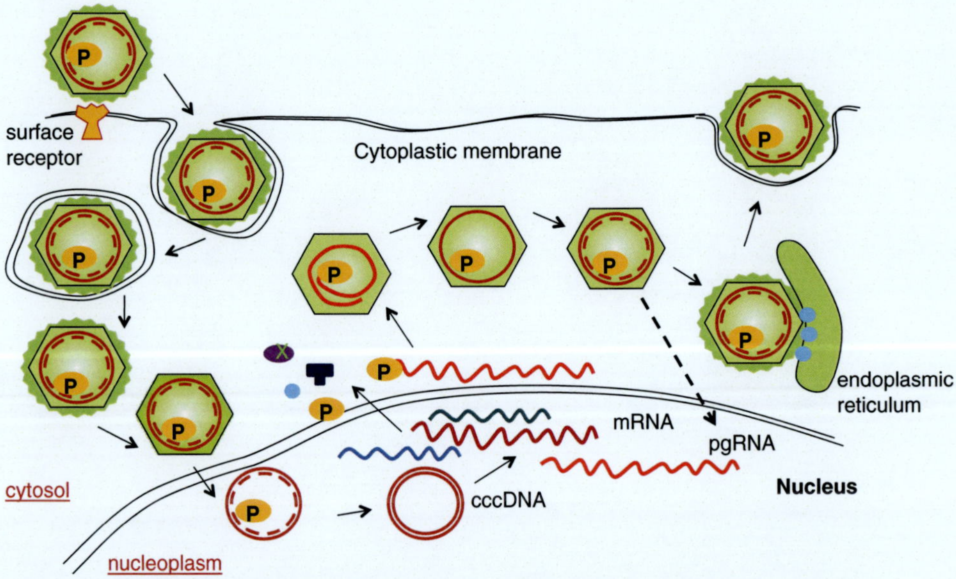

Figure 6.4 Schematic of the HBV replication life cycle. The viral entry into human hepatocytes is initiated by the Pre-S1 domain of HBsAg binding to a receptor that is normally involved in bile acid transport in the liver. After being transferred into the cytosol, the nucleocapsid is disassembled, leading to the release of the genomic material. Viral replication starts with the formation of a covalently closed circular DNA (cccDNA) in the nucleoplasm; the cccDNA templates RNA transcription and subsequently protein synthesis. Nucleocapsid (NC) assembly is driven by strong electrostatic interactions between capsid proteins and pregenome (pg) RNA. Inside NC, the polymerase (P) translates pgRNA first into a single-stranded (ss) DNA, and then a partially double-stranded (ds) (>50%) DNA. The matured capsid is then enveloped with surface proteins in the endoplasmic reticulum (ER) or migrates back to the nucleoplasm for cccDNA amplification.

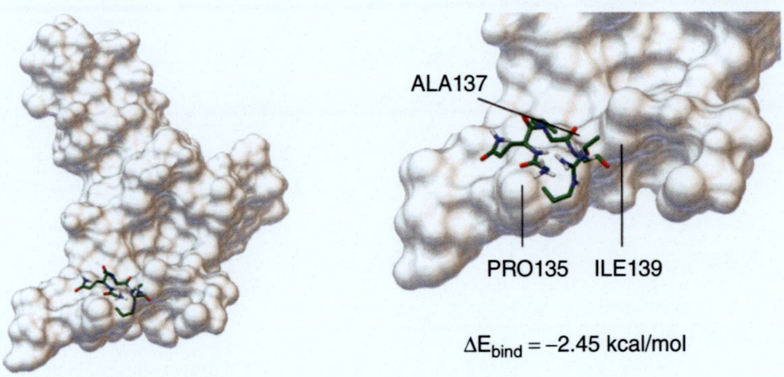

Figure 6.23 Docking of sub-fragment P4-3 onto HBcAg. Shown here are the docked conformation for P4-3 (101 – 103 aa of L-HBs) and the approximated binding energy.

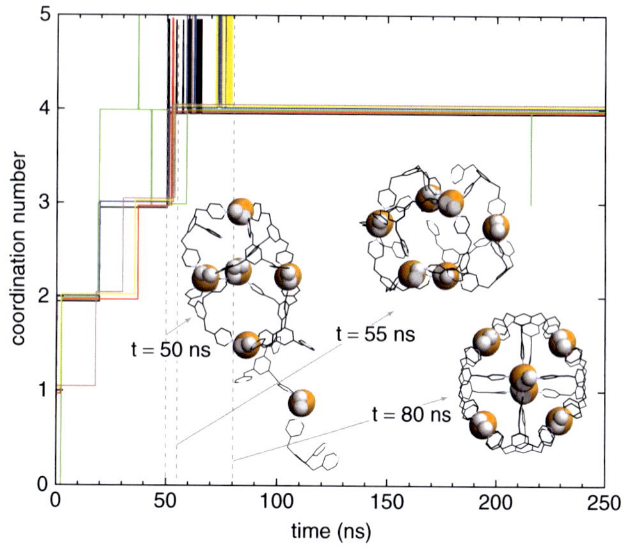

Figure 7.6 Time variations of the coordination numbers of the six Pd(II) within the completed M_6L_8 nanosphere. Each of the six plots is vertically shifted for clarity. Snapshots at the simulation times of 50, 55 and 80 ns are additionally shown in the figure.

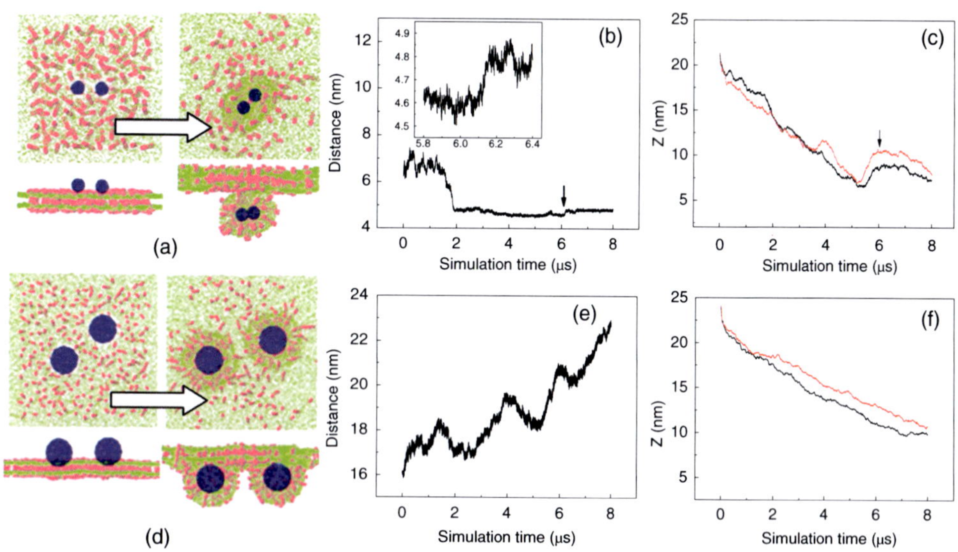

Figure 8.3 Endocytosis of two identical NPs. Two smaller NPs of 3.8 nm were placed on a membrane of 38.7 nm × 38.7 nm (a–c), while two larger NPs of 9.0 nm were placed on a membrane of 51.7 nm × 51.7 nm (d–f). (a, d) show the initial and final structure of endocytosis of two identical NPs. (b, e) show the evolution of the distance between two NPs. (c, f) show the evolution of NP positions along the membrane normal direction. The initial inter-NP distances are 6.46 nm (a–c) and 15.8 nm (d–f), respectively. Reproduced from Ref. [15] with permission from the American Chemical Society.

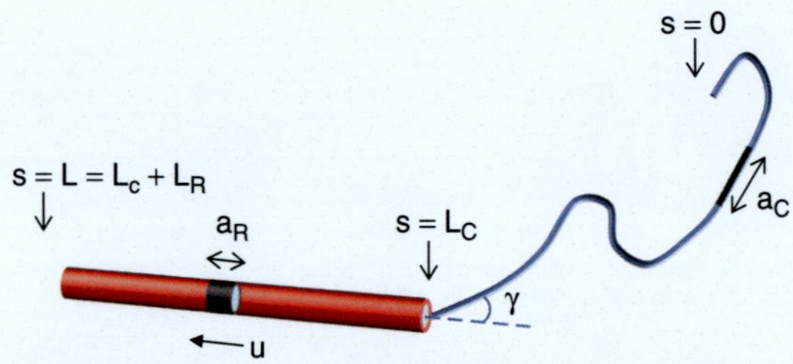

Figure 9.1 A rod–coil block copolymer of total contour length L. Basic segmental volumes are considered for lengths a_C and a_R on coil and rod blocks, respectively.

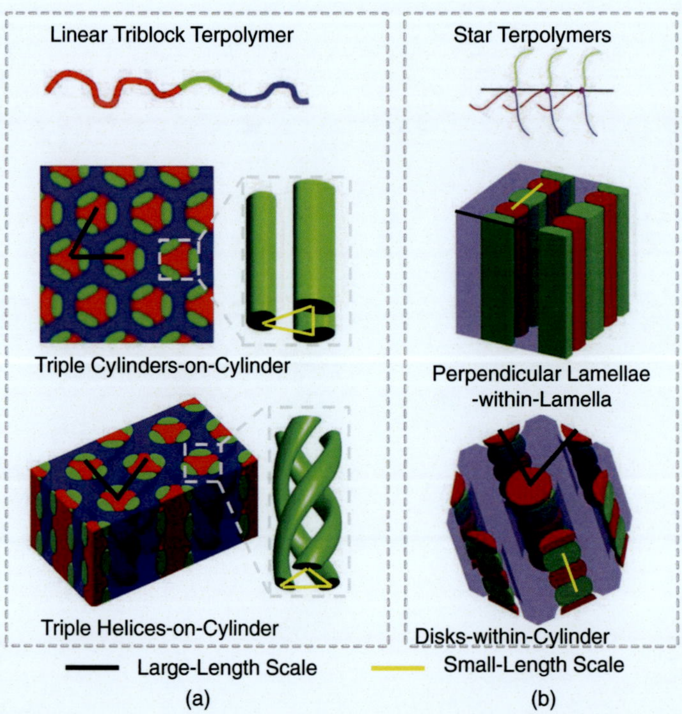

Figure 10.1 (a) Representative hierarchical structures self-assembled from ABC linear triblock terpolymers. Adapted with permission from Ref. [49]. Copyright 2012 American Chemical Society. (b) Representative hierarchical structures self-assembled from ABC star terpolymers. Adapted with permission from Ref. [60]. Copyright 2013 American Chemical Society.

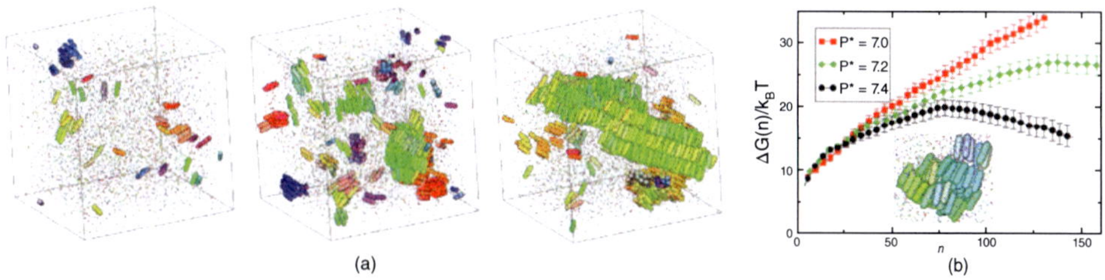

Figure 11.19 (a) Configurations for spontaneous crystal nucleation from a typical molecular dynamics trajectory at $P^* = 7.6$ and $t/\tau_{MD} = 0$, 1000 and 3000 (from left to right). Isotropic-like particles are drawn 10 times smaller than their actual size. (b) Gibbs free energy $\Delta G(n)$ as a function of the number of rods n in the crystalline cluster at pressure $P^* = 7.0$, 7.2 and 7.4. Inset: a typical configuration of a critical cluster ($n = 81$) at $P^* = 7.4$.

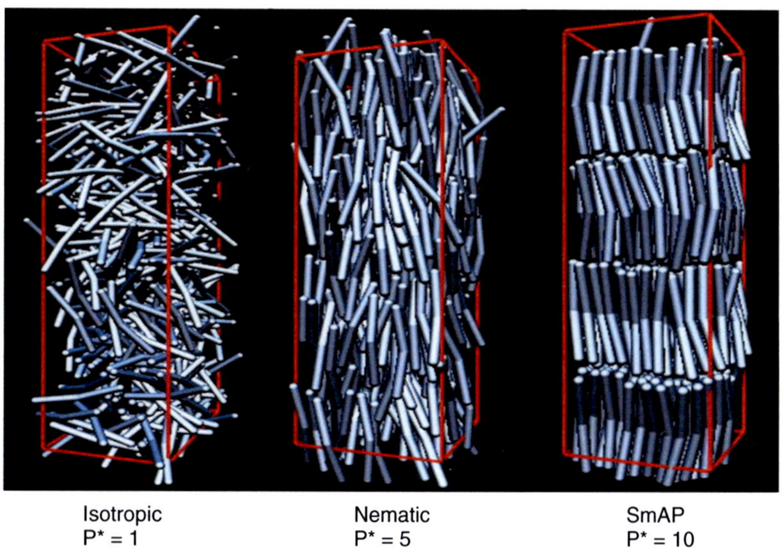

Figure 12.13 Final configurations from Monte Carlo simulations of $N = 400$ bent-core molecules with opening angle $\psi = 165°$ as a function of reduced pressure P^* ($P^* = \beta P v_0$, where v_0 is the volume of a zigzag molecule and P is the actual pressure). From left to right: isotropic phase ($P^* = 1$), nematic ($P^* = 5$) and polar smectic A ($P^* = 10$).

5

Simulation Models of Soft Janus and Patchy Particles

Zhan-Wei Li[1], Zhao-Yan Sun[1] and Zhong-Yuan Lu[2]

[1] *State Key Laboratory of Polymer Physics and Chemistry, Changchun Institute of Applied Chemistry, Chinese Academy of Sciences, China*
[2] *State Key Laboratory of Supramolecular Structure and Materials, Institute of Theoretical Chemistry, Jilin University, Changchun, China*

5.1 Introduction

Surface-anisotropic colloidal building blocks have brought an almost unbelievable revolution in materials science [1–13]. As the simplest case of colloidal particles with surface anisotropy, Janus particles have attracted considerable attention due to their unique anisotropic feature and the possibility of generating fascinating hierarchical nanostructures [12–20]. In fact, the anisotropic surface coverage of spherical particles is not restricted to a 50 : 50 ratio, and the number of the patches patterned on the surface of a spherical particle is also not restricted to one. So patchy particles with tunable number, size, direction, and geometrical arrangement of the patches provide a richer palette for the design and creation of desired aggregate structures [7, 11, 13, 17]. Up to now, a large variety of supracolloidal structures including biomimetic helices [18, 21–24], low-coordinated two-dimensional (2D) [20, 25–28] and three-dimensional (3D) [29–33] open lattices, and quasicrystal [34, 35], have been achieved in experiments and simulations through the self-assembly of surface-anisotropic Janus and patchy particles. Furthermore, due to the specificity and directionality of Janus and patchy particles that mimic the key features of atoms and molecules, Janus and patchy particles represent useful model systems to understand several long-standing fundamental questions, mainly concerned with the formation of glasses [36–39], the crystallization of proteins [40–42], and the collective behavior of living systems [43, 44].

According to their different elastic moduli, surface-anisotropic Janus and patchy particles may be classified into two major categories: hard Janus and patchy particles, and soft Janus and patchy particles. So far, hard Janus and patchy particles have been synthesized by various methods [7, 14, 15, 45–47], including templating, colloidal assembly, particle lithography, glancing-angle deposition, and so on. Unlike conventional hard and non-deformable Janus and patchy particles, soft Janus and patchy particles are flexible and deformable, and have been realized as Janus and patchy

Self-Assembling Systems: Theory and Simulation, First Edition. Edited by Li-Tang Yan.

micelles from the self-assembly of block copolymers [14, 45, 48–55], single-patch Janus microgels by atom transfer radical polymerization or a Pickering emulsion-based method [56, 57], and single-patch Janus dendrimers [58, 59] and hyperbranched polymers [60] by divergent or convergent approaches. The softness and deformability of soft Janus and patchy particles can be quantitatively described by the elastic modulus [61–64], and tuned by varying the cross-linking density [48, 65], the grafting density [56], the number of primary branches, the density of radial branches [58, 59], and so on. Thus, soft Janus and patchy particles play an important role in bridging the gap between ultra-soft block copolymers and hard Janus and patchy particles [55, 61–64, 66, 67], and exhibit fascinating aggregation behaviors into a variety of striking new self-assembled structures, and accordingly show novel aggregation mechanisms [13, 54, 55, 60–64]. Due to their soft and deformable characteristics, soft Janus and patchy particles bring new excitement to materials science, and will attract more attention on the self-assembly of novel and designable soft functional materials [53–55, 58, 60].

Soft Janus and patchy particles mostly consist of a large number of monomeric units. Their sizes typically range from about 10 nm to 1000 nm, and even up to the micrometer scale [53–55, 58, 64, 67]; the length scales of their aggregated structures will be much larger. Thus, the aggregation phenomena of soft Janus and patchy particles involve both the microscopic and mesoscopic scales. It is an extremely challenging task and also a more feasible strategy to gain insight into the aggregation behavior and mechanism of soft Janus and patchy particles via computer simulation. For hard patchy particles, several general patchy models [8], including rigid-body patchy models [3, 68, 69], spot-like patchy models [70–73], and extended Kern and Kern-inspired patchy models [17, 74–79], have been successfully developed. Because of the complex architectures of soft Janus and patchy particles, mesoscopic coarse-grained methodologies are better choices to provide solutions to investigate the aggregation behavior of soft Janus and patchy particles [54, 55, 58, 60, 80, 81]. However, to date, there are still no general mesoscopic coarse-grained models available for investigating the aggregation behavior of various types of soft Janus and patchy particles.

In this chapter, we mainly introduce two major kinds of general and effective mesoscale models to describe the aggregation behavior of soft Janus and patchy particles: soft Janus particle models (SJPMs) and soft patchy particle models (SPPMs). SJPMs are specifically designed to study the aggregation behavior of soft one-patch, and ABA-type and BAB-type triblock Janus particles, and SPPMs are mainly developed to simulate the aggregation behavior of soft multi-patch particles with tunable number, size, direction, and geometrical arrangement of the patches. The interactions between soft Janus and patchy particles are all described by single-site soft anisotropic attractive potentials that can reflect deformable and anisotropic characteristics of soft Janus and patchy particles. The major advantage of these single-site anisotropic potentials is simple, easy to implement, and computationally efficient. Furthermore, these single-site anisotropic potentials can be used to describe various types of soft Janus and patchy particles with different degrees of particle softness. The simulation parameters in these potentials can be fixed from the experimentally measurable particle properties. Thus, our SJPMs and SPPMs can be directly mapped onto experimental systems in different conditions [61–64]. In order to improve the computational efficiency of SJPMs and SPPMs in dynamics simulations, we give the simulation algorithms that fit the compute unified device architecture (CUDA) framework of NVIDIA graphics processing units (GPUs) [64, 82]. SJPMs show good performance in simulating the systems of soft one-patch and triblock Janus particles [82]. In comparison with SJPMs, the performance of SPPMs is slightly affected by the use of the quaternion method to describe the anisotropic characteristics of patchy particles, but is still very satisfactory [64]. Therefore, our SJPMs and SPPMs provide a powerful tool to investigate the aggregation behavior of soft Janus and patchy particles over larger spatial and temporal scales [61–64].

5.2 Soft Janus Particle Models

In our SJPMs [61–63], the deformable and anisotropic characteristics of soft Janus particles are described via a single-site soft anisotropic attractive potential, inspired by the soft-particle model in dissipative particle dynamics (DPD) [83] and the Kern–Frenkel model [74]. For simplicity, we use the interaction cutoff radius (r_c) as the unit of length, $k_B T$ as the unit of energy, and choose the moment of inertia (I) and the mass (m) of the particle as the units, thus the time unit $\tau = \sqrt{mr_c^2/k_B T}$. In the following, all the variables and simulation parameters are given in reduced units, and the soft anisotropic attractive potential [61–63] is expressed as

$$U_{ij} = \begin{cases} \dfrac{\alpha_{ij}^R}{2}\left(1 - \dfrac{r_{ij}}{r_c}\right)^2 - f^{\nu}\dfrac{\alpha_{ij}^A}{2}\left[\dfrac{r_{ij}}{r_c} - \left(\dfrac{r_{ij}}{r_c}\right)^2\right] & r_{ij} \leq r_c \ (\equiv 1.0) \\ 0 & r_{ij} > r_c. \end{cases} \tag{5.1}$$

In this potential, the anisotropic factor f is mainly responsible for describing the anisotropic feature of soft Janus particles, which is different for different SJPMs. The magnitude of α_{ij}^R controls the strength of repulsion, α_{ij}^A controls the strength of attraction between the attractive patches, and ν controls the angular width of the attraction. Thus, both α_{ij}^A and ν control the flexibility of Janus particle aggregates.

5.2.1 Soft One-Patch Janus Particle Model

In the soft one-patch Janus particle model [61], the anisotropic factor f is given as

$$f = \begin{cases} \cos\dfrac{\pi\theta_i}{2\beta}\cos\dfrac{\pi\theta_j}{2\beta} & \text{if } \cos\theta_i \geq \cos\beta \text{ and } \cos\theta_j \geq \cos\beta \\ 0 & \text{otherwise.} \end{cases} \tag{5.2}$$

As illustrated in Figure 5.1a, the light gray part represents the attractive patch and the dark gray part represents the repulsive region, and the directions of the attractive patches on particle i and j are specified by unit vectors $\vec{n}_i$ and $\vec{n}_j$, respectively. θ_i is the angle between $\vec{n}_i$ and the interparticle vector $\vec{r}_{ji} = \vec{r}_j - \vec{r}_i$, and θ_j is the angle between $\vec{n}_j$ and $\vec{r}_{ij} = -\vec{r}_{ji}$. The size of the attractive patch is described by Janus balance β, which is half of the opening angle of the attractive patch, and the fraction of surface covered by the attractive patch χ is related to β by the relation $\chi = \sin^2\left(\frac{\beta}{2}\right)$ [61, 74, 78].

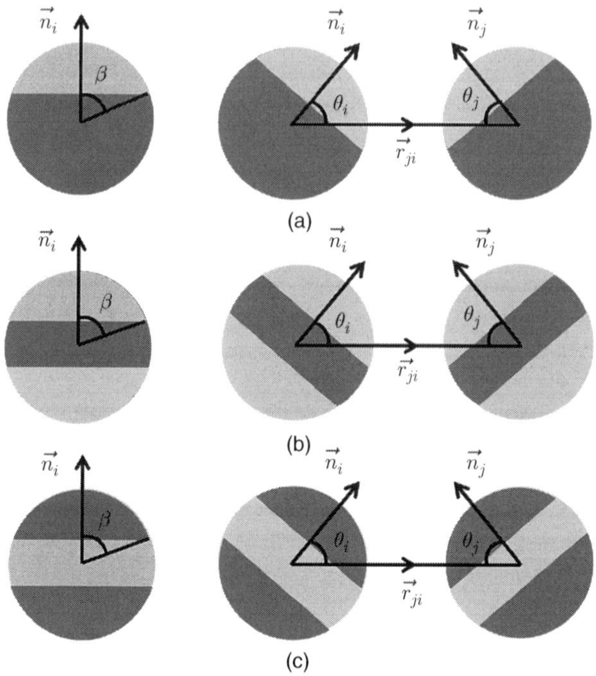

Figure 5.1 Graphical representation of soft Janus particle models. (a) Soft one-patch Janus particle model. (b) Soft ABA-type triblock Janus particle model. (c) Soft BAB-type triblock Janus particle model.

The force between two neighboring soft one-patch Janus particles $\vec{F}_{ij}$ is given by the derivation of Equation (5.1),

$$\begin{aligned}\vec{F}_{ij} &= -\frac{\partial U_{ij}}{\partial \vec{r}_{ij}}\\ &= \alpha_{ij}^{\mathrm{R}}(1-r_{ij})\frac{\vec{r}_{ij}}{r_{ij}} + \alpha_{ij}^{\mathrm{A}} f^{\nu}\left(\frac{1}{2}-r_{ij}\right)\frac{\vec{r}_{ij}}{r_{ij}} - \frac{\alpha_{ij}^{\mathrm{A}}}{2}(r_{ij}-r_{ij}^2)\\ &\quad \nu f^{\nu-1}\left(\frac{\pi}{2\beta}\sin\frac{\pi\theta_i}{2\beta}\frac{\partial\theta_i}{\partial\cos\theta_i}\frac{\partial\cos\theta_i}{\partial\vec{r}_{ij}}\cos\frac{\pi\theta_j}{2\beta}\right.\\ &\quad \left.+\frac{\pi}{2\beta}\sin\frac{\pi\theta_j}{2\beta}\frac{\partial\theta_j}{\partial\cos\theta_j}\frac{\partial\cos\theta_j}{\partial\vec{r}_{ij}}\cos\frac{\pi\theta_i}{2\beta}\right),\end{aligned} \tag{5.3}$$

where

$$\frac{\partial\theta_i}{\partial\cos\theta_i}\frac{\partial\cos\theta_i}{\partial\vec{r}_{ij}} = \begin{cases}\dfrac{1}{\sqrt{1-\cos^2\theta_i}}\left(\dfrac{\vec{n}_i}{r_{ij}} - \dfrac{(\vec{n}_i\cdot\vec{r}_{ij})\vec{r}_{ij}}{r_{ij}^3}\right) & \text{if } \cos\theta_i \geq \cos\beta,\\ 0 & \text{otherwise,}\end{cases} \tag{5.4}$$

$$\frac{\partial \theta_j}{\partial \cos\theta_j}\frac{\partial \cos\theta_j}{\partial \vec{r}_{ij}} = \begin{cases} -\dfrac{1}{\sqrt{1-\cos^2\theta_j}}\left(\dfrac{\vec{n}_j}{r_{ij}} - \dfrac{(\vec{n}_j\cdot\vec{r}_{ij})\vec{r}_{ij}}{r_{ij}^3}\right) & \text{if } \cos\theta_j \geq \cos\beta, \\ 0 & \text{otherwise.} \end{cases} \tag{5.5}$$

The force responsible for the rotation of soft one-patch Janus particle $\vec{g}_i$ is given as

$$\begin{aligned} \vec{g}_i &= -\frac{\partial U_{ij}}{\partial \vec{n}_i} \\ &= -\frac{\pi\alpha_{ij}^{\mathrm{A}}}{4\beta}(r_{ij}-r_{ij}^2)\nu f^{\nu-1}\sin\frac{\pi\theta_i}{2\beta}\frac{1}{\sqrt{1-\cos^2\theta_i}}\cos\frac{\pi\theta_j}{2\beta}\frac{\vec{r}_i}{r_{ij}}. \end{aligned} \tag{5.6}$$

Note that r_{c} is omitted in the above derivations since we have defined $r_{\mathrm{c}} \equiv 1.0$ in Equation (5.1). In the following, r_{c} will be always taken as 1.0 unless otherwise declared.

5.2.2 Soft ABA-Type Triblock Janus Particle Model

In the soft ABA-type triblock Janus particle model [62], the anisotropic factor f is given as

$$f = \begin{cases} \cos\dfrac{\pi\theta_i'}{2\beta}\cos\dfrac{\pi\theta_j'}{2\beta} & \text{if } |\cos\theta_i| \geq \cos\beta \text{ and } |\cos\theta_j| \geq \cos\beta \\ 0 & \text{otherwise.} \end{cases} \tag{5.7}$$

As illustrated in Figure 5.1b, the light gray parts represent the attractive A-type patches and the dark gray part represents the repulsive B-type middle band, and the directions of the attractive patches on particles i and j are specified by unit vectors $\vec{n}_i$ and $\vec{n}_j$, respectively. θ_i is the angle between unit vector $\vec{n}_i$ and the interparticle vector $\vec{r}_{ji} = \vec{r}_j - \vec{r}_i$, and θ_j is the angle between $\vec{n}_j$ and $\vec{r}_{ij} = -\vec{r}_{ji}$, and then $\cos\theta_i = -\vec{n}_i\cdot\vec{r}_{ij}/r_{ij}$ and $\cos\theta_j = \vec{n}_j\cdot\vec{r}_{ij}/r_{ij}$. Here, $\theta_i' = \arccos(|\cos\theta_i|)$, and $\theta_j' = \arccos(|\cos\theta_j|)$. The fraction of surface covered by the attractive A-type patches χ is related to β by the relation $\chi = 2\sin^2\left(\frac{\beta}{2}\right)$ [62, 74, 78].

The anisotropic force between two ABA-type triblock Janus particles [62] is then given as

$$\begin{aligned} \vec{F}_{ij} &= -\frac{\partial U_{ij}}{\partial \vec{r}_{ij}} \\ &= \alpha_{ij}^{\mathrm{R}}(1-r_{ij})\frac{\vec{r}_{ij}}{r_{ij}} + \alpha_{ij}^{\mathrm{A}} f^{\nu}\left(\frac{1}{2}-r_{ij}\right)\frac{\vec{r}_{ij}}{r_{ij}} - \frac{\alpha_{ij}^{\mathrm{A}}}{2}(r_{ij}-r_{ij}^2)\nu f^{\nu-1} \\ &\left(\frac{\pi}{2\beta}\sin\frac{\pi\theta_i'}{2\beta}\frac{\partial\theta_i'}{\partial|\cos\theta_i|}\frac{\partial|\cos\theta_i|}{\partial\vec{r}_{ij}}\cos\frac{\pi\theta_j'}{2\beta} + \frac{\pi}{2\beta}\sin\frac{\pi\theta_j'}{2\beta}\frac{\partial\theta_j'}{\partial|\cos\theta_j|}\right. \\ &\left.\frac{\partial|\cos\theta_j|}{\partial\vec{r}_{ij}}\cos\frac{\pi\theta_i'}{2\beta}\right), \end{aligned} \tag{5.8}$$

where

$$\frac{\partial \theta'_i}{\partial |\cos\theta_i|} = \begin{cases} 0 & \text{if } |\cos\theta_i| = 1 \\ -\dfrac{1}{\sqrt{1-|\cos\theta_i|^2}} & \text{otherwise,} \end{cases} \tag{5.9}$$

$$\frac{\partial \theta'_j}{\partial |\cos\theta_j|} = \begin{cases} 0 & \text{if } |\cos\theta_j| = 1 \\ -\dfrac{1}{\sqrt{1-|\cos\theta_j|^2}} & \text{otherwise,} \end{cases} \tag{5.10}$$

$$\frac{\partial |\cos\theta_i|}{\partial \vec{r}_{ij}} = \begin{cases} -\dfrac{\vec{n}_i}{r_{ij}} + \dfrac{(\vec{n}_i \cdot \vec{r}_{ij})\vec{r}_{ij}}{r_{ij}^3} & \text{if} \cos\theta_i \geq \cos\beta \\ \dfrac{\vec{n}_i}{r_{ij}} - \dfrac{(\vec{n}_i \cdot \vec{r}_{ij})\vec{r}_{ij}}{r_{ij}^3} & \text{if} \cos\theta_i \leq -\cos\beta, \end{cases} \tag{5.11}$$

and

$$\frac{\partial |\cos\theta_j|}{\partial \vec{r}_{ij}} = \begin{cases} \dfrac{\vec{n}_j}{r_{ij}} - \dfrac{(\vec{n}_j \cdot \vec{r}_{ij})\vec{r}_{ij}}{r_{ij}^3} & \text{if} \cos\theta_j \geq \cos\beta \\ -\dfrac{\vec{n}_j}{r_{ij}} + \dfrac{(\vec{n}_j \cdot \vec{r}_{ij})\vec{r}_{ij}}{r_{ij}^3} & \text{if} \cos\theta_j \leq -\cos\beta. \end{cases} \tag{5.12}$$

The force responsible for the rotation of soft ABA-type triblock Janus particle $\vec{g}_i$ is given as

$$\begin{aligned} \vec{g}_i &= -\frac{\partial U_{ij}}{\partial \vec{n}_i} \\ &= -\frac{\pi\alpha_{ij}^{A}}{4\beta}(r_{ij} - r_{ij}^2)\nu f^{\nu-1} \sin\frac{\pi\theta'_i}{2\beta} \frac{\partial \theta'_i}{\partial |\cos\theta_i|} \frac{\partial |\cos\theta_i|}{\partial \vec{n}_i} \cos\frac{\pi\theta'_j}{2\beta}, \end{aligned} \tag{5.13}$$

where

$$\frac{\partial |\cos\theta_i|}{\partial \vec{n}_i} = \begin{cases} -\dfrac{\vec{r}_i}{r_{ij}} & \text{if} \cos\theta_i \geq \cos\beta \\ \dfrac{\vec{r}_i}{r_{ij}} & \text{if} \cos\theta_i \leq -\cos\beta. \end{cases} \tag{5.14}$$

5.2.3 Soft BAB-Type Triblock Janus Particle Model

In the soft BAB-type triblock Janus particle model [63], the anisotropic factor f is given as

$$f = \begin{cases} \cos\dfrac{\pi}{2}\left(\dfrac{\pi/2 - \theta'_i}{\pi/2 - \beta}\right) \cos\dfrac{\pi}{2}\left(\dfrac{\pi/2 - \theta'_j}{\pi/2 - \beta}\right) & \text{if } |\cos\theta_i| \leq \cos\beta \text{ and } |\cos\theta_j| \leq \cos\beta \\ 0 & \text{otherwise.} \end{cases} \tag{5.15}$$

As illustrated in Figure 5.1c, the dark gray parts represent the repulsive B-type patches and the light gray part represents the attractive A-type middle band, and the directions of the repulsive patches on particles i and j are specified by unit vectors $\vec{n}_i$ and $\vec{n}_j$, respectively. θ_i is the angle between $\vec{n}_i$ and the interparticle vector $\vec{r}_{ji} = \vec{r}_j - \vec{r}_i$, and θ_j is the angle between $\vec{n}_j$ and $\vec{r}_{ij} = -\vec{r}_{ji}$, and then $\cos\theta_i = -\vec{n}_i \cdot \vec{r}_{ij}/r_{ij}$ and $\cos\theta_j = \vec{n}_j \cdot \vec{r}_{ij}/r_{ij}$. Here, $\theta'_i = \arccos(|\cos\theta_i|)$, and $\theta'_j = \arccos(|\cos\theta_j|)$. The fraction of surface covered by the attractive A-type middle band χ is related to β by the relation $\chi = 1 - 2\sin^2\left(\frac{\beta}{2}\right)$ [63, 74, 78].

The anisotropic force between two BAB-type triblock Janus particles is then given as

$$\begin{aligned}\vec{F}_{ij} &= -\frac{\partial U_{ij}}{\partial \vec{r}_{ij}}\\ &= \alpha_{ij}^{\mathrm{R}}(1 - r_{ij})\frac{\vec{r}_{ij}}{r_{ij}} + \alpha_{ij}^{\mathrm{A}} f^{\nu}\left(\frac{1}{2} - r_{ij}\right)\frac{\vec{r}_{ij}}{r_{ij}} - \frac{\alpha_{ij}^{\mathrm{A}}}{2}(r_{ij} - r_{ij}^2)\nu f^{\nu-1}\\ &\left[\frac{\pi}{\pi - 2\beta}\sin\frac{\pi}{2}\left(\frac{\pi/2 - \theta'_i}{\pi/2 - \beta}\right)\frac{1}{\sqrt{1 - |\cos\theta_i|^2}}\frac{\partial|\cos\theta_i|}{\partial\vec{r}_{ij}}\cos\frac{\pi}{2}\left(\frac{\pi/2 - \theta'_j}{\pi/2 - \beta}\right)\right.\\ &\left.+\frac{\pi}{\pi - 2\beta}\sin\frac{\pi}{2}\left(\frac{\pi/2 - \theta'_j}{\pi/2 - \beta}\right)\frac{1}{\sqrt{1 - |\cos\theta_j|^2}}\frac{\partial|\cos\theta_j|}{\partial\vec{r}_{ij}}\cos\frac{\pi}{2}\left(\frac{\pi/2 - \theta'_i}{\pi/2 - \beta}\right)\right],\end{aligned} \tag{5.16}$$

where

$$\frac{\partial|\cos\theta_i|}{\partial\vec{r}_{ij}} = \begin{cases} -\dfrac{\vec{n}_i}{r_{ij}} + \dfrac{(\vec{n}_i \cdot \vec{r}_{ij})\vec{r}_{ij}}{r_{ij}^3} & \text{if } 0 \le \cos\theta_i \le \cos\beta \\ \dfrac{\vec{n}_i}{r_{ij}} - \dfrac{(\vec{n}_i \cdot \vec{r}_{ij})\vec{r}_{ij}}{r_{ij}^3} & \text{if } -\cos\beta \le \cos\theta_i < 0, \end{cases} \tag{5.17}$$

and

$$\frac{\partial|\cos\theta_j|}{\partial\vec{r}_{ij}} = \begin{cases} \dfrac{\vec{n}_j}{r_{ij}} - \dfrac{(\vec{n}_j \cdot \vec{r}_{ij})\vec{r}_{ij}}{r_{ij}^3} & \text{if } 0 \le \cos\theta_j \le \cos\beta \\ -\dfrac{\vec{n}_j}{r_{ij}} + \dfrac{(\vec{n}_j \cdot \vec{r}_{ij})\vec{r}_{ij}}{r_{ij}^3} & \text{if } -\cos\beta \le \cos\theta_j < 0. \end{cases} \tag{5.18}$$

The force responsible for the rotation of soft BAB-type triblock Janus particle $\vec{g}_i$ is given as

$$\begin{aligned}\vec{g}_i &= -\frac{\partial U_{ij}}{\partial \vec{n}_i}\\ &= -\frac{\pi\alpha_{ij}^{\mathrm{A}}}{2(\pi - 2\beta)}(r_{ij} - r_{ij}^2)\nu f^{\nu-1}\sin\frac{\pi}{2}\left(\frac{\pi/2 - \theta'_i}{\pi/2 - \beta}\right)\frac{1}{\sqrt{1 - |\cos\theta_i|^2}}\end{aligned}$$

$$\frac{\partial|\cos\theta_i|}{\partial\vec{n}_i}\cos\frac{\pi}{2}\left(\frac{\pi/2-\theta_j'}{\pi/2-\beta}\right), \tag{5.19}$$

where

$$\frac{\partial|\cos\theta_i|}{\partial\vec{n}_i}=\begin{cases}-\frac{\vec{r}_{ij}}{r_{ij}} & \text{if } 0\le\cos\theta_j\le\cos\beta\\ \frac{\vec{r}_{ij}}{r_{ij}} & \text{if } -\cos\beta\le\cos\theta_j<0.\end{cases} \tag{5.20}$$

5.2.4 Integration Algorithm

For these SJPMs, the translational displacements of soft Janus particles follow Newton's equations of motion,

$$\dot{\vec{r}}_i=\vec{v}_i, \tag{5.21}$$

$$\dot{\vec{v}}_i=\frac{\vec{F}_i}{m_i}, \tag{5.22}$$

and the equations of rotational motion of soft Janus particles can be written as [62, 84]

$$\dot{\vec{n}}_i=\vec{u}_i, \tag{5.23}$$

$$\dot{\vec{u}}_i=\vec{g}_i^{\perp}/I+\lambda\vec{n}_i. \tag{5.24}$$

Here, Equation (5.23) simply defines $\vec{u}_i$ as the time derivative of the orientation $\vec{n}_i$. Physically, the first term in Equation (5.24) corresponds to the perpendicular component of $\vec{g}_i$, responsible for rotation of the particle, where

$$\vec{g}_i^{\perp}=\vec{g}_i-(\vec{g}_i\cdot\vec{n}_i)\vec{n}_i. \tag{5.25}$$

The second term in Equation (5.24) corresponds to the force $\lambda\vec{n}_i$ along the vector $\vec{n}_i$, which constrains the vector length to be a constant of the motion [84]. The equations of both translational and rotational motions in Equations (5.21)–(5.24) are numerically integrated via a half-step leap-frog algorithm [84, 85]. In the rotational motion, a full step in the integration algorithm is advanced through the equation

$$\vec{u}_i(t+\frac{1}{2}\delta t)=\vec{u}_i(t-\frac{1}{2}\delta t)+\delta t\vec{g}_i^{\perp}(t)/I-2\left[\vec{u}_i(t-\frac{1}{2}\delta t)\cdot\vec{n}_i(t)\right]\vec{n}_i(t). \tag{5.26}$$

The step is completed using

$$\vec{n}_i(t+\delta t)=\vec{n}_i(t)+\delta t\vec{u}_i\left(t+\frac{1}{2}\delta t\right). \tag{5.27}$$

In SJPMs, the simulations are performed in NVT conditions (the number of particles, the volume, and the temperature are constant). The weak coupling Berendsen thermostat is used to control the temperature at a target value [61–63, 85]. For the solution systems of soft Janus particles, the solvent particles are explicitly presented in the simulations. The solute–solvent and solvent–solvent interactions follow the first term of Equation (5.1). For a system of N particles, the number of Janus particles is $N_J=N\times\Phi$, and the number of spherical solvent particles is $N_s=N\times(1-\Phi)$ (Φ is the concentration of soft Janus particles in solution).

5.3 Soft Patchy Particle Models

5.3.1 The Model

In our soft patchy particle models (SPPMs) [64], the deformable and anisotropic characteristics of soft patchy particles are also described via a single-site soft anisotropic potential analogous to the one used in SJPMs. This anisotropic potential is expressed as

$$U_{ij} = \begin{cases} \dfrac{\alpha_{ij}^{\mathrm{R}} d_{ij}}{2}\left(1 - \dfrac{r_{ij}}{d_{ij}}\right)^2 - \displaystyle\sum_{\kappa=1}^{M_i}\sum_{\lambda=1}^{M_j} f^{\nu}(\vec{n}_i^{\kappa}, \vec{n}_j^{\lambda}, \vec{r}_{ij}) \dfrac{\alpha_{ij}^{\mathrm{A}} d_{ij}}{2}\left[\dfrac{r_{ij}}{d_{ij}} - \left(\dfrac{r_{ij}}{d_{ij}}\right)^2\right] & r_{ij} \le d_{ij} \\ 0 & r_{ij} > d_{ij}, \end{cases} \tag{5.28}$$

where

$$f\left(\vec{n}_i^{\kappa}, \vec{n}_j^{\lambda}, \vec{r}_{ij}\right) = \begin{cases} \cos\dfrac{\pi\theta_i^{\kappa}}{2\theta_m^{\kappa}} \cos\dfrac{\pi\theta_j^{\lambda}}{2\theta_m^{\lambda}} & \text{if } \cos\theta_i^{\kappa} \ge \cos\theta_m^{\kappa} \text{ and } \cos\theta_j^{\lambda} \ge \cos\theta_m^{\lambda} \\ 0 & \text{otherwise.} \end{cases} \tag{5.29}$$

Here, r_{ij} is the distance between the centers of particles i and j, d_i and d_j are the diameters of particles i and j, respectively, and $d_{ij} = (d_i + d_j)/2$. So the effect of the size polydispersity of patchy particles can be considered in SPPMs. But for simplicity, we normally choose $d_{ij} = d_i = d_j \equiv 1.0$ as the unit of length, define the interaction cutoff radius $r_c \equiv d_{ij}$, use $k_B T$ as the unit of energy, and choose the mass m_i of the particle as the unit, thus the time unit $\tau = \sqrt{m_i d_{ij}^2 / k_B T}$ [61–64]. All the variables and parameters in SPPMs are given in reduced units. As illustrated in Figure 5.2, the light gray parts of the particle surface represent the attractive patches, M_i and M_j are used to describe the number of the attractive patches of particles i and j, the directions of the attractive patches κ ($\kappa = 1, \ldots, M_i$) and λ ($\lambda = 1, \ldots, M_j$) on particles i and j are specified by patch vectors $\vec{n}_i^{\kappa}$ and $\vec{n}_j^{\lambda}$, respectively, both of which are unit vectors. θ_i^{κ} is the angle between $\vec{n}_i^{\kappa}$ and the interparticle vector $\vec{r}_{ji} = \vec{r}_j - \vec{r}_i$, and θ_j^{λ} is the angle between $\vec{n}_j^{\lambda}$ and $\vec{r}_{ij}$ ($\vec{r}_{ij} = -\vec{r}_{ji}$), and then $\cos\theta_i^{\kappa} = -\vec{n}_i^{\kappa} \cdot \vec{r}_{ij}/r_{ij}$ and $\cos\theta_j^{\lambda} = \vec{n}_j^{\lambda} \cdot \vec{r}_{ij}/r_{ij}$. Thus, $\theta_i^{\kappa} = \arccos(\cos\theta_i^{\kappa}) = \arccos(-\vec{n}_i^{\kappa} \cdot \vec{r}_{ij}/r_{ij})$, and $\theta_j^{\lambda} = \arccos(\cos\theta_j^{\lambda}) = \arccos(\vec{n}_j^{\lambda} \cdot \vec{r}_{ij}/r_{ij})$. The sizes of the attractive patches κ and λ are described by θ_m^{κ} and θ_m^{λ}, which are half of the opening angle of the attractive patches (i.e. the semi-angular widths of the patches). The fraction of surface of particle i covered by the M_i attractive patches, χ, is related to θ_m^{κ} by the relation $\chi = \sum_{\kappa=1}^{M_i} \sin^2\left(\frac{\theta_m^{\kappa}}{2}\right)$ [64, 74, 78]. As described above, due to the simplicity and

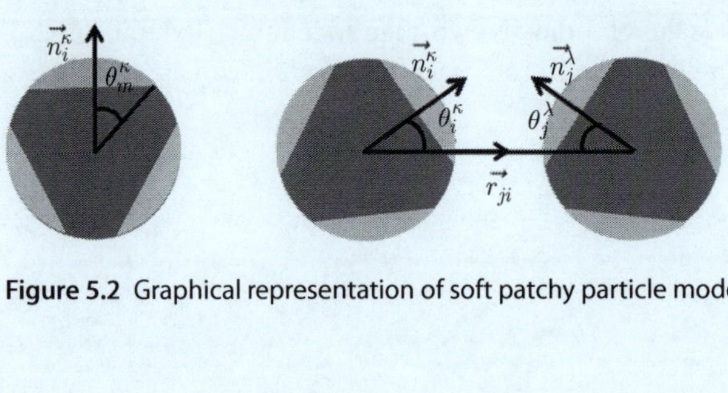

Figure 5.2 Graphical representation of soft patchy particle models.

efficiency of this single-site anisotropic potential, it is very easy and convenient to obtain a wide range of patchy particle models with different anisotropies, simply by changing the number M_i, size θ_m^κ, direction $\vec{n}_i^\kappa$, and geometrical arrangement of the patches of particle i.

5.3.2 Integration Algorithm

In dynamics simulations, all patchy particles have position $\vec{r}_i$, velocity $\vec{v}_i$, angular momentum $\vec{L}_i$, angular velocity $\vec{\omega}_i$, normalized quaternion $\vec{q}_i$, mass m_i, moment of inertia tensor $\vec{I}_i$, and number of the attractive patches M_i, and all attractive patches have the direction $\vec{n}_i^\kappa$ and size θ_m^κ. Here, the normalized quaternion $\vec{q}_i = (q_{i,0}, q_{i,1}, q_{i,2}, q_{i,3})$ is introduced in order to describe the orientation of patchy particle i, and generate a minimal representation of the rotation matrix from the body frame to the space frame [64, 69, 84, 86]

$$\vec{e}^{\mathrm{s}} = \vec{A}(\vec{q}_i)\vec{e}^{\mathrm{b}}, \tag{5.30}$$

with

$$\vec{A}(\vec{q}_i) = \begin{pmatrix} q_{i,0}^2 + q_{i,1}^2 - q_{i,2}^2 - q_{i,3}^2 & 2(q_{i,1}q_{i,2} - q_{i,0}q_{i,3}) & 2(q_{i,1}q_{i,3} + q_{i,0}q_{i,2}) \\ 2(q_{i,1}q_{i,2} + q_{i,0}q_{i,3}) & q_{i,0}^2 - q_{i,1}^2 + q_{i,2}^2 - q_{i,3}^2 & 2(q_{i,2}q_{i,3} - q_{i,0}q_{i,1}) \\ 2(q_{i,1}q_{i,3} - q_{i,0}q_{i,2}) & 2(q_{i,2}q_{i,3} + q_{i,0}q_{i,1}) & q_{i,0}^2 - q_{i,1}^2 - q_{i,2}^2 + q_{i,3}^2 \end{pmatrix}, \tag{5.31}$$

and

$$\begin{aligned} q_{i,0} &= \cos\frac{\theta}{2}\cos\left(\frac{\phi+\psi}{2}\right), \quad & q_{i,1} &= \sin\frac{\theta}{2}\cos\left(\frac{\phi-\psi}{2}\right), \\ q_{i,2} &= \sin\frac{\theta}{2}\sin\left(\frac{\phi-\psi}{2}\right), \quad & q_{i,3} &= \cos\frac{\theta}{2}\sin\left(\frac{\phi+\psi}{2}\right), \end{aligned} \tag{5.32}$$

where θ, ϕ, and ψ are the Euler angles in the standard convention. In the body frame, a patchy particle's center of mass is at the origin and the moment of inertia tensor $\vec{I}_i^{\mathrm{b}}$ is diagonal. For spherical patchy particles with homogeneous mass distribution mainly considered in the present model, the moment of inertia tensor $\vec{I}_i^{\mathrm{b}}$ is taken as a unit diagonal matrix for the sake of simplicity. Although the space-fixed direction of the attractive patches $\vec{n}_i^\kappa$ varies with time in the simulations, the body-fixed direction $\vec{n}_i^{\kappa\mathrm{b}}$ will not change provided that $\vec{n}_i^{\kappa\mathrm{b}}$ is specified in the body frame at the beginning of the simulations. The transformation of the directions from body-fixed frame to space-fixed frame is handled in a manner analogous to Equation (5.30),

$$\vec{n}_i^\kappa = \vec{A}(\vec{q}_i)\vec{n}_i^{\kappa\mathrm{b}}. \tag{5.33}$$

The motion of patchy particles is governed by Newtonian mechanics [64, 69, 84, 86], and the equations of motion of the particles in the space frame are given as

$$\dot{\vec{r}}_i = \vec{v}_i, \tag{5.34}$$

$$\dot{\vec{v}}_i = \frac{\vec{F}_i}{m_i}, \tag{5.35}$$

$$\dot{\vec{L}}_i = \vec{\tau}_i, \tag{5.36}$$

$$\dot{\vec{q}}_i = \frac{1}{2}\vec{S}(\vec{q}_i)\vec{\omega}_i, \tag{5.37}$$

where $\vec{F}_i = \sum_j \vec{F}_{ij}$ and $\vec{\tau}_i = \sum_j \vec{\tau}_{ij}$ are the force and torque acting on patchy particle i due to all its direct neighbors, respectively. In Equation (5.37),

$$\vec{\omega}_i = (0, \omega_i^x, \omega_i^y, \omega_i^z) = \vec{A}(\vec{q}_i)(\vec{I}_i^{\mathrm{b}})^{-1}\vec{A}^T(\vec{q}_i)\vec{L}_i, \tag{5.38}$$

and

$$\vec{S}(\vec{q}_i) = \begin{pmatrix} q_{i,0} & -q_{i,1} & -q_{i,2} & -q_{i,3} \\ q_{i,1} & q_{i,0} & q_{i,3} & -q_{i,2} \\ q_{i,2} & -q_{i,3} & q_{i,0} & q_{i,1} \\ q_{i,3} & q_{i,2} & -q_{i,1} & q_{i,0} \end{pmatrix}. \tag{5.39}$$

The force between two neighboring patchy particles $\vec{F}_{ij}$ is given by the derivation of Equation (5.28),

$$\begin{aligned} \vec{F}_{ij} &= -\frac{\partial U_{ij}}{\partial \vec{r}_{ij}} \\ &= \alpha_{ij}^{\mathrm{R}}\left(1 - \frac{r_{ij}}{d_{ij}}\right)\frac{\vec{r}_{ij}}{r_{ij}} + \sum_{\kappa=1}^{M_i}\sum_{\lambda=1}^{M_j}\left\{ \alpha_{ij}^{\mathrm{A}} f^{\nu}\left(\vec{n}_i^{\kappa}, \vec{n}_j^{\lambda}, \vec{r}_{ij}\right)\left(\frac{1}{2} - \frac{r_{ij}}{d_{ij}}\right)\frac{\vec{r}_{ij}}{r_{ij}} \right. \\ &\quad - \frac{\alpha_{ij}^{\mathrm{A}}}{2}\left[\frac{r_{ij}}{d_{ij}} - \left(\frac{r_{ij}}{d_{ij}}\right)^2\right] \nu f^{\nu-1}\left(\vec{n}_i^{\kappa}, \vec{n}_j^{\lambda}, \vec{r}_{ij}\right)\left(\frac{\pi}{2\theta_m^{\kappa}} \sin\frac{\pi\theta_i^{\kappa}}{2\theta_m^{\kappa}} \frac{\partial \theta_i^{\kappa}}{\partial \cos\theta_i^{\kappa}} \right. \\ &\quad \left.\left. \frac{\partial \cos\theta_i^{\kappa}}{\partial \vec{r}_{ij}} \cos\frac{\pi\theta_j^{\lambda}}{2\theta_m^{\lambda}} + \frac{\pi}{2\theta_m^{\lambda}} \sin\frac{\pi\theta_j^{\lambda}}{2\theta_m^{\lambda}} \frac{\partial \theta_j^{\lambda}}{\partial \cos\theta_j^{\lambda}} \frac{\partial \cos\theta_j^{\lambda}}{\partial \vec{r}_{ij}} \cos\frac{\pi\theta_i^{\kappa}}{2\theta_m^{\kappa}}\right)\right\}, \end{aligned} \tag{5.40}$$

where

$$\frac{\partial \theta_i^{\kappa}}{\partial \cos\theta_i^{\kappa}} = \begin{cases} 0 & \text{if } \cos^2\theta_i^{\kappa} = 1 \\ -\dfrac{1}{\sqrt{1-\cos^2\theta_i^{\kappa}}} & \text{otherwise,} \end{cases} \tag{5.41}$$

$$\frac{\partial \theta_j^{\lambda}}{\partial \cos\theta_j^{\lambda}} = \begin{cases} 0 & \text{if } \cos^2\theta_j^{\lambda} = 1 \\ -\dfrac{1}{\sqrt{1-\cos^2\theta_j^{\lambda}}} & \text{otherwise,} \end{cases} \tag{5.42}$$

$$\frac{\partial \cos\theta_i^{\kappa}}{\partial \vec{r}_{ij}} = -\frac{\vec{n}_i^{\kappa}}{r_{ij}} + \frac{(\vec{n}_i^{\kappa}\cdot\vec{r}_{ij})\vec{r}_{ij}}{r_{ij}^3}, \tag{5.43}$$

and

$$\frac{\partial \cos\theta_j^{\lambda}}{\partial \vec{r}_{ij}} = \frac{\vec{n}_j^{\lambda}}{r_{ij}} - \frac{(\vec{n}_j^{\lambda}\cdot\vec{r}_{ij})\vec{r}_{ij}}{r_{ij}^3}. \tag{5.44}$$

The torque $\vec{\tau}_{ij}$ acting on patchy particle i due to its neighboring particle j is given by

$$\begin{aligned}\vec{\tau}_{ij} &= \sum_{\kappa=1}^{M_i} -\frac{\partial U_{ij}}{\partial \vec{n}_i^{\kappa}} \\ &= \sum_{\kappa=1}^{M_i}\sum_{\lambda=1}^{M_j} \frac{\pi\alpha_{ij}^{\mathrm{A}} d_{ij}}{4\theta_m^{\kappa}} \left[\frac{r_{ij}}{d_{ij}} - \left(\frac{r_{ij}}{d_{ij}}\right)^2\right] \nu f^{\nu-1}(\vec{n}_i^{\kappa}, \vec{n}_j^{\lambda}, \vec{r}_{ij}) \sin\frac{\pi\theta_i^{\kappa}}{2\theta_m^{\kappa}} \\ &\quad \frac{\partial\theta_i^{\kappa}}{\partial\cos\theta_i^{\kappa}} \cos\frac{\pi\theta_j^{\lambda}}{2\theta_m^{\lambda}} \frac{\vec{r}_i}{r_{ij}}. \end{aligned} \tag{5.45}$$

The equations of motion of patchy particles in Equations (5.34)–(5.37) are numerically integrated via a velocity-Verlet-like algorithm [64, 69, 84]. First, the velocity and angular momentum at time $t + \frac{1}{2}\delta t$, and the position and orientation at time $t + \delta t$, are calculated using the equations

$$\vec{v}_i\left(t + \frac{1}{2}\delta t\right) = \vec{v}_i(t) + \frac{1}{2}\delta t \frac{\vec{F}_i(t)}{m_i}, \tag{5.46}$$

$$\vec{r}_i(t+\delta t) = \vec{r}_i(t) + \delta t \vec{v}_i\left(t + \frac{1}{2}\delta t\right), \tag{5.47}$$

$$\vec{L}_i\left(t + \frac{1}{2}\delta t\right) = \vec{L}_i(t) + \frac{1}{2}\delta t \vec{\tau}_i(t), \tag{5.48}$$

$$\vec{q}_i(t+\delta t) = Q\left(\vec{q}_i(t), \delta t, \vec{\omega}_i\left(t + \frac{1}{2}\delta t\right)\right), \tag{5.49}$$

where the angular velocity $\vec{\omega}_i\left(t + \frac{1}{2}\delta t\right)$ is calculated via Equation (5.38), and the function Q is an application of the Richardson method [69] to reduce the error in integrating Equation (5.37). The Richardson method to update quaternion $\vec{q}_i$ from t to $t + \delta t$ is implemented as follows:

$$\vec{\omega}_i'\left(t + \frac{1}{2}\delta t\right) = \vec{A}(\vec{q}_i(t))\left(\vec{I}_i^{\mathrm{b}}\right)^{-1}\vec{A}^T(\vec{q}_i(t))\vec{L}_i\left(t + \frac{1}{2}\delta t\right), \tag{5.50}$$

$$\dot{\vec{q}}_i'\left(t + \frac{1}{2}\delta t\right) = \frac{1}{2}\vec{S}(\vec{q}_i(t))\vec{\omega}_i'\left(t + \frac{1}{2}\delta t\right), \tag{5.51}$$

$$\vec{q}_i'(t+\delta t) = \vec{q}_i(t) + \delta t \dot{\vec{q}}_i'\left(t + \frac{1}{2}\delta t\right), \tag{5.52}$$

$$\vec{q}_i'\left(t + \frac{1}{2}\delta t\right) = \vec{q}_i(t) + \frac{1}{2}\delta t \dot{\vec{q}}_i'\left(t + \frac{1}{2}\delta t\right), \tag{5.53}$$

$$\vec{\omega}_i'\left(t + \frac{1}{2}\delta t\right) = \vec{A}\left(\vec{q}_i'\left(t + \frac{1}{2}\delta t\right)\right)\left(\vec{I}_i^{\mathrm{b}}\right)^{-1}\vec{A}^T\left(\vec{q}_i'\left(t + \frac{1}{2}\delta t\right)\right)\vec{L}_i\left(t + \frac{1}{2}\delta t\right), \tag{5.54}$$

$$\dot{\vec{q}}_i'\left(t + \frac{1}{2}\delta t\right) = \frac{1}{2}\vec{S}\left(\vec{q}_i'\left(t + \frac{1}{2}\delta t\right)\right)\vec{\omega}_i'\left(t + \frac{1}{2}\delta t\right), \tag{5.55}$$

$$\vec{q}_i'(t+\delta t) = \vec{q}_i'\left(t + \frac{1}{2}\delta t\right) + \frac{1}{2}\delta t \dot{\vec{q}}_i'\left(t + \frac{1}{2}\delta t\right), \tag{5.56}$$

$$\vec{q}_i(t+\delta t) = 2\vec{q}_i'(t+\delta t) - \vec{q}_i'(t+\delta t). \tag{5.57}$$

To preserve the constraint $q_{i,0}^2 + q_{i,1}^2 + q_{i,2}^2 + q_{i,3}^2 = 1$, all the quaternions in Equations (5.50)–(5.57) should be renormalized after being updated [84]. The force $\vec{F}_i$

and torque $\vec{\tau}_i$ at time $t + \delta t$ are then calculated based on the position and orientation at time $t + \delta t$, and the velocity and angular momentum are advanced fully to $t + \delta t$ using

$$\vec{v}_i(t+\delta t) = \vec{v}_i\left(t + \frac{1}{2}\delta t\right) + \frac{1}{2}\delta t \frac{\vec{F}_i(t+\delta t)}{m_i}, \tag{5.58}$$

$$\vec{L}_i(t+\delta t) = \vec{L}_i\left(t + \frac{1}{2}\delta t\right) + \frac{1}{2}\delta t \vec{\tau}_i(t+\delta t). \tag{5.59}$$

In SPPMs, the simulations are performed in the NVT ensemble [64, 87]. The solvent particles are explicitly considered in the simulations. The solute–solvent and solvent–solvent interactions follow the first term of Equation (5.28). The equations of motion of the solvent particles are numerically integrated via the standard velocity-Verlet algorithm [64, 84]. For a system of N particles, the number of patchy particles is $N_\text{p} = N \times \Phi$, and the number of spherical solvent particles is $N_\text{s} = N \times (1 - \Phi)$ (Φ is the concentration of patchy particles in solution). Then, the data structures required to implement the above velocity-Verlet-like algorithm consist of the following quantities for each patchy particle: N_p, m_i, $\vec{I}_i^\text{b}$, $\vec{r}_i$, $\vec{v}_i$, $\vec{L}_i$, $\vec{\omega}_i$, $\vec{q}_i$, $\vec{n}_i^{\kappa\text{b}}$, $\vec{F}_i$, and $\vec{\tau}_i$.

5.4 Physical Meanings of the Simulation Parameters in Our Models

Our soft Janus and patchy particle models can be directly mapped onto experimental systems in different ambient conditions [61–64]. As given in Refs. [61–64], if we choose the anisotropic factor $f = 1$ and $r_\text{c} = d_{ij} \equiv 1.0$, the anisotropic potentials in Equations (5.1) and (5.28) will become an isotropic potential with a shallow attractive well, $U_{ij} = \frac{\alpha_{ij}^\text{R}}{2}(1 - r_{ij})^2 - \frac{\alpha_{ij}^\text{A}}{2}(r_{ij} - r_{ij}^2)$. The parameter α_{ij}^R is related to the linear elastic modulus E of the particle by $\alpha_{ij}^\text{R} = \pi E d_\text{eff}^2/6$ [88, 89]. Here, d_eff is the effective diameter of the soft Janus or patchy particle, and can be estimated by $d_\text{eff} = (\alpha_{ij}^\text{R} + \alpha_{ij}^\text{A}/2)/(\alpha_{ij}^\text{R} + \alpha_{ij}^\text{A})$. If δ is defined as the range of attraction related to the effective diameter d_eff, and $(1 + \delta)d_\text{eff} = d_{ij}$, then δ is also related to α_{ij}^R and α_{ij}^A by $\delta = \alpha_{ij}^\text{A}/(2\alpha_{ij}^\text{R} + \alpha_{ij}^\text{A})$. The energy minimum of the attractive potential at $r_{ij} = d_\text{eff}$ gives the adhesion energy $G = -U_{ij}^\text{min} = \alpha_{ij}^\text{A}(1 - d_\text{eff})/4$, which determines the association strength between Janus or patchy particles and can be tuned by altering the salt concentration, pH, or temperature in experiments [61–64, 88]. Thus, the simulation parameters α_{ij}^R and α_{ij}^A can be fixed from experimentally measurable particle properties including the elastic modulus E, the effective diameter d_eff, and the adhesion energy G.

5.5 GPU Acceleration

Due to their high computational performance, graphics processing units (GPUs) are becoming an increasingly important element of computer simulation [64, 69, 82, 90–94]. Several highly optimized GPU-accelerated molecular dynamics (MD) packages have been designed for investigating polymeric systems efficiently, such as GROMACS [90], LAMMPS [91], AMBER [92, 93], HOOMD-blue [69], and GALAMOST [82]. In

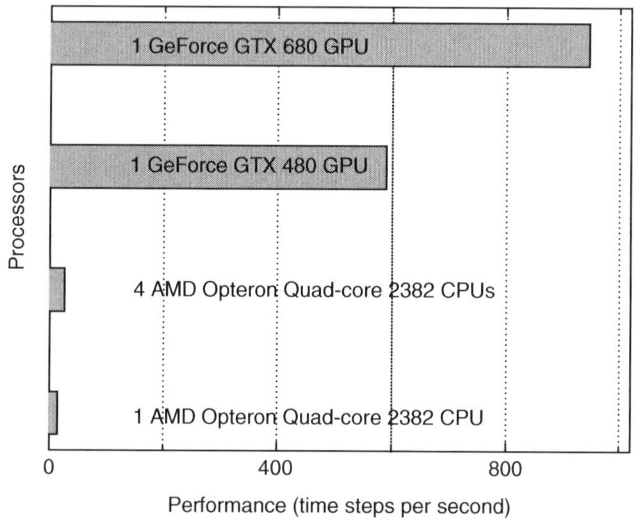

Figure 5.3 Comparison of the performance of the soft ABA-type triblock Janus particle model implemented in GALAMOST with the corresponding CPU code. The performance data are measured by time steps per second. All simulated systems have identical simulation settings with the number of ABA-type triblock Janus particles, $N_J = 24\,000$, $\alpha_{ij}^{R} = 396$, $\alpha_{ij}^{A} = 88$ ($G \approx 2.00\,k_BT$), $\nu = 1/2$, and $\beta = 45°$.

order to improve the computational efficiency of our SJPMs and SPPMs in dynamics simulations, we have given the simulation algorithms in Refs. [64, 82], which fit the compute unified device architecture (CUDA) framework of NVIDIA GPUs. As shown in Refs. [64, 82], the performance of our SJPMs and SPPMs implemented in the GALAMOST package is very satisfactory. For example, we compare the performance of our SJPMs implemented in GALAMOST with the corresponding CPU code by simulating the ordered packing of soft ABA-tpye triblock Janus systems with 24 000 Janus particles [62], as illustrated in Figure 5.3. The performance of our SJPMs implemented in GALAMOST running on a single GPU is far beyond the performance of the corresponding CPU code running on four parallel CPU cores. Therefore, our SJPMs and SPPMs provide a powerful tool to investigate the aggregation behavior of soft Janus and patchy particles over larger spatial and temporal scales.

5.6 Self-Assembly of Soft Janus and Patchy Particles

5.6.1 Self-Assembly of Soft One-Patch Janus Particles

Due to their deformable and anisotropic characteristics, soft one-patch Janus particles hide many surprises in the design and fabrication of hierarchically self-assembled structures [61]. As shown in Figure 5.4, by properly tuning the Janus balance β and the strength of attraction between attractive patches α_{ij}^{A} (i.e. the adhesion energy G), soft one-patch Janus particles can reversibly self-assemble into a number of fascinating hierarchical structures in dilute solutions, such as micelles, worm-like strings, single helices, double helices, bilayers, and complex supermicelles [61], which may have great potential in nanotechnology and biotechnology. As suggested in Ref. [61], the

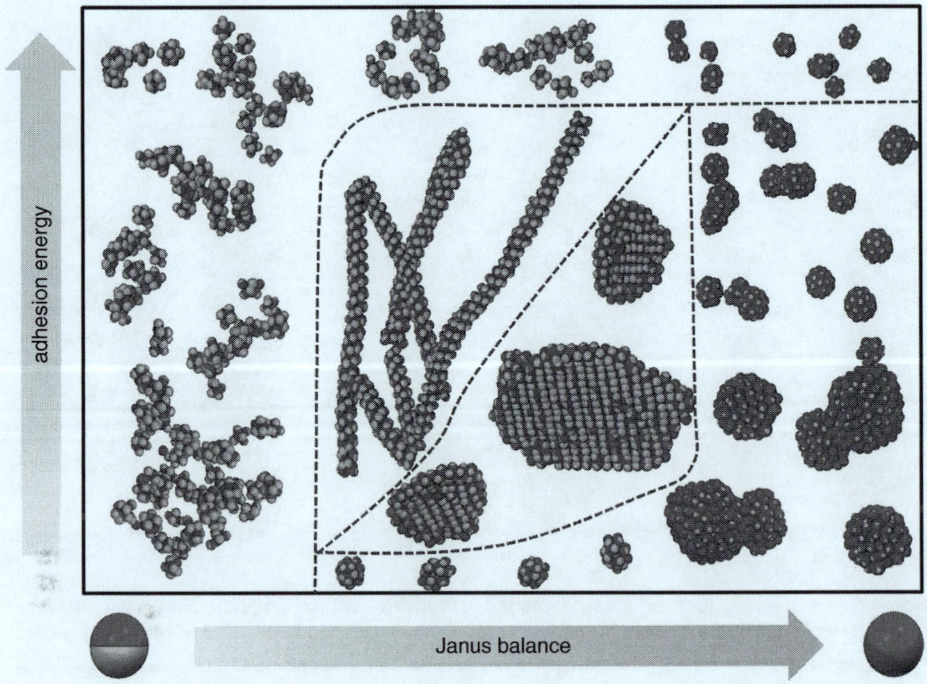

Figure 5.4 Representative equilibrium superstructures self-assembled from soft one-patch Janus particles by properly tuning the Janus balance β and the strength of attraction between attractive patches α_{ij}^{A} while keeping $\alpha_{ij}^{R} = 396$, $\phi = 5\%$, and $\nu = 1/2$. *(See insert for color representation of the figure.)*.

distribution $S(n)$ of the number of nearest neighbors per Janus particle n and the distribution $P(\cos\gamma)$ of $\cos\gamma = \vec{n}_i \cdot \vec{n}_j$ for all pairs of contacting Janus particles can be used to quantitatively characterize the structural details of different self-assembled structures. $S(n) = N_{\mathrm{p}}(n)/N_{\mathrm{solute}}$, where $N_{\mathrm{p}}(n)$ is the number of Janus particles with the number of nearest neighbors n, and N_{solute} is the total number of Janus particles in the system. The distribution $P(\cos\gamma)$ is expected to show well-defined peaks for an ordered state and to be flat in a completely disordered state [61]. Therefore, it is clear that soft one-patch Janus particles provide a rich palette for the design of novel and exciting hierarchical superstructures.

5.6.2 The Role of Particle Softness in Self-Assembling Different Supracolloidal Helices

The design and fabrication of biomimetic helical structures by simple and experimentally realizable surface-anisotropic colloidal particles remain among the current grand challenges in materials science. As the simplest case of surface-anisotropic particles, soft one-patch Janus particles exhibit remarkable ability to self-assemble into different desired helical structures [61, 95]. As presented in Ref. [61], for softer Janus particles whose deformability is higher, the range of anisotropic attraction is longer,

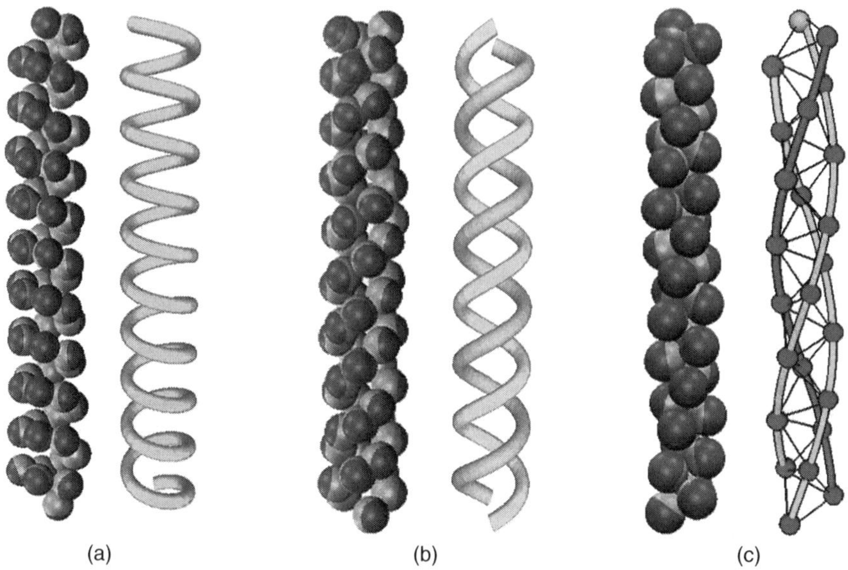

Figure 5.5 Biomimetic supracolloidal helices self-assembled from soft one-patch Janus particles by tuning particle softness. (a) Single helices ($\alpha_{ij}^{R} = 396$, $\alpha_{ij}^{A} = 308$ ($G \approx 17.00\,k_BT$), $\beta = 115°$). (b) Double helices ($\alpha_{ij}^{R} = 396$, $\alpha_{ij}^{A} = 242$ ($G \approx 11.50\,k_BT$), $\beta = 120°$). (c) Bernal spirals described as a stack of face-sharing tetrahedra ($\alpha_{ij}^{R} = 39\,996$, $\alpha_{ij}^{A} = 4420$ ($G \approx 55.00\,k_BT$), $\beta = 80°$).

and then the number of nearest neighbors per Janus particle is not limited to six. Thus, single helices (Figure 5.5a) and double helices (Figure 5.5b) with more nearest neighbors are observed. With decreasing softness of Janus particles (i.e. Janus particles become harder), the range of anisotropic attraction becomes shorter, and the number of nearest neighbors will not exceed six. Thereby, Bernal spirals described as a stack of face-sharing tetrahedra in Figure 5.5c are obtained through the self-assembly of harder Janus particles [95]. These building blocks in our simulations are sufficiently simple, and the experimental implementation of soft one-patch Janus particles with tunable softness is within the reach of today's experimental capabilities [13, 48, 51–53, 56, 58, 65]. Therefore, soft one-patch Janus particles with tunable softness hold significant promise for the design and fabrication of biomimetic helical structures [61, 95].

5.6.3 Self-Assembly of Soft ABA-Type Triblock Janus Particles

As presented in Ref. [62], the aggregation behavior of soft ABA-type triblock Janus particles mainly depends on the Janus balance β and the flexibility of Janus particle aggregates (the strength of attraction α_{ij}^{A} and the angular width of attraction ν). By examining the influence of β, α_{ij}^{A}, and ν, some ordered packing structures including hexagonally packed columnar structure and body-centered tetragonal structure are observed. Furthermore, typical aggregated structures, such as thread-like structures, network structures, and pyrochlore lattice structures (i.e., the 3D generalization of the 2D kagome lattices) are also observed by the self-assembly of soft ABA-type triblock Janus particles in dilute

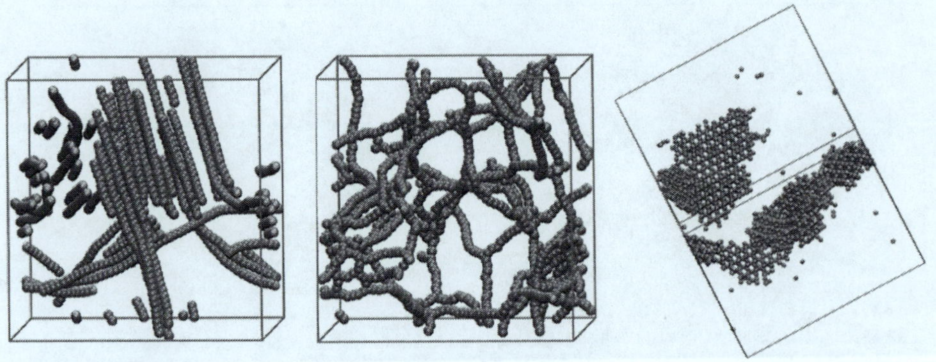

Figure 5.6 Typical equilibrium structures self-assembled from soft ABA-type triblock Janus particles with different Janus balance while keeping $\alpha_{ij}^{R} = 396$, $\alpha_{ij}^{A} = 220$ ($G \approx 10.00\,k_BT$), $\phi = 5\%$, and $\nu = 1/2$: thread-like structures ($\beta = 30°$), network structures ($\beta = 60°$), and pyrochlore lattice structures ($\beta = 65°$).

solutions, as shown in Figure 5.6. Therefore, soft ABA-type triblock Janus particles can not only provide the building blocks for the design and creation of various ordered structures, but also serve as very useful model systems to understand some fundamental questions such as polymerization and gelation in polymer science.

5.6.4 Template-Free Fabrication of Two-Dimensional Exotic Nanostructures through the Self-Assembly of Soft BAB-Type Triblock Janus Particles

The design and creation of two-dimensional (2D) well-ordered nanostructures by a facile template-free fabrication strategy remain major scientific and technological challenges. Recently, we have introduced a new concept to achieve template-free fabrication of diverse 2D nanostructures by resorting to the deformable and anisotropic characteristics of soft BAB-type triblock Janus particles [63]. As shown in Figure 5.1c, these particles are spherical and symmetrically decorated with two repulsive patches separated by an attractive middle band. As can be seen in Figure 5.7, the attraction strength between attractive middle bands α_{ij}^{A} (i.e. the adhesion energy G) and the Janus balance β coordinate to generate a number of fascinating 2D ordered non-close-packed (ncp) and close-packed nanostructures, such as hexagonal, square, and honeycomb-like ncp arrays, and hexagonally close-packed (hcp) arrays. Interestingly, the 2D hexagonal and square ncp nanostructures can further close to form vesicles with hexagonal and square ncp surface lattices, respectively. The relative stability of these 2D nanostructures can be efficiently evaluated through calculating their free energy differences with respect to the reference nanostructure [63, 87, 96, 97]. As suggested in Ref. [63], the free energy differences per particle of these 2D nanostructures are low enough to allow spontaneous formation of these ordered nanostructures, and it is possible to observe reversible structural transitions between them over a convenient timescale. It is clear that the interplay between particle softness and controllable directional attraction plays a dominant role in creating these nanostructures, which may lead to promising applications such as coating curved substrates and soft-electronic/optical devices. Therefore, our

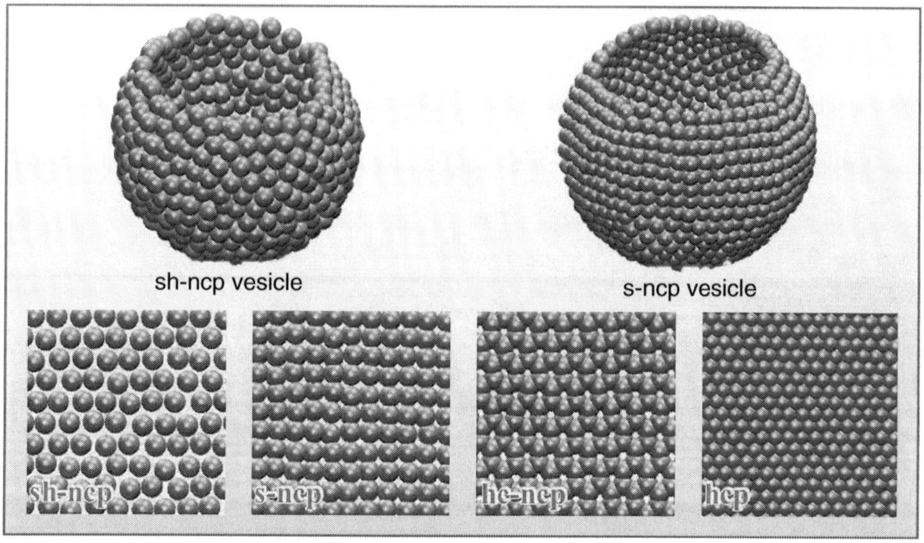

Figure 5.7 Representative non-close-packed (ncp) and close-packed nanostructures self-assembled from soft BAB-type triblock Janus particles, including 2D single-layer hexagonal ncp (sh-ncp), square ncp (s-ncp), honeycomb-like ncp (hc-ncp), and hexagonal close-packed (hcp) arrays, and vesicles with sh-ncp and s-ncp surface lattices, at $\alpha_{ij}^{R} = 396$, $\phi = 5\%$, and $\nu = 1/2$.

work paves the way for the experimental realization of 2D well-ordered nanostructures, and offers a new direction for the creation of novel colloidal nanostructures without any template [63].

5.6.5 Self-Assembly of Soft Multi-Patch Particles

With the aid of SPPMs, the self-assembly behavior of soft multi-patch particles with tunable number, size, direction, and geometrical arrangement of the patches can also be effectively investigated. Apparently, 2D and 3D open lattices can be obtained from the self-assembly of soft three-patch and four-patch particles with regular triangular and tetrahedral arrangement of the patches, respectively. In the body-fixed frame, the patch vectors of the three-patch particle in Figure 5.8a are $\vec{n}_i^1 = (0, 0, 1)$, $\vec{n}_i^2 = (0.8660, 0, -0.5)$, and $\vec{n}_i^3 = (-0.8660, 0, -0.5)$, and the patch vectors of the four-patch particle in Figure 5.8b are $\vec{n}_i^1 = (0, 0, 1)$, $\vec{n}_i^2 = (0.8165, -0.4714, -0.3333)$, $\vec{n}_i^3 = (-0.8165, -0.4714, -0.3333)$, and $\vec{n}_i^4 = (0, 0.9428, -0.3333)$. To obtain well-ordered self-assembly structures, the annealing method [62, 64, 98] is adopted in the simulations. As shown in Figure 5.8a, soft three-patch particles that mimic sp^2 hybridized atomic orbitals [6], self-assemble into a graphene-like two-dimensional structure with a hexagonal honeycomb lattice, which accommodates a large potential in opto-electronic applications [99, 100]. Soft four-patch particles with a regular tetrahedral patch arrangement are able to stabilize the diamond lattice structure in Figure 5.8b [54, 55], which has attracted considerable attention in the field of patchy particles [70, 77, 79, 101] due to their potential applications in photonics and biosensing [102, 103].

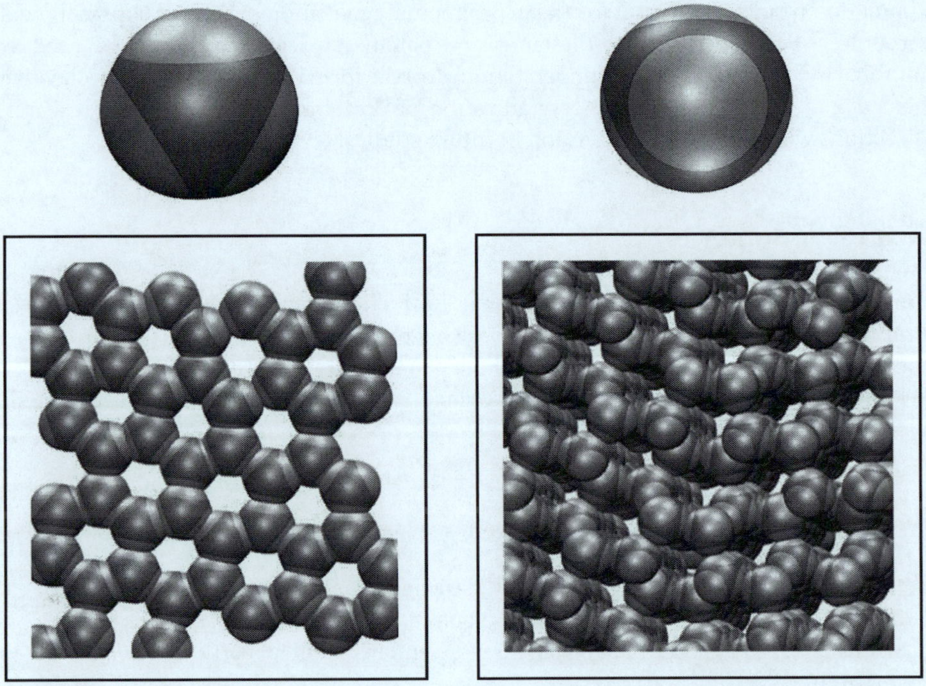

(a) (b)

Figure 5.8 (a) Graphene-like two-dimensional structure with a hexagonal honeycomb lattice self-assembled from soft three-patch particles with regular triangular arrangement of the patches. (b) Diamond lattice structure self-assembled from soft four-patch particles with regular tetrahedral arrangement of the patches. In the simulations, $\alpha_{ij}^{R} = 396$, $\alpha_{ij}^{A} = 220$ ($G \approx 10.00\,k_B T$), $\nu = 0.5$, and $\theta_m^{\kappa} = 45°$. *(See insert for color representation of the figure.).*

5.7 Conclusions

In this chapter, we have introduced two kinds of general and effective mesoscale models, which are capable of investigating the aggregation behavior and mechanism of various types of soft Janus and patchy particles. The simulation models and algorithms are presented in detail, and all these models can be directly mapped onto experimental systems in different conditions. We have also summarized possible structures obtained from the self-assembly of soft one-patch, and ABA-type and BAB-type triblock Janus particles [61–63, 95] and soft multi-patch particles [64]. A number of fascinating ordered structures have been achieved by utilizing the deformable and anisotropic characteristics of soft Janus and patchy particles, for example, biomimetic helical structures (single helices, double helices, and Bernal spirals), low-coordinated 2D (non-close-packed hexagonal, square, and honeycomb-like arrays and graphene-like structures) and 3D (pyrochlore lattices and diamond lattices) structures. Therefore, soft Janus and patchy particles with deformable and anisotropic characteristics open up new avenues for the design and creation of desired supracolloidal structures. Our soft Janus and patchy particle models provide a powerful tool to investigate the aggregation behavior of soft Janus and patchy particles over larger spatial and temporal scales. These simulation results are helpful for the experimental realization of various supracolloidal structures. Furthermore, the soft Janus and patchy particle models may be used as minimal models to deepen our understanding on the formation of glasses, the collective behavior of living systems, the crystallization of proteins, and so on. These research directions will receive more attention in future studies.

Acknowledgments

This work is subsidized by the National Basic Research Program of China (973 Program, 2012CB821500). The authors also appreciate the financial support from the National Science Foundation of China (21474110, 21474111, 21222407). Z.Y.L. acknowledges the support of the Jilin Province Science and Technology Development Plan (20140519004JH).

References

1. Frenkel, D. (2002) Playing tricks with designer "atoms." *Science*, **296**, 65.
2. Glotzer, S.C. (2004) Some assembly required. *Science*, **306**, 419.
3. Zhang, Z. and Glotzer, S.C. (2004) Self-assembly of patchy particles. *Nano Lett.*, **4**, 1407.
4. Glotzer, S.C. and Solomon, M.J. (2007) Anisotropy of building blocks and their assembly into complex structures. *Nat. Mater.*, **6**, 557.
5. Jones, M.R. and Mirkin, C.A. (2012) Materials science: Self-assembly gets new direction. *Nature*, **491**, 42.
6. Wang, Y., Wang, Y., Breed, D.R., Manoharan, V.N., Feng, L., Hollingsworth, A.D., Weck, M., and Pine, D.J. (2012) Colloids with valence and specific directional bonding. *Nature*, **491**, 51.
7. Pawar, A.B. and Kretzschmar, I. (2010) Fabrication, assembly, and application of patchy particles. *Macromol. Rapid Commun.*, **31**, 150.
8. Bianchi, E., Blaak, R., and Likos, C.N. (2011) Patchy colloids: state of the art and perspectives. *Phys. Chem. Chem. Phys.*, **13**, 6397.
9. Lee, K.J., Yoon, J., and Lahann, J. (2011) Recent advances with anisotropic particles. *Curr. Opin. Colloid Interface Sci.*, **16**, 195.
10. Li, F., Josephson, D.P., and Stein, A. (2011) Colloidal assembly: The road from particles to colloidal molecules and crystals. *Angew. Chem. Int. Ed.*, **50**, 360.
11. Yi, G.R., Pine, D.J., and Sacanna, S. (2013) Recent progress on patchy colloids and their self-assembly. *J. Phys.: Condens. Matter*, **25**, 193101.
12. Hu, J., Zhou, S., Sun, Y., Fang, X., and Wu, L. (2012) Fabrication, properties and applications of Janus particles. *Chem. Soc. Rev.*, **41**, 4356.
13. Walther, A. and Müller, A.H.E. (2013) Janus particles: Synthesis, self-assembly, physical properties, and applications. *Chem. Rev.*, **113**, 5194.
14. Walther, A. and Müller, A.H.E. (2008) Janus particles. *Soft Matter*, **4**, 663.
15. Du, J. and O'Reilly, R.K. (2011) Anisotropic particles with patchy, multicompartment and Janus architectures: Preparation and application. *Chem. Soc. Rev.*, **40**, 2402.

16. Lattuada, M. and Hatton, T.A. (2011) Synthesis, properties and applications of Janus nanoparticles. *Nano Today*, **6**, 286.
17. Sciortino, F., Giacometti, A., and Pastore, G. (2009) Phase diagram of Janus particles. *Phys. Rev. Lett.*, **103**, 237801.
18. Chen, Q., Whitmer, J.K., Jiang, S., Bae, S.C., Luijten, E., and Granick, S. (2011) Supracolloidal reaction kinetics of Janus spheres. *Science*, **331**, 199.
19. Yan, J., Bloom, M., Bae, S.C., Luijten, E., and Granick, S. (2012) Linking synchronization to self-assembly using magnetic Janus colloids. *Nature*, **491** (7425), 578.
20. Romano, F. and Sciortino, F. (2011) Colloidal self-assembly: Patchy from the bottom up. *Nat. Mater.*, **10**, 171.
21. Morgan, J.W.R., Chakrabarti, D., Dorsaz, N., and Wales, D.J. (2013) Designing a bernal spiral from patchy colloids. *ACS Nano*, **7**, 1246.
22. Olesen, S.W., Fejer, S.N., Chakrabarti, D., and Wales, D.J. (2013) A left-handed building block self-assembles into right- and left-handed helices. *RSC Adv.*, **3**, 12905.
23. Fejer, S.N., Chakrabarti, D., Kusumaatmaja, H., and Wales, D.J. (2014) Design principles for Bernal spirals and helices with tunable pitch. *Nanoscale*, **6**, 9448.
24. Guo, R., Mao, J., Xie, X.M., and Yan, L.T. (2014) Predictive supracolloidal helices from patchy particles. *Sci. Rep.*, **4**, 7021.
25. Chen, Q., Bae, S.C., and Granick, S. (2011) Directed self-assembly of a colloidal kagome lattice. *Nature*, **469**, 381.
26. Chen, Q., Diesel, E., Whitmer, J.K., Bae, S.C., Luijten, E., and Granick, S. (2011) Triblock colloids for directed self-assembly. *J. Am. Chem. Soc.*, **133**, 7725.
27. Romano, F. and Sciortino, F. (2011) Two-dimensional assembly of triblock Janus particles into crystal phases in the two bond per patch limit. *Soft Matter*, **7**, 5799.
28. Chapela, G.A., Guzmán, O., Martínez-González, J.A., Díaz-Leyva, P., and Quintana-H, J. (2014) Self-assembly of kagome lattices, entangled webs and linear fibers with vibrating patchy particles in two dimensions. *Soft Matter*, **10**, 9167.
29. Chen, Q., Bae, S.C., and Granick, S. (2012) Staged self-assembly of colloidal metastructures. *J. Am. Chem. Soc.*, **134**, 11080.
30. Cates, M.E. (2013) Patchy colloids: Entropy stabilizes open crystals. *Nat. Mater.*, **12**, 179.
31. Mao, X. (2013) Entropic effects in the self-assembly of open lattices from patchy particles. *Phys. Rev. E*, **87**, 062319.
32. Mao, X., Chen, Q., and Granick, S. (2013) Entropy favours open colloidal lattices. *Nat. Mater.*, **12**, 217.
33. Romano, F. and Sciortino, F. (2012) Patterning symmetry in the rational design of colloidal crystals. *Nat. Commun.*, **3**, 975.
34. van der Linden, M.N., Doye, J.P.K., and Louis, A.A. (2012) Formation of dodecagonal quasicrystals in two-dimensional systems of patchy particles. *J. Chem. Phys.*, **136**, 054904.
35. Reinhardt, A., Romano, F., and Doye, J.P.K. (2013) Computing phase diagrams for a quasicrystal-forming patchy-particle system. *Phys. Rev. Lett.*, **110**, 255503.
36. Jiang, S., Yan, J., Whitmer, J.K., Anthony, S.M., Luijten, E., and Granick, S. (2014) Orientationally glassy crystals of Janus spheres. *Phys. Rev. Lett.*, **112**, 218301.

37. Smallenburg, F., Leibler, L., and Sciortino, F. (2013) Patchy particle model for vitrimers. *Phys. Rev. Lett.*, **111**, 188002.
38. Smallenburg, F. and Sciortino, F. (2013) Liquids more stable than crystals in particles with limited valence and flexible bonds. *Nat. Phys.*, **9**, 554.
39. Romano, F., Russo, J., and Tanaka, H. (2014) Influence of patch-size variability on the crystallization of tetrahedral patchy particles. *Phys. Rev. Lett.*, **113**, 138303.
40. Doye, J.P.K., Louis, A.A., Lin, I.C., Allen, L.R., Noya, E.G., Wilber, A.W., Kok, H.C., and Lyus, R. (2007) Controlling crystallization and its absence: Proteins, colloids and patchy models. *Phys. Chem. Chem. Phys.*, **9**, 2197.
41. Liu, H., Kumar, S.K., and Douglas, J.F. (2009) Self-assembly-induced protein crystallization. *Phys. Rev. Lett.*, **103**, 018101.
42. Fusco, D. and Charbonneau, P. (2013) Crystallization of asymmetric patchy models for globular proteins in solution. *Phys. Rev. E*, **88**, 012721.
43. Huisman, B.A.H., Bolhuis, P.G., and Fasolino, A. (2008) Phase transition to bundles of flexible supramolecular polymers. *Phys. Rev. Lett.*, **100**, 188301.
44. Yang, Y., Meyer, R.B., and Hagan, M.F. (2010) Self-limited self-assembly of chiral filaments. *Phys. Rev. Lett.*, **104**, 258102.
45. Wurm, F. and Kilbinger, A.F.M. (2009) Polymeric Janus particles. *Angew. Chem. Int. Ed.*, **48**, 8412.
46. Nie, Z., Li, W., Seo, M., Xu, S., and Kumacheva, E. (2006) Janus and ternary particles generated by microfluidic synthesis: Design, synthesis, and self-assembly. *J. Am. Chem. Soc.*, **128**, 9408.
47. Lattuada, M. and Hatton, T.A. (2007) Preparation and controlled self-assembly of Janus magnetic nanoparticles. *J. Am. Chem. Soc.*, **129**, 12878.
48. Erhardt, R., Zhang, M., Böker, A., Zettl, H., Abetz, C., Frederik, P., Krausch, G., Abetz, V., and Müller, A.H.E. (2003) Amphiphilic Janus micelles with polystyrene and poly(methacrylic acid) hemispheres. *J. Am. Chem. Soc.*, **125**, 3260.
49. Srinivas, G. and Pitera, J.W. (2008) Soft patchy nanoparticles from solution-phase self-assembly of binary diblock copolymers. *Nano Lett.*, **8**, 611.
50. Holmes, R.M. and Williams, D.R.M. (2011) Single-chain asymmetric block copolymers in poor solvents: Candidates for patchy colloids. *Macromolecules*, **44**, 6172.
51. Gröschel, A.H., Schacher, F.H., Schmalz, H., Borisov, O.V., Zhulina, E.B., Walther, A., and Müller, A.H.E. (2012) Precise hierarchical self-assembly of multicompartment micelles. *Nat. Commun.*, **3**, 710.
52. Gröschel, A.H., Walther, A., Löbling, T.I., Schmelz, J., Hanisch, A., Schmalz, H., and Müller, A.H.E. (2012) Facile, solution-based synthesis of soft, nanoscale Janus particles with tunable Janus balance. *J. Am. Chem. Soc.*, **134**, 13850.
53. Gröschel, A.H., Walther, A., Löbling, T.I., Schacher, F.H., Schmalz, H., and Müller, A.H.E. (2013) Guided hierarchical co-assembly of soft patchy nanoparticles. *Nature*, **503**, 247.
54. Capone, B., Coluzza, I., LoVerso, F., Likos, C.N., and Blaak, R. (2012) Telechelic star polymers as self-assembling units from the molecular to the macroscopic scale. *Phys. Rev. Lett.*, **109**, 238301.
55. Capone, B., Coluzza, I., Blaak, R., Verso, F.L., and Likos, C.N. (2013) Hierarchical self-assembly of telechelic star polymers: From soft patchy particles to gels and diamond crystals. *New J. Phys.*, **15**, 095002.

56. Bradley, M. and Rowe, J. (2009) Cluster formation of Janus polymer microgels. *Soft Matter*, **5**, 3114.
57. Suzuki, D., Tsuji, S., and Kawaguchi, H. (2007) Janus microgels prepared by surfactant-free Pickering emulsion-based modification and their self-assembly. *J. Am. Chem. Soc.*, **129**, 8088.
58. Percec, V., Wilson, D.A., Leowanawat, P., Wilson, C.J., Hughes, A.D., Kaucher, M.S., Hammer, D.A., Levine, D.H., Kim, A.J., Bates, F.S., Davis, K.P., Lodge, T.P., Klein, M.L., DeVane, R.H., Aqad, E., Rosen, B.M., Argintaru, A.O., Sienkowska, M.J., Rissanen, K., Nummelin, S., and Ropponen, J. (2010) Self-assembly of Janus dendrimers into uniform dendrimersomes and other complex architectures. *Science*, **328**, 1009.
59. Peterca, M., Percec, V., Leowanawat, P., and Bertin, A. (2011) Predicting the size and properties of dendrimersomes from the lamellar structure of their amphiphilic Janus dendrimers. *J. Am. Chem. Soc.*, **133**, 20507.
60. Liu, Y., Yu, C., Jin, H., Jiang, B., Zhu, X., Zhou, Y., Lu, Z., and Yan, D. (2013) A supramolecular Janus hyperbranched polymer and its photoresponsive self-assembly of vesicles with narrow size distribution. *J. Am. Chem. Soc.*, **135**, 4765.
61. Li, Z.W., Lu, Z.Y., Sun, Z.Y., and An, L.J. (2012) Model, self-assembly structures, and phase diagram of soft Janus particles. *Soft Matter*, **8**, 6693.
62. Li, Z.W., Lu, Z.Y., Zhu, Y.L., Sun, Z.Y., and An, L.J. (2013) A simulation model for soft triblock Janus particles and their ordered packing. *RSC Adv.*, **3**, 813.
63. Li, Z.W., Lu, Z.Y., and Sun, Z.Y. (2014) Soft Janus particles: Ideal building blocks for template-free fabrication of two-dimensional exotic nanostructures. *Soft Matter*, **10**, 5472.
64. Li, Z.W., Zhu, Y.L., Lu, Z.Y., and Sun, Z.Y. (2016) A versatile model for soft patchy particles with various patch arrangements. *Soft Matter*, **12**, 741.
65. Angell, C.A. and Ueno, K. (2009) Materials science: Soft is strong. *Nature*, **462**, 45.
66. Likos, C.N. (2006) Soft matter with soft particles. *Soft Matter*, **2**, 478.
67. Heyes, D.M. and Brańka, A.C. (2009) Interactions between microgel particles. *Soft Matter*, **5**, 2681.
68. Yan, L.T., Popp, N., Ghosh, S.K., and Böker, A. (2010) Self-assembly of Janus nanoparticles in diblock copolymers. *ACS Nano*, **4**, 913.
69. Nguyen, T.D., Phillips, C.L., Anderson, J.A., and Glotzer, S.C. (2011) Rigid body constraints realized in massively parallel molecular dynamics on graphics processing units. *Comput. Phys. Commun.*, **182**, 2307.
70. Zhang, Z., Keys, A.S., Chen, T., and Glotzer, S.C. (2005) Self-assembly of patchy particles into diamond structures through molecular mimicry. *Langmuir*, **21**, 11547.
71. Bianchi, E., Largo, J., Tartaglia, P., Zaccarelli, E., and Sciortino, F. (2006) Phase diagram of patchy colloids: Towards empty liquids. *Phys. Rev. Lett.*, **97**, 168301.
72. Sciortino, F., Bianchi, E., Douglas, J.F., and Tartaglia, P. (2007) Self-assembly of patchy particles into polymer chains: A parameter-free comparison between Wertheim theory and Monte Carlo simulation. *J. Chem. Phys.*, **126**, 194903.
73. Romano, F., Tartaglia, P., and Sciortino, F. (2007) Gas-liquid phase coexistence in a tetrahedral patchy particle model. *J. Phys.: Condens. Matter*, **19**, 322101.

74. Kern, N. and Frenkel, D. (2003) Fluid–fluid coexistence in colloidal systems with short-ranged strongly directional attraction. *J. Chem. Phys.*, **118**, 9882.

75. Giacometti, A., Lado, F., Largo, J., Pastore, G., and Sciortino, F. (2009) Phase diagram and structural properties of a simple model for one-patch particles. *J. Chem. Phys.*, **131**, 174114.

76. Giacometti, A., Lado, F., Largo, J., Pastore, G., and Sciortino, F. (2010) Effects of patch size and number within a simple model of patchy colloids. *J. Chem. Phys.*, **132**, 174110.

77. Romano, F., Sanz, E., and Sciortino, F. (2010) Phase diagram of a tetrahedral patchy particle model for different interaction ranges. *J. Chem. Phys.*, **132**, 184501.

78. Sciortino, F., Giacometti, A., and Pastore, G. (2010) A numerical study of one-patch colloidal particles: From square-well to Janus. *Phys. Chem. Chem. Phys.*, **12**, 11869.

79. Romano, F., Sanz, E., and Sciortino, F. (2011) Crystallization of tetrahedral patchy particles in silico. *J. Chem. Phys.*, **134**, 174502.

80. Klein, M.L. and Shinoda, W. (2008) Large-scale molecular dynamics simulations of self-assembling systems. *Science*, **321**, 798.

81. Bianchi, E., Capone, B., Kahl, G., and Likos, C.N. (2015) Soft-patchy nanoparticles: modeling and self-organization. *Faraday Discuss.*, DOI: 10.1039/C4FD00271G.

82. Zhu, Y.L., Liu, H., Li, Z.W., Qian, H.J., Milano, G., and Lu, Z.Y. (2013) GALAMOST: GPU-accelerated large-scale molecular simulation toolkit. *J. Comput. Chem.*, **34**, 2197.

83. Groot, R.D. and Warren, P.B. (1997) Dissipative particle dynamics: Bridging the gap between atomistic and mesoscopic simulation. *J. Chem. Phys.*, **107**, 4423.

84. Allen, M.P. and Tildesley, D.J. (1987) *Computer Simulation of Liquids*, Clarendon Press, Oxford.

85. Berendsen, H.J.C., Postma, J.P.M., van Gunsteren, W.F., DiNola, A., and Haak, J.R. (1984) Molecular dynamics with coupling to an external bath. *J. Chem. Phys.*, **81**, 3684.

86. Miller, T.F., Eleftheriou, M., Pattnaik, P., Ndirango, A., Newns, D., and Martyna, G.J. (2002) Symplectic quaternion scheme for biophysical molecular dynamics. *J. Chem. Phys.*, **116**, 8649.

87. Frenkel, D. and Smit, B. (2002) *Understanding Molecular Simulations*, 2nd edition, Academic Press, Waltham, MA.

88. Groot, R.D. and Stoyanov, S.D. (2008) Mesoscopic model for colloidal particles, powders, and granular solids. *Phys. Rev. E*, **78**, 051403.

89. Groot, R.D. and Stoyanov, S.D. (2010) Equation of state of surface-adsorbing colloids. *Soft Matter*, **6**, 1682.

90. Berendsen, H.J.C., van der Spoel, D., and van Drunen, R. (1995) GROMACS: A message-passing parallel molecular dynamics implementation. *J. Comput. Phys.*, **91**, 43.

91. Plimpton, S. (1995) Fast parallel algorithms for short-range molecular dynamics. *J. Comput. Phys.*, **117**, 1.

92. Götz, A.W., Williamson, M.J., Xu, D., Poole, D., Grand, S.L., and Walker, R.C. (2012) Routine microsecond molecular dynamics simulations with AMBER on GPUs. 1. Generalized Born. *J. Chem. Theory Comput.*, **8**, 1542.

93. Salomon-Ferrer, R., Götz, A.W., Poole, D., Grand, S.L., and Walker, R.C. (2013) Routine microsecond molecular dynamics simulations with AMBER on GPUs. 2. Explicit solvent particle mesh Ewald. *J. Chem. Theory Comput.*, **9**, 3878.

94. Rovigatti, L., Šulc, P., Reguly, I.Z., and Romano, F. (2015) A comparison between parallelization approaches in molecular dynamics simulations on GPUs. *J. Comput. Chem.*, **36**, 1.

95. Zou, Q.Z., Li, Z.W., Lu, Z.Y., and Sun, Z.Y. (2016) Supracolloidal helices from soft Janus particles by tuning the particle softness. *Nanoscale*, **8**, 4070.

96. Müller, M. and Daoulas, K.C. (2008) Calculating the free energy of self-assembled structures by thermodynamic integration. *J. Chem. Phys.*, **128**, 024903.

97. Norizoe, Y., Daoulas, K.C., and Müller, M. (2010) Measuring excess free energies of self-assembled membrane structures. *Faraday Discuss.*, **144**, 369.

98. Zhang, J., Lu, Z.Y., and Sun, Z.Y. (2013) Self-assembly structures of amphiphilic multiblock copolymer in dilute solution. *Soft Matter*, **9**, 1947.

99. Allen, M.J., Tung, V.C., and Kaner, R.B. (2010) Honeycomb carbon: A review of graphene. *Chem. Rev.*, **110**, 132.

100. Xu, M., Liang, T., Shi, M., and Chen, H. (2013) Graphene-like two-dimensional materials. *Chem. Rev.*, **113**, 3766.

101. Noya, E.G., Vega, C., Doye, J.P.K., and Louis, A.A. (2010) The stability of a crystal with diamond structure for patchy particles with tetrahedral symmetry. *J. Chem. Phys.*, **132**, 234511.

102. Hartl, A., Schmich, E., Garrido, J.A., Hernando, J., Catharino, S.C.R., Walter, S., Feulner, P., Kromka, A., Steinmuller, D., and Stutzmann, M. (2004) Protein-modified nanocrystalline diamond thin films for biosensor applications. *Nat. Mater.*, **3**, 736.

103. Aharonovich, I., Greentree, A.D., and Prawer, S. (2011) Diamond photonics. *Nat. Photon.*, **5**, 397.

6

Molecular Models for Hepatitis B Virus Capsid Formation, Maturation, and Envelopment

Jehoon Kim[1] *and Jianzhong Wu*[2]

[1] *Department of Chemical Engineering, Massachusetts Institute of Technology, Cambridge, USA*
[2] *Department of Chemical and Environmental Engineering, University of California Riverside, USA*

6.1 Introduction

Hepatitis B virus (HBV) is an infectious pathogen affecting many people throughout the world [1]. According to the World Health Organization (WHO) [2], over 240 million people alive today are chronically infected with HBV, placing them at high risk of liver cirrhosis (scarring), hepatocellular carcinoma (liver cancer), and liver failure [3]. The prevalence of chronic HBV infection is particularly high in sub-Saharan Africa and East Asia (between 5% and 10% of the adult population). Even in the US, more than one million people are chronically infected with HBV, and up to 40 000 Americans are becoming HBV carriers each year [4]. The virus is responsible for up to 80% of all cases of hepatocellular carcinoma worldwide, second only to tobacco among known human carcinogens. Every year, there are over 4 million acute clinical cases of HBV infection globally, and about 780 000 persons die from HBV-induced chronic active hepatitis, cirrhosis, or primary liver cancer.

Although HBV infection is now preventable through vaccination, existing vaccines have no therapeutic effect on the chronically infected population. Current treatment is mostly based on antiviral cytokine interferon alpha (IFN-α) and reverse transcriptase inhibitors [5, 6]. Interferon therapy is effective in about 20%–40% of patients with rather unpredictable efficacy, and long-term treatment is often accompanied by strong side effects. The HBV-polymerase has been a well-recognized target for inhibiting the viral replication and several nucleos(t)ide-analogs (e.g., tenofovir and entecavir) have been developed as antiviral agents [7–10]. These drugs are effective in suppressing HBV replication with little toxicity and adverse effect. However, current treatment does not cure hepatitis B infection completely in most people and requires sustained medication to prevent the virus rebound. Tremendous studies have been devoted to improving the drug efficacy and overcoming potential resistance [11, 12]. Because HBV mutates fast in a changing environment, drug resistance is almost inevitable for any therapy that targets a specific motif of the viral genome or the reverse transcriptase. Alternative

Self-Assembling Systems: Theory and Simulation, First Edition. Edited by Li-Tang Yan.

antiviral strategies have been pursued in recent years including non-interferon immune enhancers and therapeutics improving the vaccine efficiency [1].

Development of more effective therapeutic strategies for HBV infection hinges on a better understanding of the viral structure and the morphogenesis from a system's perspective. As shown schematically in Figure 6.1, the HBV virion (*viz.*, "Dane particle" [13]) is a double-shelled particle made of an icosahedral nucleocapsid (NC) and a spherical shell of lipoproteins [14–16]. The whole particle measures 42 nm in the outer diameter, enclosing a 36 nm-diameter protein capsid and a 26 nm-diameter genomic core. The envelope is composed of the HBV surface proteins embedded in a lipid bilayer (Figure 6.2). These lipoproteins, known as HBV surface antigen (HBsAg), are conventionally used as a target of diagnosis for the viral infection.

The envelope proteins exist in three interrelated forms derived from a single open reading frame of the HBV genome. These proteins are distinguished according to their sizes: small (S), middle (M), and large (L). The S protein (S-HBs) is the most abundant; it consists of 226 amino-acid (aa) residues that can be divided into four trans-membrane domains (TM1-4). M and L proteins are augmented with the extended N-terminal domains (Pre-S1 and Pre-S2) after a common S domain. Each M-protein (M-HBs) is formed by the addition of Pre-S2 domain (55 aa) to the N-terminus of S-HBs, and

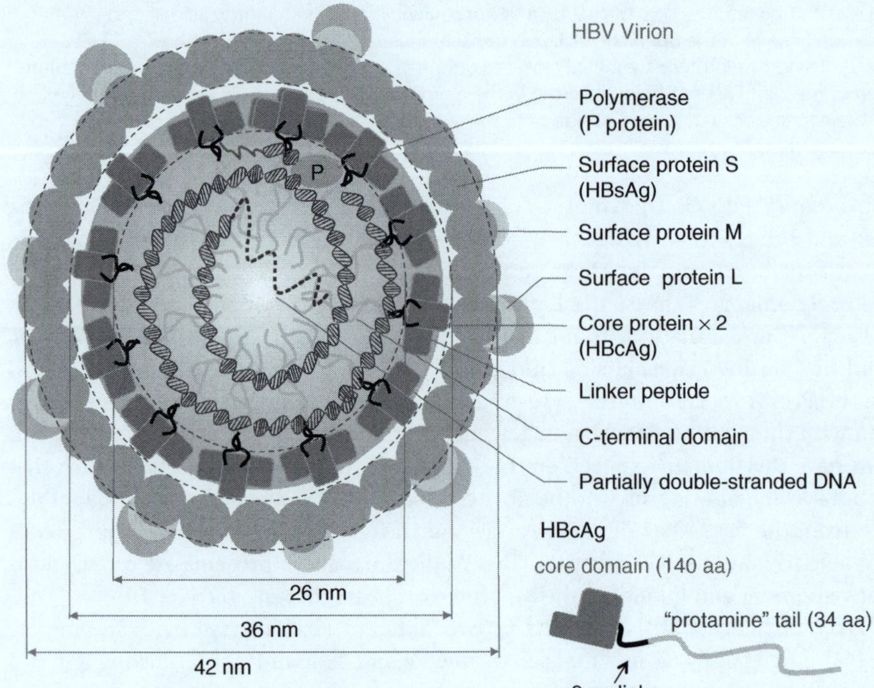

Figure 6.1 A schematic representation of an HBV virion. The virus has a double-shelled structure containing an icosahedral nucleocapsid, about 36 nm in outer diameter, coated with three interrelated lipoproteins (S, M, L) known as the HBV surface antigens (HBsAg). Inside the nucleocapsid is a partially double strained DNA in strong binding with the highly flexible and positively charged C-terminal domains of the capsid proteins.

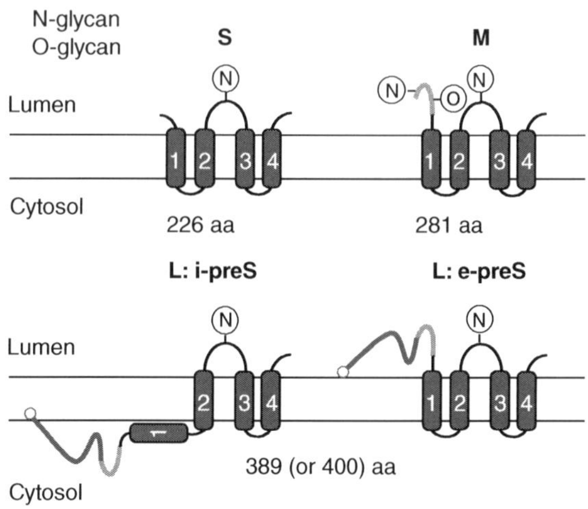

Figure 6.2 HBV lipoproteins (S, M, L) share a common S domain with different N-terminal extensions. The S protein consists of 226 amino acid (aa) residues; it is mostly imbedded in the lipid bilayer and can be divided into four trans-membrane (TM) regions. The M protein is N-terminally extended from the S protein with the Pre-S2 domain (with 55 aa). The L protein carries an additional N-terminal Pre-S1 domain (gray), which consists of 108 or 119 aa depending on the genotype. The N-terminus of the L protein is anchored into the lipid bilayer by a 14-carbon chain of saturated fatty acid added by myristylation (shown as the open circle). Upon translational membrane insertion, the Pre-S1 and Pre-S2 regions of the L protein are initially located on the cytosolic side of the endoplasmic reticulum (ER) membrane with TM1 not being inserted in the membrane (L: i-preS). In HBV virions, about half of the L proteins translocate the Pre-S region onto the viral surface (L: e-preS). Reproduced from Ref. [17] with permission from Springer.

the L-protein (L-HBs) is an extension of both Pre-S1 (108 aa or 119 depending on the strains) and Pre-S2 to S-HBs. The S proteins are mostly embedded in the lipid and represent a major component of the viral envelope.

The Pre-S domains of the M and L proteins are flexible and not embedded in the lipid membrane. After translation, about half of the Pre-S domains are on the cytoplasmic side and half on the cisternal side (lumen) of the endoplasmic reticulum (ER) membrane. The Pre-S1 located on the cytosolic side (outer) is required for surface proteins binding with the viral nucleocapsids. Recently, it has been shown that about 50% of L proteins have this domain exposed on the surface of infectious virus particles such that it functions as the binding site for the host cell receptor [18]. In particular, lipopeptides derived from the first 48 aa of Pre-S1 are essential for HBV entering human liver cells and are able to inhibit HBV infection [18]. While the S and L proteins are required for HBV envelopment and infection, the M-protein appears dispensable [15, 19].

The HBV nucleocapsid (NC) exists in two different sizes *in vivo*, with the majority (>90%) having a $T = 4$ icosahedral symmetry and a small fraction having a $T = 3$ symmetry [20]. The T4 capsid is composed of 120 capsomers, i.e., dimers of the capsid protein known as hepatitis B core antigen (HBcAg) or core proteins [21]. These capsomers are arranged symmetrically on the triangulation lattices of the icosahedral surfaces (Figure 6.5). The HBV capsids are fenestrated, with hydrophilic pores 1.4 nm in diameter around the 3-fold axes; hydrophobic oval pores, about 2.4 nm long and 1.8 nm wide, at the 2-fold axes; and smaller pores 0.3 nm in diameter at the positions of the

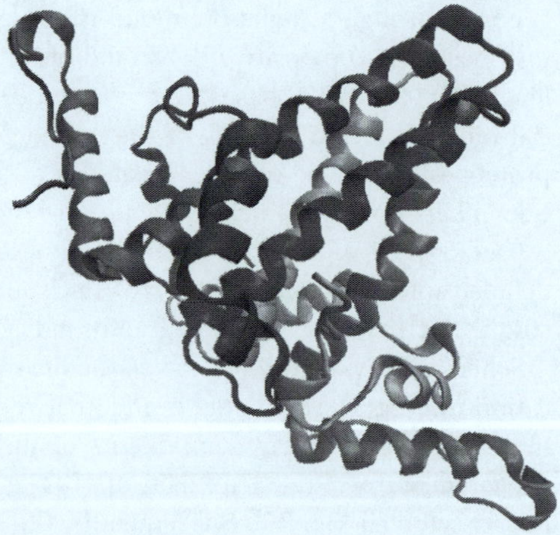

Figure 6.3 The structure of the N-terminal region of the HBV capsid protein (HBcAg 1–149) in its dimer form (i.e., capsomer). The capsid dimer has a T-shaped structure stabilized by strong hydrophobic attractions between α-helical hairpins.

5-fold icosahedral axes [22]. These pores make nucleotides and small chemical species directly accessible to the capsid lumen.

In wild type (WT) HBV, each capsid protein comprises 183 aa residues, which is separated into an N-terminal assembly domain (aa 1–140), a nonapeptide linker (aa 140–149), and a highly basic, protamine-like C-terminal domain (CTD, aa 150–183). The assembly domain consists of the first 140 aa residues from the N terminus and exists as a dimer (i.e., capsomer) right after synthesis. Figure 6.3 shows the dimer structure as determined from high resolution cryo-EM [23] and X-ray crystallography [24] measurements. Each monomer consisting of a protruding α-helical hairpin (aa 50–110), with the lower part surrounded by three shorter helixes (aa 13–17, aa 27–43, and aa 112–127), an irregular N-terminal region (aa 1–12), a proline-rich loop (aa 128–136), and an extended strand (aa 137–140). The C-terminus of the capsid protein is not seen in the cryo-EM images. It includes the flexible nonapeptide linker (aa 141–149) and the CTD tail that is rich in positively charged arginine residues (aa 150–183). The linker and CTD tail are highly flexible and may be located on either side of the capsid [25, 26].

The capsid proteins are dimerized right after synthesis, driven by the strong association of two α-helical hairpins into a compact four-helix bundle [27]. The dimer has a T-shaped form stabilized by hydrophobic attractions between α-helical hairpins and a disulfide linking between C61 of each monomer. Recent cryo-EM structure shows that the disulfide bond is not indispensable for the capsomer formation [25]. While mutations or truncations at the assembly domain impair capsomer formation, CTD chains play a critical role in HBV morphogenesis, particularly during NC formation, RNA packaging, and DNA synthesis.

Inside, each T4 HBV capsid contains a genome that is made of a relaxed circular (rc) [28], partially double-stranded (ds) DNA, approximately 3200 base pairs (bp) in length. The negative (−) strand of dsDNA, i.e., the one with opposite polarity to the mRNA, encodes four overlapping reading frames for the synthesis of HBV surface proteins, capsid (or core) proteins, X proteins, and the viral polymerase (P). While the multifunctional P protein (832 or 854 aa depending on the strains) is indispensable for viral replication, the X protein (154 aa) plays an important role modulating the remodeling of the HBV genome through acetylation, mono-methylation, and phosphorylation of DNA-bound histones [29]. The positive (+) strand of the DNA genome is incomplete and varies in length (1800 to 2700 bases). As revealed in cryo-EM images [30], the genomic material binds to the CTD tails of the capsid proteins, resulting in a distinctive band of protein – nucleic acid complex located between radii 9 – 12.5 nm inside the NC. The peptide linkers provide radial connections of the CTD tails to the inner surface of the viral capsid [31]. Because of the flexibility of these highly hydrophilic macromolecules, more detailed internal structures of HBV NCs are not amenable to conventional structural studies such as X-ray crystallography and cryo-EM.

Schematically, Figure 6.4 shows the life cycle of HBV infection [32]. The viral replication undergoes sequential events that require spatially and temporally coordinated interactions of the key components of the infectious virion [33]. The viral entry is triggered by the Pre-S1 domain of HBsAg binding to a receptor at the surface of a hepatocyte cell [34, 35]. Subsequently, the entire virion is internalized by membrane fusion [16]. After entering the liver cell, the surface proteins are uncoated by the lipid molecules while the nucleocapsid (NC) migrates into the cytosol of the infected cell. The NC particle is somehow destabilized, possibly due to the change of local environment

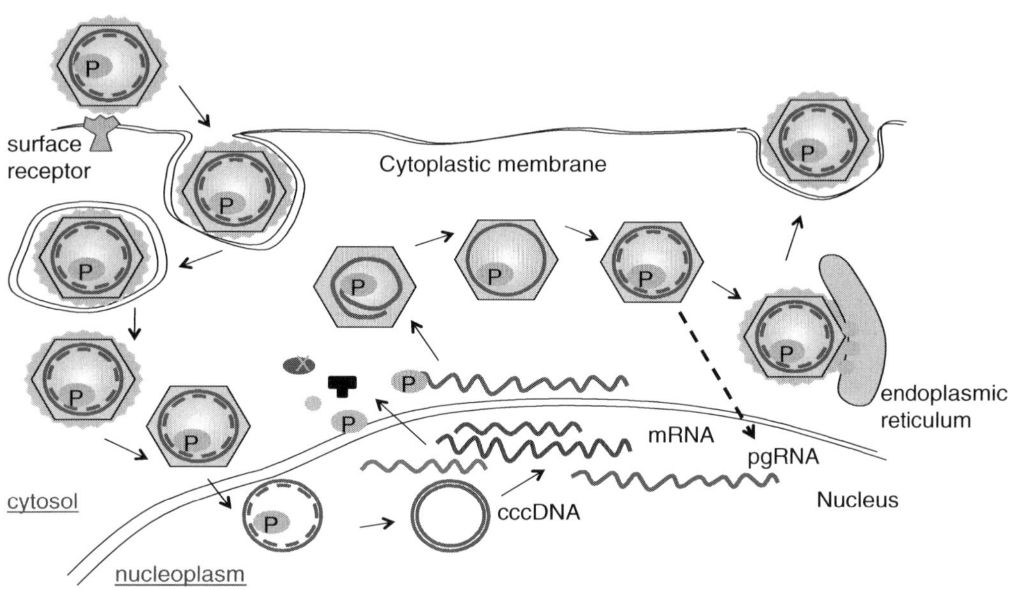

Figure 6.4 Schematic of the HBV replication life cycle. The viral entry into human hepatocytes is initiated by the Pre-S1 domain of HBsAg binding to a receptor that is normally involved in bile acid transport in the liver. After being transferred into the cytosol, the nucleocapsid is disassembled, leading to the release of the genomic material. Viral replication starts with the formation of a covalently closed circular DNA (cccDNA) in the nucleoplasm; the cccDNA templates RNA transcription and subsequently protein synthesis. Nucleocapsid (NC) assembly is driven by strong electrostatic interactions between capsid proteins and pregenome (pg) RNA. Inside NC, the polymerase (P) translates pgRNA first into a single-stranded (ss) DNA, and then a partially double-stranded (ds) (>50%) DNA. The matured capsid is then enveloped with surface proteins in the endoplasmic reticulum (ER) or migrates back to the nucleoplasm for cccDNA amplification. *(See insert for color representation of the figure.).*

(e.g., pH decreases) during the transport of the viral genome to the cellular nucleus. The detailed mechanisms for NC disassembly are not yet well understood. The conformational change to the nucleocapsid results in deproteinization of rc-DNA and triggers the nuclear localization signal (NLS) for subsequent rc-DNA nuclear import [36]. Within the nucleoplasm, the rc-DNA is first converted into a plasmid-like, covalently closed circular (ccc) DNA [37]. Clearance of cccDNA remains challenging for existing antiviral drugs. In recent years, it has been a major target for the development of new therapeutic strategies [1].

cccDNA serves as the transcriptional template to produce multiple copies of pregenomic (pg) and subgenomic RNA chains through RNA polymerase II of the host cell [38]. After the transcriptions, the mRNAs are exported into the cytoplasm of the infected cell and translated into the HBV surface, core, X, and P proteins. In different reading frames, the pgRNA (about 3500 nucleotides in total length) acts as the mRNA for the synthesis of both the core and P proteins. Self-assembly of HBV capsids occurs when the core proteins accumulate above a certain threshold level, concomitant to the formation of P protein and pgRNA complexes that are encapsidated in the progeny NCs [39, 40].

Inside a newly formed (immature) NC particle, the pgRNA serves as the template for the synthesis of a complete (−) strand DNA [41, 42]. While the pgRNA is degraded during the (−) DNA synthesis, the (+) strand DNA is formed with the (−) strand DNA as the template. For reasons yet unknown, the synthesis of (+) strand DNA is incomplete during the replication process. Recent experiments uncovered a critical role of electrostatic interactions and CTD phosphorylation in NC maturation [43–45]. However, the molecular basis for this essential regulation remains obscure. Depending on the concentration of envelope proteins, the matured NCs can be transported back to the nucleoplasm for intracellular amplification of cccDNA or bud into the ER membrane of the host cell for envelopment with the viral surface proteins [15]. An intricate aspect of the hepadnavirus life cycle is that only nucleocapsids containing dsDNA and a large excess of empty capsids can be coated with envelope proteins [46]. Nether single-stranded RNA nor DNA containing capsids can bind with the surface proteins or be transported into the nucleoplasm.

In addition to the virions and empty capsids, HBV replication results in abundant subviral spherical and filamentous particles (SVP) that contain no viral genome. The spherical SVP, about 22 nm in diameter, are predominately composed of S-HBs, and the filaments are formed because of addition of L-HBs [47]. While SVP outnumber virions in the serum of HBV patients by orders of magnitude, their biological functions and the pathway of self-assembly are not yet known. It has been suggested that SVP contribute to a persistent infection by enhancing immune tolerance [48]. SVP induce shared immunological responses with the HBV virions, and thus can be used as effective immunogens for antiviral treatments. For example, the current (the second generation) human vaccine for HBV was developed by expression of S-HBs particles in yeast [49].

Capsid formation, genome maturation and viral envelopment represent three critical steps in the HBV life cycle. A theoretical description of such processes from a molecular perspective may help better understand viral replication and morphogenesis, leading to fresh insights into the experimental studies of antiviral strategies. The main purpose of this review is to summarize our recent theoretical work concerning the microscopic structure and physiochemical properties of HBV at the various stages of replication.

Specifically, we introduce molecular thermodynamic models for predicting the stability of HBV capsids either with or without loading nucleic materials and for describing key biomolecular forces underlying genome encapsidation and envelopment. We show that HBV capsids are stabilized due to a delicate balance of hydrophobic and electrostatic interactions and thus their stability is sensitive to capsid mutation and solution conditions. The theoretical model is used to examine how the electrostatic interaction between the packaged genome and the highly charged C-terminal domains (CTD) of capsid proteins regulates the nucleocapsid formation. New biophysical schemes are proposed to regulate HBV replication by controlling the charge status and positions of CTD tails. We also investigate probable interaction sites between the Pre-S1 domain of L-HBs and the capsid core proteins (HBc) based on a combination of molecular dynamic (MD) simulations and molecular docking. By addressing fundamental issues related to capsid assembly, genome packaging, and viral maturation, we expect that the thermodynamic models will contribute to a better understanding of various biophysical processes of the HBV life cycle and development of novel antiviral strategies for ultimate eradication of HBV.

6.2 Molecular Thermodynamics of Capsid Formation

HBV infects only primary hepatocytes from humans, chimpanzees, and a primate-like animal called the treeshrew (*Tupaia belangeri*). Toward the development of a more effective treatment of HBV infection, it is critically important to understand the thermodynamic stability of the viral capsids at different stages of replication [50, 51]. Because of the limited availability of primary hepatocytes from permissive hosts, it is extremely difficult to experimentally investigate the energetic profiles of the capsid assembly under an intracellular environment. *In vivo* assembly of HBV is further hampered by the small amount of virus produced in an infected human cell, which has been estimated on the order of 1 – 10 virions per cell per day [52, 53]. Nonetheless, *in vitro* assembly of HBV capsids devoid of CTD and nucleic acids has been analyzed in detail [54, 55]. By describing the self-assembly process in terms of a cascade of low-order association reactions, Zlotnick and coworkers found that the intermediate concentrations were extremely low and the kinetic of self-assembly followed essentially a one-step process [56]. The single-step model allows for the quantification of the thermodynamic stability of empty HBV capsids on the basis of the concentrations of the capsomers and the fully assembled capsids at various solution conditions [57].

Whereas existing experimental analyses of capsid stability were mostly performed on empty capsids devoid of any nucleic materials or the CTD tails, there have been a number of investigations into the role of CTD in HBV replication, such as its relations with the genome size [58] as well as the unknown functions associated with phosphorylation [44]. Several experimental studies have been reported on the *in vitro* assembly of empty or RNA-containing HBV capsids with the WT CTD [59–61]. When compared to that for empty capsids without CTD, the assembly of the full-length capsids occurs at relatively high salt concentration (≥ 0.25 M) to avoid strong electrostatic repulsion among the basic tails. Whereas favorable electrostatic attractions between CTD and RNA chains are expected to make NC more stable than corresponding empty capsids [54, 59–61], NC particles undergo significant stability changes during the replication cycle. It has been suggested that NCs with dsDNA are relatively unstable in comparison to those containing ssRNA or ssDNA [45, 62]. Further evaluation of the NC stability is very desirable, in particular with respect to different genome contents and characteristics.

In recent years, the assembly of viral capsids has been subject to extensive molecular simulation studies (e.g. [63–65]). Whereas simulation advances the fundamentals of viral assembly and dynamics [65–67], the theoretical analysis is mostly based on simple model systems within a relatively narrow range of the parameter space. Simulation of the life cycle of a real virus is computationally challenging, if not impossible, due to the large number of underlying biomacromolecules, complicated intracellular environment, and long duration pertinent to the self-assembly pathways [64]. Little information is currently available for the capsid stability under the physiological conditions.

Built upon previous developments [68–71], we have recently formulated a thermodynamic stability model specific to HBV nucleocapsids [72]. Using coarse-grained models to represent the key viral components, we derived analytical expressions for intermolecular interactions upon self-assembly and complexation of capsid proteins and genome. Although the coarse-grained model captures major non-specific interactions important for the viral formation, by necessity it does not account for the detailed site-specific interactions that are also part of the self-assembly. Overall, the molecular thermodynamic model is able to quantify the free energies of capsid formation for both empty capsids and nucleocapsids in good agreement with experimental data.

6.2.1 Energetics of Viral Assembly

Capsid formation is entropy driven associated with the reorganization of water molecules for hydrophobic interactions and the release of massive counterions due to the binding of the nucleic acids with oppositely charged polypeptides. Experimental investigations on the formation of viral capsids and polyelectrolyte complexes render an excellent starting point for interrogating the molecular driving forces of the capsid assembly processes. Despite decades of theoretical efforts from a fundamental perspective, a microscopic approach is yet to be established to quantify hydrophobic and polyelectrolyte complex interactions. Fortunately, a number of semi-empirical models are readily available to capture the entropic effects responsible for the virus assembly. We use such models in this work to describe hydrophobic and polyelectrolyte interactions underlying the formation of empty and RNA-loaded HBV capsids. While the discussion here is mainly focused on HBV, we expect that the procedures are similarly applicable to capsid formation for other viruses.

6.2.1.1 Rigid Capsids

It was proposed first by Crick and Watson that many small viruses have an icosahedral capsid with near perfect arrangement of protein subunits [73]. The lattice structure gives the repeating subunits an identical environment and allows for a good estimation of the formation free energy simply based on the contact value of the protein–protein interactions [57]. While a thermodynamic model has been proposed before for the formation of empty HBV capsids from the assembly domain of the core proteins (*viz.*, HBcAg without CTD) [74], the icosahedral lattice was not explicitly used in the previous work.

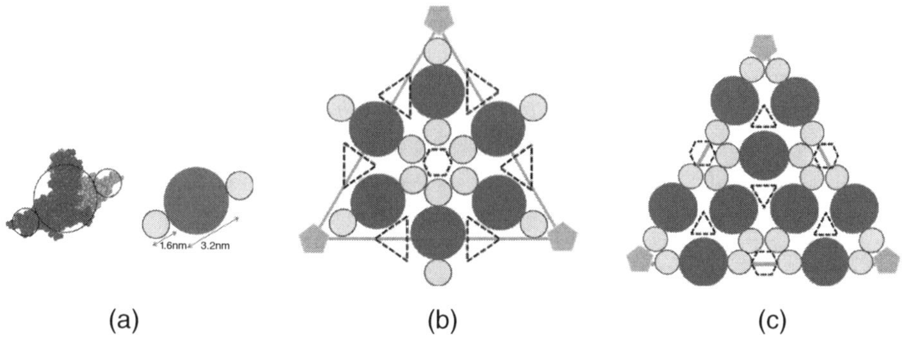

Figure 6.5 A course-grained model for the arrangement of HBV capsomers on T3 and T4 capsids. (a) Each capsomer consists of two core proteins (HBcAg without CTD), depicted as three tangentially connected spheres with their radii determined from the protein crystal structure. (b) A lattice model for the arrangement of capsid proteins at a triangular facet of a T3 capsid. (c) The lattice model for a T4 capsid. In both cases, the major capsid holes are shown as hexagons and triangles.

In our coarse-grained representation of an HBV capsid, we assume that each capsomer has a rigid structure that can be described by three tangentially connected spheres mimicking the topology of HBcAg determined from experiments. Figure 6.5 shows schematically the lattice model for the arrangement of HBV capsomers at the triangular facets of T3 and T4 capsids. The diameters of these coarse-grained segments are selected such that they approximately reproduce the protein volume as determined from the crystal structure [24]. The large sphere of the capsomer, 3.2 nm in diameter, corresponds to the 4-helix bundle of each HBcAg subunit. The large sphere is flanged between two small spheres, each 1.6 nm in diameter, representing the base part of the HBcAg assembly domain (see Figure 6.3). According to the crystalline structure of the HBV capsids and the icosahedral symmetry [75], each subunit has four nearest neighbors in both T3 and T4 capsids, joined through hydrophobic contacts.

The Helmholtz free energy of capsid formation corresponds to the reversible work to bring the subunits from infinitely apart to the T3 or T4 icosahedral lattice. Approximately, the free energy can be decomposed into contributions due to hydrophobic (F_{hyd}), van der Waals (F_{vdw}), and electrostatic (F_{ele}) interactions:

$$\Delta F_0 = F_{hyd} + F_{vdw} + F_{ele}. \tag{6.1}$$

As has been well documented [76], the hydrophobic interaction is short-ranged and can be related to the solvent accessible surface area of the protein subunits. While both van der Waals and electrostatic interactions are longer-ranged, depending on the protein–protein separations and the salt concentration. These longer-ranged potentials are relatively insignificant for the formation of empty HBV capsids because of the low charge density of the assembly domain of the core proteins and the small Hamaker constant for proteins in an aqueous solution [51].

The coarse-grained model for the subunits shown in Figure 6.5 allows us to estimate the hydrophobic free energy without explicitly accounting for the protein surface area. Because the subunits are in contact only through the small spheres, the total hydrophobic free energy of association corresponds to a summation of their contact hydrophobic potentials. According to a recent work on hydrophobic interactions [77], the potential of mean force between two hydrophobic spheres of diameter $\sigma = 1.6\,\text{nm}$ at contact can be estimated from the overlapping volume of the hydration layers:

$$\beta w_{hyd} = -A\sigma_w^2(2\sigma_w + 3_\sigma), \tag{6.2}$$

where $\sigma_W = 2.8$ Å denotes the diameter of a water molecule, and A is a constant reflecting the difference between the entropy of water molecules within the solvation shell and that in the bulk [77]. Based on the solubility of small hydrocarbons in water over the temperature range of interest in this work, we find that parameter A (in units of nm^{-3}) can be linearly correlated with the absolute temperature:

$$A = 0.067T + 4.481. \tag{6.3}$$

Substituting the expression for A into Equation (6.2) gives the reduced hydrophobic free energy for each inter-dimer contact,

$$\beta w_{hyd} = -0.028T - 1.883, \tag{6.4}$$

where $\beta = 1/(k_B\,T)$, k_B is the Boltzmann constant, and T is the absolute temperature.

According to Equation (6.4), the hydrophobic free energy between the neighboring dimers is around −6 kcal/mol at room temperature ($T = 298\,\text{K}$). The value is close to that estimated by Ceres and Zlotnick [57] and comparable to the binding free energy used in molecular simulation of capsid assembly [78]. In a T4 HBV capsid, our coarse-grained model predicts $120 \times 4/2 = 240$ direct contacts between small hydrophobic spheres, which yields a hydrophobic free energy of $F_{hyd} = 240w_{hyd}$. For a T3 capsid, the hydrophobic free energy is $F_{hyd} = 180w_{hyd}$.

We can predict the van der Waals energy between the protein subunits using the standard equations from the colloids literature [79]. Because the solvent-mediated van der Waals force between small spheres is insignificant in comparison to the hydrophobic potential, we may estimate the van der Waals energy from that between the large spheres. For spherical particles much larger than a water molecule, the van der Waals free energy is given by

$$w_{vdw} = -\frac{H\sigma_L}{24d}. \tag{6.5}$$

In Equation (6.5), $\sigma_L = 3.2$ nm is the diameter of the large sphere in our coarse-grained model of the capsomers, H denotes the Hamaker constant, and $d = 0.8$ nm is the surface-to-surface distance between the large spheres. For proteins in water, the Hamaker constant is not subject to large variations [80]. In this work, we assume $H \approx 4 \times 10^{-21}$ J, independent of temperature. Using this Hamaker constant, we calculate the van der Waals energy for the attraction between two neighboring capsomers, about −0.1 kcal/mol. As expected, the contribution of van der Waals energy to the free energy of capsid formation is much smaller than that of the hydrophobic interaction. Nevertheless, each big sphere interacts with four nearest neighbors and, in a T4 HBV capsid, the total van der Waals free energy is $F_{vdw} \approx -40H$. For a T3 capsid, the total van der Waals free energy is $F_{vdw} \approx -30H$.

We follow Kegel and van der Schoot [70] to estimate the electrostatic repulsion among the subunits. In this model, the electrostatic charges are assumed to distribute uniformly kat the surface of a spherical capsid shell. The free energy of surface charging is given by

$$F_{ele} = A_c\,.\,\sigma^2\,.\,l_B\,.\,\kappa^{-1}.\,k_B T, \tag{6.6}$$

where A_c represents the total charged surface area, σ is the charge density, l_B stands for the Bjerrum length, $\kappa^{-1} = 1/\sqrt{8\pi l_B \rho_s}$ is the Debye screening length, and ρ_s is the overall number density of ions in the electrolyte solution. The Bjerrum length depends only on temperature and the solvent dielectric constant:

$$l_B = e^2 \,/\, (4\pi\varepsilon_0\varepsilon_r k_B T), \tag{6.7}$$

where e is the unit charge, ε_0 is the vacuum permittivity, and ε_r is the solvent dielectric constant. Over the temperature range considered in this work (273 K to 318 K), the water dielectric constant can be accurately represented by an empirical quadratic function of the absolute temperature (in Kelvin) [81]:

$$\varepsilon_r(T) = 249.21 - 0.79069T + (0.72997 \times 10^{-3})T^2. \tag{6.8}$$

We estimate the effective charge density from the number of charged residues in the region of α-helical hairpins (50–110 aa) of the capsid proteins. This region forms spikes at the capsid surface and is depicted as a large sphere in our dimer model. At the physiological condition, each native HBV core protein contains eight charged residues in the α-helical hairpin domain [24]. The same number was used in an earlier work for the capsid charge without CTD [70]. We assume that these charges are uniformly distributed at a spherical surface of $R = 15$ nm in radius, approximately equal to the inner radius of the T4 capsid. From the spherical cavity model, we can estimate the parameters σ and A_c.

6.2.1.2 Nucleocapsids

We may extend the coarse-grained model discussed above to formation of nucleocapsids by including additional contributions due to the CTD tails and nucleic acids. Both T3 and T4 capsids may contain ssRNA, ssDNA, or dsDNA, depending on the specific stage of HBV replication. For simplicity, here we consider only T4 capsids containing ssRNA. Similar models can be applied to capsids containing either ssDNA or dsDNA.

We assume that, at the physiological condition, every arginine residue carries one positive charge, every glutamate residue or phosphorylated serine (155th, 162th, and 170th residue) carries one negative charge, and all other residues from the CTD are electrostatically neutral. Accordingly, the CTD chain of a wild type capsid has 16 positive (valence $Z_i = 1$), 4 negative ($Z_i = -1$), and 14 neutral ($Z_i = 0$) amino-acid residues. To describe the molecular excluded volume effects, we describe the CTD chains as tangentially connected hard spheres, each sphere standing for one aa residue. These coarse-grained segments have the same size (with diameter $a_c = 0.5$ nm) but different electrostatic valences. Similarly, the confined RNA chain is represented by a tangentially connected chain of charged spheres with identical diameter and valence, $a_r = 0.75$ nm and $Z_r = -1$, respectively. While recent theoretical studies suggest the importance of the secondary structure of RNA in the formation of nucleocapsids [83, 84], it is known that RNA encapsidation in HBV is not sensitive to the genome origin and that nucleocapsid formation is mainly driven by strong and non-specific RNA–CTD interactions [60]. As a result, we may assume that the RNA sequence, and, subsequently, specific forms of the secondary structure, have little effect on capsid formation. In both wild type and *E. coli* regenerated T4 capsids, the number of RNA segments is about 3400, and the RNA content is slightly less in a T3 nucleocapsid [54].

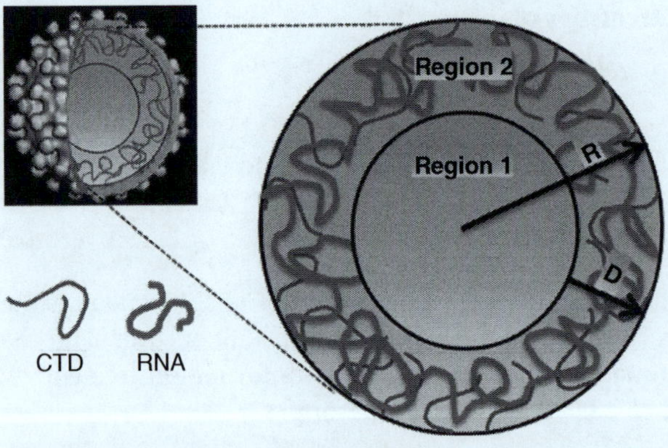

Figure 6.6 A schematic representation of an HBV nucleocapsid containing CTD and RNA chains. Here, D denotes the thickness of Region 2, and $R \sim 12.6$ nm is the radius at the capsid interior wall. The electron-microscopy image of an HBV capsid is adapted from the literature [82].

In comparison to empty capsids, nucleocapsid formation includes an additional free energy due to the change in the conformations of encapsidated chains, the molecular excluded volume effects, the entropy of mixing among macromolecules and small ions, and electrostatic interactions. All these contributions can be estimated from the polymer model shown schematically in Figure 6.6. According to this model, the capsid confining effect is represented by a spherical cavity of radius R. The inner space of the capsid is divided into two regions: Region 1 is free of polymers, and Region 2 consists of the nucleic acids associated with the CTD chains. The division between these two regions, i.e., the thickness of the complex region D, can be determined by minimization of the complexation free energy,

$$\frac{\partial F_{plex}(D)}{\partial D} = 0. \tag{6.9}$$

The electrolyte solution of small ions in Region 1 is fully equilibrated with the surroundings and makes negligible contribution to the free energy of capsid formation. For RNA and CTD chains in Region 2, the concentrations of cations and anions, $\rho_2{}^+$ and $\rho_2{}^-$, are defined by the Donnan equilibrium:

$$\rho_2{}^+ \rho_2{}^- = \rho_0{}^2, \tag{6.10}$$

$$\rho_2{}^+ - \rho_2{}^- - \rho_p = 0, \tag{6.11}$$

where ρ_p stands for the total charge density of RNA and CTD chains, and ρ_0 denotes the overall number densities of small ions in Region 1. The total charge density of the biomacromolecules is related to their molecular characteristics and the volume of Region 2:

$$V_2 = 4\pi(R^2D - RD^2 + D^3/3). \tag{6.12}$$

The free energy of complex formation reflects the difference between the free energy of the confined polymer chains and those corresponding to isolated RNA and CTD chains at infinite dilution. Approximately, this free energy includes contributions due to the changes in polymer elastic energies (*elas*), segment excluded volume effects (*exc*), electrostatic interactions, and the entropy of mixing [85]:

$$F_{plex} = \Delta F_r^{\,elas} + \Delta F_c^{\,elas} + \Delta F^{exc} + \Delta F^{els}. \tag{6.13}$$

We calculate the change in elastic energy for transferring the RNA from an isolated state to that in the spherical shell from the freely rotating chain model,

$$\beta \Delta F_r^{\,elas} = \frac{\pi^2 n_r a_r^{\,2}}{6D^2}, \tag{6.14}$$

where $n_r \approx 3400$ is the number of nucleotide segments in the RNA chain, and a_r is the monomer diameter. In Equation (6.14), the proportionality constant accords with the ground-state dominance approximation [86]. Similarly, the conformation free energy associated with the confinement of CTD tails is obtained from [87]

$$\beta \Delta F_c^{\,elas} = \frac{3}{2} N_t \left(\frac{R_e^{\,2}}{D^2} + \frac{D^2}{R_e^{\,2}} \right), \tag{6.15}$$

where $N_t = 240$ is the number of CTD tails in each T4 capsid, and R_e is the average end-to-end distance for a free CTD chain. Assuming that each CTD tail is a freely rotating chain with a bond angle of 116°, we have $R^2{}_e \approx 2.56 n_c a_c^2$ where $n_c = 34$ is the number of amino-acid residues and a_c is the segment diameter.

At infinite dilution, the excluded volume is negligible for CTD and RNA chains. In that case, the excluded-volume free energy can be estimated from the modified Flory theory for the polyelectrolyte complex [74]:

$$\beta \Delta F^{exc} = \frac{\nu_r N_r^{\,2} + \nu_{rc} N_r N_c + \nu_c N^2{}_c}{4\pi R^2 D}, \tag{6.16}$$

where ν stands for segmental excluded volume. In principle, ν depends on both the segment size and electrostatic interactions [84, 88, 89]. Here we use the “bare” excluded volumes of the polymers on the right-hand side of Equation (6.16), i.e., $\nu_r = 4\pi a_r^{\,2}/3$, $\nu_c = 4\pi a_c^{\,3}/3$, and $\nu_{rc} = 4\pi\{(a_r + a_c)/2\}^3/3$ for RNA, CTD, and RNA–CTD interactions, respectively. As detailed in the following, we account for the charge effects along with their Coulomb interactions with the small ions.

The excess free energy due to electrostatic interactions and entropy of mixing can be calculated from the Overbeek–Voorn (OV) theory for polyelectrolyte complexes [90],

$$\beta F^{els} = -\frac{V_2 \kappa_2^{\,3}}{12\pi} + \sum_i N_i \ln \phi_i, \tag{6.17}$$

where κ_2 represents the Debye screening parameter (in Region 2), N_i stands for the number of RNA or CTD chains or salt ions, and $\phi_i = n_i \nu_i N_i / V_2$ represents the volume fraction. For simplicity, we assume that $n_i = 1$ for salt ions, and that the ionic molar volume is the same as that for water molecules, $\nu_i = 18\,\text{cm}^3/\text{mol}$. For RNA and CTD chains, the volume of individual segments is related to the segment diameter.

Because the OV theory includes the electrostatic energy and the entropy of mixing among small ions, we must subtract from Equation (6.17) the free energy corresponding to the background electrolyte occupying the same volume and the free energies of RNA and CTD chains at infinite dilution. In calculating the background free energy, the salt concentration is the same as that of the bulk and the polymer concentration vanishes. To use the OV theory for RNA and CTD chains at infinite dilution, we assume that the electrostatic energy and entropy of mixing are the same as those corresponding to a confined chain in a spherical cage with cavity radius the same as the polymer gyration radius. According to the literature [91–94], the radius of gyration for a single RNA chain depends on its chain length and the surrounding salt concentration,

$$R_{g,r} = 5.84 \times n_r^{1/3} C_s^{-1/8}, \tag{6.18}$$

where R_g has units of Å and C_s is in M. We calculate the radius of gyration for each CTD chain according to the freely rotating chain model with a bond angle of 116°:

$$R_{g,c} = (0.43 n_c a_c^2/6)^{1/2}. \tag{6.19}$$

We use different models for RNA/CTD in energetic and geometric characterizations because the former is divided into several interrelated contributions (e.g., elastic, electrostatic, etc.) while the latter (e.g., the radius of gyration) represents the overall behavior of polymer chains.

6.2.2 Thermodynamics of Capsid Formation and Stability

The stability of HBV capsids has been analyzed by *in vitro* experiments either with or without the genomic materials. In both cases, the intermediate structures are insignificant, suggesting that the thermodynamics of capsid formation can be described in terms of a one-step process similar to that commonly used for micellization. While micelles often show a broad polydispersity in terms of size and shape, viral capsids adopt a nearly uniform size due to the quasi-equivalence of the capsid proteins arranged in a lattice-like structure

6.2.2.1 Stability of CTD-Free Empty Capsids

As mentioned above, *in vitro* formation of empty HBV capsids free of CTD can be described by the single-step model, i.e., in terms of a quasi-chemical reaction for the core protein dimers [95],

$$\mathrm{N} \times \mathrm{Dimer} \leftrightarrow \mathrm{Capsid}, \tag{6.20}$$

Where $N = 90$ and 120 for T3 and T4 capsids, respectively [75]. In a dilute solution, the equilibrium dimer concentration (D_0) and that of the fully assembled capsids (C) satisfy the mass action law

$$K_C = [\mathrm{C}]/[\mathrm{D}_0]^N, \tag{6.21}$$

where K_C denotes the apparent equilibrium constant. In terms of the mole fractions of dimers (X_{D0}) and capsids (X_C), the equilibrium constant is dimensionless:

$$K_0 = \frac{X_C}{X^N{}_{D_0}}. \tag{6.22}$$

We can find the connection between the two equilibrium constants by converting the molar concentrations into the corresponding mole fractions:

$$[\mathrm{D}_0] = [\mathrm{W}] \cdot X_{D0},$$
$$[\mathrm{C}] = [\mathrm{W}] \cdot X_C, \tag{6.23}$$

where [W] is the molarity of pure water. Substituting Equation (6.23) into (6.22) gives

$$\ln K_0 = \ln K_c + (N-1)\ \ln[\mathrm{W}]. \tag{6.24}$$

The free energy of capsid formation is related to the dimensionless equilibrium constant

$$\Delta F_0 = -k_B T\ \ln K_0. \tag{6.25}$$

According to Equation (6.25), the free energy of capsid formation reflects the difference between the chemical potential of the empty capsid and that of the CTD-truncated protein dimers at the standard conditions, i.e., in their corresponding hypothetical pure states [70]. With the concentrations of the core-protein dimers and the fully assembled capsids measured at various equilibrium conditions, we can quantify the equilibrium constant and subsequently the free energy of capsid formation [51].

We compare in Figure 6.7 the theoretical predictions with *in vitro* experiments for the equilibrium constants of empty T4 capsids over a large range of temperatures and salt concentrations [57]. Here, the symbols represent the experimental data from the *in vitro* formation of CTD-free empty T4 HBV capsids, and the lines are from our theoretical predictions. Clearly, the coarse-grained model captures both the temperature and salinity effects on the capsid stability. In light of the simplicity of the theoretical model and noticeable scattering in the experimental data, the quantitative agreement between the theory and experiment is satisfactory. Our theoretical results further support the conjecture that capsid formation is a one-step process, mainly driven by hydrophobic interactions [57]. In addition, we find that the contact hydrophobic free energy can be quantitatively described with the solvation shell model [77].

It has been shown that the self-assembly of empty HBV capsids from CTD-truncated core proteins is entropy driven, primarily due to inter-dimer hydrophobic interactions [57]. If one divides the free energy of capsid formation into the contact values between nearest neighboring dimers, the equilibrium constant provides an indirect measure of the pair association constant [51],

$$K_{contact} = (NK_0/j^{N-1})^{2/cN}, \tag{6.26}$$

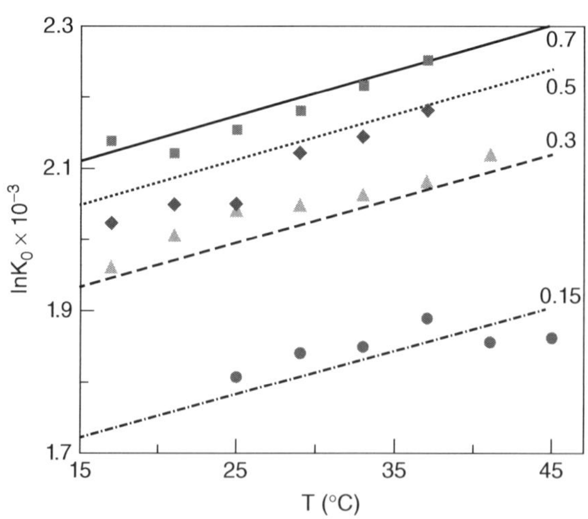

Figure 6.7 The equilibrium constant for the formation of CTD-free empty T4 capsids. Symbols are experimental data [57] and lines are theoretical predictions. The salt concentration (M) is denoted beside each line.

where j represents the degeneracy of the capsid with N dimers, and c stands for the number of contacts per dimer. The association free energy per contact (for each dime pair) is then given by

$$\Delta F_{contact} = -k_B T \ \ln \ K_{contact}. \tag{6.27}$$

Experiments for *in vitro* assembly of empty HBV capsids indicate that the average value for the contact free energy is between −3 and −4 kcal/mol, which is "*surprisingly*" small in comparison to alternative estimations based on the dimer structure (~ −10 kcal/mol) [57]. The discrepancy was attributed to multiple causes including possible changes in the protein structure due to the self-assembly, the pairwise additivity assumption for the binding free energy, or the imprecise calculation of the hydrophobic interactions. As shown in Figure 6.7, our thermodynamic model faithfully captures the equilibrium constant for the assembly of empty T4 in all conditions.

Whereas hydrophobic attractions are mainly responsible for empty capsid formation, it is worthwhile mentioning that electrostatic interactions and van der Waals forces are also needed to attain a quantitative performance of the theoretical model. For example, Figure 6.8 shows the equilibrium constant versus temperature at 0.15 M salt condition. Excluding van der Waals association free energy (F_{vdw}) would result in a systematic deviation from the experimental values. The sensitivity of the numerical results is related, in part, to the partial cancelation of the electrostatic and hydrophobic interactions, and to the large number of proteins for the formation of a capsid. Besides, the sensitivity also reflects the exponential relationship between the formation free energy and the equilibrium constant.

It has been shown before that, at higher temperature or salt concentration, the HBV capsids become more stable and the increased stability can be rationalized in terms of hydrophobic and electrostatic interactions between capsid proteins [57, 70, 96]. Our numerical analysis confirms the dependence of protein–protein interactions on both temperature and ionic strength [66, 71, 97]. It is worthwhile noting that other effects may also influence the capsid satiability. For example, the increase in ionic strength might lead to the conformation change of the capsid protein, thereby enhancing the capsid

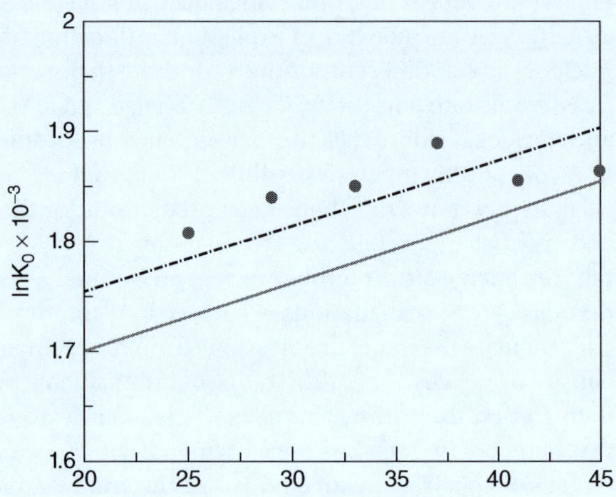

Figure 6.8 The effect of van der Waals force on capsid stability. The symbols and the dashed line are the same as those shown in Figure 6.7, and the solid line represents ln K_0 without considering van der Waals interactions.

stability [57]. Besides, chemicals such as zinc ions or heteroaryl dihydro pyrimidines may also alter the conformation of the capsid protein, preventing their proper assembly [98–101]. Recent experiments indicate that mutation on certain residues of the capsid protein modulates the tendency of capsid assembly [102–105].

6.2.2.2 Stability of Nucleocapsids

Our thermodynamic analysis can be similarly applied to the assembly of HBV nucleocapsids with native core protein dimers in the presence of either RNA or DNA. In this case, the quasi-chemical reaction becomes

$$\text{N} \times \text{Dimer (D)} + \text{Nucleic Acids (NA)} \leftrightarrow \text{Nucleocapsid (NC)}, \qquad (6.28)$$

and the corresponding equation for the mass action law is

$$K_{NC} = \frac{[\text{NC}]}{[\text{D}]^N[\text{NA}]}. \qquad (6.29)$$

In a dilute solution, the apparent equilibrium constant is related to the free energy of capsid formation,

$$\beta\Delta F = -\ln K_{NC} - N\ln[\text{W}]. \qquad (6.30)$$

Assuming that NC capsids maintain the structure integrity for both the core-protein assembly domain as well as the capsid shell, we may divide the free energy of nucleocapsid formation in terms of that corresponding to the formation of a CTD-free empty capsid, designated as ΔF_0, and a contribution due to CTD–RNA/DNA interactions and encapsidation of biomacromolecules:

$$\Delta F = \Delta F_0 + F_{plex}, \qquad (6.31)$$

where F_{plex} denotes the free energy of polyelectrolyte complexation between CTD and nucleic acids. Interestingly, complex formation between oppositely charged polyelectrolytes is also entropy driven, primarily due to the release of a large number of counterions associated with the electrostatic binding, i.e., between the positively charged CTD chains and negatively charged nucleic acids [106]. Whereas *in vitro* assemblies of both empty and RNA-containing nucleocapsids have been recently demonstrated [60, 61, 107, 108], we are unaware of explicit quantification of the formation free energy for HBV nucleocapsid at different solution conditions. Because native HBV core proteins are able to assemble into empty capsids both *in vitro* and *in vivo*, the competition between empty and nucleocapsids makes the experimental procedure for studying nucleocapsid formation conceivably much more difficult than that for empty capsids.

Figure 6.9 shows our theoretical predictions for the encapsidation free energy of RNA and CTD chains versus the thickness of the polyelectrolyte complex, D. As expected, the free energy shows a minimum at an optimal D, which varies with the solution conditions. In these calculations, we use $R = 13$ nm for the inner radius of the HBV capsid, and calculate the ion concentrations from the Donnan equilibrium with polymer charge density $\rho_p = 520/V_2$. For the range of bulk salt concentrations (0.1 ~ 0.7 M) considered in this work, the optimal complex thickness lies around 10.5 ~ 11 nm, much larger than that detected in cryo-EM measurements.

Nucleocapsids are expected to be thermodynamically more stable than the corresponding empty capsids owing to the strong electrostatic attraction between RNA

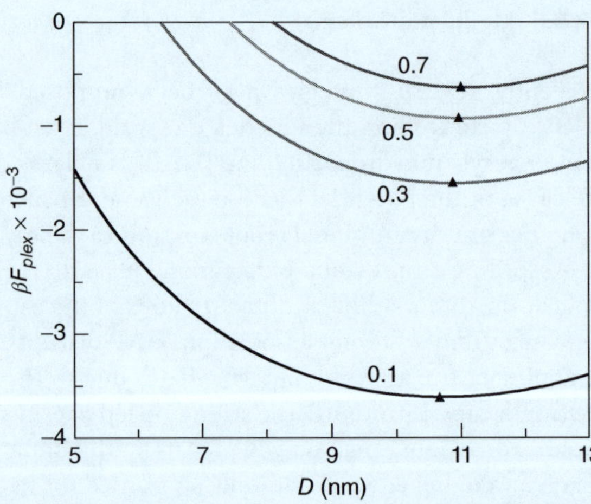

Figure 6.9 The free energy of encapsidation for CTD and RNA chains as a function of the thickness of the complex layer. The salt concentration (M) is denoted beside each line, and the minimum points are indicated by triangle symbols.

and oppositely charged CTD chains. To validate this hypothesis, Figure 6.10 shows the effects of salt concentration on the total free energy of CTD/RNA encapsidation. For all conditions considered in this work, F_{plex} is negative, indicating that RNA encapsidation indeed stabilizes the T4 capsid. As expected, the complex formation free energy becomes less negative as the salt concentration increases. Experimental studies confirmed that the nucleocapsids are more stable than the corresponding empty capsids [54, 61]. The enhanced stability was also suggested by a previous theoretical investigation [109]. Figure 6.10 shows that the encapsidation free energy approaches zero at sufficiently high salt concentration (C_s). In agreement with experiments, the stabilizing effect of complex formation is most effective when C_s is relatively small.

We may explain the overall salt effect on the nucleocapsid stability as a balance of protein–protein and RNA–CTD interactions. While salt ions reduce the electrostatic repulsion between capsid proteins, they also abate the electrostatic binding energy between CTD and RNA chains. As a result, nucleocapsid assembly would be most favorable at an intermediate salt condition. Indeed, an intermediate salt condition (0.25 ~ 0.35 M) must be applied for *in vitro* nucleocapsid assembly (e.g., [61, 82]).

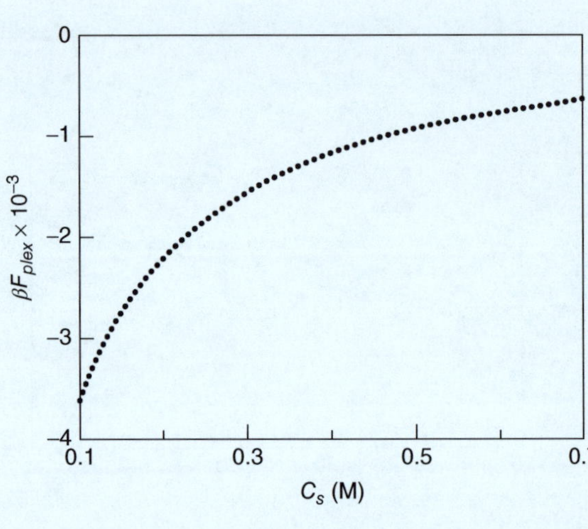

Figure 6.10 The encapsidation free energy for CTD and RNA chains in a T4 capsid (βF_{plex}) as a function of the salt concentration (C_S).

6.2.3 Modulation Effects

Recently, several strategies have been proposed to design new drug molecules for HBV by controlling the kinetics of capsid assembly and stability [110]. For example, heteroacryl- dihydropyrimidine (HAP) is able to disrupt capsid formation [99–101]. Because of the potential therapeutic values of such molecules, it is worthwhile testing whether or not our model is able to capture their effects on HBV capsid formation.

According to previous publications [99–101], HAP molecules are able to intercalate at the dimer–dimer contact regions of the capsid proteins, and HAP intercalation enhances dimer–dimer association. HAP disrupts capsid formation mainly by kinetic trapping of the assembly intermediates due to the increased hydrophobic association between capsid proteins. The strengthened capsid association is due to the filling of the void hydrophobic regions while keeping the dimer–dimer distance unchanged [101]. Such a wedging effect is naturally accounted for in our hydrophobic model for the capsomers association. As shown in Figures 6.5b and 6.5c, the modulation effect would be most significant if the size and shape of the hydrophobic molecule could match those corresponding to the hydrophobic pores of two-fold symmetry or triangular wedges between dimer–dimer contact regions.

We may also predict a scenario such that hydrophilic molecules are able to bind the dimer contact regions of the capsid proteins [111]. In this case, the association energy would be decreased due to the reduction of the dimer contact area. Figure 6.11 shows that a reduction of the hydrophobic patch area of the protein dimers by 10% would decrease the capsid stability by ~200 in the ln K value. For comparison, we consider the effect of electrostatic interactions on the capsid stability by increasing either the charged surface area or the surface charge density. Figure 6.11 indicates that the influence is much less significant in comparison to the hydrophobic effect. According to our thermodynamic model, an efficient HBV “modulator” should bind specifically to the base part of the HBcAg assembly domain while enhancing repulsion among the protein

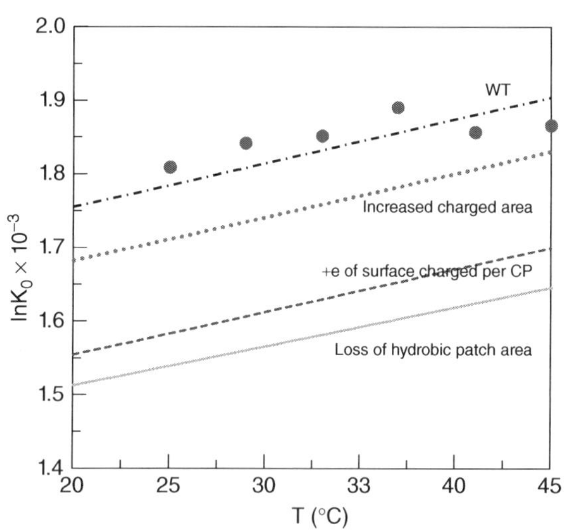

Figure 6.11 Hydrophobic and electrostatic effects on the equilibrium constant of empty HBV capsids. The solid line corresponds to the theoretical results if there is a 10% reduction of the buried hydrophobic area at each dimer contact; the dotted line is for a 10% increase of the capsid charged area; and the dashed line is for an increase of one unit charge for each capsid protein. For comparison, also shown in this figure are the equilibrium constants of wild type capsids from theory (WT, dash-dotted line) and experiment (circle points) at 0.15 M.

subunits, such as an amphiphilic molecule with a hydrophobic binding domain associated with a charged group or a hydrophilic chain. The hypothesis may be easily validated with future experiments.

6.2.4 T3/T4 Dimorphism

T3 and T4 capsids form concurrently during both *in vivo* and *in vitro* syntheses of HBV capsids. The composition of dimorphism reflects the difference in the free energy of capsid formation. Because each dimer maintains the same number of contact neighbors in both T3 and T4 capsids (see Figure 6.5), the association energy between dimer units has little influence on T3 and T4 dimorphism except that the number of dimers is 90 in T3 instead of 120 in T4. However, the size of an empty T3 capsid is slightly larger than would be expected according to the degree of association. The average radius of a T4 capsid is around 15 nm, and that of the T3 capsid is 13.3 nm [55]. Accordingly, the ratio of spherical surface area is $13.3^2 : 15^2 = 0.786 : 1$, which is slightly larger than the ratio for the number of capsomers (0.75 : 1). Such a loosely assembled structure can be depicted in our capsid model by setting two small units at each dimer contact slightly farther apart, i.e., by setting the distance between small units to 1.5 nm for a T3 capsid instead of 1.4 nm as for the T4 capsid.

Table 6.1 presents the hydrophobic contact energy and the electrostatic repulsion energy calculated from our thermodynamic model. Due to the reduced hydrophobic potential, the free energy for T3 capsid formation per each monomer ($-4\,k_B T$) is significantly less negative that that of the T4 form ($-7.7\,k_B T$). The higher energy state for T3 capsids explains the prevalence of T4 capsids at both *in vivo* and *in vitro* conditions [55].

Table 6.1 Free energy of self-assembly for T3 and T4 capsids.

	T4 (Cp149*, HBeAg**)	T3 (Cp149*)	T3 (HBeAg**)
Number of monomers (n)	240	180	180
R_{ave}	15	13.3*	12.9**
F_{cap}	−1856.19	−725.99	−1540.13
F_{cap}/n	−7.73	−4.03	−8.55
$F_{hyd+vdw}$	−2592.44	−1252.77	−2100.08
$F_{hyd+vdw}/n$	−10.80	−6.96	−11.67
F_{ele}	736.25	526.78	559.95
F_{ele}/n	3.07	2.93	3.11

The average radius (R_{ave}) is obtained from Zlotnick et al. for Cp149 capsid* [55], and from Wingfield et al. for HBeAg capsid** [54]. Unit of R is nm. The free energy of capsid formation, F_{cap}, and its division into hydrophobic and van der Waals interactions, $F_{hyd+vdw}$, and electrostatic interaction F_{ele} are compared with the corresponding values per monomer for both T4 and T3 capsids. Here the unit of energy is $k_B T$, and the solution condition is 37 °C and 0.15 M salt.

The higher energy for T3 capsids agrees with an earlier analysis of the capsid stability in vacuum [112].

It has been suggested that the capsid protein plays an important role in T4/T3 dimorphism [54, 55]. Without CTD, T4 capsids are more prevalent than T3 capsids. However, truncation of more than 34 C-terminal residues inverts the relative composition, i.e., T3 capsids may be preferred over the T4 form at the same assembly condition [54, 55]. We demonstrated above that the coarse-grained model provides a good description of empty capsid formation without CTD (CP149). For comparison, we may also quantify another truncation case where only 3–144 aa residues of the wild type capsid protein are maintained. Such a system was studied many years ago by Wingfield et al. [54]. Following their data, we estimate that the mean radius of the T3 capsid is 12.9 nm, while that for T4 is 15.1 nm. Because the ratio of the spherical surface area is now slightly smaller than the dimer number ratio, the T3 capsid is more tightly assembled in comparison to a conventional T4 capsid. In a T3 capsid, the contact distance between nearest dimers (r) is about 1.38 nm. Although the change of r is small (about 10% of the small unit overlap distance), the contact free energy per each monomer increases up to $-8.6\,k_BT$, explaining the prevalence of the T3 capsids (see Table 6.1). As discussed in the following, the dimorphism of the HBV capsid is also sensitive to the variation of the C-terminal domain of the capsid proteins.

We have also evaluated the effects of RNA–CTD interaction on the dimorphism of HBV nucleocapsids. It is well known that T3 nucleocapsids are rare (<5%) in both *in vivo* or *in vitro* assembly [113]. To check the dimorphic probability using our nucleocapsid model, we estimate the pgRNA and CTD encapsidation free energy in a way similar to that for T4 nucleocapsids. For T3 nucleocapsids, we set the radius of the inside shell to 11 nm, considering that the capsid thickness is about 2 nm. Whereas the encapsidation free energy on a T4 capsid is to be obtained in a self-consistent way with respect to an optimal thickness D for an RNA–CTD complex region, the CTD–RNA association in a T3 capsid tends to spread into the entire capsid lumen without reaching a free energy minimum. The absence of a polymer-free Region 1 is probably because the excluded volume effect is too large to package pgRNA into a narrow T3 nucleocapsid. A recent study on *in vitro* encapsidation of nanoparticles also indicates that the volumetric restriction makes T3 nucleocapsids less stable than T4 nucleocapsids [114]. For pgRNA encapsidation in WT T3 nucleocapsids, our model predicts that the free energy of CTD–RNA complex formation, $F_{plex} = -1863\,k_BT$ at 37 °C with 0.15 M salt condition. At the same condition, $F_{plex} = -4582\,k_BT$ for a T4 nucleocapsid. Because the stabilizing effect by RNA–CTD complex formation is much smaller for T3 nucleocapsids, we ascertain that the energetics in genome encapsidation results in the prevalence of T4 nucleocapsids.

6.3 Electrostatics of Genome Packaging

Recent mutagenesis studies indicate that pgRNA packaging is strongly correlated with the net charge of the CTD tails, leading to the so-called "charge balance hypothesis" [59, 115, 116]. A partial truncation of the flexible, highly charged CTD tails reduces the amount of encapsidated RNA, and restoring the arginine residues to the truncated CTD increases the nucleic content. A positive correlation between the C-terminal tail length and the RNA content was confirmed by cryo-EM measurements [117].

While the charge balance hypothesis entails a stoichiometry-like correlation between basic residues and their binding partner of nucleic acids, a more quantitative description of the charge–charge relation is difficult to establish because electrostatic interaction between macromolecules is affected not only by the net charges of the macromolecules but also by the solution conditions such as salt concentration, identities of low molecular weight ions, pH, and temperature [104]. Besides, the charge distribution at the CTD tails may also play an important role. For example, it is unclear whether all arginine residues at the C-terminus make equal contributions to the electrostatic interaction with encapsidated RNA or DNA during the HBV life cycle [116].

The regulative role of electrostatic interactions in capsid formation has been identified before for many ss-RNA/DNA viruses [118–126]. Like HBV, genome packaging is intimately affiliated with electrostatic interactions between the nucleic acids and oppositely charged, flexible domains of the capsid proteins [127]. Several theoretical studies suggest that viral packaging conforms with a fixed charge ratio between nucleic acids and capsid proteins [128–132]. A conserved ratio ($N/Q = 1.61$) of the genome size (N) to the overall charge of capsid proteins (Q) was observed for more than 30 wild type viruses. While the universality of a linear relation between capsid charge and genome size has been partially justified [68], its applicability to all viruses was challenged by more recent theoretical studies [46, 62]. Genome packaging involves other mechanisms such as the Donnan equilibrium [46] and specific effects including the genome structure [83, 91] that may lead to nonlinearity or different charge ratios in complexation between macromolecules of opposite charges.

6.3.1 Thermodynamics of RNA Encapsidation

To a certain degree, the thermodynamics of RNA encapsidation may be considered as an equilibrium partition of a long RNA chain between the inner and outer space of the capsid. Figure 6.12 depicts, schematically, a packaged chain at the early stage of assembly. The exposed portion of the RNA chain is eventually degraded by intracellular nuclease [115, 133], and the inside portion defines the optimal size of the genome.

The thermodynamic potential for RNA fragments both inside and outside the capsid can be written as

$$F = F_{cap} + F_{out}, \tag{6.32}$$

where F_{cap} accounts for the free energy of the entire nucleocapsids, and F_{out} represents the energy of the exposed RNA in the local cellular environment outside capsid. At equilibrium, the optimal genome size minimizes the overall potential, i.e.,

$$\frac{\partial F}{\partial N_{RNA}} = 0, \tag{6.33}$$

where N_{RNA} stands for the number of nucleotides of the packaged RNA chain.

We may estimate the free energy of RNA outside the capsid from the reversible work to insert the unpackaged RNA segments into the cellular environment. The free energy of RNA insertion corresponds to the volumetric work to create a cavity in a cellular milieu ($\Pi \approx 5$ bar [134, 135]),

$$F_{out} = \Pi v_R (N_{pgRNA} - N_{RNA}), \tag{6.34}$$

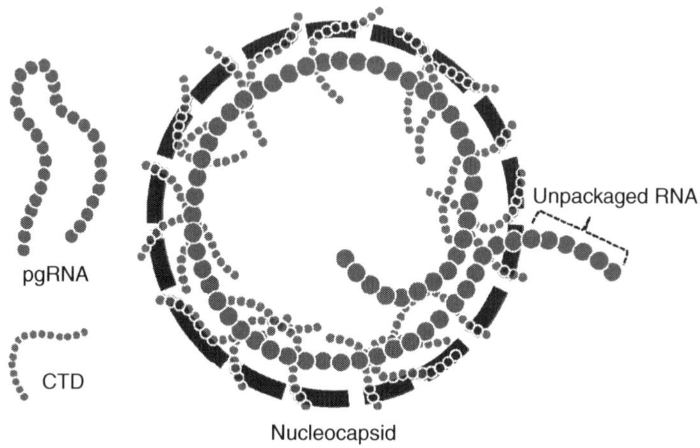

Figure 6.12 Schematic of RNA packaging in an HBV nucleocapsid. The optimal genome size can be determined from a partial encapsidation of the RNA chain that minimizes the thermodynamic potential of the entire nucleocapsid in an aqueous electrolyte solution. The unpackaged RNA fragments are subjected to degradation by intracellular nucleases.

where ν_R stands for the excluded volume of each RNA segment, and N_{pgRNA} is the number of segments of the entire pgRNA chain. Because the free energy for the outside part of the RNA is a linear function of N_{RNA}, its derivative with respect to N_{RNA}, i.e., the thermodynamic potential, is independent of the size of the unpackaged RNA. It has been shown that, under the cellular osmotic pressure, the contribution to the total thermodynamic potential due to the exposed RNA is relatively minor [136].

The thermodynamic potential within the nucleocapsid accounts for the genome–capsid interactions with the participation of salt ions. The corresponding semi-grand potential (F_{cap}) depends on the compositions of the RNA and CTD chains, the chemical potentials of the small ions, and the interaction potentials between those components and the capsid shell:

$$\begin{aligned} F_{cap} = & \sum_{\alpha=\pm ions} \int d\mathbf{r}\rho_\alpha(\mathbf{r})[\ln \rho_\alpha(\mathbf{r}) - 1] \\ & + \sum_{\alpha=\pm ions} \int d\mathbf{r}\rho_\alpha(\mathbf{r})[\psi_\alpha(\mathbf{r}) - \mu_\alpha] \\ & + \sum_{M=R,T} \int d\mathbf{R}\rho_M(\mathbf{R})\, \psi_M(\mathbf{R}) - F^{ex}, \end{aligned} \tag{6.35}$$

where subscripts α and M denote salt ions and macromolecules (R = RNA, T = CTD tail); $d\mathbf{R} = d\mathbf{r}_1 d\mathbf{r}_2 \cdots d\mathbf{r}_M$ stands for a set of differentials; μ_α is the chemical potentials of salt ions; $\psi_M(\mathbf{R}) = \sum_{i=1}^{M} \psi_p(\mathbf{r}_i)$ represents the external potential for the polyion, while $\psi_p(\mathbf{r}_i)$ is that of each segment; $\psi_\alpha(\mathbf{r})$ is the external potential of small salt ions; $\rho_\alpha(\mathbf{r})$ and $\rho_M(\mathbf{R})$ are the density profiles of small ions and polymers, respectively; and F^{ex} is the excess potential due to intermolecular interactions. With an appropriate model for polymers and small ions, both excess Helmholtz energy and molecular density profiles can be calculated from classical density functional theory (DFT). We have demonstrated in our earlier works the theoretical performance of DFT

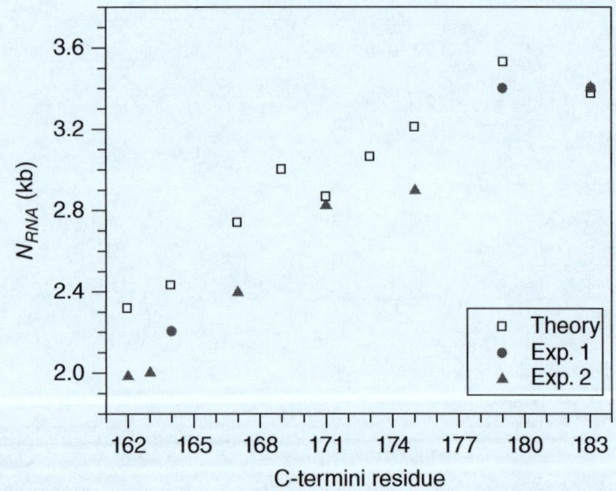

Figure 6.13 Theoretical predictions (black open square) and experimental data (Exp. 1 [115], circle; Exp. 2 [141], triangle) for the optimal RNA length (kb or kilo-base) in the wild type (WT) and mutant HBV nucleocapsids.

for inhomogeneous polyelectrolyte systems [137, 138] and its successful applications to HBV nucleocapsids [136, 139, 140].

6.3.2 The Optimal Genome Size of an HBV Nucleocapsid

The thermodynamic model for RNA encapsidation has been used to identify the optimal RNA length for both WT and mutants of HBV. It has been shown that a newly formed WT HBV nucleocapsid contains a single copy of pgRNA, which consists of about 3400 nucleotides. A similar amount of bacterial ssRNA can be encapsidated by expression of the full-length capsid proteins in *E. coli* [61]. When the CTD tails are truncated, however, the capsid contains a smaller RNA chain.

Figure 6.13 shows the theoretical predictions for the RNA size in the WT and mutant nucleocapsids with different degrees of CTD truncation. The number of residues in the core protein (183~162) is used to denote the WT and C-terminally truncated mutants. The theoretical data for the optimal genome size are compared with mutagenesis studies in *E. coli* systems [115, 141]. In general, the theoretical predictions correspond well with those from experiments. The predicted values are slightly higher than the experimental data probably because our model determines the optimal genome size without considering the polymerase and other proteins that are also encapsidated [40]. Although these proteins make little contribution to the overall charge of the HBV capsid in comparison to the core proteins, they occupy certain capsid volume influencing the total amount of packaged RNA. For WT and mutants with low degree CTD truncations, nevertheless, the theoretical predictions are excellent. By contrast, the calculated RNA size for mutants with shorter CTD tail (e.g., with C-termini residue of 162 or 164) exhibits more significant deviation (10%~20%) from the experimental data. Such a corroborated result confirms that RNA packaging in HBV is mostly regulated by nonspecific interactions between RNA and the CTD tails.

6.3.3 Charge Balance between Packaged RNA and CTD Tails

Our DFT calculations support the lack of a universal stoichiometric relation in HBV genome packaging. To illustrate, Figure 6.14 presents the packaged RNA size (N_{RNA}) as a

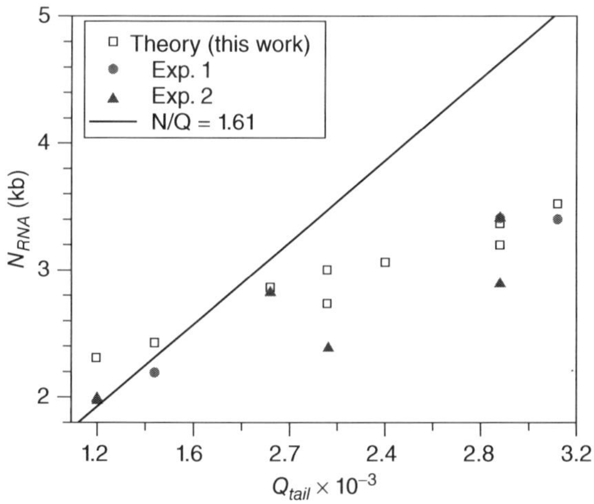

Figure 6.14 Correlation between the RNA size (N_{RNA}) and the net charge of C-terminal tails (Q_{tail}). The solid line ($N/Q = 1.61$) was proposed by Belyi and Muthukumar [68].

function of the net charge of CTD tails (Q_{tail}) of the HBV core proteins. Both the theoretical predictions and experiment data for the optimal genome size (shown in Figure 6.13) are assorted according to Q_{tail} for WT and mutants. Although there is an overall trend that N_{RNA} increases with Q_{tail}, both experimental data and theoretical predictions show that the charge ratio (N_{RNA}/Q_{tail}) is not constant. For example, two mutants have the same Q_{tail} but noticeably different N_{RNA} values (e.g., 167 and 169).

A number of thermodynamic models have been proposed to explain the existence of an optimal charge ratio between nucleic acids and capsid proteins. For example, Hu et al. argued that the optimal RNA size is dependent on the contour length of peptide tails, and obtained a fixed charge ratio of $N/Q = 2$ [33]. Different charge ratios (<1, 1, and 2) have also been predicted with the assumption of a uniform capsid charge placed at the inner capsid surface [128, 131, 132]. Most notably, Belyi and Muthukumar determined $N/Q \sim 1.6$ based on the counterion-condensation theory for RNA and peptide tails [68]. The linear charge relation is compared in Figure 6.14 together with our model and experimental data. Clearly, neither the WT nor mutant HBV nucleocapsids follow the "universal" charge ratio. Instead, the DFT calculations show that the charge ratio varies from 1.1 to 1.8 depending on specific mutants. According to a recent experiment, the charge ratio universality is also not valid for brome mosaic virus (BMV) [62]. The number of charges in capsid protein arms is not a sole factor in controlling the RNA packaging. Our DFT calculations on HBV packaging draw a similar conclusion that the additional controlling factors beyond the net charge effect should be considered. As indicated before [136], the excluded volume and correlation effects of macromolecules and salt ions contribute in the RNA packaging process. A proper consideration of such effects allows us to quantify the equilibrium constants for the capsid formation and obtain an optimal genome size in good agreement with experiments.

The theoretical results shown in Figure 6.14 were obtained by assuming that three serine residues (155th, 162nd, and 170th aa) were phosphorylated for the WT virus and that each phosphorylated serine carried a negative unit charge. Although the bacterial capsids are not phosphorylated, we assume that *E. coli* can make up to the total amount of negative charge equivalent to that in mammalian capsids by packaging multiple RNA chains [60]. In other words, the lack of phosphorylation is compensated by the difference between small RNAs and pgRNA. Whereas these serine residues are most relevant to phosphorylation *in vivo*, it has been reported that serine residues at the 176th and 178th aa may also be phosphorylated [142]. If that is the case, the optimal genome size would be reduced by ~170 bases in comparison to that for the WT virus. If all five serine sites are phosphorylated, the RNA size is longer (by ~70 bases) than that of the C-terminal truncated mutant 169 with regular phosphorylation, although these mutants have the same net charge (Q_{tail}) for the CTD tails. Apparently, such a difference is not accounted for by the universal charge ratio or charge balance hypothesis. The difference is somewhat expected because, as discussed below, the additional phosphorylation sites are located around the C-terminal end region and mostly exposed at the capsid surface.

6.4 Dynamic Structure of HBV Nucleocapsids

A balanced electrostatic interaction between the CTD tails and the encapsidated nucleic acid is required for many viral activities including capsid stability, assembly, RNA encapsidation, and DNA replication [115, 143]. However, the HBV capsid undergoes drastic changes in both electrostatic charge and packing density as the pgRNA is reversely transcribed to a partially double-strained DNA. Several important questions may be raised from a biophysical perspective: Does the nucleocapsid preserve an internal charge balance with regard to the capsid evolving stages? How does the capsid structure, in particular distributions of the protamine tails and the overall packing density, evolve during the reverse transcription? What are the internal properties of nucleocapsids, such as the electrostatic potential distribution, osmotic pressure, and forces on the capsid inner surface? What are the roles of ions in charge regulation? Addressing all these questions calls for more detailed information on the distributions of all charged components, including nucleic acids, protein tails, and ions, within the nucleocapsid.

6.4.1 Structure of WT and Mutant Nucleocapsids

As mentioned before, the HBV capsid is fenestrated with symmetrical pores allowing for mass transport and extrusion of the CTD tails [15, 26, 40]. The CTD exposure affects genome packaging, nuclear entry, as well as capsid envelopment and egress [58, 144, 145]. In addition, it has been recognized that the CTD tails are involved in signaling capsid maturation [37, 145, 146]. Because CTD exposure alters the surface properties of the NC particle, its location at different stages of replication may differentiate the mature and immature NCs.

Figure 6.15 presents the theoretical predictions for the radial distributions of RNA and CTD segments at different levels of CTD truncation. As indicated in a previous report [136], RNA segments exhibit extremely high contact density at the inner capsid surface. Immediately following the RNA layer are tethered CTD segments, with the rest forming a brush-like structure. For mutants with truncation over five residues, we find the radial distributions of capsid components are virtually identical to our previous predictions from a confined capsid model [136]. Noticeably, however, some CTD segments approach outside the capsid region for both WT and CP179 mutant.

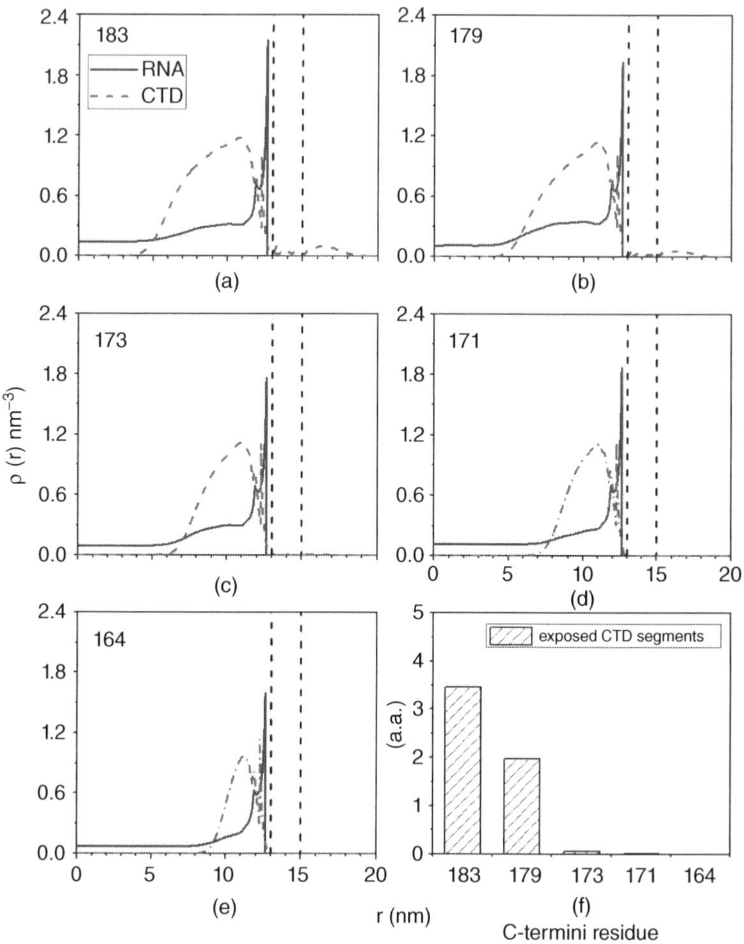

Figure 6.15 Radial distributions of RNA (solid line) and CTD chains (dashed line) in WT and mutant HBV nucleocapsids [147]. Here, *r* stands for the radial distance from the capsid center; the perpendicular dashed lines indicate the position of the capsid shell. (f) The average number of amino acid (aa) residues from the CTD tails located outside the capsid wall for WT and mutants.

The last panel of Figure 6.15 compares the average number of exposed CTD residues for different mutants. For WT nucleocapsids, 3–4 CTD residues per capsid protein are exposed in the outside region. Because these residues do not participate in the complex formation with encapsidated RNA, we expect about 10% of CTD segments are dispensable for genome packaging. The exposure of CTD tails in wild type HBV virus is consistent with previous biochemical analysis showing their accessibility to proteases and antibodies [148, 149]. It has also been confirmed by recent cryo-EM structures [62, 150]. Because our theoretical calculations assume a spherical symmetry for CTD distributions, we are not able to distinguish different holes of the capsid shell responsible for the CTD exposure.

6.4.2 The Location of CTD Residues

In order to clarify how CTD chains regulate genome evolvement as well as intracellular trafficking of HBV capsids [146, 151], it is imperative to determine their spatial distribution during different stages of replication. Regrettably, experimental detection of the CTD structure has been elusive. Whereas the assembly domain of HBV capsids maintains a relatively rigid structure, CTD chains are highly flexible and their structure is changing in response to the evolving genome [37, 149, 151–154]. Recently, the serine–arginine protein kinase (SRPK) was utilized to analyze the location of CTD tails in empty and RNA-containing HBV capsids [155]. The *in vitro* experiments for the binding affinity of SRPK with HBV capsids indicate that the CTD tails are exposed at the surface when the capsid does not contain any nucleic materials. Conversely, the CTD tails are mainly distributed inside the capsid for the RNA-containing nucleocapsids [155]. However, the CTD chains were not visible through a conventional microscopy such as cryo-EM [22, 156]. Only recent studies capture a trace amount of CTD chains through reconstructed EM images [150, 154].

HBV capsids expressed in *E. coli* contain the bacterial RNA instead of pgRNA [151, 154, 155, 157, 158]. The *in vitro* assembly of HBV capsids typically occurs at an electrolyte concentration much larger than that corresponding to the physiological condition. The high salinity is required not only to avoid capsid aggregation but also to maintain the thermodynamic stability of the reassembled capsids [158]. While DNA-containing viral particles are available from expression in mammalian cells at physiological conditions, a discrete separation of the capsids at each maturation state is currently impracticable. Detection of the capsid structure is also complicated by the intracellular environmental effects [108].

The theoretical framework discussed above allows us to determine the distributions of individual residues of the CTDs at different stages of replication. Figure 6.16 shows the accumulated number of linker-CTD residues as a function of the radial distance r from the capsid center. At each radial distance, we calculate the accumulated number of CTD residues per capsid protein (i.e., it changes from 0 to 42 as r increases from 0 at the capsid center to 22 at its fullest extent outside the capsid surface). For an empty capsid (Figure 6.16a), the accumulated number of CTD residues is about 29 at the inner capsid wall ($r = 13$ nm) and 30 at the outer surface ($r = 15$ nm). Accordingly, 30 residues are on average located inside the empty capsid, and 12 residues are exposed outside. However, for RNA-containing NC (Figure 6.16b), the accumulated number of linker-CTD residues is about 40 at the outer surface ($r = 15$ nm), indicating that only two CTD residues are distributed outside the capsid ($r > 15$ nm). At $r = 19$ nm from the center of the NC, the accumulated number of linker-CTD residues approaches 42. For the empty capsid (Figure 6.16a), CTD residues are distributed further from capsid center (up to $r = 22$ nm). We have also considered the distribution data for charged and neutral amino-acid groups but found not much variation between the empty and RNA-containing capsids.

Figure 6.17 compares the exposure ratio of individual CTD segments. A segment is "100% of exposure" if it is fully distributed in the outer region of the capsid, i.e., $r \geq 17$ nm. We assume that the residues in the outer region are able to fully interact with surrounding molecules like cellular enzymes. For convenience, we

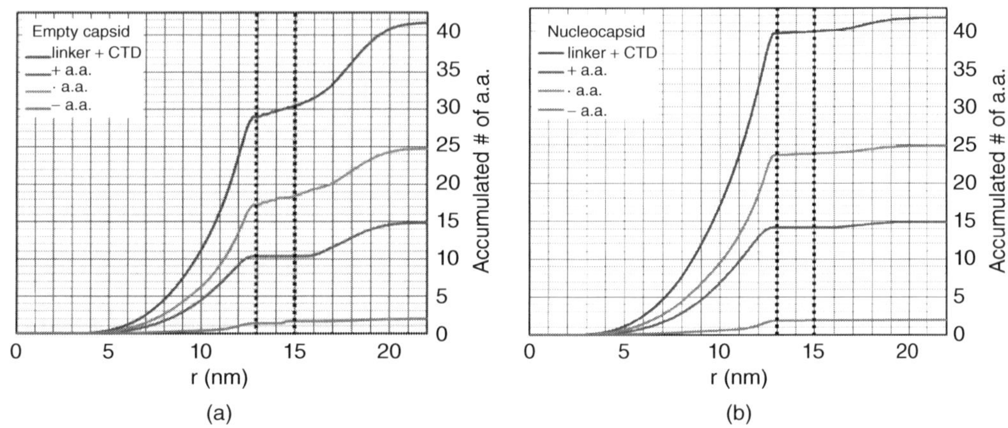

Figure 6.16 The accumulated number of CTD residues for empty (a) and RNA-containing (b) HBV capsids. The accumulated number is obtained from a radial integration of the density profiles for the linker-CTD residues and normalized by the total number of CTD tails ($N_t = 240$). In other words, the *y*-axis corresponds to the summed number of residues per CTD tail up to the radial distance (r) from the capsid center. As r increases, the summed value approaches 42, which is the total number of traced residues for each tail. The accumulated values for charged (+ and −) and neutral (·) residue groups are shown separately.

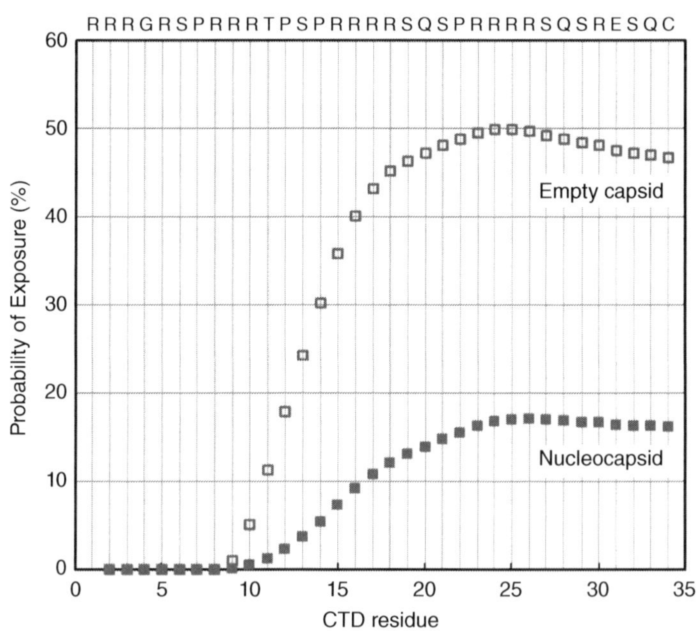

Figure 6.17 The ratio of the CTD residues exposed to the outer region of HBV capsids. The CTD residues exposed in the region $r \geq 17$ nm are counted for both the empty capsid and nucleocapsid. Each residue is ranked from the tethering site (1) to the end segment (34).

rank each residue according to its position at the CTD chain, i.e., residue 1 represents the amino acid tethered at the inner capsid surface, and residue 34 is the C-terminal residue. As expected, the empty capsid shows higher overall CTD exposure compared to the NC. Assuming that the RNA content is mainly responsible for the deviation between the empty capsid and the NC, we find that the attraction between CTDs and RNA internalizes more than half of exposed CTD segments in the empty capsid. In the latter case, the exposed ratio peaks around 21st–26th residues (^{170}Ser-^{171}Pro-^{172}Arg-^{173}Arg-^{174}Arg-^{175}Arg), coinciding with a typical binding motif for SRPK [159, 160].

Schematically, Figure 6.18 depicts the distribution of CTD residues for the empty capsid according to our theoretical calculations. Because the end residues have a slightly lower exposure ratio than the 21st–26th residues, a CTD tail must somehow bend in the middle, giving it a hook-like shape. The bent structure increases the possibility of the motif to contact with the SRPK. We find that about 30% of the residues present in such a motif are within reach of the SRPK kinase. *In vitro* study by Chen et al. indicates the transient CTD location in HBV capsids [82]. It revealed the association of SRPK on the outer surface of the empty capsid, mediated by the enzyme-binding motif in the CTD. The same assay was conducted for the NC-containing RNA genome, but in that case the enzyme binding was mostly inhibited. The SRPK binding demonstrated

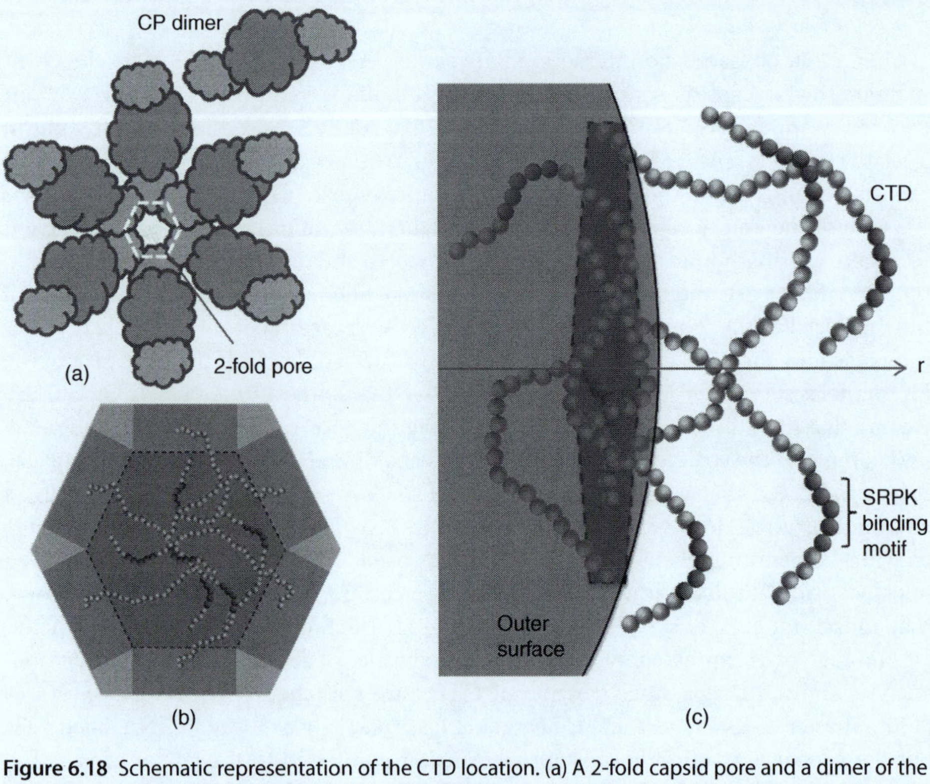

Figure 6.18 Schematic representation of the CTD location. (a) A 2-fold capsid pore and a dimer of the capsid core protein [64]. (b) Six CTD tails in each 2-fold pore. (c) CTD tails of the empty capsid. Here, the tails are distributed both inside and outside the empty capsid through the 2-fold hole.

unequivocally that the surface characteristics of the capsid change with the CTD location. Corresponding to the experimental approach, our theoretical results capture the transient feature of the capsid surface, rendering additional evidence of the interaction of CTD with SRPK. Although the primary kinase responsible for HBV phosphorylation is still under active investigation, SRPK has been a well-speculated candidate [142, 161]. SRPK is able to fulfill the *in vitro* phosphorylation of HBV CPs [142]. Moreover, the binding of this enzyme to every 2-fold pore of an HBV empty capsid has been reported [155].

The density distributions of individual CTD segments in HBV capsids suggest that a specific motif, 21st–26th CTD residues of the CTD, is located in the outer region ($r \geq 17$ nm) with the highest probability. Such a serine/arginine repeats domain serves as an ideal substrate for SRPK binding and nuclear localization signals (NLS) [162, 163]. At the physiological condition ($C_s = 0.14M$), the exposure fraction of the SRPK binding motif is expected to be ~30% of that corresponding to the empty capsid. A recent *in vitro* HBV study demonstrated that the empty capsid was decorated with SRPK proteins at all 2-fold capsid pores. However, RNA-containing nucleocapsids (RNA-NCs) failed in that binding assay, indicating that CTDs were not accessible to SRPK [155]. Regarding the exposure of the kinase binding motif, the deviation between the empty capsid and the NC has been confirmed by our DFT calculations. In the presence of the RNA, the exposure fraction of the SRPK binding motif of the 21st–26th CTD segments is reduced to ~5%.

Chen et al. observed no interaction between RNA-NCs and SRPKs from the chromatography of capsids sample into the SRPK bound column [155]. However, our theoretical model predicts that CTD chains are partially exposed in RNA-NCs but in a relatively much smaller portion in comparison to empty capsids. One explanation for this discrepancy is that the enzyme binding requires substantial substrate contact as for the case of the empty capsids. For RNA-NCs, the target motif exposure is insufficient to retain capsids bound on the column. It is worth mentioning that, in experiment, the RNA-NCs were obtained from an *E. coli* system, filled with the host RNA. Although the bacterial RNA mimic the authentic pgRNA regarding its content [22], their sequence and chemical characteristics are surely different [116]. If the RNA size in the NC model is set to have a slightly larger value than that of the authentic pgRNA (3.5 kb), we find that the CTD exposure ratio decreases significantly (data not shown). Rabe et al. [37] compared the structures of *E-coli*-derived capsids and authentic immature capsids. It was indicated that a certain portion of CTDs was exposed to outside RNA-NCs obtained *in vivo*. However, NCs obtained from *E. coli* were not affected by outside enzymes, suggesting their CTDs were not exposed. But even for an *E-coli*-derived nucleocapsid with RNA, it was speculated that a certain portion of CTDs are exposed. Vanlandschoot el al. observed the CTD-mediated attachment of nucleocapsids to glycosaminoglycans expressed on the plasma membranes of cells [85]. A recent structural analysis showed that a small portion of CTDs are stretched out through NC pores [150]. These results suggest that CTDs are at least partially exposed in RNA-filled NCs but the exposure fraction is substantially smaller compared to the empty capsid. The distinct structures of capsids suggest that the CTD location, specifically the distribution of the kinase target motifs, can be modulated in response to the encapsidated genome.

6.4.3 Implication of the CTD Exposure

Several recent investigations indicate the transient exposure of CTDs on the capsid surface [24, 146, 149, 151, 164, 165]. It has been postulated that the exposed CTDs may regulate the post-translational process of the HBV core, i.e., trafficking into nucleus, envelopment, and secretion. Kann et al. determined the fraction of the exposed CTDs for NCs in different maturation stages, and suggested the CTD-associated signal modulates the capsid delivery into the cellular nucleus [166]. Ning et al. observed that the secreted HBV particles contained either empty capsids or NCs with dsDNA, while the immature NCs, i.e., those filled with pgRNA or single-stranded DNA, were excluded from secretion [108]. Accordingly, a hypothesis was set such that the immature NCs negatively regulate the trafficking process [108, 154, 155].

Our theoretical model supports the mechanism of genome-regulated exposure of CTDs. Although we are not describing the whole process of HBV replication, a substantial structural change of CTD implies its functional correlation with the maturation signaling. Our model predicts that about ten residues per CTD tail (34 residues) are exposed outside the empty capsid. The exposure of 30% of CTD segments modifies the surface characteristics and affects the capsid trafficking. For empty capsids, the CTD tails have been suggested to extrude into far space from the capsid center, so that the outermost reachable r is $\sim 19\,\text{nm}$ for RNA-NC but $\sim 22\,\text{nm}$ for the empty capsid. Such a structural deviation between empty and RNA-filled capsid supports the hypothesis that the degree of CTD exposure may trigger selective selection upon the post-translational process [108, 154, 155]. The hypothesis, specifically the rationale on the transient CTD structure, was endorsed by experiments [154, 155]. In supporting those observations, our model gives the evidence on the CTD exposure and accessibility into the outer capsid space.

6.4.4 The Effect of Phosphorylation of Capsid Structure

Formation of HBV capsids and pgRNA reverse transcription are strongly correlated with phosphorylation and dephosphorylation of serine residues at the CTD of the core protein [44, 167, 168]. Phosphorylation at this domain is essential for packaging of pgRNA and for at least the initial stage of DNA synthesis. However, matured nucleocapsids are fully dephosphorylated and mutation on the phosphate sites may block synthesis of DNA [43, 115, 169]. Because phosphorylation affects the capsid charge and its affinity to the genomic nucleotides, reversible phosphorylation and dephosphorylation of the HBV core proteins provide an effective means to regulate electrostatic and other non-specific interactions within the HBV nucleocapsid. A quantitative understanding of the phosphorylation/dephosphorylation mechanism will therefore allow us to identify promising biophysical schemes to halt HBV replication by controlling the positions and charge status of the CTD chains.

Each CTD chain has three phosphorylation sites (S155, S162, and S170) important for RNA encapsidation and reverse transcription [144, 170]. For the wild type virus, the CTD chains are phosphorylated at the early stage of replication (with pgRNA or intermediate DNA) [145], but dephosphorylated for matured capsids (with dsDNA) [44, 171]. Replacing three serine residues with charged amino acids affects pgRNA encapsidation and DNA synthesis. For example, a phosphorylated analog, with one serine residue

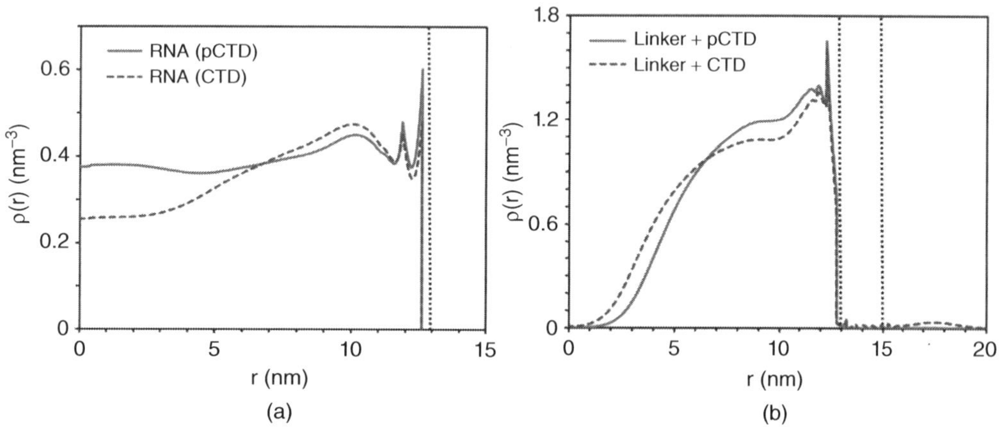

Figure 6.19 Effect of phosphorylation on the distributions of RNA and linker-CTD tails. (a) RNA distribution in phosphorylated (pCTD) and unphosphorylated (CTD) capsids. (b) The distributions of phosphorylated (pCTD) and unphosphorylated CTD tails.

replaced by glutamic acid, is able to package the pregenomic RNA but fails to support viral DNA replication [144]. Meanwhile, substantial reduction of RNA packaging was observed by substitution of serine with aspartic acid [168, 172].

Phosphorylation effects can be studied with DFT calculations simply by setting the valence of those three serine residues from 0 to -1 for the phosphorylated case. The addition of negative charges affects both the RNA distribution and the exposure of CTD chains. Figure 6.19 shows that the RNA structure inside the capsid varies significantly in response to CTD phosphorylation. Compared to the unphosphorylated case, the RNA segments become more uniformly distributed and are positioned closer to the capsid surface. CTD phosphorylation also transduces the RNA distribution to have a relatively higher peak near the capsid surface. Besides, phosphorylation reduces the extent of CTD exposure outside the capsid. Because the addition of negative charge reduces intra-chain electrostatic repulsion, the CTD brush is slightly collapsed in comparison to the unphosphorylated brush.

It was postulated that HBV carries serine residues in different phosphorylation states during the process of the capsid assembly and pgRNA reverse transcription. Specifically, in a duck hepatitis B virus (DHBV), the capsid proteins [64] were in phosphorylated form upon the capsid assembly. However, they were dephosphorylated for the mature NC filled with rcDNA [44, 173]. The CTD in phosphorylation state is believed to control the pgRNA encapsidation as well as the reverse transcription. Mutation of phosphorylation sites on CTD to mimic unphosphorylation hinders RNA packaging [168, 172, 174]. Similar mutation studies indicate that CTD phosphorylation affects DNA synthesis [145, 172, 174].

Our DFT calculations show that CTD phosphorylation induces RNA segments to be more localized toward inner capsid surface (Figure 6.19). The alteration in RNA structure was observed by Wang et al. [154]. In the phosphorylation-mimic case, the pgRNA inside the capsid shows more ordered structure than that in an *E-coli*-derived nucleocapsid. Our theoretical analysis indicates that such a structural ordering of RNA is correlated with the CTD location. In phosphorylated state, more CTD segments are localized at the region around $7 \sim 12$ nm from the capsid center, and the inner shell (< 7 nm) distribution of the phosphorylated CTDs is relatively depleted. It is expected that the RNA–CTD interaction would be reduced because of negative charges added to the CTD by the phosphorylation. Figure 6.19 shows such reduced complex formation between RNA and CTDs. At the inner region ($r < 7$ nm), the density profile of RNA for the phosphorylated case is higher than that for the unphosphorylated one. However, the corresponding densities of CTD segments are inverted in that region, thus the unphosphorylated CTD chains show higher segmental density than phosphorylated CTD chains. In other words, the CTD chains stay relatively apart from the RNA when they gain additional negative charges from phosphorylation. Accordingly, phosphorylation results in higher RNA density close to the inner surface of the capsid and it maintains monotonic radial distribution except near the inner wall. By contrast, RNA exhibits a more inhomogeneous distribution if the CTD chains are unphosphorylated.

6.5 Capsid Envelopment with Surface Proteins

HBV virions are generated from matured nucleocapsids after coating with surface proteins pre-existing in the ER membrane [48, 175]. As indicated above (see Figure 6.2), the Pre-S domain of the L-HBs plays a critical role in the envelopment of nucleocapsids [176], as well as in signaling cell binding with the hepatocyte receptors [18]. To mediate those interactions, L-HBs take alternative topologies for binding with the capsid proteins and hepatocyte receptors, with the Pre-S domain located at the inside and outside of the viral membrane, respectively. The internal Pre-S domain also mediates the interaction between the surface protein and cytosolic HBV nucleocapsids [177, 178].

Mutagenesis studies reveal candidate residues in HBcAg that interact specifically with the envelope proteins. Those residues may be located at the tip of the HBcAg spikes [179] or around the base of the 4-helix bundle where the C-terminus residues are located close to the capsid pores [180]. However, it remains unclear how the surface and the capsid proteins are engaged in an HBV virion. One key question is whether there are specific interactions between HBsAg and HBcAg. While electron microscopy measurements reveal a non-icosahedral arrangement of HBsAg in the envelope [31], Seitz et al. proposed a distinct packaging rule for the surface proteins, supporting their specific interactions with the nucleocapsid [181].

We may consider the putative interactions of HBcAg with the Pre-S domain of HBsAg as a special type of regulatory cellular processes that are mediated by protein–peptide interactions [182]. Although high-resolution analysis, such as NMR spectroscopy and X-ray crystallography, uncovers the atomic structure of many protein–peptide complexes, the structural information for the putative protein–peptide pair in HBV is unavailable. While it is experimentally challenging to detect the HBV surface–core protein recognition and binding processes, theoretical modeling and computation provide a valuable alternative.

We have identified probable interactions between the Pre-S1 domain of L-HBs and the core protein HBcAg using MD simulation and molecular docking methods. In MD simulations, we select 12 peptide fragments from the Pre-S1 domain in L-HBs (Figure 6.20b). Each peptide is initially placed at a random position near the HBcAg monomer. Parallel MD runs are then applied to each peptide to locate the binding

```
1 MGGWSSKPRQ GMGTNLSVPN PLGFFPDHQL DPAFGANSNN PDWDFNPNKD HWPEAHQVGA
  GAFGPGFTPP HGGLLGWSPQ AQGILTTVPV APPPASTNRQ SGRQPTPISP PLRDSHPQAM
  QWNSTTFHQA LLDPKVRGLY FPAGGSSSGT VNPVPTTASP ISSIFSRTGD PAPN 174
```

(a)

```
P1  (89-96) : PVAPPPAS            P4  (101-108): QSGRQPTPIS
P2 (93-100) : PPASTNRQ            P4-1 (99-102): RQSG
P3 (97-104) : TNRQSGRQ            P4-2(101-104): SGRQ
P4(101-108) : SGRQPTPI            P4-3(103-106): RQPT
P5(105-112) : PTPISPPL            P4-4(105-108): PTPI
P6(109-116) : SPPLRDSH            P4-5(107-110): PISP
P7(113-120) : RDSHPQAM                  (c)
P8(117-124) : PQAMQWNS
P9(121-128) : QWNSTTFH
P10(125-132) : TTFHQALL
P11(129-136) : QALLDPKV
P12(133-140) : DPKVRGLY
        (b)
```

Figure 6.20 The sequence of amino-acid (aa) residues in the Pre-S domain of the large HBV surface protein (L-HBs). (a) N-terminal residues (1 – 174 aa) of L-HBs. (b) 8-mer peptides (P1~P12) used in MD simulations. (c) 4-mer peptides (P4) used for molecular docking.

site. In most cases, the peptide moves around the protein without making any specific binding with the protein. However, the P4 peptide is able to bind with the core protein in a short simulation time [102], and remains associated until the end of the simulation [177].

Figure 6.21 shows two snapshots of the P4 peptide–core protein conformation from MD simulation. Within about 2 ns time frame, the peptide lands on the protein surface and the bound state is preserved over the entire duration of the simulation. The peptide approaches from a random position (Figure 6.21a), and ends up in a fixed bound state with the core protein (Figure 6.21b). The binding domain is located at residues

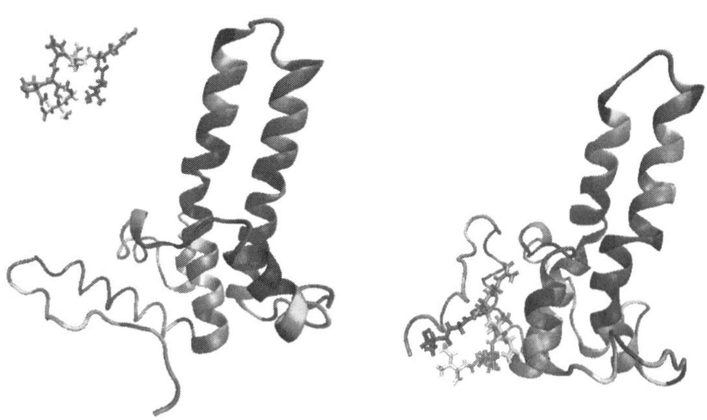

(a) (b)

Figure 6.21 Snapshots from MD simulation for P4 peptide binding to the HBcAG protein. (a) Initial conformation with P4 randomly placed around HBcAg. (b) The P4-core structure after 2 ns of simulation.

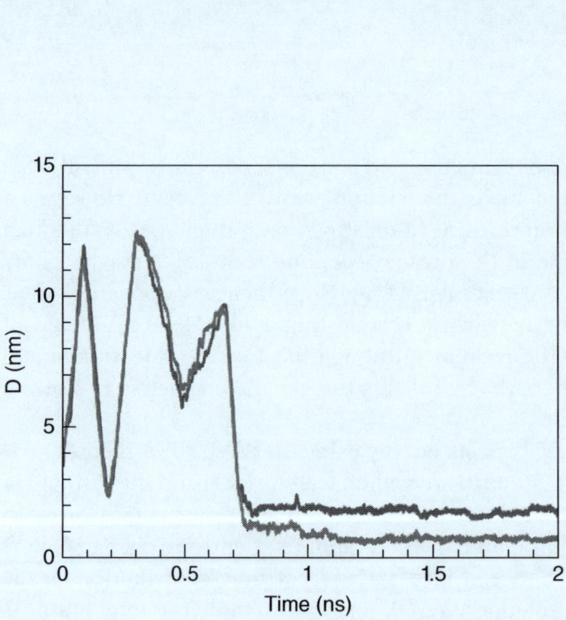

(a)

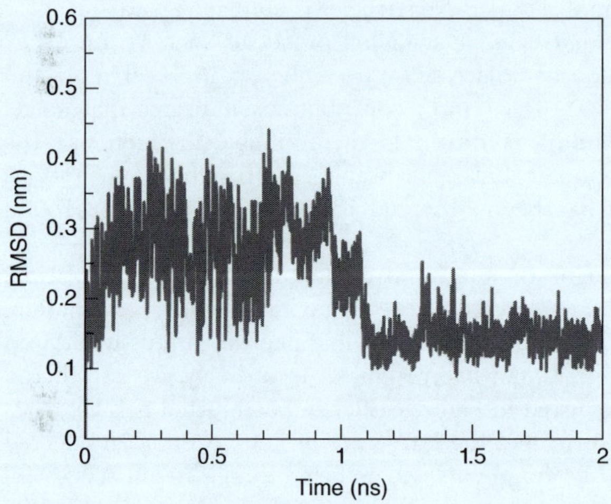

(b)

Figure 6.22 Evolution of P4–HBc association. (a) Change of the distance between the averaged centers of the P4 peptide and two binding domains (130–134 aa; 135–139 aa) from HBcAg. (b) Root mean square distribution (RMSD) of the backbone atoms in the P4 peptide.

135–139 of the capsid protein. According to the 3D protein structure, these residues are located at the initial portion of the linker domain of HBcAg and form a relatively flexible loop.

Figure 6.22a shows the change in distance between the center of mass for the P4 peptide and that of the putative binding domain (135–139) in HBcAg during the MD simulation. At the early stage of simulation, the P4–HBc separation oscillates at a frequency much lower than that of RMSD because its conformation fluctuates quite fast before reaching the binding domain of the core protein. At ~ 0.8 ns, however, the P4–HBc distance converges within about 2 nm. In comparison with the RMSD data (Figure 6.22b), we see that the peptide binds to the protein around 1 ns before its conformation converges into a stable form at ~ 1.1 ns. At ~ 1.5 ns, the peptide RMSD reduces to approximately 1.0 Å due to its association with the protein. However, a certain range of structural fluctuation remains, indicating that it might not be the final binding state. The equilibrated state of the protein–peptide complex is reached only after a long simulation, on the order of several μs [183]. Nevertheless, it appears evident that there exists a selective binding between a certain fragment (P4, 101–108 aa) of the Pre-S1 domain in L-HBs and HBc protein. Although this binding is identified at a relatively short time frame, we expect that it reflects the general characteristics of the specific peptide–protein association.

We may determine the P4–HBcAg binding energy using the MM/PBSA method. For that purpose, we run additional MD steps in an explicit water model and the results are shown in Table 6.2. Approximately, the internal energy (ΔE_{tot}) for the peptide–protein complex formation is ~ −36 kcal/mol, and the corresponding entropy change ($T\Delta S$) is ~ −29 kcal/mol. The difference between internal and entropic contributions leads to the peptide–protein binding free energy (ΔG), −6.67 kcal/mol. The interaction is relatively weak in comparison to that typical for peptide–protein binding (for ~ 8-mer peptide, it is about −10 kcal/mol [184]). Nevertheless, similar ligand–receptor association has been shown in other cellular signaling processes such as the T-cell response [185, 186]. A comparison of polar ($\Delta U_{Coul} + \Delta W_{PB} = 13.5$ kcal/mol) and non-polar ($\Delta U_{VDW} + \Delta W_{np} = -49.18$ kcal/mol) contributions indicates that, overall, the surface–core protein binding is driven by hydrophobic interactions. The combination of favorable hydrophobic but unfavorable electrostatic contributions to protein–peptide associations has been observed in many promiscuous protein recognitions [28, 147, 187].

To confirm the molecular recognition mechanism identified by MD simulation, we have also conducted the docking analysis for the P4 peptide–HBcAg association. It has been suggested that a probable docking of flexible peptide is mostly allowed for a short peptide composed of less than five residues. Accordingly, we considered 4-mer sub-fragments of the P4, as listed in Figure 6.20c, and conducted parallel trials of the docking procedure to the surface of the HBV core protein. Among 20 selected grid positions for each of five sub-P4 fragments, we detected three pairs of successful docking with negative binding energy. Interestingly, one of them is the case of P4-3 docked on the domain of 135–139 aa from the HBV core protein. This complex formation is clearly overlapped with the P4–HBcAg binding conformation predicted by our MD simulation. As seen in Figure 6.23, the P4-3 peptide, representing 103–106 aa in L-HBs, is supposed to be well adopted into a narrow loop region of HBcAg that is close to the linker domain of the core protein.

Table 6.2 The binding energy (kcal/mol) for the core protein–peptide interactions. Here, ΔU_{VDW} is the van der Waals energy; ΔU_{Coul} is the electrostatic energy; ΔW_{PB} and ΔW_{np} are the polar and non-polar contributions to the solvation energy; ΔE_{tot} is the total interaction energy, which is the sum of ΔE_{polar} and ΔE_{np}; $T\Delta S$ is the entropic energy term; ΔG is the binding free energy.

ΔU_{VDW}	ΔU_{Coul}	ΔW_{PB}	ΔW_{np}	ΔE_{tot}	$T\Delta S$	ΔG
−43.37	−292.19	305.69	−5.81	−35.66	−28.99	−6.67

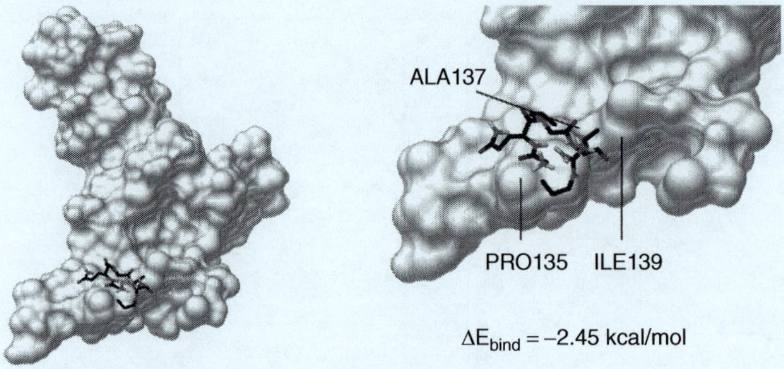

Figure 6.23 Docking of sub-fragment P4-3 onto HBcAg. Shown here are the docked conformation for P4-3 (101–103 aa of L-HBs) and the approximated binding energy. *(See insert for color representation of the figure.).*

We recall that only the mature nucleocapsids (with dsDNA) can be enveloped while the immature nucleocapsids (with ssRNA or ssDNA) are excluded in the envelopment. The probable binding between L-HBsAg and HBcAg suggests a possible scenario explaining the maturation-dependent selection mechanism for the formation of HBV virions. MD simulations and the docking analysis indicate that the recognition is mediated by 135–139 aa domain in HBcAg, which is located near the flexible C-terminal region. Such binding can be inhibited by the exposure of the C-terminal domain (CTD, 150–183 aa) of HBc. Although the CTD structure is not yet experimentally available, the equilibrium structure of the HBV nucleocapsid including the CTD chain location can be determined by the DFT calculation [140, 188].

Figure 6.24 shows the DFT predictions for the radial distributions of CTD residues in three different capsids: the mature form with dsDNA, immature form with ssRNA, and empty capsid without any genome content. The CTD chains tend to be associated more strongly with dsDNA with virtually all residues residing inside the capsid. On the other hand, a marginal portion of CTD segments is located beyond the inner capsid wall in the immature capsid. For empty capsids, an even greater portion of CTD residues are extruding outside the capsid region. Such a varied CTD positioning with respect to the genome content strongly suggests that the selective mechanism in the envelopment depends on the CTD location. When the CTD chains are totally internalized, i.e., in mature nucleocapsids, the putative core–envelope interaction should not be disturbed. However, if the CTD chains are placed close to the inside capsid wall or block the pore area on the capsid, then the contact of L-HBs to the binding site in HBc (135–139 aa) could easily be inhibited. Such a scenario may help to explain why only the empty and mature nucleocapsid can be enveloped. It is possible that the secretion of matured and empty capsids goes through different mechanisms. The requirements from core or envelope proteins for secreting empty capsids are still being investigated.

6.6 Summary and Outlook

In this review, we highlight our recent work on the thermodynamic basis of capsid formation, maturation, and envelopment in the life cycle of the hepatitis B virus

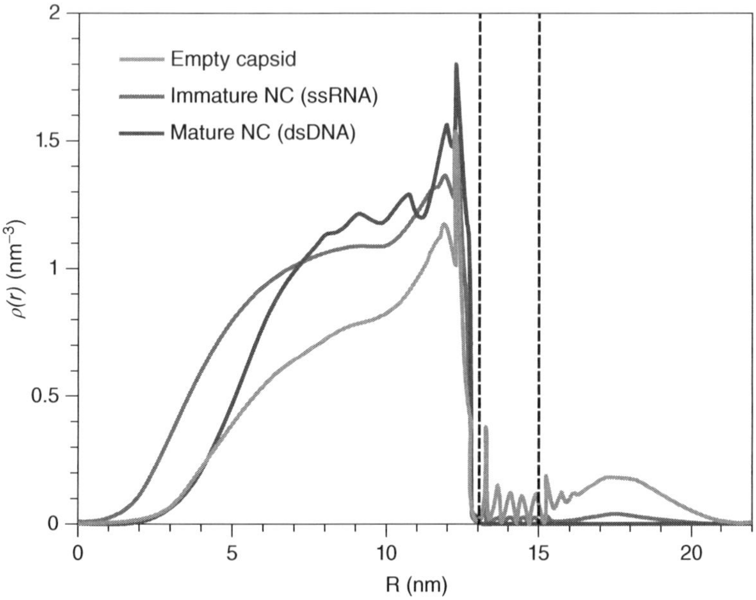

Figure 6.24 Radial distributions of CTD chains at different stages of maturation. Density profiles [$\rho(r)$] of the CTD residues are compared for the mature (with dsDNA), immature (with ssRNA), and empty capsids. Perpendicular dashed lines show the location of the capsid wall.

(HBV). Our objective is to provide a quantitative understanding of the viral replication processes and help development of promising antiviral strategies. A systematic investigation is conducted by modeling physiochemical processes at different scales from macroscopic range to atomic level, from which the thermodynamic basis on the capsid assembly, genome packaging, dynamic capsid structure, and capsid envelopment are elaborated. These theoretical models not only unveil the assembly pathways and replication mechanisms but also provide quantitative descriptions of molecular interactions and guidelines for designing new experiments.

Self-assembly of capsid proteins and genome encapsidation are two critical steps in the life cycle of most plant and animal viruses. A theoretical description of such processes from a physiochemical perspective may help better understand viral replication and morphogenesis, and thus provide fresh insights into the experimental studies of antiviral strategies. The molecular thermodynamic models developed in this work can be used to predict the stability of HBV capsids either with or without loading nucleic materials. With the key components represented by coarse-grained models, the theoretical predictions are in excellent agreement with experimental data for the formation free energies of empty T4 capsids over a broad range of temperature and ion concentrations. The theoretical model also predicts T3/T4 dimorphism in good agreement with the capsid formation for *in vivo* and *in vitro* conditions. In addition, we have studied the stability of the viral particles in response to physiological cellular conditions with the explicit consideration of the hydrophobic association of capsid subunits, electrostatic interactions, molecular excluded volume effects, entropy of mixing, and conformational changes of the biomolecular species. Because the course-grained model captures the essential features of the HBV nucleocapsid stability revealed by recent experiments, we expect that it can also be used to unravel HBV stability after cell entry and at different stages of capsid maturation.

In addition, we studied the equilibrium properties of HBV capsids using the classical density functional theory (DFT). DFT has the advantage of numerical efficiency compared to simulation methods, and provides direct information on thermodynamic properties with precise evaluation of various non-bonded molecular interactions beyond mean field approximations. We show that, in combination with coarse-grained polymer models, the DFT is able to quantify structure and thermodynamic properties of empty or genome-containing nucleocapsids. DFT calculations predict the optimal RNA lengths in quantitative agreement with experiments on the wild type (WT) virus and C-terminally truncated mutants. It is shown that the encapsidated genome and the flexible tails of core proteins undergo structural transitions with regard to viral maturing steps or in response to phosphorylation of the capsid proteins. We find that both nucleocapsid formation and genome packaging strongly depend on the charge density of CTDs. Moreover, theoretical results for the distributions of amino acid residues in CTD tails in different replication stages unveil the signature of mature nucleocapsids that differentiate them from the immature ones, suggesting that the CTD tails may be utilized to regulate the post-transcriptional process of the viral particles.

The combined analysis of MD simulations using both implicit and explicit water models as well as molecular docking reveals specific peptide–protein binding underlying the HBsAg–HBcAg association. The theoretical results suggest that the specific binding of Pre-S domain of L-HBsAg with HBcAg arises from recognition between protein domains consisting of 101st–108th residues from L-HBsAG and 135th–139th residues from HBcAg. A theoretical analysis based on the molecular mechanics/Poisson–Boltzmann surface area (MM/PBSA) method indicates that the approximated binding energy is −6.7 kcal/mol, which is weaker than that for typical cellular receptor ligand binding. The capsid–envelope association is enhanced by multi-body cooperative effects and may be regulated by the spatial location of flexible C-termini of HBcAg. The theoretical model provides further insights into the selective envelopment of HBV capsids at different degrees of maturation. We found the domains of 101–108 aa in L-HBs and 135–139 aa in HBc are critical in this molecular recognition. The L-HBs–HBc association has a lower binding free energy in comparison with typical cellular receptor–ligand binding, indicating that envelopment is a cooperatively driven multi-body process. Considering that the binding site is close to the C-terminal domain (CTD) of HBcAg, we expect that the core–envelope association is closely coordinated with the nucleocapsid maturation. In immature nucleocapsids, the putative docking of L-HBsAg to HBcAg is blocked because the CTD tails are partially exposed and cover the capsid surface. The dynamic nature of the CTD distribution explains the selective envelopment of HBV nucleocapsids during virion formation.

Acknowledgments

The authors are grateful to Dr Jianming Hu at Penn State College of Medicine and Dr Rex Hjelm at the Los Alamos National Laboratory for careful reading of this manuscript and helpful comments. For financial support, we thank the National Institutes of Health (R21-AI077532) and the National Science Foundation (NSF-CBET-0852353). The allocation of supercomputer hours from the National Energy Research Scientific Computing Center (NERSC) is also gratefully acknowledged.

References

1. Ahmed, M., F. Wang, A. Levin, C. Le, Y. Eltayebi, M. Houghton, L. Tyrrell, and K. Barakat, Targeting the Achilles heel of the hepatitis B virus: a review of current treatments against covalently closed circular DNA. *Drug Discov Today*, 2015. **20**(5): 548–561.
2. WHO, *Hepatitis B: Fact sheet N°204* 2015, World Health Organization.
3. Trepo, C., H.L.Y. Chan, and A. Lok, Hepatitis B virus infection. *Lancet*, 2014. **384**(9959): 2053–2063.
4. Ott, J.J., G.A. Stevens, J. Groeger, and S.T. Wiersma, Global epidemiology of hepatitis B virus infection: New estimates of age-specific HBsAg seroprevalence and endemicity. *Vaccine*, 2012. **30**(12): 2212–2219.
5. Liaw, Y.F., N. Leung, R. Guan, G.K.K. Lau, I. Merican, G. McCaughan, E. Gane, J.H. Kao, M. Omata, and A.-P.C.U. Work, Asian-Pacific consensus statement on the management of chronic hepatitis B: a 2005 update. *Liver International*, 2005. **25**(3): 472–489.
6. Orlando, R., M. Foggia, A.E. Maraolo, S. Mascolo, G. Palmiero, O. Tambaro, and G. Tosone, Prevention of hepatitis B virus infection: From the past to the future. *European Journal of Clinical Microbiology & Infectious Diseases*, 2015. **34**(6): 1059–1070.
7. Chu, C.K., T.W. Ma, K. Shanhmuganathan, C.G. Wang, Y.J. Xiang, S.B. Pai, G.Q. Yao, J.P. Sommadossi, and Y.C. Cheng, Use of 2'-fluoro-5-methyl-beta-l-arabinofuranosyluracil as a novel antiviral agent for hepatitis-B virus and epstein-barr-virus. *Antimicrobial Agents and Chemotherapy*, 1995. **39**(4): 979–981.
8. de Man, R.A., L.M.M. Wolters, F. Nevens, D. Chua, M. Sherman, C.L. Lai, A. Gadano, Y. Lee, F. Mazzotta, N. Thomas, and D. DeHertogh, Safety and efficacy of oral entecavir given for 28 days in patients with chronic hepatitis B virus infection. *Hepatology*, 2001. **34**(3): 578–582.
9. Dienstag, J.L., R.P. Perrillo, E.R. Schiff, M. Bartholomew, C. Vicary, and M. Rubin, A preliminary trial of lamivudine for chronic hepatitis-B infection. *New England Journal of Medicine*, 1995. **333**(25): 1657–1661.
10. Lai, C.L., S.G. Lim, N.A. Brown, X.J. Zhou, D.M. Lloyd, Y.M. Lee, M.F. Yuen, G.C. Chao, and M.W. Myers, A dose-finding study of once-daily oral telbivudine in HBeAg-positive patients with chronic hepatitis B virus infection. *Hepatology*, 2004. **40**(3): 719–726.
11. Chong, Y.H. and C.K. Chu, Understanding the unique mechanism of L-FMAU (clevudine) against hepatitis B virus: Molecular dynamics studies. *Bioorganic & Medicinal Chemistry Letters*, 2002. **12**(23): 3459–3462.
12. Yadav, V. and C.K. Chu, Molecular mechanisms of adefovir sensitivity and resistance in HBV polymerase mutants: A molecular dynamics study. *Bioorganic & Medicinal Chemistry Letters*, 2004. **14**(16): 4313–4317.

13. Dane, D.S., C.H. Cameron, and M. Briggs, Virus-like particles in serum of patients with Australia-antigen-associated hepatitis. *Lancet*, 1970. **1**(7649): 695–698.
14. Seeger, C. and W.S. Mason, Hepatitis B virus biology. *Microbiology and Molecular Biology Reviews*, 2000. **64**(1): 51–68.
15. Bruss, V., Hepatitis B virus morphogenesis. *World Journal of Gastroenterology*, 2007. **13**(1): 65–73.
16. Beck, J. and M. Nassal, Hepatitis B virus replication. *World Journal of Gastroenterology*, 2007. **13**(1): 48–64.
17. Prange, R., Host factors involved in hepatitis B virus maturation, assembly, and egress. *Medical Microbiology and Immunology*, 2012. **201**(4): 449–461.
18. Glebe, D. and S. Urban, Viral and cellular determinants involved in hepadnaviral entry. *World Journal of Gastroenterology*, 2007. **13**(1): 22–38.
19. Bruss, V. and D. Ganem, The role of envelope proteins in hepatitis-B virus assembly. *Proceedings of the National Academy of Sciences of the United States of America*, 1991. **88**(3): 1059–1063.
20. Zlotnick, A., Viruses and the physics of soft condensed matter. *Proceedings of the National Academy of Sciences of the United States of America*, 2004. **101**(44): 15549–15550.
21. Zlotnick A, V.B., Tan Z, Lewellyn E, Turner W, Francis S., Core protein: A pleiotropic keystone in the HBV lifecycle. *Antiviral Research*, 2015. **121**: 82–93.
22. Zlotnick, A., N. Cheng, S.J. Stahl, J.F. Conway, A.C. Steven, and P.T. Wingfield, Localization of the C terminus of the assembly domain of hepatitis B virus capsid protein: Implications for morphogenesis and organization of encapsidated RNA. *Proceedings of the National Academy of Sciences of the United States of America*, 1997. **94**(18): 9556–9561.
23. Bottcher, B., S.A. Wynne, and R.A. Crowther, Determination of the fold of the core protein of hepatitis B virus by electron cryomicroscopy. *Nature*, 1997. **386**(6620): 88–91.
24. Wynne, S.A., R.A. Crowther, and A.G.W. Leslie, The crystal structure of the human hepatitis B virus capsid. *Molecular Cell*, 1999. **3**(6): 771–780.
25. Yu, X.K., L. Jin, J. Jih, C.H. Shih, and Z.H. Zhou, 3.5 angstrom cryoEM structure of hepatitis B virus core assembled from full-length core protein. *Plos One*, 2013. **8**(9): e69729.
26. Watts, N.R., J.F. Conway, N.Q. Cheng, S.J. Stahl, D.M. Belnap, A.C. Steven, and P.T. Wingfield, The morphogenic linker peptide of HBV capsid protein forms a mobile array on the interior surface. *Embo Journal*, 2002. **21**(5): 876–884.
27. Conway, J.F., N. Cheng, A. Zlotnick, P.T. Wingfield, S.J. Stahl, and A.C. Steven, Visualization of a 4-helix bundle in the hepatitis B virus capsid by cryo-electron microscopy. *Nature*, 1997. **386**(6620): 91–94.
28. Basdevant, N., H. Weinstein, and M. Ceruso, Thermodynamic basis for promiscuity and selectivity in protein-protein interactions: PDZ domains, a case study. *Journal of the American Chemical Society*, 2006. **128**(39): 12766–12777.
29. Luo, L., S. Chen, Q. Gong, N. Luo, Y. Lei, J.J. Guo, and S. He, Hepatitis B virus X protein modulates remodelling of minichromosomes related to hepatitis B virus replication in HepG2 cells. *International Journal of Molecular Medicine*, 2013. **31**(1): 197–204.

30. Roseman, A.M., J.A. Berriman, S.A. Wynne, P.J.G. Butler, and R.A. Crowther, A structural model for maturation of the hepatitis B virus core. *Proceedings of the National Academy of Sciences of the United States of America*, 2005. **102**(44): 15821–15826.
31. Dryden, K.A., S.F. Wieland, C. Whitten-Bauer, J.L. Gerin, F.V. Chisari, and M. Yeager, Native hepatitis B virions and capsids visualized by electron cryomicroscopy. *Molecular Cell*, 2006. **22**(6): 843–850.
32. Hu, J., *HBV Virology and Replication*. 2015, Springer.
33. Nassal, M., Hepatitis B viruses: Reverse transcription a different way. *Virus Research*, 2008. **134**(1–2): 235–249.
34. Watashi, K., S. Urban, W.H. Li, and T. Wakita, NTCP and beyond: Opening the door to unveil hepatitis B virus entry. *International Journal of Molecular Sciences*, 2014. **15**(2): 2892–2905.
35. Yan, H., G.C. Zhong, G.W. Xu, W.H. He, Z.Y. Jing, Z.C. Gao, Y. Huang, Y.H. Qi, B. Peng, H.M. Wang, L.R. Fu, M. Song, P. Chen, W.Q. Gao, B.J. Ren, Y.Y. Sun, T. Cai, X.F. Feng, J.H. Sui, and W.H. Li, Sodium taurocholate cotransporting polypeptide is a functional receptor for human hepatitis B and D virus. *Elife*, 2012. **1**: e00049.
36. Guo, H.T., C.X. Xu, T.L. Zhou, T.M. Block, and J.T. Guo, Characterization of the host factors required for hepadnavirus covalently closed circular (ccc) DNA formation. *Plos One*, 2012. **7**(8): e43270.
37. Rabe, B., A. Vlachou, N. Pante, A. Helenius, and M. Kann, Nuclear import of hepatitis B virus capsids and release of the viral genome. *Proceedings of the National Academy of Sciences of the United States of America*, 2003. **100**(17): 9849–9854.
38. Hu, J. and C. Seeger, Hepadnavirus genome replication and persistence. *Cold Spring Harb Perspect Med*, 2015. **5**(7): 10.1101/cshperspect.a021386.
39. Ryu, D.K., S. Kim, and W.S. Ryu, Hepatitis B virus polymerase suppresses translation of pregenomic RNA via a mechanism involving its interaction with 5' stem-loop structure. *Virology*, 2008. **373**(1): 112–123.
40. Bartenschlager, R. and H. Schaller, Hepadnaviral assembly is initiated by polymerase binding to the encapsidation signal in the viral RNA genome. *Embo Journal*, 1992. **11**: 3413–3420.
41. Liu, N., R. Tian, and D.D. Loeb, Base pairing among three cis-acting sequences contributes to template switching during hepadnavirus reverse transcription. *Proceedings of the National Academy of Sciences of the United States of America*, 2003. **100**(4): 1984–1989.
42. Lewellyn, E.B. and D.D. Loeb, Base pairing between cis-acting sequences contributes to template switching during plus-strand DNA synthesis in human hepatitis B virus. *Journal of Virology*, 2007. **81**(12): 6207–6215.
43. Basagoudanavar, S.H., D.H. Perlman, and J.M. Hu, Regulation of hepadnavirus reverse transcription by dynamic nucleocapsid phosphorylation. *Journal of Virology*, 2007. **81**(4): 1641–1649.
44. Perlman, D.H., E.A. Berg, P.B. O'Connor, C.E. Costello, and J.M. Hu, Reverse transcription-associated dephosphorylation of hepadnavirus nucleocapsids. *Proceedings of the National Academy of Sciences of the United States of America*, 2005. **102**(25): 9020–9025.
45. Cui, X., L. Ludgate, X.J. Ning, and J.M. Hu, Maturation-associated destabilization of hepatitis B virus nucleocapsid. *Journal of Virology*, 2013. **87**(21): 11494–11503.

46. Ning, X.J., D. Nguyen, L. Mentzer, C. Adams, H. Lee, R. Ashley, S. Hafenstein, and J.M. Hu, Secretion of genome-free hepatitis B virus: Single strand blocking model for virion morphogenesis of para-retrovirus. *Plos Pathogens*, 2011. 7(9): e1002255.

47. Gilbert, R.J.C., L. Beales, D. Blond, M.N. Simon, B.Y. Lin, F.V. Chisari, D.I. Stuart, and D.J. Rowlands, Hepatitis B small surface antigen particles are octahedral. *Proceedings of the National Academy of Sciences of the United States of America*, 2005. **102**(41): 14783–14788.

48. Ganem, D., Assembly of hepadnaviral virions and subviral particles. *Current Topics in Microbiology and Immunology*, 1991. **168**: 61–83.

49. McAleer, W.J., E.B. Buynak, R.Z. Maigetter, D.E. Wampler, W.J. Miller, and M.R. Hilleman, Human hepatitis-B vaccine from recombinant yeast. *Nature*, 1984. **307**(5947): 178–180.

50. Mateu, M.G., Assembly, stability and dynamics of virus capsids. *Archives of Biochemistry and Biophysics*, 2013. **531**(1–2): 65–79.

51. Katen, S. and A. Zlotnick, The thermodynamics of virus capsid assembly. *Methods in Enzymology*, 2009. **455**: 395–417.

52. Nowak, M.A., S. Bonhoeffer, A.M. Hill, R. Boehme, H.C. Thomas, and H. McDade, Viral dynamics in hepatitis B virus infection. *Proceedings of the National Academy of Sciences of the United States of America*, 1996. **93**(9): 4398–4402.

53. Yang, D.R., C.H. Zuo, X.H. Wang, X.H. Meng, B.B. Xue, N.L. Liu, R. Yu, Y.W. Qin, Y.M. Gao, Q.P. Wang, J. Hu, L. Wang, Z.B. Zhou, B. Liu, D.M. Tan, Y. Guan, and H.Z. Zhu, Complete replication of hepatitis B virus and hepatitis C virus in a newly developed hepatoma cell line. *Proceedings of the National Academy of Sciences of the United States of America*, 2014. **111**(13): E1264–E1273.

54. Wingfield, P.T., S.J. Stahl, R.W. Williams, and A.C. Steven, Hepatitis core antigen produced in *Escherichia coli*: subunit composition, conformational analysis, and in vitro capsid assembly. *Biochemistry*, 1995. **34**: 4919–4932.

55. Zlotnick, A., N. Cheng, J.F. Conway, F.P. Booy, A.C. Steven, S.J. Stahl, and P.T. Wingfield, Dimorphism of hepatitis B virus capsids is strongly influenced by the C-terminus of the capsid protein. *Biochemistry*, 1996. **35**(23): 7412–7421.

56. Zlotnick, A., J.M. Johnson, P.W. Wingfield, S.J. Stahl, and D. Endres, A theoretical model successfully identifies features of hepatitis B virus capsid assembly. *Biochemistry*, 1999. **38**(44): 14644–14652.

57. Ceres, P. and A. Zlotnick, Weak protein–protein interactions are sufficient to drive assembly of hepatitis B virus capsids. *Biochemistry*, 2002. **41**(39): 11525–11531.

58. Nassal, M., The arginine-rich domain of the hepatitis B virus core protein is required for pregenome encapsidation and productive viral positive-strand DNA synthesis but not for virus assembly. *Journal of Virology*, 1992. **66**: 4107–4116

59. Chua, P.K., F.-M. Tang, J.-Y. Huang, C.-S. Suen, and C. Shih, Testing the balanced electrostatic interaction hypothesis of hepatitis B virus DNA synthesis by using an *in vivo* charge rebalance approach. *Journal of Virology*, 2010. **84**(5): 2340–2351.

60. Newman, M., P.K. Chua, F.M. Tang, P.Y. Su, and C. Shih, Testing an electrostatic interaction hypothesis of hepatitis B virus capsid stability by using an in vitro capsid disassembly/reassembly system. *Journal of Virology*, 2009. **83**(20): 10616–10626.

61. Porterfield, J.Z., M.S. Dhason, D.D. Loeb, M. Nassal, S.J. Stray, and A. Zlotnick, Full-length hepatitis B virus core protein packages viral and heterologous

RNA with similarly high levels of cooperativity. *Journal of Virology*, 2010. **84**: 7174–7184.

62. Dhason, M.S., J.C.Y. Wang, M.F. Hagan, and A. Zlotnick, Differential assembly of hepatitis B virus core protein on single- and double-stranded nucleic acid suggest the dsDNA-filled core is spring-loaded. *Virology*, 2012. **430**(1): 20–29.
63. Yu, N. and M.F. Hagan, Simulations of HIV capsid protein dimerization reveal the effect of chemistry and topography on the mechanism of hydrophobic protein association. *Biophysical Journal*, 2012. **103**(6): 1363–1369.
64. Freddolino, P.L., A.S. Arkhipov, S.B. Larson, A. McPherson, and K. Schulten, Molecular dynamics simulations of the complete satellite tobacco mosaic virus. *Structure*, 2006. **14**(3): 437–449.
65. Hagan, M.F., Modeling viral capsid assembly. *Advances in Chemical Physics*, 2013. **155**:1–68.
66. Siber, A., A.L. Bozic, and R. Podgornik, Energies and pressures in viruses: Contribution of nonspecific electrostatic interactions. *Physical Chemistry Chemical Physics*, 2012. **14**(11): 3746–3765.
67. Zhang, R., E. Wernersson, and P. Linse, Icosahedral capsid formation by capsomer subunits and a semiflexible polyion. *Rsc Advances*, 2013. **3**(47): 25258–25267.
68. Belyi, V.A. and M. Muthukumar, Electrostatic origin of the genome packing in viruses. *Proc. Natl. Acad. Sci. U.S.A.*, 2006. **103**: 17174–17178.
69. Bruinsma, R.F., W.M. Gelbart, D. Reguera, J. Rudnick, and R. Zandi, Viral self-assembly as a thermodynamic process. *Physical Review Letters*, 2003. **90**: 248101.
70. Kegel, W.K. and P. van der Schoot, Competing hydrophobic and screened-Coulomb interactions in hepatitis B virus capsid assembly. *Biophysical Journal*, 2004. **86**(6): 3905–3913
71. Siber, A. and R. Podgornik, Nonspecific interactions in spontaneous assembly of empty versus functional single-stranded RNA viruses. *Physical Review E*, 2008. **78**: 051915.
72. Kim, J. and J.Z. Wu, A molecular thermodynamic model for the stability of hepatitis B capsids. *Journal of Chemical Physics*, 2014. **140**(23): 235101.
73. Crick, F.H.C. and J.D. Watson, Structure of small viruses. *Nature*, 1956. **177**(4506): 473–475.
74. van der Schoot, P. and R. Bruinsma, Electrostatics and the assembly of an RNA virus. *Physical Review E*, 2005. **71**: 061928.
75. Crowther, R.A., N.A. Kiselev, B. Bottcher, J.A. Berriman, G.P. Borisova, V. Ose, and P. Pumpens, Three-dimensional structure of hepatitis B virus core particles determined by electron cryomicroscopy. *Cell*, 1994. **77**: 943–950.
76. Baldwin, R.L., The new view of hydrophobic free energy. *Febs Letters*, 2013. **587**(8): 1062–1066.
77. Wu, J. and J.M. Prausnitz, Pairwise-additive hydrophobic effect for alkanes in water. *Proc. Natl. Acad. Sci. U.S.A.*, 2008. **105**: 9512–9515.
78. Hagan, M.F., Controlling viral capsid assembly with templating. *Physical Review E*, 2008. **77**: 051904.
79. Israelachvili, J.N., *Intermolecular and Surface Forces*. 3rd ed. 2011, Burlington, MA: Academic Press.

80. Farnum, M. and C. Zukoski, Effect of glycerol on the interactions and solubility of bovine pancreatic trypsin inhibitor. *Biophysical Journal*, 1999. **76**(5): 2716–2726.

81. Weast, R.C., *Handbook of Chemistry and Physics*. 1984, Boca Ranton, FL: CRC Press.

82. Chen, C., J.C.-Y. Wang, and A. Zlotnick, A kinase chaperones hepatitis B virus capsid assembly and captures capsid dynamics in vitro. *Plos Pathogens*, 2011. **7**: e1002388.

83. Perlmutter, J.D., C. Qiao, and M.F. Hagan, Viral genome structures are optimal for capsid assembly. *eLife*, 2013. **2**: e00632.

84. van der Schoot, P. and R. Zandi, Impact of the topology of viral RNAs on their encapsulation by virus coat proteins. *Journal of Biological Physics*, 2013. **39**: 289–299.

85. Vanlandschoot, P., F. Van Houtte, B. Serruys, and G. Leroux-Roels, The arginine-rich carboxy-terminal domain of the hepatitis B virus core protein mediates attachment of nucleocapsids to cell-surface-expressed heparan sulfate. *Journal of General Virology*, 2005. **86**: 75–84.

86. Casassa, E.F. and Y. Tagami, An equilibrium theory for exclusion chromatography of branched and linear polymer chains. *Macromolecules*, 1969. **2**: 14–26.

87. Prinsen, P., P. van der Schoot, W.M. Gelbart, and C.M. Knobler, Multishell structures of virus coat proteins. *Journal of Physical Chemistry B*, 2010. **114**: 5522–5533.

88. Netz, R.R. and D. Andelman, Neutral and charged polymers at interfaces. *Physics Reports-Review Section of Physics Letters*, 2003. **380**: 1–95.

89. Zandi, R. and P. van der Schoot, Size regulation of ss-RNA viruses. *Biophysical Journal*, 2009. **96**: 9–20

90. Overbeek, J.T.G. and M.J. Voorn, Phase separation in polyelectrolyte solutions. Theory of complex coacervation. *J Cell and Comp. Physiol.*, 1957. **49**: 7–26.

91. Yoffe, A.M., P. Prinsen, A. Gopal, C.M. Knobler, W.M. Gelbart, and A. Ben-Shaul, Predicting the sizes of large RNA molecules. *Proc. Natl. Acad. Sci. U.S.A.*, 2008. **105**: 16153–16158.

92. Hyeon, C., R.I. Dima, and D. Thirumalai, Size, shape, and flexibility of RNA structures. *Journal of Chemical Physics*, 2006. **125**: 194905.

93. Carrillo, J.-M.Y. and A.V. Dobrynin, Polyelectrolytes in salt solutions: Molecular dynamics simulations. *Macromolecules*, 2011. **44**: 5798–5816.

94. Caliskan, G., C. Hyeon, U. Perez-Salas, R.M. Briber, S.A. Woodson, and D. Thirumalai, Persistence length changes dramatically as RNA folds. *Physical Review Letters*, 2005. **95**: 268303.

95. Zlotnick, A., P. Ceres, S. Singh, and J.M. Johnson, A small molecule inhibits and misdirects assembly of hepatitis B virus capsids. *Journal of Virology*, 2002. **76**: 4848–4854.

96. Kegel, W.K. and P. van der Schoot, Physical regulation of the self-assembly of tobacco mosaic virus coat protein. *Biophysical Journal*, 2006. **91**(4): 1501–1512.

97. Siber, A. and R. Podgornik, Role of electrostatic interactions in the assembly of empty spherical viral capsids. *Physical Review E*, 2007. **76**(6).

98. Stray, S.J., P. Ceres, and A. Zlotnick, Zinc ions trigger conformational change and oligomerization of hepatitis B virus capsid protein. *Biochemistry*, 2004. **43**(31): 9989–9998.

99. Stray, S.J., C.R. Bourne, S. Punna, W.G. Lewis, M.G. Finn, and A. Zlotnick, A heteroaryldihydropyrimidine activates and can misdirect hepatitis B virus capsid assembly. *Proceedings of the National Academy of Sciences of the United States of America*, 2005. **102**(23): 8138–8143.

100. Katen, S.P., S.R. Chirapu, M.G. Finn, and A. Zlotnick, Trapping of hepatitis B virus capsid assembly intermediates by phenylpropenamide assembly accelerators. *Acs Chemical Biology*, 2010. **5**(12): 1125–1136.

101. Bourne, C.R., M.G. Finn, and A. Zlotnick, Global structural changes in hepatitis B virus capsids induced by the assembly effector HAP1. *Journal of Virology*, 2006. **80**(22): 11055–11061.

102. Alexander, C.G., M.C. Jurgens, D.A. Shepherd, S.M.V. Freund, A.E. Ashcroft, and N. Ferguson, Thermodynamic origins of protein folding, allostery, and capsid formation in the human hepatitis B virus core protein. *Proceedings of the National Academy of Sciences of the United States of America*, 2013. **110**(30): E2782–E2791.

103. Ceres, P., S.J. Stray, and A. Zlotnick, Hepatitis B virus capsid assembly is enhanced by naturally occurring mutation F97L. *Journal of Virology*, 2004. **78**(17): 9538–9543.

104. Newman, M., F.M. Suk, M. Cajimat, P.M. Chua, and C. Shih, Stability and morphology comparisons of self-assembled virus-like particles from wild-type and mutant human hepatitis B virus capsid proteins. *Journal of Virology*, 2003. **77**(24): 12950–12960.

105. Packianathan, C., S.P. Katen, C.E. Dann, and A. Zlotnick, Conformational changes in the hepatitis B virus core protein are consistent with a role for allostery in virus assembly. *Journal of Virology*, 2010. **84**(3): 1607–1615.

106. de Vries, R. and M.C. Stuart, Theory and simulations of macroion complexation. *Current Opinion in Colloid & Interface Science*, 2006. **11**(5): 295–301.

107. Yu, X., L. Jin, J. Jih, C. Shih, and Z.H. Zhou, 3.5A cryoEM structure of hepatitis B virus core assembled from full-length core protein. *PLoS One*, 2013. **8**(9): e69729.

108. Ning, X., D. Nguyen, L. Mentzer, C. Adams, H. Lee, R. Ashley, S. Hafenstein, and J. Hu, Secretion of genome-free hepatitis B virus: Single strand blocking model for virion morphogenesis of para-retrovirus. *Plos Pathogens*, 2011. 7(9): e1002255.

109. Ting, C.L., J. Wu, and Z.-G. Wang, Thermodynamic basis for the genome to capsid charge relationship in viral encapsidation. *Proc. Natl. Acad. Sci. U.S.A.*, 2011. **108**: 16986–16991.

110. Campagna, M.R., F. Liu, R.C. Mao, C. Mills, D.W. Cai, F. Guo, X.S. Zhao, H. Ye, A. Cuconati, H.T. Guo, J.H. Chang, X.D. Xu, T.M. Block, and J.T. Guo, Sulfamoylbenzamide derivatives inhibit the assembly of hepatitis B virus nucleocapsids. *Journal of Virology*, 2013. **87**(12): 6931–6942.

111. Tan, Z.N., K. Pionek, N. Unchwaniwala, M.L. Maguire, D.D. Loeb, and A. Zlotnick, The interface between hepatitis B virus capsid proteins affects self-assembly, pregenomic RNA packaging, and reverse transcription. *Journal of Virology*, 2015. **89**(6): 3275–3284.

112. Uetrecht, C., C. Versluis, N.R. Watts, P.T. Wingfield, A.C. Steven, and A.J.R. Heck, Stability and shape of hepatitis B virus capsids in vacuo. *Angewandte Chemie-International Edition*, 2008. **47**: 6247–6251.

113. Kenney, J.M., C.H. Vonbonsdorff, M. Nassal, and S.D. Fuller, Evolutionary conservation in the hepatitis B virus core structure: comparison of human and duck cores. *Structure*, 1995. **3**: 1009–1019.

114. He, L., Z. Porterfield, P. van der Schoot, A. Zlotnick, and B. Dragnea, Hepatitis virus capsid polymorph stability depends on encapsulated cargo size. *Acs Nano*, 2013. **7**(10): 8447–8454.

115. Le Pogam, S., P.K. Chua, M. Newman, and C. Shih, Exposure of RNA templates and encapsidation of spliced viral RNA are influenced by the arginine-rich domain of human hepatitis B virus core antigen (HBcAg 165-173). *Journal of Virology*, 2005. **79**(3): 1871–1887.

116. Newman, M., P.K. Chua, F.-M. Tang, P.-Y. Su, and C. Shih, Testing an electrostatic interaction hypothesis of hepatitis B virus capsid stability by using an *in vitro* capsid disassembly/reassembly system. *Journal of Virology*, 2009. **83**(20): 10616–10626.

117. Liu, S., J. He, C. Shih, K. Li, A. Dai, Z.H. Zhou, and J. Zhang, Structural comparisons of hepatitis B core antigen particles with different C-terminal lengths. *Virus Research*, 2010. **149**(2): 241–244.

118. Bringas, R., Folding and assembly of hepatitis B virus core protein: A new model proposal. *Journal of Structural Biology*, 1997. **118**(3): 189–196.

119. Dong, X.F., P. Natarajan, M. Tihova, J.E. Johnson, and A. Schneemann, Particle polymorphism caused by deletion of a peptide molecular switch in a quasiequivalent icosahedral virus. *Journal of Virology*, 1998. **72**(7): 6024–6033.

120. Kaplan, I.B., L. Zhang, and P. Palukaitis, Characterization of cucumber mosaic virus: V. Cell-to-cell movement requires capsid protein but not virions. *Virology*, 1998. **246**(2): 221–231.

121. Marshall, D. and A. Schneemann, Specific packaging of nodaviral RNA2 requires the N-terminus of the capsid protein. *Virology*, 2001. **285**(1): 165–175.

122. Schmitz, I. and A.L.N. Rao, Deletions in the conserved amino-terminal basic arm of cucumber mosaic virus coat protein disrupt virion assembly but do not abolish infectivity and cell-to-cell movement. *Virology*, 1998. **248**(2): 323–331.

123. Schneemann, A. and D. Marshall, Specific encapsidation of nodavirus RNAs is mediated through the C terminus of capsid precursor protein alpha. *Journal of Virology*, 1998. **72**(11): 8738–8746.

124. Scott, K.P., M.J. Farmer, D.J. Robinson, L. Torrance, and A.F. Murant, Comparison of the coat protein of groundnut rosette assistor virus with those of other luteoviruses. *Annals of Applied Biology*, 1996. **128**(1): 77–83.

125. Smith, T.J., E. Chase, T. Schmidt, and K.L. Perry, The structure of cucumber mosaic virus and comparison to cowpea chlorotic mottle virus. *Journal of Virology*, 2000. **74**(16): 7578–7586.

126. Tihova, M., K.A. Dryden, T.V.L. Le, S.C. Harvey, J.E. Johnson, M. Yeager, and A. Schneemann, Nodavirus coat protein imposes dodecahedral RNA structure independent of nucleotide sequence and length. *Journal of Virology*, 2004. **78**(6): 2897–2905.

127. Hagan, M.F., Modeling viral capsid assembly. *Advances in Chemical Physics*, 2014. **155**: 1–67.

128. Angelescu, D.G., R. Bruinsma, and P. Linse, Monte Carlo simulations of polyelectrolytes inside viral capsids. *Physical Review E*, 2006. **73**(4): 041921.

129. Belyi, V.A. and M. Muthukumar, Electrostatic origin of the genome packing in viruses. *Proceedings of the National Academy of Sciences of the United States of America*, 2006. **103**(46): 17174–17178.
130. Hu, T., R. Zhang, and B.I. Shkovskii, Electrostatic theory of viral self-assembly. *Physica a-Statistical Mechanics and Its Applications*, 2008. **387**(12): 3059–3064.
131. Siber, A. and R. Podgornik, Nonspecific interactions in spontaneous assembly of empty versus functional single-stranded RNA viruses. *Physical Review E*, 2008. **78**(5): 051915.
132. van der Schoot, P. and R. Bruinsma, Electrostatics and the assembly of an RNA virus. *Physical Review E*, 2005. **71**(6): 061928.
133. Ostrow, K.M. and D.D. Loeb, Underrepresentation of the 3' region of the capsid pregenomic RNA of duck hepatitis B virus. *Journal of Virology*, 2004. **78**(5): 2179–2186.
134. Jiang, H. and S.X. Sun, Cellular pressure and volume regulation and implications for cell mechanics. *Biophysical Journal*, 2013. **105**(3): 609–619.
135. Tinevez, J.-Y., U. Schulze, G. Salbreux, J. Roensch, J.-F. Joanny, and E. Paluch, Role of cortical tension in bleb growth. *Proceedings of the National Academy of Sciences of the United States of America*, 2009. **106**(44): 18581–18586.
136. Jiang, T., Z.-G. Wang, and J. Wu, Electrostatic regulation of genome packaging in human hepatitis B virus. *Biophysical Journal*, 2009. **96**(8): 3065–3073.
137. Jiang, T., Z. Li, and J. Wu, Structure and swelling of grafted polyelectrolytes: Predictions from a nonlocal density functional theory. *Macromolecules*, 2007. **40**(2): 334–343.
138. Li, Z.D. and J.Z. Wu, Density functional theory for polyelectrolytes near oppositely charged surfaces. *Physical Review Letters*, 2006. **96**(4): 048302.
139. Kim, J. and J. Wu, A theoretical study of SRPK interaction with the flexible domains of hepatitis B capsids. *Biophysical Journal*, 2014. **107**(6): 1453–1461.
140. Meng, D., R.P. Hjelm, J. Hu, and J. Wu, A theoretical model for the dynamic structure of hepatitis B nucleocapsid. *Biophysical Journal*, 2011. **101**(10): 2476–2484.
141. Sominskaya, I., D. Skrastina, I. Petrovskis, A. Dishlers, I. Berza, M. Mihailova, J. Jansons, I. Akopjana, I. Stahovska, D. Dreilina, V. Ose, and P. Pumpens, A VLP library of C-terminally truncated hepatitis B core proteins: Correlation of RNA encapsidation with a Th1/Th2 switch in the immune responses of mice. *Plos One*, 2013. **8**(9): e75938.
142. Daub, H., S. Blencke, P. Habenberger, A. Kurtenbach, J. Dennenmoser, J. Wissing, A. Ullrich, and M. Cotten, Identification of SRPK1 and SRPK2 as the major cellular protein kinases phosphorylating hepatitis B virus core protein. *Journal of Virology*, 2002. **76**(16): 8124–8137.
143. Chu, T.H., A.T. Liou, P.Y. Su, H.N. Wu, and C.H. Shih, Nucleic acid chaperone activity associated with the arginine-rich domain of human hepatitis B virus core protein. *Journal of Virology*, 2014. **88**(5): 2530–2543.
144. Lan, Y.T., J. Li, W.Y. Liao, and J.H. Ou, Roles of the three major phosphorylation sites of hepatitis B virus core protein in viral replication. *Virology*, 1999. **259**(2): 342–348.
145. Lewellyn, E.B. and D.D. Loeb, Serine phosphoacceptor sites within the core protein of hepatitis B virus contribute to genome replication pleiotropically. *Plos One*, 2011. **6**(2): e17202.

146. Schmitz, A., A. Schwarz, M. Foss, L. Zhou, B. Rabe, J. Hoellenriegel, M. Stoeber, N. Pante, and M. Kann, Nucleoporin 153 arrests the nuclear import of hepatitis B virus capsids in the nuclear basket. *Plos Pathogens*, 2010. **6**(1): e1000741.

147. Huang, Y.-m.M., M. Kang, and C.-e.A. Chang, Mechanistic insights into phosphopeptide-BRCT domain association: Preorganization, flexibility, and phosphate recognition. *Journal of Physical Chemistry B*, 2012. **116**(34): 10247–10258.

148. Gallina, A., F. Bonelli, L. Zentilin, G. Rindi, M. Muttini, and G. Milanesi, A recombinant hepatitis B core antigen polypeptide with the protamine-like domain deleted self-assembles into capsid particles but fails to bind nucleic acids. *Journal of Virology*, 1989. **63**(11): 4645–4652.

149. Seifer, M. and D.N. Standring, A protease-sensitive hinge linking the two domains of the hepatitis B virus core protein is exposed on the viral capsid surface. *Journal of Virology*, 1994. **68**(9): 5548–5555.

150. Yu, X., L. Jin, J. Jih, C. Shih, and Z.H. Zhou, 3.5 angstrom cryoEM structure of hepatitis B virus core assembled from full-length core protein. *Plos One*, 2013. **8**(9): e69729.

151. Kann, M., B. Sodeik, A. Vlachou, W.H. Gerlich, and A. Helenius, Phosphorylation-dependent binding of hepatitis B virus core particles to the nuclear pore complex. *Journal of Cell Biology*, 1999. **145**(1): 45–55.

152. Schlicht, H.J., R. Bartenschlager, and H. Schaller, The duck hepatitis B virus core protein contains a highly phosphorylated C terminus that is essential for replication but not for RNA packaging. *Journal of Virology*, 1989. **63**(7): 2995–3000.

153. Guo, H., R. Mao, T.M. Block, and J.-T. Guo, Production and function of the cytoplasmic deproteinized relaxed circular DNA of hepadnaviruses. *Journal of Virology*, 2010. **84**(1): 387–396.

154. Wang, J.C.Y., M.S. Dhason, and A. Zlotnick, Structural organization of pregenomic RNA and the carboxy-terminal domain of the capsid protein of hepatitis B virus. *Plos Pathogens*, 2012. **8**(9): e1002919.

155. Chen, C., J.C.-Y. Wang, and A. Zlotnick, A kinase chaperones hepatitis B virus capsid assembly and captures capsid dynamics *in vitro*. *Plos Pathogens*, 2011. **7**(11): e1002388.

156. Roseman, A.M., O. Borschukova, J.A. Berriman, S.A. Wynne, P. Pumpens, and R.A. Crowther, Structures of hepatitis B virus cores presenting a model epitope and their complexes with antibodies. *Journal of Molecular Biology*, 2012. **423**(1): 63–78.

157. Wingfield, P.T., S.J. Stahl, R.W. Williams, and A.C. Steven, Hepatitis core antigen produced in *Escherichia coli*: Subunit composition, conformational analysis, and *in vitro* capsid assembly. *Biochemistry*, 1995. **34**(15): 4919–4932.

158. Porterfield, J.Z., M.S. Dhason, D.D. Loeb, M. Nassal, S.J. Stray, and A. Zlotnick, Full-length hepatitis B virus core protein packages viral and heterologous RNA with similarly high levels of cooperativity. *Journal of Virology*, 2010. **84**(14): 7174–7184.

159. Gui, J.F., W.S. Lane, and X.D. Fu, A serine kinase regulates intracellular-localization of splicing factors in the cell cycle. *Nature*, 1994. **369**(6482): 678–682.

160. Colwill, K., L.L. Feng, J.M. Yeakley, G.D. Gish, J.F. Caceres, T. Pawson, and X.D. Fu, SRPK1 and Clk/Sty protein kinases show distinct substrate specificities for serine/arginine-rich splicing factors. *Journal of Biological Chemistry*, 1996. **271**(40): 24569–24575.

161. Zheng, Y.Y., X.D. Fu, and J.H.J. Ou, Suppression of hepatitis B virus replication by SRPK1 and SRPK2 via a pathway independent of the phosphorylation of the viral core protein. *Virology*, 2005. **342**(1): 150–158.
162. Li, H.-C., E.-Y. Huang, P.-Y. Su, S.-Y. Wu, C.-C. Yang, Y.-S. Lin, W.-C. Chang, and C. Shih, Nuclear export and import of human hepatitis B virus capsid protein and particles. *Plos Pathogens*, 2010. **6**(10): e1001162.
163. Yeh, C.T., Y.F. Liaw, and J.H. Ou, The arginine-rich domain of hepatitis-B virus precore and core proteins contains a signal for nuclear transport. *Journal of Virology*, 1990. **64**(12): 6141–6147.
164. Yu, M.S. and J. Summers, Phosphorylation of the duck hepatitis B virus capsid protein associated with conformational changes in the C terminus. *Journal of Virology*, 1994. **68**(5): 2965–2969.
165. Hilmer, J.K., A. Zlotnick, and B. Bothner, Conformational equilibria and rates of localized motion within hepatitis B virus capsids. *Journal of Molecular Biology*, 2008. **375**(2): 581–594.
166. Kann, M., A. Schmitz, and B. Rabe, Intracellular transport of hepatitis B virus. *World Journal of Gastroenterology*, 2007. **13**(1): 39–47.
167. Machida, A., H. Ohnuma, F. Tsuda, A. Yoshikawa, Y. Hoshi, T. Tanaka, S. Kishimoto, Y. Akahane, Y. Miyakawa, and M. Mayumi, Phosphorylation in the carboxyl-terminal domain of the capsid protein of hepatitis-B virus: Evaluation with a monoclonal-antibody. *Journal of Virology*, 1991. **65**(11): 6024–6030.
168. Gazina, E.V., J.E. Fielding, B. Lin, and D.A. Anderson, Core protein phosphorylation modulates pregenomic RNA encapsidation to different extents in human and duck hepatitis B viruses. *Journal of Virology*, 2000. **74**(10): 4721–4728.
169. Liu, K., L. Ludgate, Z.H. Yuan, and J.M. Hu, Regulation of multiple stages of hepadnavirus replication by the carboxyl-terminal domain of viral core protein in trans. *Journal of Virology*, 2015. **89**(5): 2918–2930.
170. Nassal, M., The arginine-rich domain of the hepatitis B virus core protein is required for pregenome encapsidation and productive viral positive-strand DNA synthesis but not for virus assembly. *Journal of Virology*, 1992. **66**(7): 4107–4116.
171. Basagoudanavar, S.H., D.H. Perlman, and J. Hu, Regulation of hepadnavirus reverse transcription by dynamic nucleocapsid phosphorylation. *Journal of Virology*, 2007. **81**(4): 1641–1649.
172. Kock, J., M. Nassal, K. Deres, H.E. Blum, and F. von Weizsacker, Hepatitis B virus nucleocapsids formed by carboxy-terminally mutated core proteins contain spliced viral genomes but lack full-size DNA. *Journal of Virology*, 2004. **78**(24): 13812–13818.
173. Pugh, J., A. Zweidler, and J. Summers, Characterization of the major duck hepatitis B virus core particle protein. *Journal of Virology*, 1989. **63**(3): 1371–1376.
174. Melegari, M., S.K. Wolf, and R.J. Schneider, Hepatitis B virus DNA replication is coordinated by core protein serine phosphorylation and HBx expression. *Journal of Virology*, 2005. **79**(15): 9810–9820.
175. Huovila, A.P.J., A.M. Eder, and S.D. Fuller, Hepatitis-B surface-antigen assemblies in a post-ER, pre-Golgi compartment. *Journal of Cell Biology*, 1992. **118**(6): 1305–1320.

176. Bruss, V., A short linear sequence in the pre-S domain of the large hepatitis B virus envelope protein required for virion formation. *Journal of Virology*, 1997. **71**(12): 9350–9357.

177. Bruss, V. and K. Vieluf, Functions of the internal pre-S domain of the large surface protein in hepatitis-B virus particle morphogenesis. *Journal of Virology*, 1995. **69**(11): 6652–6657.

178. Poisson, F., A. Severac, C. Hourioux, A. Goudeau, and P. Roingeard, Both pre-S1 and S domains of hepatitis B virus envelope proteins interact with the core particle. *Virology*, 1997. **228**(1): 115–120.

179. Bottcher, B., N. Tsuji, H. Takahashi, M.R. Dyson, S. Zhao, R.A. Crowther, and K. Murray, Peptides that block hepatitis B virus assembly: Analysis by cryomicroscopy, mutagenesis and transfection. *Embo Journal*, 1998. **17**(23): 6839–6845.

180. Ponsel, D. and V. Bruss, Mapping of amino acid side chains on the surface of hepatitis B virus capsids required for envelopment and virion formation. *Journal of Virology*, 2003. **77**(1): 416–422.

181. Seitz, S., S. Urban, C. Antoni, and B. Boettcher, Cryo-electron microscopy of hepatitis B virions reveals variability in envelope capsid interactions. *Embo Journal*, 2007. **26**(18): 4160–4167.

182. Schittl, B. and V. Bruss, Mutational profiling of the variability of individual amino acid positions in the hepatitis B virus matrix domain. *Virology*, 2014. **458**: 183–189.

183. Shan, Y., E.T. Kim, M.P. Eastwood, R.O. Dror, M.A. Seeliger, and D.E. Shaw, How does a drug molecule find its target binding site? *Journal of the American Chemical Society*, 2011. **133**(24): 9181–9183.

184. Logean, A., A. Sette, and D. Rognan, Customized versus universal scoring functions: Application to class I MHC-peptide binding free energy predictions. *Bioorganic & Medicinal Chemistry Letters*, 2001. **11**(5): 675–679.

185. Corse, E., R.A. Gottschalk, and J.P. Allison, Strength of TCR-peptide/MHC interactions and *in vivo* T cell responses. *Journal of Immunology*, 2011. **186**(9): 5039–5045.

186. Germain, R.N. and I. Stefanova, The dynamics of T cell receptor signaling: Complex orchestration and the key roles of tempo and cooperation. *Annual Review of Immunology*, 1999. **17**: 467–522.

187. Hou, T., K. Chen, W.A. McLaughlin, B. Lu, and W. Wang, Computational analysis and prediction of the binding motif and protein interacting partners of the Abl SH3 domain. *Plos Computational Biology*, 2006. **2**(1): 46-55.

188. Kim, J. and J.Z. Wu, A theoretical study of SRPK interaction with the flexible domains of hepatitis B capsids. *Biophysical Journal*, 2014. **107**(6): 1453–1461.

7

Simulation Studies of Metal–Ligand Self-Assembly

Makoto Yoneya

National Institute of Advanced Industrial Science and Technology, Tsukuba, Japan

7.1 Introduction

In biological systems, hydrogen bonds are known to play a crucial role in the structure formation of a DNA helix, protein alpha-helix and beta-sheet, among others [1]. Compared to this hydrogen-bond-directed self-assembly, coordination-bond-directed self-assembly is less pronounced in biological systems, but has been extensively studied recently in synthetic chemistry [2, 3]. In these coordination systems, both finite supramolecular complexes and infinite polymer networks are possible via self-assembly with the appropriate metal–ligand system design [3, 4]. Well-known examples are the finite spherical complex $Pd(II)_{12}L_{24}$, which is composed of 12 palladium ions [Pd(II)] and 24 bent ligands (L in Figure 7.1a) [4], and the two-dimensional (2-D)-like infinite network which is composed of cadmium ions and linear ligands (4,4′-bipyridine, Figure 7.1b) [5].

The former $M_{12}L_{24}$ cages are most extensively studied as finite supramolecular complexes [6], and proven to be functional molecular cages to encapsulate perfluoroalkanes [7], polymerize methyl methacrylate segments [8] and protect a small protein [9]. The latter-type 2-D-like or 3-D infinite coordination polymer networks, which are also known as metal–organic frameworks (MOFs), have been extensively studied over the last decade for various applications, e.g., gas storage, gas separation, and catalysis [10, 11]. Understanding the mechanism of these coordination-directed self-assemblies is important for designing new functional materials. However, only a limited number of experimental studies have been reported to date on the self-assembly process in both finite supramolecular complexes and infinite polymer networks [12–15].

Computational studies can play a significant role in this area of research [3]. Generally, the time scale of the metal–ligand self-assembly is far beyond the accessible time scale of common molecular dynamics (MD) simulations. Therefore we must speed up the self-assembly using some types of coarse-grained modeling to fill the gap between simulations and real reaction systems.

In this chapter, we will introduce MD simulation studies of the metal–ligand self-assembly, which begins with our modeling method.

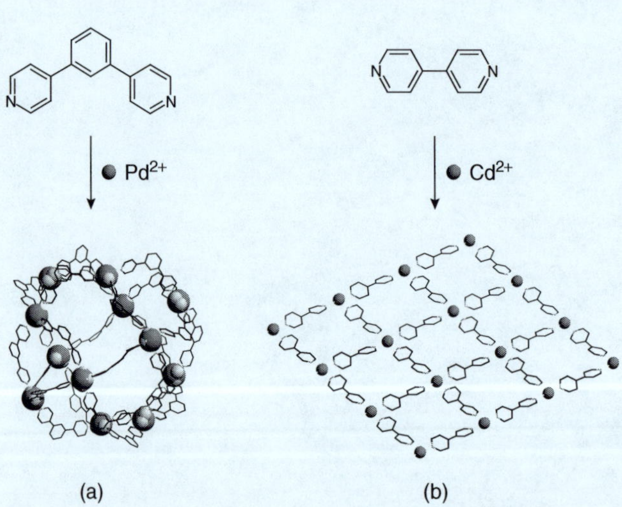

Figure 7.1 (a) 4,4′-benzene-1,3-diyldipyridine (L) and (b) 4,4′-bipyridine (bpy).

7.2 Modeling Metal–Ligand Self-Assembly

7.2.1 Modeling Metals, Ligands and their Interactions

We began with the relatively straightforward part of the modeling, i.e., metals and ligands. For the metal-atom non-bonded interaction parameters, literature values [16] or values in the universal force field (UFF) [17] were applied, and the van der Waals bond length was scaled to reproduce the metal–ligand nitrogen distance from the X-ray diffraction study. For ligand molecules, we applied the flexible united- or all-atom model with the exception that all bond-stretching degrees of freedom were constrained to the equilibrium bond lengths. For the inter- and intramolecular interactions, a general AMBER force field (GAFF) [18] was used. In the united-atom model, the CH and CH_2 united-atom parameters from a reoptimized united-atom force field [19] were used. Restrained electrostatic potential (RESP) charges [20], which were obtained using *ab initio* molecular orbital calculations with the B3LYP/6-31G(d) level in the Gaussian03 program [21], were used for the atomic charges of the ligands.

Next, the metal–ligand coordination interaction is to be modeled. Various methods have been proposed to model the coordination of metal ions and ligands [22]. For example, the bonded model uses covalent bonds between metal ions and coordinating sites of ligands to maintain the specific symmetry of the coordinations. For our purposes, this bonded model did not apply because we wanted to simulate spontaneous coordination bond formations and breaks over the course of the simulations. In this case, we applied a non-bonded model, which maintains the coordination with the electrostatic and van der Waals interactions. More specifically, we applied the cationic dummy atom (CaDA) model, which was originally applied to zinc complex proteins [23]. The zinc CaDA model uses four identical dummy atoms that are tetrahedrally attached to the zinc ions and evenly transfers the atomic charge of the zinc divalent cation to the

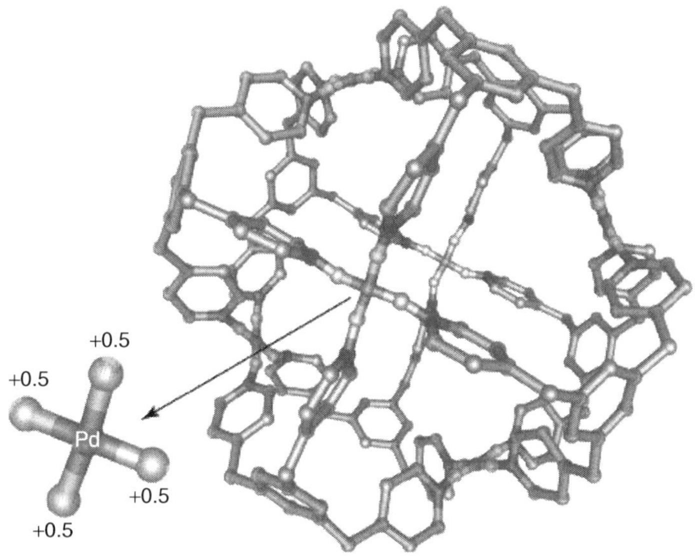

Figure 7.2 Pd(II) model using cationic dummy atoms (CaDA) and M_6L_8 nanosphere. The Pd(II) CaDA model uses four identical dummy atoms, which are coplanarly attached to the Pd ions, and evenly transfers the atomic charge of the Pd(II) cation to the four dummy atoms.

four dummy atoms. Similar to this zinc model, the Pd(II) CaDA model uses four identical dummy atoms, which are coplanarly attached to Pd, as shown in Figure 7.2. The metal-dummy atom distance was set to 0.09 nm as in the original CaDA paper [23].

The metal–ligand binding in the CaDA model is purely electrostatic and possibly too simple to represent the actual coordination interaction. We compared the calculated metal–ligand binding energy between the present model and the *ab initio* MO calculations. The results are summarized in Table 7.1. The binding energy was obtained as the dissociation energy of a pyridine (py) molecule from the Pd(II)(py)$_4$ complex. The MP2/6-31G* level optimized geometry of Pd(II)(py)$_4$ was used to calculate the metal–ligand binding energy. The DGDZVP basis set [24] was used for Pd. The binding energy value with the present force-field calculation was between those of the standard Hartree–Fock (HF) and high-level MP2 methods; we think that it is sufficient for the present study.

Table 7.1 Calculated metal–ligand binding energy (kcal/mol)

	HF	MP2	Present force-field
			−52.8
6-31G*	−49.65	−61.18	
6-311G*	−49.20	−61.26	
6-311G**	−49.10	−61.16	
cc-pVDZ	−48.79	−60.61	
aug-cc-pVDZ	−48.41	−65.05	

7.2.2 Modeling Solvents

Next, we explain the method that we used to speed up the assembly process of the model metal–ligand system in a solution.

As discussed in Section 7.1, we introduce the methods to fill the time scale gap between the actual and the simulated systems. In this study, we accelerated the assembly of metals and ligands by implicitly treating the solvent as a continuum medium. There are various types of implicit-solvent models [25]. Here, we applied a combination of three methods: Langevin (stochastic) dynamics, reaction fields and coarse-grained potentials. To account for the dynamic (heat-bath) effects of the solvent molecules, Langevin dynamics (LD) [26], which adds a friction and random force to the conventional MD, was applied. To represent the electrostatic effects of the solvent, the generalized reaction field method [27], which extends the "classical" reaction field method to ionic systems, was used. In this method, the relative dielectric constant beyond the distance parameter for the complete neutralization (screening) can be separately specified as the far field value ϵ_{rf} from the near-field value ϵ_r. Similar to the electrostatic effects, the van der Waals interactions among the ligands are also screened by the solvation to effectively be short-range repulsive ones. We applied a Weeks–Chandler–Andersen (WCA) potential [28] for this short-range repulsive potential. The WCA potential was truncated at the repulsive part of the common Lennard-Jones potential and shifted as follows:

$$V(r_{ij}) = \begin{cases} 4\epsilon_{ij}\left[\left(\frac{\sigma_{ij}}{r_{ij}}\right)^{12} - \left(\frac{\sigma_{ij}}{r_{ij}}\right)^{6}\right] + \epsilon_{ij} & r_{ij} < 2^{1/6}\sigma_{ij} \\ 0 & r_{ij} \geq 2^{1/6}\sigma_{ij}, \end{cases} \quad (7.1)$$

where r_{ij} is the distance between the atoms, and ϵ and σ are Lennard-Jones potential parameters. Such a combination of the Coulomb interaction with an effective dielectric constant and the WCA potential interaction was used as a simple implicit-solvent model in the ionic-system study [29]. A combination of the LD heat-bath and WCA potential is also used in the well-known Kremer–Grest model for polymers in a good solvent [30].

We did not include counter anions (e.g., NO_3^-) in our model. To realize the metal–ligand self-assembly within the available computational time, we should enhance the probability of coordination formation between the Pd(II) ions and the ligands. Inclusion of the counter ions reduces this probability because these ions would compete with the ligands for the metal ions. In reality, the metal–ligand self-assembly takes a much longer time than that in our model because the ligand coordination requires time to dissociate the solvents or counter anions from metal ions. Our model (without real solvents and counter anions) overestimates the coordination rate of the metal ions to the ligands. Because of the overestimation, our model is certainly a poor model to evaluate an accurate coordination rate. In principle, the time scale in any coarse-grained model is not the "real" time scale, but the "coarse-grained" time scale because of the acceleration dynamics, which is the case for our model. However, this overestimation effectively accelerates the coordination rate of the metals to the ligands and enables our model to realize the metal–ligand self-assembly within the available computational time.

To study the metal–ligand self-assembly, we should consider the elements that play important roles in the self-assembly process, but we think that it is not necessary to use the most realistic possible model. Currently, we do not have evidence showing that the dissociation process of the solvents or anions from metal ions plays such an active role. Thus, we did not include the actual solvents and anions in our current model to propose the "minimum" model that realizes the metal–ligand self-assembly.

7.2.3 Computational Methods

Finally, we explain the computation methods. Trajectories were produced using the MD program package GROMACS (version 4) [31] with modification of the non-bonded interaction routine to use the WCA potential. For the LD time integration, we used the leapfrog stochastic dynamics integrator [32] and LINCS bond constraint [33], where the time step was either 4 fs for the all-atom model or 5 fs for the united-atom model because of the stability of the LINCS algorithm [34].

The charge-group-based twin-range 0.65 nm van der Waals and 1.4 nm electrostatic cutoff distances [35] were applied to the non-bonded interactions. In contrast to the common setting for the latter value (1.4 nm) [36], the former value (0.65 nm) is shorter than the common value because we used the short-ranged WCA potential. The simulation temperature was maintained by coupling to a stochastic thermostat with a time constant $\tau_t = 0.1$ ps via LD. The LD friction coefficient for each atom was specified as the GROMACS default value, i.e., mass/τ_t.

7.3 Self-Assembly of Supramolecular Coordination Complex

7.3.1 Self-Assembly of M_6L_8 Spherical Complex [37]

We first tried one of the simplest examples of the spherical complex, M_6L_8, a nanosphere of octahedral symmetry with six palladium (Pd) ions and eight tridentate ligands (ligand **1** in Figure 7.3) [38]. The M_6L_8 complex can be readily formed from a 4 : 3 mixture of ligand **1** and $Pd(NO_3)_2$ in dimethyl sulfoxide (DMSO) at 90°C. The formation process was directly monitored by nuclear magnetic resonance spectroscopy in DMSO – d_6 solvent, which shows the completed complexation reaction within 5 min [38]. Additionally, no peaks other than those due to the M_6L_8 complex were observed, which establishes the monodisperse assembly and remarkable thermodynamic stability of the complex. The

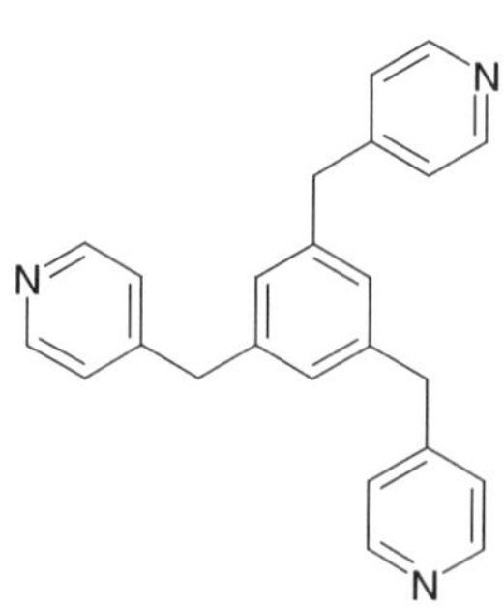

Figure 7.3 Pyridine-capped tridentate ligand **1**.

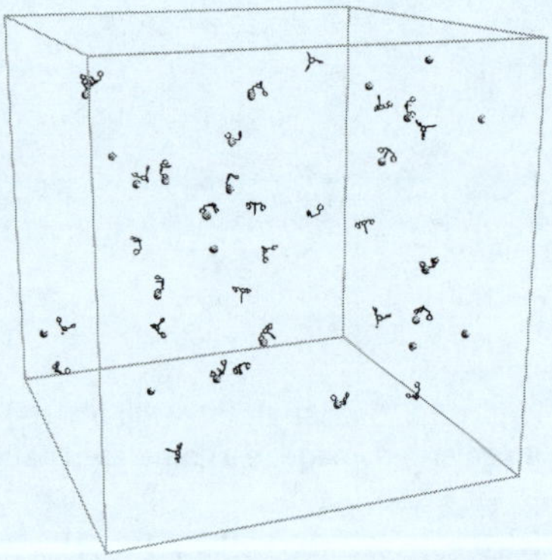

Figure 7.4 Initial structure made by random placement of 24 Pd(II) and 48 ligand **1** models (which corresponds to four M_6L_8 nanospheres) in the cubic simulation box with a volume of (20 nm)^3.

structure of M_6L_8 was determined by X-ray diffraction (XRD) [38], which reveals that the six Pd(II) occupy the apices of the octahedral array and the trigonal faces of the octahedron are capped by the tripodal non-planar ligand **1** in a truncated fashion, leading to a near-spherical shape.

We tried a relatively small system to clarify the acceleration factors in the MD simulation, using the model in the previous section. The initial structure was made by random placement of 24 Pd(II) and 48 ligand **1** models (which corresponds to four M_6L_8 nanospheres) in the cubic simulation box with a volume of (20 nm)^3, as shown in Figure 7.4. The initial concentration is the factor that affects the formation process of the M_6L_8 nanosphere, i.e., a long time free diffusion occurred when the initial concentration was too low and a highly random aggregation occurred when it was for too high. The concentration of the Pd(II) ions and ligands in the initial structure described above was chosen by trial and error and is about 1/234 of that in the M_6L_8 crystal state [38], but a far higher concentration than that of the experimentally prepared solution (0.01 mol/*l*).

There are many other factors that could affect the formation rate of the spherical complex. Temperature is surely another factor, but we set it to the experimental condition, i.e., 90°C (363 K). Herein, we focus on the parameter that controls the metal–ligand coordination interactions, i.e. the near-field relative dielectric constant ϵ_r. The far-field relative dielectric constant, ϵ_{rf}, was specified as 47.0 for the DMSO solvent as described in the previous section. Application of this value ($\epsilon_r = 47$) for even short-distance (near-field) charge–charge interactions are clearly inadequate because this means there are as many DMSO molecules (as the continuum medium) between these short-distance-separated charges. However, a common choice for the near-field $\epsilon_r = 1$, which corresponds to the vacuum, is also inadequate in this study, because this completely neglects the pseudo-ligand effect of the DMSO solvent. A polar solvent such as DMSO could behave like a ligand, where the resulting metal–solvent interaction enhances the metal–ligand exchange probability. This pseudo-ligand effect of the polar solvent could be parameterized as the enlarged ϵ_r rather than one that corresponds to the vacuum.

We compare the results with three different near-field dielectric constants, $\epsilon_r = 1.0$, 2.5 and 4.0, all starting with the same initial configuration. Snapshots after the 250 ns LD runs are shown in Figure 7.5 for these three ϵ_r settings. Among these, the complete M_6L_8 nanosphere formation during the same 250 ns LD simulation was only observed

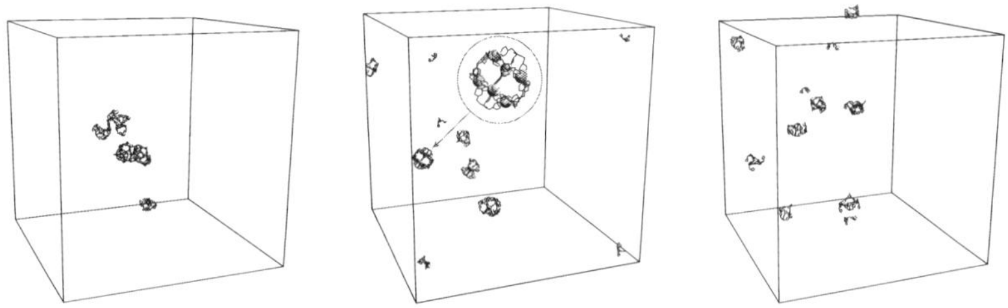

Figure 7.5 Snapshots after the 250 ns LD runs setting $\epsilon_r = 1.0$ (left), 2.5 (center), 4.0 (right). An enlarged image of a completed nanosphere is additionally shown in the central case.

with $\epsilon_r = 2.5$. For all six Pd(II) within this completed nanosphere, time variations of the coordination numbers (judged from the distance between Pd and ligand-nitrogen atoms with the threshold of 0.27 nm) of each Pd(II) are shown in Figure 7.6. The time evolutions in this figure can be divided into three stages. The period up to 50 ns was characterized by a rather monotonic increase of coordination numbers from metal–ligand assembly (an assembly stage). The subsequent period, 50–80 ns, was characterized by frequent coordination exchange via tentative coordination of more than four that evolves from a rather randomly assembled cluster to a near-spherical shape (an evolution stage). In the last period, above 80 ns, the coordination numbers were almost fixed to four and the nanosphere was stable (a fixation stage). This simulated three-stage process in Figure 7.6 corresponds well to the stages explained in the experimental study [12].

We then compare the time evolutions of the ligand exchange rates of the three cases $\epsilon_r = 1.0$, 2.5 and 4.0. We define the ligand exchange rate as the number of metal–ligand

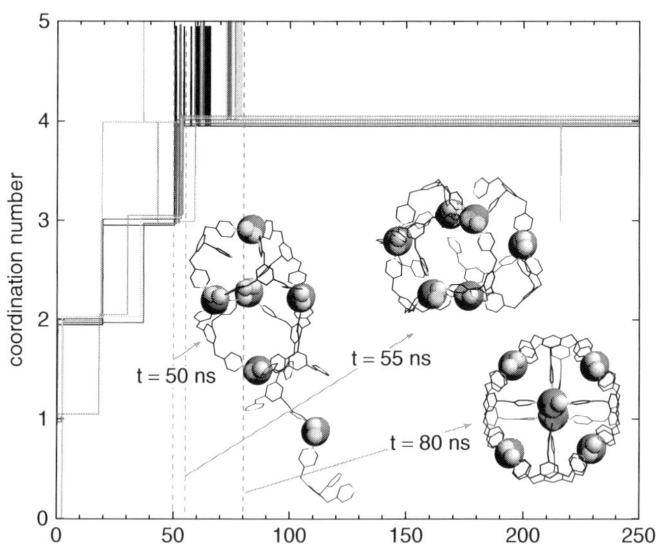

Figure 7.6 Time variations of the coordination numbers of the six Pd(II) within the completed M_6L_8 nanosphere. Each of the six plots is vertically shifted for clarity. Snapshots at the simulation times of 50, 55 and 80 ns are additionally shown in the figure. *(See insert for color representation of the figure.)*.

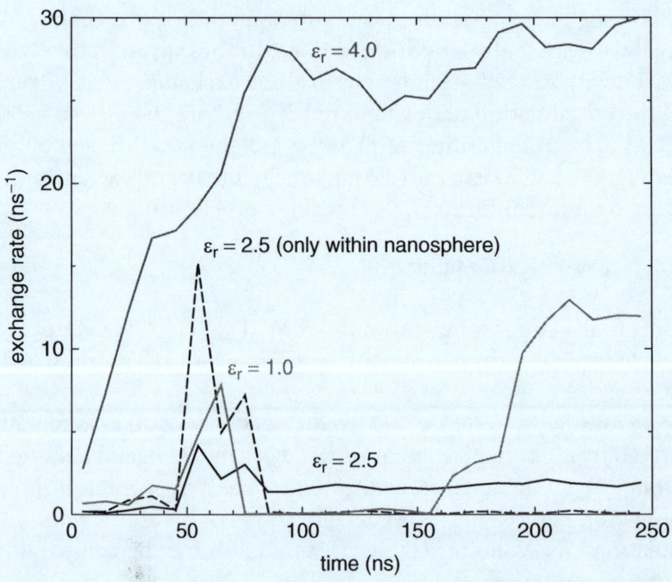

Figure 7.7 Time variations of the ligand exchange rates for the LD runs with $\epsilon_r = 1.0$, 2.5 and 4.0. The broken line corresponds to the exchange rates of the six Pd(II) ions within the completed nanosphere (instead of all the Pd(II) ions) for the LD run with $\epsilon_r = 2.5$.

pair alterations over a specified time period, τ (taken here to be 10 ns), and then average over all of the pairs in the system. The evaluated rates vs. time are shown in Figure 7.7. In this figure, the broken line corresponds to the exchange rates of the six Pd(II) ions within the completed nanosphere (instead of all the Pd(II) ions) for the LD run with $\epsilon_r = 2.5$. The broken plot clearly corresponds to Figure 7.6, i.e., the exchange rate increased in stage 2 (50 – 80 ns) and then steeply (more than two orders of magnitude) decreased to a very small exchange rate value in stage 3 (above 80 ns). We think this large rate difference between the completed nanosphere (in stage 3) and the incomplete clusters (in stage 2) corresponds to the large experimental exchange rate difference between the $M_{12}L_{24}$ nanosphere and a fractional tetra-monodentate model complex [12]. The exchange rate constants themselves are much lager (*ca* 0.1 – 10 ns^{-1} in Figure 7.7) in our simulation model than those in the experimental report (ranging from ms^{-1} to days^{-1} in Ref. [12]). The real process will depend on both the diffusivity of the fragments in the DMSO solution and the ligand exchange rate. We think the acceleration in our current model was realized not by the fastened diffusivity, but by the enhanced ligand exchange rates described above. Actually, we found that the diffusion constants in the implicit solvent model with the Langevin dynamics (this study) and those in the explicit DMSO model [39] solvents with the ordinary MD are the same order. Moreover, the dependence of the diffusivity on the simulated self-assembly was diminished in our study because of the much enhanced initial system concentration in comparison with the real experimental system.

From the results described above, we believe that our accelerated simulation correctly follows the self-assembly process of the real reaction systems when we properly select the simulation parameters, e.g., ϵ_r. This ϵ_r can be regarded as a parameter that scales the metal–ligand binding energy, because Coulomb potentials, $q_iq_j/(4\pi\epsilon r_{ij})$, are the only source of the (metal–ligand) attractive interactions. In the current model, the three-state process was seen with $\epsilon_r = 2.0-2.5$. In contrast, the exchange rate was very large when $\epsilon_r = 4.0$, and the coordination bonds tend to break before the dimers grow into larger clusters, i.e., clustering rate limiting. When $\epsilon_r = 1.0$, the Pd(II)-ligand binding energy is relatively strong; thus, it takes a rather long time to break and exchange the coordination bonds to develop new conformers, i.e., breaking rate limiting.

7.3.2 Self-Assembly of $M_{12}L_{24}$ Spherical Complex [40]

Next, we applied our model to the larger spherical complex $M_{12}L_{24}$, which is a nanocage with 12 Pd(II) ions and 24 pyridine-capped bidentate ligands [4, 41] as shown in Figure 7.8 for example.

To compare with the previous M_6L_8 case, we first used a relatively small system with an initial structure generated by random placement of 48 Pd(II) and 96 ligand **a** models (which corresponds to four $M_{12}L^a_{24}$ nanocages) with the same solvent condition [37]. The initial structure is shown in Figure 7.9a. The concentrations of the Pd(II) ions and ligands in the cubic simulation box (volume = 30 nm^3) were chosen to be comparable to the M_6L_8 system in our previous study.

Snapshots after the 500 ns Langevin dynamics (LD) [42] runs at the simulation temperature of 343 K (70°C, the experimental reaction conditions [4]) are shown in Figure 7.9b. As shown in the enlarged image in Figure 7.10, a complete $M_{12}L^a_{24}$ nanocage was successfully formed during this LD run. For all 12 Pd(II) within this completed nanocage in Figure 7.10, the time variations of the coordination numbers (judged from the distance between Pd and ligand-nitrogen atoms with the specified threshold [37]) of each Pd(II) are shown in Figure 7.11. The time evolutions in this figure can be divided

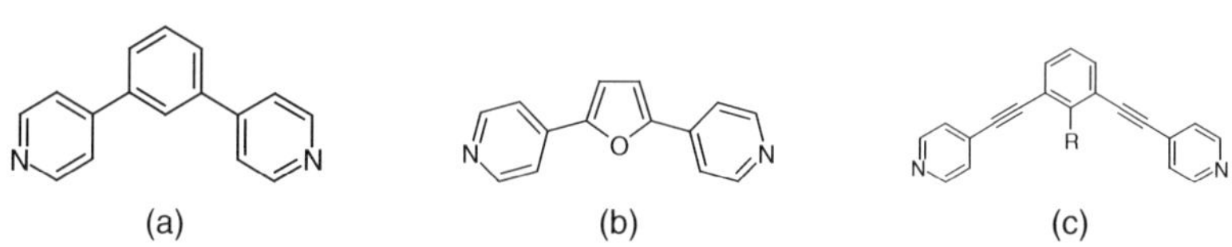

Figure 7.8 Pyridine-capped banana-shaped bidentate ligands **a**, **b** and **c** (R = OCH_3).

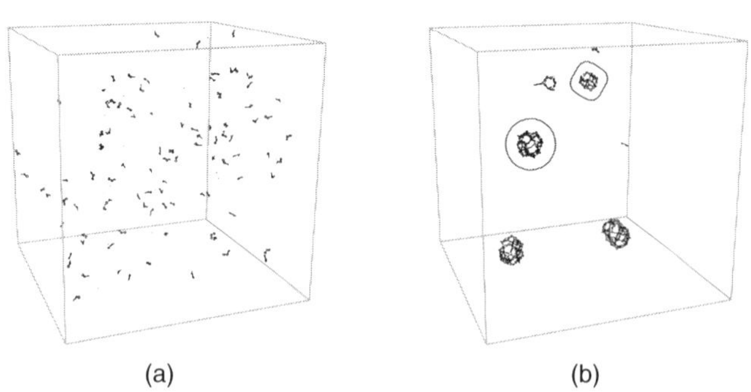

Figure 7.9 a) Initial structure generated by random placement of 48 Pd(II) and 96 ligand **a** models in the cubic simulation box (volume = 30 nm^3). (b) Snapshot after a 500 ns LD run with $M_{12}L^a_{24}$ and $M_6L^a_{12}$ nanocages in the circle and square, respectively.

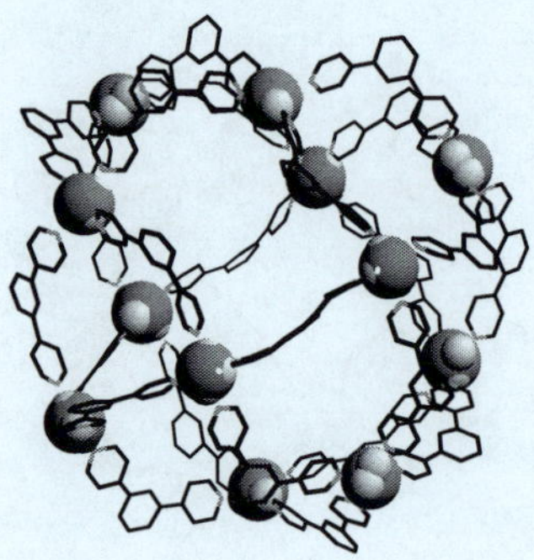

Figure 7.10 Enlarged view of the $M_{12}L^{a}_{24}$ nanocage from the snapshot after the 500 ns LD run in Figure 7.9b.

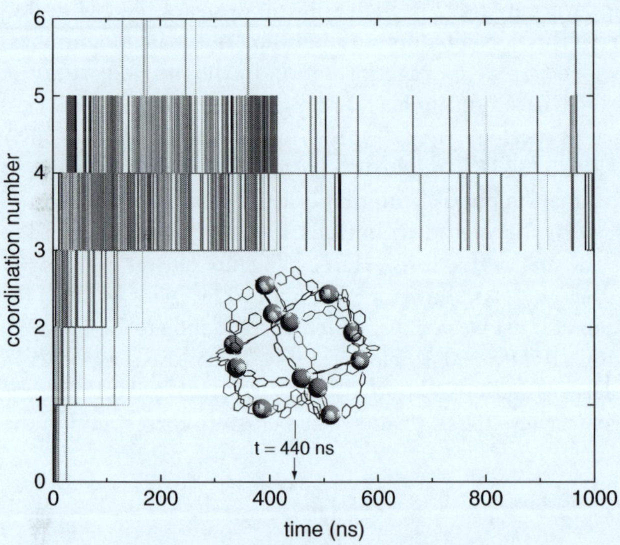

Figure 7.11 Time variations of the coordination numbers of the six Pd(II) within the completed $M_{12}L^{a}_{24}$ nanocage. Snapshots at the simulation time of 440 ns are additionally shown in the figure.

into three stages: an assembly stage up to *ca* 100 ns, an evolution stage for 100–500 ns and a fixation stage above 500 ns. The simulated three-stage process in Figure 7.11 corresponds well to the stages found in our previous simulations on the M_6L_8 system and that explained in the experimental study on the $M_{12}L_{24}$ system [12].

We also tried completely parallel simulations for the furan-cored ligand **b** shown in Figure 7.8. With this ligand **b**, we also observed the spontaneous formation of the $M_{12}L^{b}_{24}$ nanocage as shown in Figure 7.12. From the results described above, we are convinced that our model, which was originally developed for the M_6L_8 system, is capable of demonstrating the self-assembly of the larger $M_{12}L_{24}$ system for two of the ligand types shown in Figure 7.8.

Next, we enlarged the system size to see the cluster size distribution evolution during the $M_{12}L_{24}$ nanocage formation. We constructed a system eight times larger than the first system (with 384 Pd(II) ions and 768 ligand **a** models, which corresponds to 32 $M_{12}L^{a}_{24}$ nanocages) by doubly stacking the MD cells in each *xyz* direction. As in our

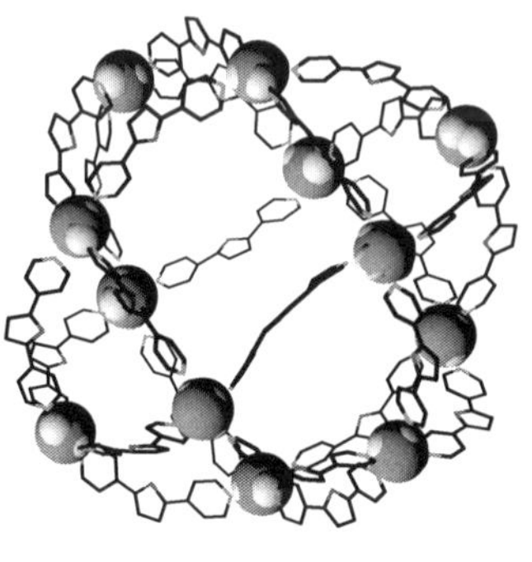

Figure 7.12 $M_{12}L^{b}_{24}$ nanocage from the snapshot after the 500 ns LD run.

previous study [37], we also incorporated 192 explicit solvent atoms of the CH united atom type into the above system. This solvent atom number (half of the Pd ion number) was the optimum number to speed up M_6L_8 cage formation in the previous study. A random initial structure was constructed through a 50 ns high-temperature (473 K) LD run using uncharged Pd(II)-CaDA models. The LD run with the original Pd(II)-CaDA model was then performed for 1 μs at 343 K. The cluster size distributions were analyzed based on the ligand-metal coordination bonds from the simulated trajectories.

Figure 7.13 shows the ligand cluster size distributions as a function of time; the vertical axis corresponds to the mass fractions. Here, a ligand cluster size of 24 corresponds to the completed $M_{12}L^{a}_{24}$. As shown in this figure, the mass fraction for 24-sized clusters at the final step (1.0 μs) was *ca* 0.1, which corresponds to 3–4 $M_{12}L^{a}_{24}$ cage assemblies, compared with 1.0 for a 100% yield assembly. In contrast, we observe peaks at 12, 16 and 18 for the larger fraction at approximately 0.14, 0.1 and 0.15 (corresponds to 4–9 nanocages), respectively. These smaller-sized clusters correspond to the

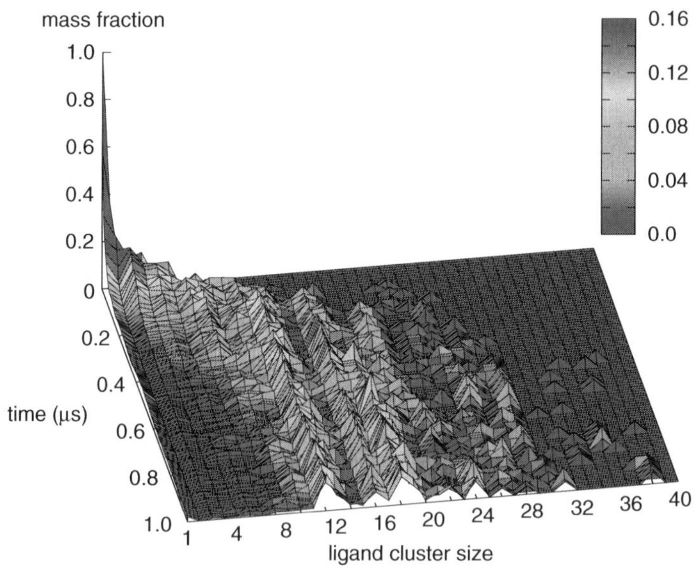

Figure 7.13 Cluster size distributions of the ligand **a** as a function of time. The vertical axis corresponds to the mass fractions.

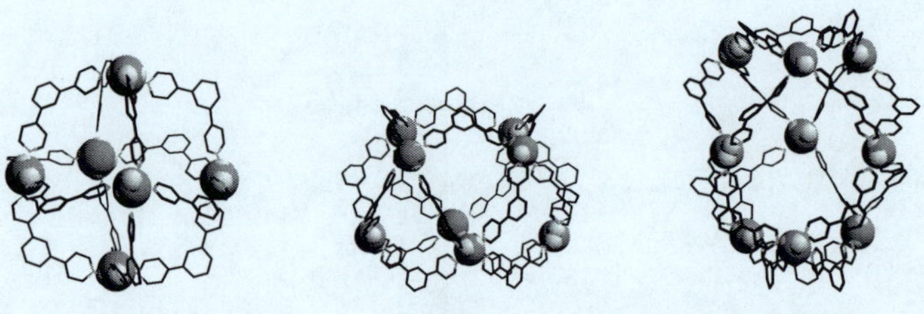

Figure 7.14 (a) $M_6L^a_{12}$, (b) $M_8L^a_{16}$ and (c) $M_9L^a_{18}$ cages from the snapshot after the 1 μs LD run shown in Figure 7.13.

lower-nuclearity complexes with specific symmetries. A peak for the ligand cluster size 12 corresponds to the cube shape $M_6L^a_{12}$ cage (shown in Figure 7.14a) whose formation was reported in a real reaction system of bidentate ligand with acetylene spacers (ligand **c** in Figure 7.8) using platinum (Pt) instead of the Pd ion [41]. The latter peaks at 16 and 18 correspond to the $M_8L^a_{16}$ and $M_9L^a_{18}$ nanocages shown in Figures 7.14b and 7.14c, respectively. These two cages are composed of eight three-ligand membered rings as in $M_{12}L^a_{24}$, but two and three (instead of six for the $M_{12}L^a_{24}$) four-ligand membered rings in the $M_8L^a_{16}$ and $M_9L^a_{18}$ nanocages, respectively. Recently, these two forms, M_8L_{16} and M_9L_{18}, were experimentally confirmed as metastable intermediate structures that exist during the real self-assembly process [13]. In these smaller nanocages ($M_6L^a_{12}$, $M_8L^a_{16}$ and $M_9L^a_{18}$), the coordination network geometries around the Pd(II) ions were deformed compared with $M_{12}L^a_{24}$. These geometry distortions were not so large as being toward tetrahedral geometry from the original coplanar one, but relatively small out-of-plane and adjacent angle deformations between the coordinated pyridines. The results above show that the larger $M_{12}L_{24}$ self-assembly process noticeably involves the trapping structures with lower-nuclearity in contrast to the smaller M_6L_8 cage.

We also tried corresponding simulations with the furan-cored ligand **b**. Figure 7.15 shows the ligand cluster size distributions as a function of time, which corresponds to Figure 7.13 for the ligand **a**. The mass fraction for 24-sized clusters at the final step (around 0.1) is similar to that in Figure 7.13 for the ligand **a**. However, the peaks for the smaller-sized clusters (at 12, 16 and 18) were much suppressed compared to Figure 7.13. It is highly probable that the slightly larger bend angle (*ca* 127° [43]) of the furan-cored ligand **b** would make those smaller-sized clusters energetically unfavorable compared to ligand **a**. This shows that the trapping at the lower nuclearities sensitively depends on the bend angle of ligands, and a slightly larger bend angle (compared to the geometrically optimum angle for the final structure, here 120° for $M_{12}L_{24}$ [44]) would be effective to suppress the trapping.

As mentioned above, in the bidentate ligand real reaction system using Pt(II) instead of the Pd(II) ion, the deformed $Pt(II)_6L^c_{12}$ cube, instead of the $Pt(II)_{12}L^c_{24}$ sphere, was observed [41]. It is proposed that the angle between the pyridyl coordination vectors (bend angle) determines which structures is observed experimentally [3, 44, 45]. Indeed, the formation of an M_6L_{12} cube using the ligands with a 90° bend angle [46] follows the proposed hypothesis. However, $Pt(II)_6L^c_{12}$ vs. $Pd(II)_{12}L^c_{24}$ formation from the same ligand **c** (with bend angle $\sim$ 120°) is not predicted by the hypothesis [43]. It is said that

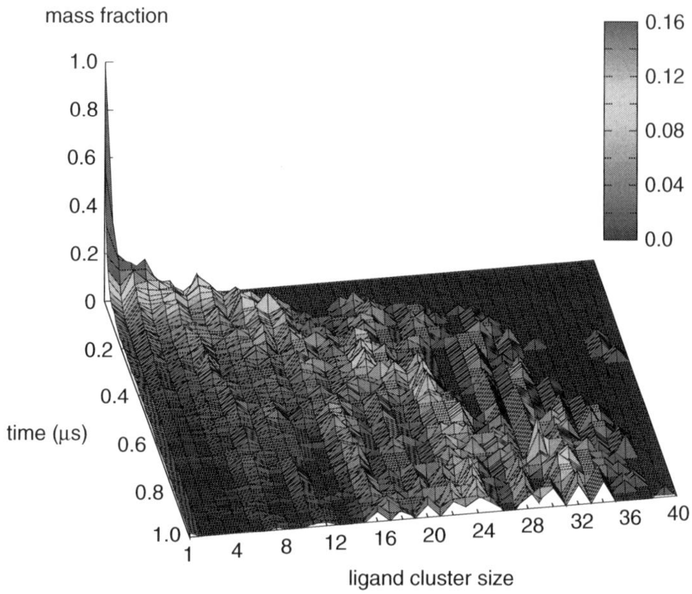

Figure 7.15 Cluster size distributions of the ligand **b** as a function of time. The vertical axis corresponds to the mass fractions.

the stronger metal–ligand binding energy by the Pt(II) ion kinetically stabilizes the metastable complex $Pt(II)_6L^c_{12}$ even with the deformation energy penalties [41].

7.4 Self-Assembly of Metal–Organic Frameworks [47]

In this section, we apply our previous model developed for the self-assembly of a finite complex to infinite MOFs.

7.4.1 Self-Assembly of 2D-like MOF

We first tried to simulate 2-D-like MOF self-assembly using our previous simulation model for $Pd(II)_{12}L_{24}$ by simply changing the ligand from the bent form in $Pd(II)_{12}L_{24}$ (Figure 7.1a) to the linear form 4,4′-bpy (Figure 7.1b) in DMSO. The DMSO solvent was treated implicitly as a continuum medium with a far-field relative dielectric constant of 47.0, as in the previous cases. The simulation temperature was maintained at room temperature (293 K) by coupling to a stochastic thermostat [37, 40]. The initial structure was made by random placement of 384 Pd(II) ions and 768 4,4′-bipyridine (bpy) ligands in the cubic simulation cell with a volume of (60^3 nm^3), as shown in Figure 7.16a. We note that the all-atom ligand model and the coarse-grained solvent model without atomic solvent was used in this section.

The simulated time evolution of the total number of coordination bonds (judged from the distance between the Pd atoms and the nitrogen atoms of the ligands $Pd-N_{bpy}$ using the appropriate threshold [37]) is shown in Figure 7.17 with a solid black line. It shows saturating behavior after *ca* 0.2 μs, which implies that the coordination bonding rates

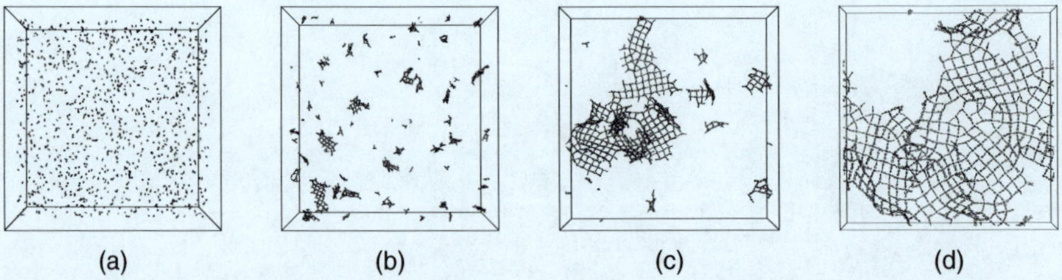

Figure 7.16 (a) Initial structure and snapshot after (b) 0.4 μs, (c) 0.6 μs and (d) 0.8 μs of the [Pd(II)(4, 4′-bpy)$_2$] system simulation viewing along the z-axis (hydrogen atoms are omitted for clarity). The bounding box corresponds to the simulation box with 3-D periodic boundary conditions. The box sizes and shapes are different after 0.4 μs.

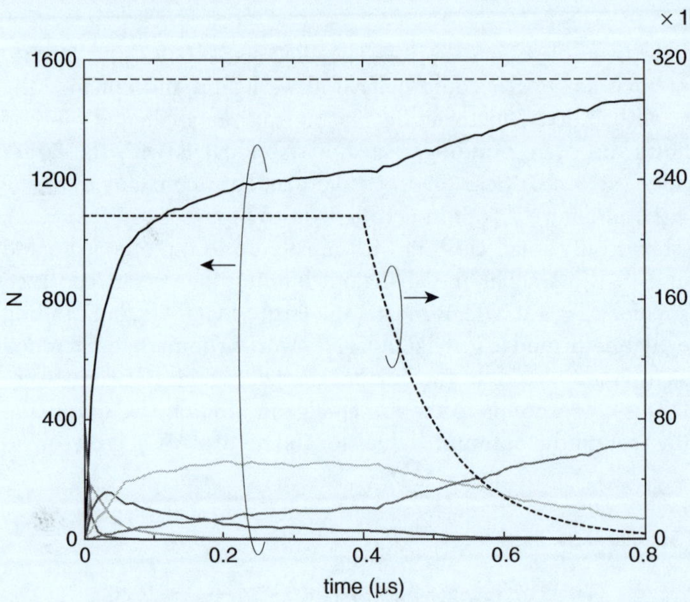

Figure 7.17 Time evolution of the number of coordination bonds (black line) and number of zero- to four-coordinated (gray lines) Pd(II) and the simulation box volume (black dashed line) in the self-assembly process of the [Pd(II)(4, 4′-bpy)$_2$] system. The dashed horizontal line at $N = 1536 = 2 \times 768$ corresponds to the full-bonding limit of the total coordination bonds.

are strongly dependent on the number density of the unbound pyridine rings. Then, to increase the bonding rates, we shrunk the simulation box volume (*ca* −1.0%/ns) after 0.4 μs, as shown in the box volume change in Figure 7.17 with a black dashed line. With this box shrinkage, the number of four-coordinated Pd(II) was increased and the total number of coordination bonds approached its full-bonding value, as shown in Figure 7.17.

Here, we note the effect of the metal–ligand binding strength upon the MOF self-assembly. In our model, the metal–ligand binding was modeled with Coulomb potentials, $q_i q_j/(4\pi\epsilon_r r_{ij})$, and the near-field relative dielectric constant ϵ_r can be regarded as a scaling parameter of the binding energy. As described previously,

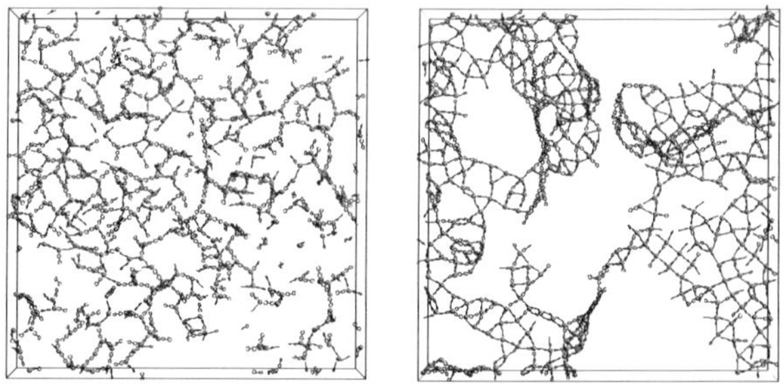

Figure 7.18 Snapshot after 0.8 μs of the [Pd(II)(4, 4′-bpy)$_2$] system simulation with (a) $\epsilon_r = 4.0$ and (b) $\epsilon_r = 1.0$ (hydrogen atoms are omitted for clarity).

the far-field relative dielectric constant (ϵ_{rf}) was specified as 47.0 for the DMSO solvent. A polar solvent such as DMSO could behave like a ligand and enhance the metal–ligand exchange probability (effectively decreasing the metal–ligand binding energy). Figure 7.18 shows the corresponding snapshots after 0.8 μs with the larger ($\epsilon_r = 4.0$) and the smaller ($\epsilon_r = 1.0$) near-field relative dielectric constant than the previous setting of $\epsilon_r = 2.5$ in Figure 7.16d. In comparison to Figure 7.16d ($\epsilon_r = 2.5$), Figure 7.18a ($\epsilon_r = 4.0$) shows only small clusters with mostly up to three-coordinated Pd(II) complexes because of the smaller metal–ligand binding energy. In contrast, Figure 7.18b with the smaller $\epsilon_r = 1.0$, which means the larger metal–ligand binding energy, shows more irregular deformed (e.g., triangular) networks compare to the much regular square networks in Figure 7.16d. In the current model, regular networks were obtained with $\epsilon_r \sim 2-3$. This corresponds to the metal–ligand binding strength being within a window of ±20% around the optimum values for the regular MOF growth.

7.4.2 Self-Assembly of 3D-like MOF

Next, we considered the 3-D extension of the 2-D-like MOF described above using octahedral metal centers. With the use of [9]ane-S_3 capped hexacoordinated ruthenium ions and 4,4′-bpys, the formation of a supramolecular cube $\{([9]\text{ane-S}_3)\text{Ru(II)}\}_8(4,4'\text{-bpy})_{12}$ in nitromethane was reported [48]. Removal of the capping groups may give an infinite 3-D cubic network, because an analogous octahedrally coordinated pyridine (py) complex Ru(II)(py)$_6$ has previously been reported [49]. However, no such 3-D MOFs consisting of only the octahedral centers and the 4,4′-bpys have been reported to date [3]. We then tried the corresponding simulation for the system.

The Ru(II) ion's nonbonding parameters were taken from the universal force field (UFF) [17], and the van der Waals bond length (originally 0.2963 nm) was scaled by 0.78 to reproduce the Ru–$N_{py,bpy}$ distance (*ca* 0.21 nm) from the XRD studies [48, 49]. The atomic charge of the Ru(II) evenly transferred to the dummy atoms that were attached octahedrally to the Ru(II) atom. The solvent, nitromethane, was treated implicitly as a continuum medium with the far- and near-field relative dielectric constant of 36.2 and 1.75, respectively. The initial structure (Figure 7.19a) was made by random placement

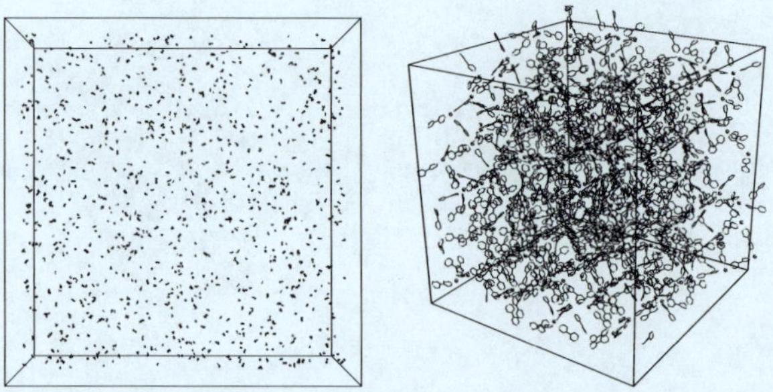

(a) (b)

Figure 7.19 (a) Initial structure and snapshot after (b) 1.0 μs of the [Ru(II)(4, 4′-bpy)$_3$] system simulation (hydrogen atoms are omitted for clarity). Scales (box sizes) are different in these snapshots.

of 256 Ru(II) ions and 768 4,4′-bpy ligands in the cubic simulation box with the same volume (60^3 nm^3) as in the previous system.

The time evolution of the total number of coordination bonds and the number of zero- to six-coordinated Ru(II) is shown in Figure 7.20a. Though we applied the same box shrinkage rate after 0.4 μs as in the previous case, up to five-coordinated Ru(II) complexes were obtained without any six-coordinated ones. Additionally, the total coordination bond number was saturated at much lower values, and the snapshot after 1.0 μs (Figure 7.19b) shows disordered non-regular networks. This result may be consistent with the lack of experimental reports of 3-D MOFs consisting of Ru(II) ions and 4,4′-bpys noted above.

Then, we tried to identify the key design factor to realize regular 3-D MOFs using the current simulation model as a tool. The clue for this was the previously reported 3-D cubic MOF composed of silver (Ag) ions and pyrazine (pyz) ligands [50]. In this system, relatively long Ag–N$_{pyz}$ distances of *ca* 0.25 nm were obtained by XRD measurements [50]. This length is larger than the reported Ru–N$_{bpy}$ distance of *ca* 0.21 nm [48],

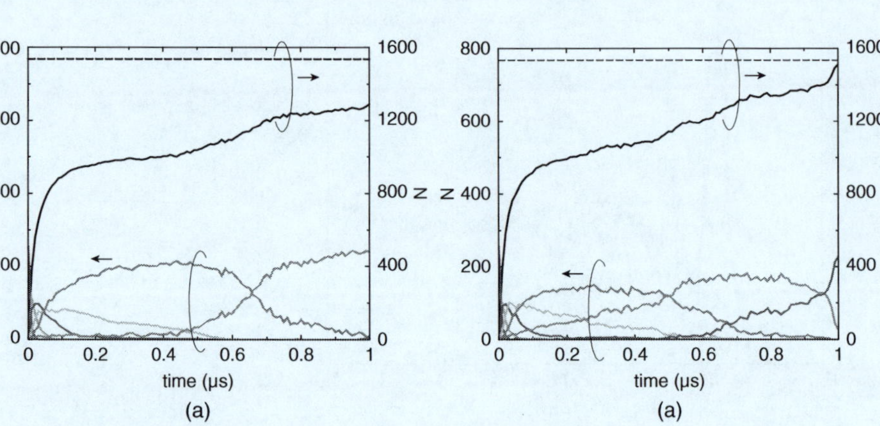

(a) (a)

Figure 7.20 (a) Time evolution of the number of coordination bonds (black line) and number of zero- to six-coordinated (gray lines) Ru(II) in the self-assembly process of the [Ru(II)(4, 4′-bpy)$_3$] system. (b) Corresponding time evolution with the Ru–N distance enlargement.

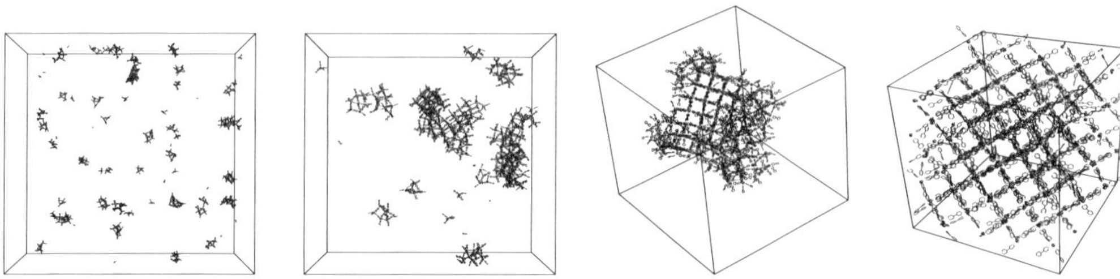

Figure 7.21 Snapshot after (a) 0.4 μs, (b) 0.6 μs, (c) 0.8 μs and (d) 1.0 μs of the enlarged Ru–N distance [Ru(II)(4, 4′-bpy)$_3$] system simulation (hydrogen atoms are omitted for clarity). Scales (box sizes) are different in these snapshots.

which we had used to fit the Ru(II) van der Waals bond length parameter in the simulation above. Because the octahedrally coordinated structure with six 4,4′-bpys is sterically more crowded than the planar four-coordination center, a larger $M-N_{bpy}$ distance would be required [51].

To validate this, we resized the Ru(II) van der Waals bond length parameter to the $Ru-N_{bpy}$ distance of *ca* 0.255 nm with a scaling factor of 1.2 instead of the previous value of 0.78. The corresponding time evolution of the total number of coordination bonds from zero- to six-coordinated Ru(II) is shown in Figure 7.20b. With this enlarged $Ru-N_{bpy}$ distance, six-coordinated Ru(II) became the dominant species and the total number of coordination bonds approached the full-bonding values at $t = 1.0\ \mu s$, as shown in Figure 7.20b. Snapshots in Figure 7.21 show the spontaneous growth process of the 3-D MOF. The marked peaks in the radial distribution function (RDF) in Figure 7.22 with the enlarged Ru–N distance (black line) show the high structural

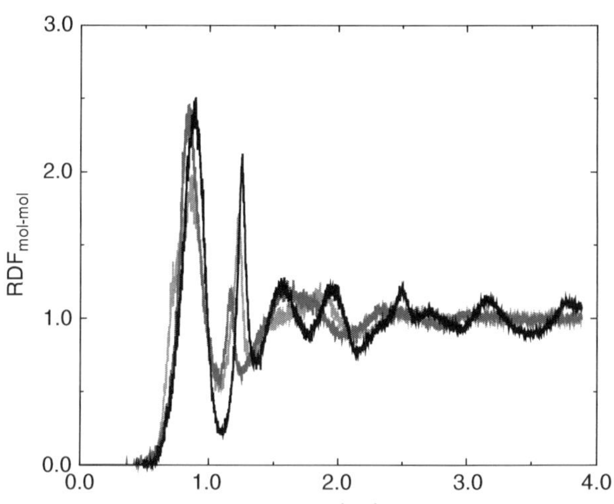

Figure 7.22 The 4,4′-bpy center of molecular mass radial distribution functions averaged over the last 100 ps without (gray line) and with (black line) the Ru–N distance enlargement and with the Ru–N distance enlargement and also the smaller $\epsilon_r = 1.2$ (light gray line).

regularity of the obtained 3-D MOF. Also, RDF for the case without the Ru–N distance enlargement (gray line) flattens out around the value of one over a distance larger than 3 nm, which corresponds to the structureless uniform distributions of the 4,4′-bpys. We also tried using the Ru–N_{bpy} distance of *ca* 0.24 nm and found that it was insufficient to obtain the regular MOF.

Here, we note again the effect of the metal–ligand binding strength upon this 3-D MOF case. The near-field relative dielectric constant $\epsilon_r = 1.75$ applied above was the optimum value we found for the regular 3-D MOF self-assembly. With a smaller value, e.g., $\epsilon_r = 1.2$ (corresponding to the larger metal–ligand binding energy), we obtained the RDF (light gray line) in Figure 7.22 that flattens out again as for the case without the Ru–N distance enlargement (gray line), even with the enlarged Ru–N distance. This could be the result of too strong metal–ligand binding. In contrast, too weak metal–ligand binding gives no MOF growth; it should be within a window between appropriate values again in this 3-D MOF case.

7.5 Conclusion and Outlook

We are convinced that our model can realize the self-assembly of the finite spherical complexes M_6L_8 and $M_{12}L_{24}$ in a distinct three-stage process, which is well correlated with the experimental results [12]. We found that the difference in life times (or ligand exchange rate) of the smaller-sized incomplete clusters and complete spherical complexes is crucially important in this self-assembly.

The above result is consistent with the findings in the highly simplified model study of the growth of the virus capsid [52]. These characteristics of the size–lifetime relations, which are accomplished by growth reversibility in the assembly process (i.e., breakup is even more likely than growth in smaller-sized clusters), are noted as the crucial factors for capsid self-assembly [52]. We found identical importance using a much more direct (realistic) model than the capsid model and completely different inter-subunit interactions from those in the capsid model. These correspondences strongly support the general importance of the marked life time difference and growth reversibility in the supramolecular self-assembly.

Our results also show that the appropriate strength of the metal–ligand binding is definitely important in both finite supramolecular complexes and infinite polymer networks. It should not be too weak or too strong for regular structure formation. The metal–ligand binding strength depends on the characteristics of the metals, ligands and solvents (and counter anions, etc.). For example, the dielectric constant ϵ_r of solvents can affect the Coulombic binding energy $q_i q_j/(4\pi\epsilon_r r_{ij})$ between the metals and the ligands. Then, the selection of solvent dielectric characteristics is an important factor for adjusting the metal–ligand binding strength. The solvent dielectric constant that directly affects the metal–ligand binding strength is the near-field dielectric constant ϵ_r in our model. The effective value of this near-field dielectric constant ϵ_r is roughly proportional to the solvent macroscopic dielectric constant (far-field value ϵ_{rf} in our model), but may vary with the molecular characteristics of the solvents (such as the size, affinity to the metal, etc.). These molecular characteristics of the solvents are also important factors for appropriate solvent selection.

In future studies, the effect of the temperature would be interesting to investigate. Generally, the self-assembly process strongly depends on the synthesis temperature via entropic effects [53]. The thermal energy also contributes as an activation energy to disconnect irregularly bonded networks, which may lead to reconnection to regular networks. Thus, a stronger-binding-energy system may require a higher temperature. A study of these temperature effects will be notably important and should be conducted in future work. We consider that the MD simulation can also be a valuable tool to study the temperature effects in metal–ligand coordination-directed self-assembly.

Acknowledgments

The studies in this chapter were collaborations with Prof. Makoto Fujita of University of Tokyo, Prof. Sota Sato of Tohoku University, Dr Tomohiko Yamaguchi, Dr Seiji Tsuzuki and Dr Masaru Aoyagi of Advanced Industrial Science and Technology.

Reproduction from Ref. [47] with permission from the PCCP Owner Societies.

References

1. Jeffrey, G.A. and Saenger, W. (2012) *Hydrogen bonding in biological structures*, Springer Science & Business Media.
2. Lehn, J.M. (2002) Toward self-organization and complex matter. *Science*, **295** (5564), 2400–2403.
3. Cook, T.R., Zheng, Y.R., and Stang, P.J. (2012) Metal–organic frameworks and self-assembled supramolecular coordination complexes: Comparing and contrasting the design, synthesis, and functionality of metal–organic materials. *Chem. Rev.*, **113** (1), 734–777.
4. Tominaga, M., Suzuki, K., Kawano, M., Kusukawa, T., Ozeki, T., Sakamoto, S., Yamaguchi, K., and Fujita, M. (2004) Finite, spherical coordination networks that self-organize from 36 small components. *Angew. Chem., Int. Ed.*, **43** (42), 5621–5625.
5. Fujita, M., Kwon, Y.J., Washizu, S., and Ogura, K. (1994) Preparation, clathration ability, and catalysis of a two-dimensional square network material composed of cadmium (II) and 4,4′-bipyridine. *J. Am. Chem. Soc.*, **116** (3), 1151–1152.
6. Harris, K., Fujita, D., and Fujita, M. (2013) Giant hollow M_nL_{2n} spherical complexes: Structure, functionalisation and applications. *Chem. Commun.*, **49**, 6703–6712.
7. Sato, S., Iida, J., Suzuki, K., Kawano, M., Ozeki, T., and Fujita, M. (2006) Fluorous nanodroplets structurally confined in an organopalladium sphere. *Science*, **313** (5791), 1273.
8. Murase, T., Sato, S., and Fujita, M. (2007) Nanometer-sized shell molecules that confine endohedral polymerizing units. *Angew. Chem., Int. Ed.*, **46** (7), 1083–1085.
9. Fujita, D., Suzuki, K., Sato, S., Yagi-Utsumi, M., Yamaguchi, Y., Mizuno, N., Kumasaka, T., Takata, M., Noda, M., Uchiyama, S., Kato, K., and Fujita, M. (2012) Protein encapsulation within synthetic molecular hosts. *Nat. Commun.*, **3**, 1093.
10. Furukawa, H., Cordova, K.E., O'Keeffe, M., and Yaghi, O.M. (2013) The chemistry and applications of metal-organic frameworks. *Science*, **341** (6149), 1230444.

11. Zhou, H.C. and Kitagawa, S. (2014) Metal–organic frameworks (MOFs). *Chem. Soc. Rev.*, **43** (16), 5415–5418.
12. Sato, S., Ishido, Y., and Fujita, M. (2009) Remarkable stabilization of $M_{12}L_{24}$ spherical frameworks through the cooperation of 48 Pd (II)–pyridine interactions. *J. Am. Chem. Soc.*, **131** (17), 6064–6065.
13. Fujita, D., Yokoyama, H., Ueda, Y., Sato, S., and Fujita, M. (2015) Geometrically restricted intermediates in the self-assembly of an m12l24 cuboctahedral complex. *Angewandte Chemie International Edition*, **54** (1), 155–158.
14. Millange, F., Medina, M.I., Guillou, N., Férey, G., Golden, K.M., and Walton, R.I. (2010) Time-resolved in situ diffraction study of the solvothermal crystallization of some prototypical metal–organic frameworks. *Angew. Chem., Int. Ed.*, **49** (4), 763–766.
15. Cravillon, J., Schröder, C., Nayuk, R., Gummel, J., Huber, K., and Wiebcke, M. (2011) Fast nucleation and growth of ZIF-8 nanocrystals monitored by time-resolved in situ small-angle and wide-angle X-ray scattering. *Angew. Chem., Int. Ed.*, **50** (35), 8067–8071.
16. Calligaris, M., Zangrando, E., Milani, B., and Marson, A. (2005) Stereochemical investigation of palladium (II) complexes with phenanthroline ligands: A molecular mechanics approach. *Eur. J. Inorg. Chem.*, **2005** (4), 704–712.
17. Rappé, A.K., Casewit, C.J., Colwell, K., Goddard III, W., and Skiff, W. (1992) UFF, a full periodic table force field for molecular mechanics and molecular dynamics simulations. *J. Am. Chem. Soc.*, **114** (25), 10024–10035.
18. Wang, J., Wolf, R.M., Caldwell, J.W., Kollman, P.A., and Case, D.A. (2004) Development and testing of a general AMBER force field. *J. Comput. Chem.*, **25**, 1157.
19. Tiberio, G., Muccioli, L., Berardi, R., and Zannoni, C. (2008) Towards in silico liquid crystals. realistic transition temperatures and physical properties for n-cyanobiphenyls *via* molecular dynamics simulations. *ChemPhysChem*, **10**, 125.
20. Bayly, C.I., Cieplak, P., Cornell, W., and Kollman, P.A. (1993) A well-behaved electrostatic potential based method using charge restraints for deriving atomic charges: The RESP model. *J. Chem. Phys.*, **97** (40), 10269.
21. Frisch, M.J., Trucks, G.W., Schlegel, H.B., Scuseria, G.E., Robb, M.A., Cheeseman, J.R., Montgomery, J.J.A., Vreven, T., Kudin, K.N., Burant, J.C., Millam, J.M., Iyengar, S.S., Tomasi, J., Barone, V., Mennucci, B., Cossi, M., Scalmani, G., Rega, N., Petersson, G.A., Nakatsuji, H., Hada, M., Ehara, M., Toyota, K., Fukuda, R., Hasegawa, J., Ishida, M., Nakajima, T., Honda, Y., Kitao, O., Nakai, H., Klene, M., Li, X., Knox, J.E., Hratchian, H.P., Cross, J.B., Bakken, V., Adamo, C., Jaramillo, J., Gomperts, R., Stratmann, R.E., Yazyev, O., Austin, A.J., Cammi, R., Pomelli, C., Ochterski, J.W., Ayala, P.Y., Morokuma, K., Voth, G.A., Salvador, P., Dannenberg, J.J., Zakrzewski, V.G., Dapprich, S., Daniels, A.D., Strain, M.C., Farkas, O., Malick, D.K., Rabuck, A.D., Raghavachari, K., Foresman, J.B., Ortiz, J.V., Cui, Q., Baboul, A.G., Clifford, S., Cioslowski, J., Stefanov, B.B., Liu, G., Liashenko, A., Piskorz, P., Komaromi, I., Martin, R.L., Fox, D.J., Keith, T., Al-Laham, M.A., Peng, C.Y., Nanayakkara, A., Challacombe, M., Gill, P.M.W., Johnson, B., Chen, W., Wong, M.W., Gonzalez, C., and Pople, J.A. (2004) *Gaussian03, Revision C.02*, Gaussian, Inc., Wallingford CT.
22. Sousa, S., Fernandes, P., and Ramos, M. (2010) Molecular dynamics simulations: Difficulties, solutions and strategies for treating metalloenzymes. *Kinet. Dyn.*, 299–330.

23. Pang, Y. (1999) Novel zinc protein molecular dynamics simulations: Steps toward antiangiogenesis for cancer treatment. *J. Mol. Model.*, **5** (10), 196–202.
24. Godbout, N., Salahub, D.R., Andzelm, J., and Wimmer, E. (1992) Optimization of Gaussian-type basis sets for local spin density functional calculations. Part I. Boron through neon, optimization technique and validation. *Canadian Journal of Chemistry*, **70** (2), 560–571.
25. Smith, P. and Pettitt, B. (1994) Modeling solvent in biomolecular systems. *The Journal of Physical Chemistry*, **98** (39), 9700–9711.
26. Yun-yu, S., Lu, W., and Van Gunsteren, W. (1988) On the approximation of solvent effects on the conformation and dynamics of cyclosporin A by stochastic dynamics simulation techniques. *Mol. Simul.*, **1** (6), 369–383.
27. Tironi, I., Sperb, R., Smith, P., and van Gunsteren, W. (1995) A generalized reaction field method for molecular dynamics simulations. *J. Chem. Phys.*, **102**, 5451.
28. Weeks, J., Chandler, D., and Andersen, H. (1971) Role of repulsive forces in determining the equilibrium structure of simple liquids. *J. Chem. Phys.*, **54** (12), 5237.
29. Hess, B., Holm, C., and van der Vegt, N. (2006) Modeling multibody effects in ionic solutions with a concentration dependent dielectric permittivity. *Physical Review Letters*, **96** (14), 147801.
30. Grest, G.S. and Kremer, K. (1986) Molecular dynamics simulation for polymers in the presence of a heat bath. *Physical Review A*, **33** (5), 3628.
31. Hess, B., Kutzner, C., Van Der Spoel, D., and Lindahl, E. (2008) GROMACS 4: Algorithms for highly efficient, load-balanced, and scalable molecular simulation. *J. Chem. Theory Comput.*, **4** (3), 435–447.
32. van Gunsteren, W. and Berendsen, H. (1988) A leap-frog algorithm for stochastic dynamics. *Mol. Simul.*, **1**, 173–185.
33. Hess, B., Bekker, H., Berendsen, H.J.C., and Fraaije, J.G.E.M. (1997) LINCS: A linear constraint solver for molecular simulations. *J. Comput. Chem.*, **18** (12), 1463.
34. Feenstra, K.A., Hess, B., and Berendsen, H.J. (1999) Improving efficiency of large timescale molecular dynamics simulations of hydrogen-rich systems. *J. Comput. Chem.*, **20**, 786–798.
35. Lindahl, E., Hess, B., and van der Spoel, D. (2001) Gromacs 3.0: A package for molecular simulation and trajectory analysis. *J. Mol. Model.*, **7** (8), 306.
36. Villa, A., Mark, A., Saracino, G., Cosentino, U., Pitea, D., Moro, G., and Salmona, M. (2006) Conformational polymorphism of the PrP106-126 peptide in different environments: A molecular dynamics study. *J. Phys. Chem. B*, **110** (3), 1423–1428.
37. Yoneya, M., Yamaguchi, T., Sato, S., and Fujita, M. (2012) Simulation of metal–ligand self-assembly into spherical complex M_6L_8. *J. Am. Chem. Soc.*, **134** (35), 14401–14407.
38. Chand, D., Biradha, K., Fujita, M., Sakamoto, S., and Yamaguchi, K. (2002) A molecular sphere of octahedral symmetry. *Chem. Commun.*, (21), 2486–2487.
39. Liu, H., Mueller-Plathe, F., and van Gunsteren, W. (1995) A force field for liquid dimethyl sulfoxide and physical properties of liquid dimethyl sulfoxide calculated using molecular dynamics simulation. *Journal of the American Chemical Society*, **117** (15), 4363–4366.
40. Yoneya, M., Tsuzuki, S., Yamaguchi, T., Sato, S., and Fujita, M. (2014) Coordination-directed self-assembly of $M_{12}L_{24}$ nanocage: Effects of kinetic trapping on the assembly process. *ACS nano*, **8** (2), 1290–1296.

41. Fujita, D., Takahashi, A., Sato, S., and Fujita, M. (2011) Self-assembly of Pt(II) spherical complexes *via* temporary labilization of the metal–ligand association in 2,2,2-trifluoroethanol. *J. Am. Chem. Soc.*, **133** (34), 13317–13319.

42. Scheraga, H.A., Khalili, M., and Liwo, A. (2007) Protein-folding dynamics: Overview of molecular simulation techniques. *Annu. Rev. Phys. Chem.*, **58**, 57–83.

43. Bunzen, J., Iwasa, J., Bonakdarzadeh, P., Numata, E., Rissanen, K., Sato, S., and Fujita, M. (2012) Self-assembly of $M_{24}L_{48}$ polyhedra based on empirical prediction. *Angew. Chem., Int. Ed.*, **51** (13), 3161–3163.

44. Williams, A.F. (2011) Ligand design for hollow spherical complexes. *Coord. Chem. Rev.*, **255** (17), 2104–2110.

45. Leininger, S., Olenyuk, B., and Stang, P.J. (2000) Self-assembly of discrete cyclic nanostructures mediated by transition metals. *Chem. Rev.*, **100** (3), 853–908.

46. Suzuki, K., Tominaga, M., Kawano, M., and Fujita, M. (2009) Self-assembly of an M_6L_{12} coordination cube. *Chem. Commun.*, (13), 1638–1640.

47. Yoneya, M., Tsuzuki, S., and Aoyagi, M. (2015) Simulation of metal–organic framework self-assembly. *Physical Chemistry Chemical Physics*, **17** (14), 8649–8652.

48. Roche, S., Haslam, C., Adams, H., Heath, S.L., and Thomas, J.A. (1998) Self-assembly of a supramolecular cube. *Chem. Commun.*, (16), 1681–1682.

49. Templeton, J.L. (1979) Hexakis (pyridine) ruthenium (II) tetrafluoroborate. molecular structure and spectroscopic properties. *J. Am. Chem. Soc.*, **101** (17), 4906–4917.

50. Carlucci, L., Ciani, G., Proserpio, D.M., and Sironi, A. (1995) Novel networks of unusually coordinated silver (I) cations: The wafer-like structure of $[Ag(pyz)_2][Ag_2(pyz)_5](PF_6)_3 \cdot 2G$ and the simple cubic frame of $[Ag(pyz)_3](SbF_6)$. *Angew. Chem., Int. Ed.*, **34** (17), 1895–1898.

51. Soldatov, D.V. and Ripmeester, J.A. (1998) Hexapyridine addition compounds of metal (II) salts (nickel and magnesium perchlorates, magnesium nitrate): Their supramolecular nature, and anion coordination of the metal cation [1]. *Supramolecular Chemistry*, **9** (3), 175–181.

52. Rapaport, D. (2008) Role of reversibility in viral capsid growth: A paradigm for self-assembly. *Physical Review Letters*, **101** (18), 186101.

53. Palma, C.A., Cecchini, M., and Samorì, P. (2012) Predicting self-assembly: from empirism to determinism. *Chemical Society Reviews*, **41** (10), 3713–3730.

8

Simulations of Cell Uptake of Nanoparticles: Membrane-Mediated Interaction, Internalization Pathways, and Cooperative Effect

Falin Tian[1], Tongtao Yue[2], Ye Li[1] and Xianren Zhang[1]

[1] State Key Laboratory of Organic–Inorganic Composites, Beijing University of Chemical Technology, China
[2] State Key Laboratory of Heavy Oil Processing, Center for Bioengineering and Biotechnology, China University of Petroleum (East China), Qingdao, China

Nanoparticles (NPs) nowadays receive intense interest for their potential applications serving as a vehicle for drug delivery as well as for their potential to cause undesirable effects on human health. A biomembrane, composed of a particular type of amphiphilic molecules, lipid molecules, is one of the most important parts of cells, participating in a number of cellular activities, like cell signal transduction and transportation. Thus, studying the interaction between the biomembrane and NPs is essential for understanding the nature of cellular life and for the safe application of nanoparticles. Recently, with the development of computer technology and simulation techniques, computer simulations have become an important tool for studying the interaction between NPs and the biomembrane. In this chapter, we summarize recent developments of computer simulation studies on the NP–membrane interaction, and particularly stress the results from dissipative particle dynamics (DPD) simulation. In particular, we concentrate on membrane-mediated interaction, internalization pathways, and cooperative effect. These studies undoubtly deepen our understanding of the mechanisms for NPs interacting with the plasma membrane, which can not only help us to understand the cytotoxicity of NPs, but also guide us to design new NPs for drug delivery.

8.1 Introduction

Nowadays, nanoparticles (NPs) have received intense interest due to their potential applications in diagnostics and therapeutics. However, the rapid development of nanotechnology not only provides a new route to use NPs as a vehicle for drug delivery, but also implies the potential to cause undesirable effects, including the potential negative impacts of nanomaterials on human health, namely, nanotoxicology.

As a natural barrier of the cell, the plasma membrane is composed of amphiphile molecules such as lipids and proteins. It is the amphiphilic nature of the lipid molecule that decides the structure of the lipid membrane, in which the hydrophobic parts

Self-Assembling Systems: Theory and Simulation, First Edition. Edited by Li-Tang Yan.

of lipids are wrapped by their hydrophilic parts. The cell membrane, serving as the interface between the cell and the surrounding environment, protects the integrity of the cell and participates in a number of bioprocesses, such as signal communication among cells and uptake of nutrients. Therefore, to understand how NPs serve as carriers for drug delivery and induce cytotoxicity it is crucial to understand the interaction between the cell membrane and NPs.

In recent years, numerous studies have investigated the interaction between NPs and the cell membrane through *in vivo* and *in vitro* experiments, theoretical analysis, and computer simulations. The studies fall into two broad categories: NP toxicity and the internalization pathways of NPs, both of which are found to be affected by several factors such as NP size, geometrical shape, surface modification, and the surrounding environment. Multiple mechanisms of interaction between the plasma membrane and NPs have been proposed.

Some experimental works studied the effect of NP size. For example, Mou et al. [1] used confocal laser scanning microscopy (CLSM) to demonstrate that the cellular uptake of the monodisperse mesoporous silica nanoparticle (MSN) depends on the NP size. They found that NPs having a size of 170, 110, 50, and 30 nm can be internalized into cells, and the efficiency of cellular uptake is highly dependent on particle size, in the order 50 nm > 30 nm > 110 nm > 280 nm > 170 nm. Similar particle size dependence was reported by Osaki et al. [2] and Jiang et al. [3]. Besides the influence of NP size, it has also been found that cellular uptake depends on NP concentration in solution. It was found that there exists a threshold concentration for NP internalization, above which the uptake efficiency increases with NP concentration [4]. In addition to the above studies, there exist also numerous studies on the interaction between the plasma membrane and NPs that predict the cellular uptake to be dependent on NP size [5–9].

Although extensive experimental studies have been carried out in order to understand the NP–membrane interaction, much less is known about the kinetic evolution of these interaction processes. Moreover, in most cases, experiments cannot provide any detailed information about the interaction mechanism at the molecular level. Therefore, it is highly desirable to study the molecular mechanism for NP–membrane interaction with the help of molecular simulations. As a result, coarse-grained molecular dynamics (CGMD) and dissipative particle dynamics (DPD) simulations have been performed to study the interaction mechanism and the corresponding interaction pathway. As an example, Li et al. [10] observed experimentally that when cells (like lung epithelial cells and macrophages) are exposed to a solution of graphene, a type of NP with a strong shape anisotropy, graphene NPs would penetrate the cell membrane. To interpret the penetration mechanism, the authors performed computer simulations to show that graphene NPs enter the cells through spontaneous penetration at edge or corner sites.

With the development of supercomputers and simulation techniques, molecular simulation has become a very useful investigative tool. Different types of simulation methods have been elaborated, e.g., Monte Carlo (MC) methods, molecular dynamics (MD), dissipative particle dynamics (DPD), Brownian dynamics (BD), etc. Each method has its advantages and shortcomings. In this chapter, we mainly focus on recent computer simulation studies on the interaction between the plasma membrane and NPs, and we particularly stress the results from DPD simulations.

8.2 *N*-Varied DPD Technique

8.2.1 Traditional DPD Method

The dissipative particle dynamics method was first introduced to simulate the hydrodynamic behavior of complex fluids [11], and proved to be especially useful in studying the mesoscale behaviors of lipid bilayer membranes [12–15]. The elementary units of DPD simulation are soft beads whose dynamics are governed by Newton's equations of motion. The force between each pair of beads contains three parts: conservative forces ($\mathbf{F}_{ij}^{\mathrm{C}}$), dissipative forces ($\mathbf{F}_{ij}^{\mathrm{D}}$), and random forces ($\mathbf{F}_{ij}^{\mathrm{R}}$). The conservative force between beads i and j, which is soft repulsive, is determined by

$$\mathbf{F}_{ij}^{\mathrm{C}} = \begin{cases} a_{ij}(1 - r_{ij})\hat{\mathbf{r}}_{ij} & r_{ij} < r_{\mathrm{C}} \\ 0 & r_{ij} > r_{\mathrm{C}} \end{cases} \tag{8.1}$$

where a_{ij} is the maximum repulsive strength, $\mathbf{r}_{ij} = \mathbf{r}_i - \mathbf{r}_j$ ($\mathbf{r}_i$ and $\mathbf{r}_j$ are the positions of beads i and j), $r_{ij} = |\mathbf{r}_{ij}|$, $\hat{\mathbf{r}}_{ij} = \mathbf{r}_{ij}/|\mathbf{r}_{ij}|$, and $r_{\mathrm{C}} = 1$ is the interaction range. The other two forces $\mathbf{F}_{ij}^{\mathrm{D}}$ and $\mathbf{F}_{ij}^{\mathrm{R}}$ are both responsible for the conservation of total momentum in the system and incorporate the effect of Brownian motion into the larger length scale.

The dissipative force and random force are

$$\mathbf{F}_{ij}^{\mathrm{D}} = -\gamma\omega^{\mathrm{D}}(r_{ij})(\hat{\mathbf{r}}_{ij}\cdot\mathbf{v}_{ij})\hat{\mathbf{r}}_{ij}, \quad \mathbf{F}_{ij}^{\mathrm{R}} = \sigma\omega^{\mathrm{R}}(r_{ij})\theta_{ij}\hat{\mathbf{r}}_{ij}, \tag{8.2}$$

where $\mathbf{v}_{ij} = \mathbf{v}_i - \mathbf{v}_j$ are the velocity differences between particles i and j, γ is the friction coefficient, σ is the noise and satisfies $\sigma^2 = 2\gamma k_{\mathrm{B}}T$, ω^{D} and ω^{R} are the weighting factors,

$$\omega^{\mathrm{D}}(r) = [\omega^{\mathrm{R}}(r)]^2 = \begin{cases} (1 - r)^2 & r < r_{\mathrm{C}} \\ 0 & r \geq r_{\mathrm{C}} \end{cases} \tag{8.3}$$

and $\theta_{ij}(t)$ is a random number distributed with

$$\langle\theta_{ij}(t)\rangle = 0, \langle\theta_{ij}(t)\theta_{kl}(t\prime)\rangle = (\delta_{ik}\delta_{jl} + \delta_{il}\delta_{jk})\delta(t - t\prime). \tag{8.4}$$

8.2.2 *N*-Varied DPD Method

A variant version of DPD, N-varied DPD, is particularly suitable for studying the NP–membrane interaction. In conventional DPD simulations, the conservation of the number of lipid molecules in a membrane patch would lead to an increase of surface tension when the flat membrane patch bends up. Therefore, large membrane deformations, like budding and invagination, are prohibited in conventional DPD simulations of flat membrane patches. However, for cellular membranes, the much larger membrane area compared to the NP size offers sufficient excess area to release the tension induced by a large membrane deformation. In order to solve this problem, we adopted the N-varied DPD simulation method introduced by Hong et al. [16] and developed by Yue et al. [17, 18]. This method offers sufficient excess area to release the tension induced by a large membrane deformation, as in a real cellular membrane.

For implementing this type of DPD simulation, the membrane patch in the simulation box is divided into two parts: the central region and the surrounding boundary region; the targeted membrane tension is maintained by monitoring the lipid number per area

(LNPA) in the boundary region. The preset membrane tension can implicitly include the effects of cytoskeletal force associated with actin polymerization, which *in vivo* is tightly related to most internalization processes. The boundary region plays a role as the reservoir of lipids, and the value of LNPA in the region (denoted by ρ^{BR}_{LNPA}) is kept constant by addition/deletion moves. In an addition/deletion move, new lipid molecules are added/deleted to the boundary region if its local lipid density significantly deviates from the given value of ρ^{BR}_{LNPA}. At the same time, a corresponding number of water beads are randomly added or deleted to keep the whole density of the simulation box constant. In practice, an addition/deletion move is often performed after 200–1500 time steps to leave enough time to propagate the membrane tension to the whole membrane.

8.3 The Interaction between NP and Membrane

The plasma membrane consists of different types of phospholipids and membrane proteins. To maintain the cell integrity, the cell membrane must be impermeable to most macromolecules and NPs. Adsorption of NPs on the membrane mainly influences the membrane in several aspects: First, the adsorbed NPs may alter other properties of the plasma membrane, such as the membrane rigidity, the membrane tension, or the diffusion coefficient of lipids on the membrane. Second, the adsorbed NPs may change global properties of the membrane: the adsorption of NPs can induce phase behavior of the multiple-component membrane. Charged NPs may trigger phase separation by interacting strongly with a special type of lipid having opposite charge, and the phase separation will affect the signal transduction crossing cells. Finally, the addition of NPs changes the structure of the cell membrane and induces various membrane responses. When NPs approach and adhere to the membrane, the addition of NPs alters the ordering of nearby lipid molecules and changes the membrane thickness, creating local membrane curvature, and even breaking the integrity of the cell membrane. Depending on the properties of NPs and membrane, different forms of NP–membrane interaction are found, including NP adhesion on the membrane, NP aggregation on the membrane, NP encapsulation inside the membrane, NP penetration (translocation) across the membrane, NP endocytosis, membrane rupture, and the formation of membrane holes. The factors that influence the interaction between NPs and membrane can be divided into different categories: the (physical and chemical) properties of NPs and those of the membrane. The effects of these factors are discussed below.

8.3.1 Membrane-Mediated Interaction between NPs

Due to its flexibility and elasticity, a lipid membrane can be easily deformed when the membrane interacts with NPs or some membrane proteins. In turn, these deformations may result in more complex effective interactions with NPs or proteins, leading to effective interactions between these “embedded inclusions” [13, 19]. In recent years, a number of simulations and experimental studies have been devoted to revealing the membrane-mediated interactions with NPs or proteins, such as transmembrane proteins or anchored proteins [13, 19, 20]. The main mechanism of membrane-mediated interactions between the inclusions had been considered in some cases as the so-called hydrophobic mismatch effect, but in other cases not.

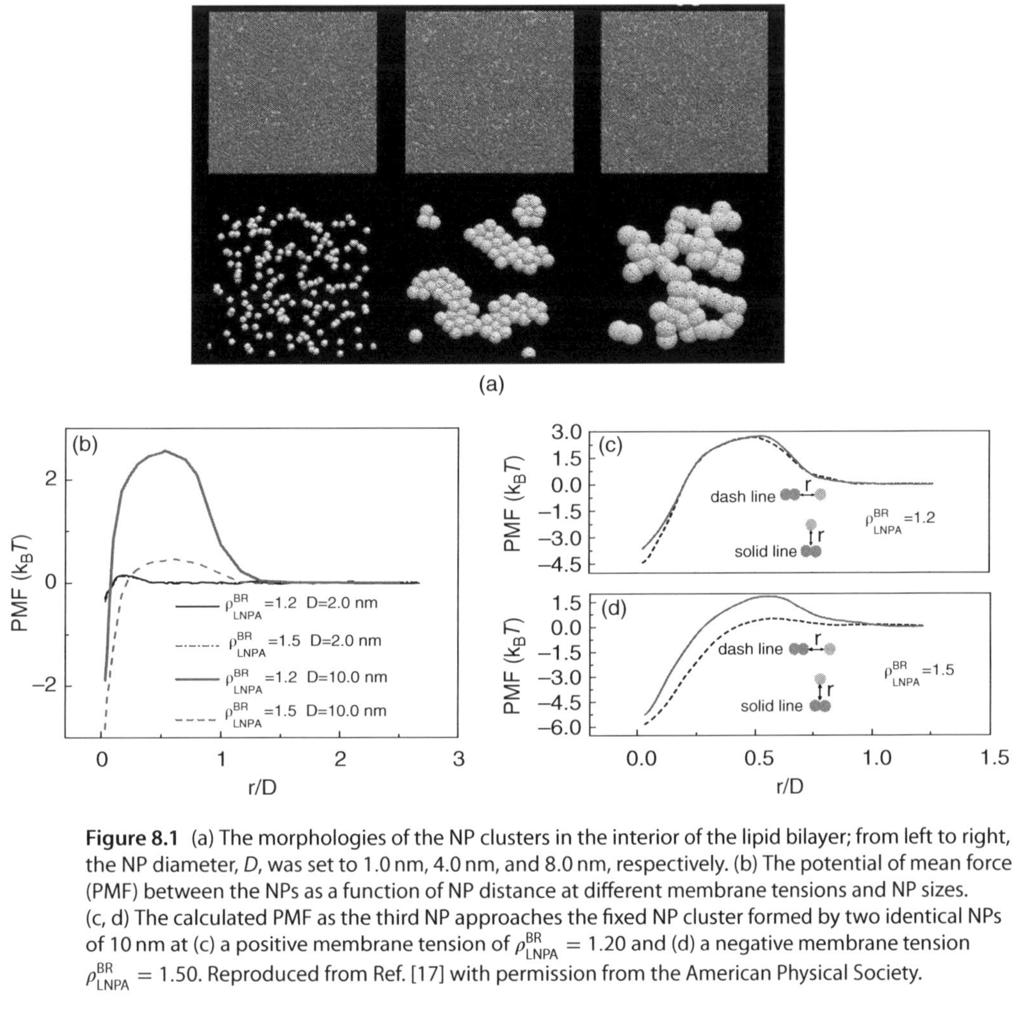

Figure 8.1 (a) The morphologies of the NP clusters in the interior of the lipid bilayer; from left to right, the NP diameter, *D*, was set to 1.0 nm, 4.0 nm, and 8.0 nm, respectively. (b) The potential of mean force (PMF) between the NPs as a function of NP distance at different membrane tensions and NP sizes. (c, d) The calculated PMF as the third NP approaches the fixed NP cluster formed by two identical NPs of 10 nm at (c) a positive membrane tension of $\rho^{BR}_{LNPA} = 1.20$ and (d) a negative membrane tension $\rho^{BR}_{LNPA} = 1.50$. Reproduced from Ref. [17] with permission from the American Physical Society.

Recently, Tian et al. [17] found through DPD simulations that hydrophobic NPs with a diameter comparable to the hydrophobic thickness of the membrane (5 nm), prefer to go into the hydrophobic compartments of the cell membrane and are capable of clustering in the interior of a cell membrane (Figure 8.1a). In contrast, NPs with diameter less than 3 nm or larger than 12 nm stay in the membrane as free monomers. Energy analysis shows that the driven force for NP aggregation comes from the membrane deformation induced by the NP insertion (Figure 8.1b). In other words, the membrane deformation mediates an effective attraction between neighboring NPs, which in turn causes NP clustering to minimize the membrane deformation. Moreover, a series of simulation runs demonstrated that the morphology of the NP clusters changes as a function of membrane tension and NP diameter. This indicates that the effective force changes with membrane tension and NP diameter (Figures 8.1c and 8.1d).

Unlike this "hydrophobic mismatch" induced by the external inclusions (such as transmembrane proteins, anchored proteins, and surface hydrophobic NPs), it was found that some charged proteins or NPs would also lead to membrane

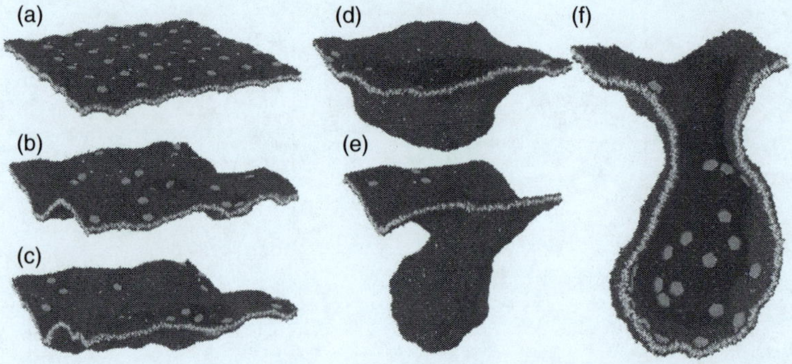

Figure 8.2 A series of simulation snapshots of membrane-curvature-induced NP attraction. The times of the simulation snapshots are: (a) 0*t*, (b) 20 000*t*, (c) 40 000*t*, (d) 50 000*t*, (e) 60 000*t*, and (f) 70 000*t*, the last corresponding to roughly 1 ms. Reproduced from Ref. [21] with permission from Nature Publication Group.

deformation and reorganization due to the attractive interaction between NP and membrane. This membrane deformation then induces an effective interaction between the neighboring adsorbed NPs in the absence of direct NP–NP interaction. Reynwar et al. [21] have performed coarse-grained simulations to investigate the membrane-curvature-mediated interactions between NPs adsorbed on a membrane. From the simulations they confirmed that when the NPs (or proteins) were adsorbed on the surface of the lipid bilayer membrane, the NP–membrane binding would curve the membrane near the NPs. Once a minimal local bending is realized, the membrane-curvature-mediated attractive NP–NP interaction would drive the NPs (proteins) to form a cluster and finally deform the flat bilayer to a vesicle (Figure 8.2).

Liang et al. [22] used self-consistent field theory to study the membrane-mediated interactions between NPs, and found that effective interactions between two NPs are mainly dominated by competition of the deformations of the different parts of the lipid bilayers as well as chain stretching of the lipid molecules sandwiched between NPs. By varying the packing density of the lipids in the membrane or changing the NP size, the authors found that the effective interactions between NPs can be efficiently modulated. Yue and Zhang [15] performed DPD simulations to investigate the interaction between two NPs during NP internalization at different NP sizes. The distance between two small NPs (initial inter-NP distances of 6.46 nm) decreases during NP internalization, indicating an effective attractive inter-NP interaction (Figure 8.3a–c). However, for large NPs (initial inter-NP distances of 15.8 nm), the two NPs would keep away from each other during the internalization (Figure 8.3d–f). This indicated that an effective interaction between neighboring NPs arises purely as a result of membrane curvature induced by NP adsorption. Due to the strong receptor–ligand interaction, the local geometry of a small NP deforms the membrane and produces local membrane curvature. When two membrane-curving NPs approach each other, the bilayer deformations overlap. Then the neighboring NPs might cluster in order to reduce the curvature energy of the membrane because the clustering could share the work needed to bend the membrane and thereby lower the stored elastic energy. The curvature-mediated interaction between small NPs is thus effectively attractive even without direct attraction between neighboring NPs (Figure 8.3b), while increasing NP

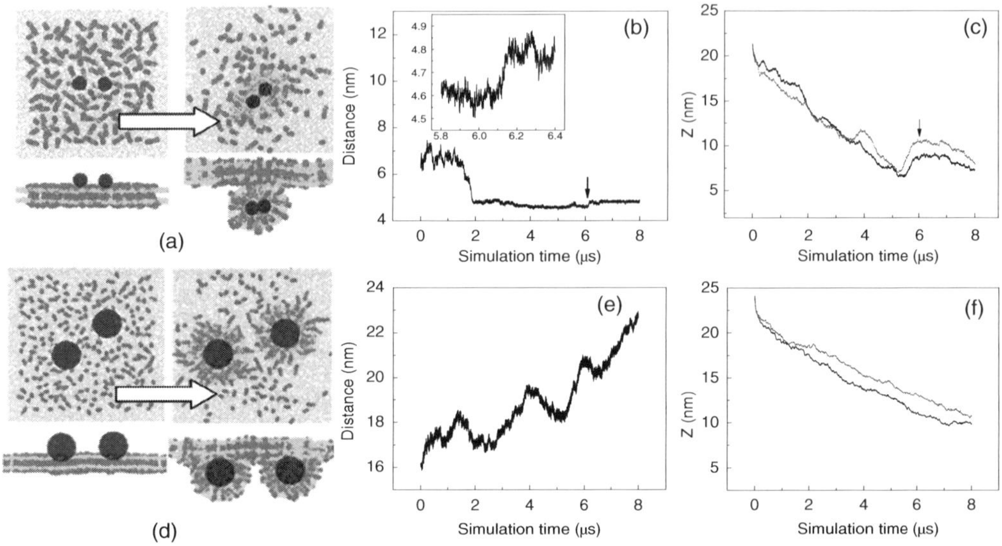

Figure 8.3 Endocytosis of two identical NPs. Two smaller NPs of 3.8 nm were placed on a membrane of 38.7 nm × 38.7 nm (a–c), while two larger NPs of 9.0 nm were placed on a membrane of 51.7 nm × 51.7 nm (d–f). (a, d) show the initial and final structure of endocytosis of two identical NPs. (b, e) show the evolution of the distance between two NPs. (c, f) show the evolution of NP positions along the membrane normal direction. The initial inter-NP distances are 6.46 nm (a–c) and 15.8 nm (d–f), respectively. Reproduced from Ref. [15] with permission from the American Chemical Society. *(See insert for color representation of the figure.)*.

size would weaken the membrane-curvature-mediated interaction because the increase in the NP size lowers the local membrane curvature. Moreover, when the NP size becomes large enough, we found that the effective interaction between neighboring NPs seems to be repulsive, as shown in Figure 8.3e. Further simulations indicated that this membrane-curvature-mediated interaction was controlled by NP size and the inter-NP distance or, equivalently, the dose of the NPs.

Chen et al. [23] and Yue et al. [24] employed computer simulation to study the interaction of hard NPs with lipid aggregates with curved surfaces (vesicle and membrane tube), and confirmed that when the NPs adhere on the curved membrane surface, they may cluster due to the membrane-induced attraction as in the situation of the flat bilayer. Moreover, it was found that the morphology of the NP clusters mediated by the membrane depends on NP size, interaction strength between the NPs and the lipid membrane, and membrane tension.

8.3.2 Internalization Pathways of the NPs

As NPs play an important role in drug delivery, more and more attention has been paid to the internalization pathways of NPs across plasma membranes. Yue and Zhang performed N-varied DPD simulations of receptor-mediated endocytosis of spherical NPs coated with ligands [18]. Depending on the membrane tension, ligand density on the NP surface, and NP size, four different membrane responses were observed: receptor-mediated endocytosis, NP adhered on the membrane surface, NP penetration, and NP-adsorption-induced membrane rupture (see Figure 8.4). This

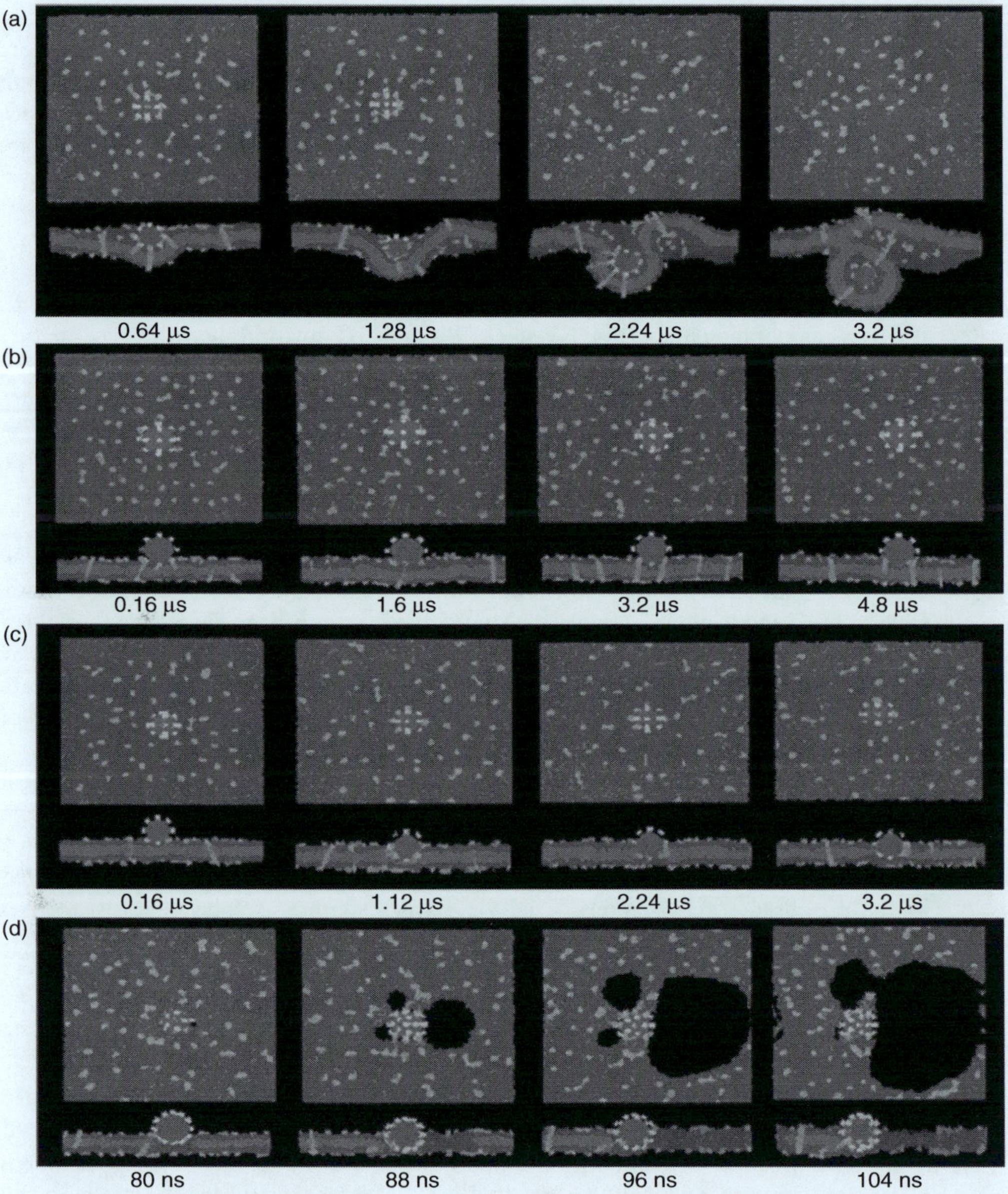

Figure 8.4 Four kinds of membrane responses to adsorption of NPs: (a) Receptor-mediated endocytosis, (b) adhesion of the NP on the membrane surface, (c) penetration of the NP into the membrane, and (d) NP-induced membrane rupture. Both top view and side view are displayed for each process. Reproduced from Ref. [18] with permission from the Royal Society of Chemistry.

study not only reveals the dynamic pathway for the NP uptake, but also demonstrates how the internalization pathway changes with NP properties and membrane tension. In the past few years, a number of works focused on the pathway of NP internalization and investigated which NP properties (such as NP size, topography, and surface modification) and membrane properties (such as membrane composition, tension) are the dominant factors affecting the NP internalization pathways.

8.3.2.1 NP Properties Affecting the NP–Membrane Interaction

With the rapid development of synthesis techniques, various NPs with different properties have appeared. At present, the following NP properties have been considered for the NP–membrane interaction: NP size, NP shape, NP hydrophilicity, surface coating and especially surface charge, and NP dose (concentration). Sometimes, one of these properties dominates the interaction between the NPs and the membrane, but more frequently there exist several properties that cooperatively determine the interaction. Below, we discuss how the NP properties affect the interaction between NPs and the plasma membrane, especially from the viewpoint of computer simulation.

NP Size NP size is an important parameter in designing suitable cell tracking and drug delivery systems, and it often determines the mechanism and rate of NP uptake. Many experimental and computer simulation studies have recently been devoted to the field of cellular uptake of NPs, and show that particle size can affect the efficiency and pathway of cellular uptake. Furthermore, several mechanisms have been proposed to indicate that the NP size relates to the extent of membrane disruption and to the structure of NP–lipid assemblies.

Using DPD simulation techniques, Yue and Zhang [18] suggested that small hydrophilic NPs with a radius smaller than 3.0 nm possibly penetrate the membrane directly, sometimes followed by membrane rupture or the formation of membrane a pore at a high membrane tension; for large NPs, they can often be partly encapsulated by the membrane or fully internalized via pathways either as "phagocytosis" (or cell eating) or as "pinocytosis" (or cell drinking).

Similarly, Lin et al. [25] performed CGMD simulations to investigate the size effect for hydrophobic NPs across the dipalmitoylphosphatidylcholine (DPPC) bilayer. The authors found that different NP sizes have different effects on the structural change of the membrane. However, no obvious energy barrier can be seen in the free energy profiles during the process of NP translocation. The same size effect was showed experimentally by Roiter et al. [26], although the critical dimension of the particles obtained from experiments is different when compared to that from simulations. Atomic-force microscopy images indicated that small NPs ($R < 22$ nm) would induce a hole in the lipid bilayer, whereas larger NPs are mostly covered with the lipid bilayer as a whole [26].

Other studies also suggested that it is thermodynamically favorable for hydrophobic NPs of size 2–8 nm to be embedded within a lipid bilayer. It is thermodynamically possible to embed hydrophobic NPs having a diameter in proximity to, or even exceeding, the thickness of the bilayer. The size effect was often interpreted theoretically with the Helfrich elastic model [27], in which the total free energy of the NP–membrane system can be decomposed into two parts: the bending energy of the membrane and the adhesion energy between NPs and the membrane. The bending energy of the membrane changes with the radius of the NPs when NPs are wrapped by the membrane. To minimize the total free energy of the NP–membrane system one can obtain theoretically the relation between the NP radius and the membrane bending as a function of membrane tension.

Topography of NPs Besides the NP size, the topography of the NPs is another important intrinsic property affecting the NP–membrane interaction, as demonstrated both experimentally and theoretically. For example, some kinds of NPs with a high aspect ratio, like carbon nanotubes or asbestos fibers, have been shown to cause length-dependent toxicity in certain cells [28].

To understand how NPs with a high aspect ratio interact with a plasma membrane, Shi et al. [28] performed both experimental and simulation studies, and found that NPs with a high aspect ratio enter cells through the tip first. It is the tip-first mechanism that leads to near-vertical entry of carbon nanotubes with end caps and prevents the long NPs from entering a cell completely. However, nanotubes without caps on their tip show a different mechanism for NP uptake. A subsequent study [29] found that the internalization pathway of spherocylindrical NPs is different when compared to spherical NPs. For a spherocylindrical NP with an initial upright docking position on the membrane plane, endocytosis proceeds through a laying-down-then-standing-up sequence (see Figure 8.5). This is similar to the conclusion by Vacha et al. [30] that passive endocytosis (i.e., not ATP driven) depends on the shape of the NPs. In their studies, CGMD simulation results demonstrated that the efficiency of the passive endocytosis is higher for spherocylindrical particles than for spherical particles of the same diameter.

The NP topography influences not only the efficiency of cellular uptake but also the uptake pathway. For NPs with different shapes, Yang and Ma [14] showed that the shape anisotropy and initial orientation of the NPs dominate the process for NP translocation across the lipid bilayer (see Figure 8.5). Li et al. [31] performed DPD simulations to show that rotation of NPs, which is found to be one of the most important mechanisms for endocytosis of shaped NPs, regulates the competition between ligand–receptor binding and membrane deformation. As a result, the whole internalization process for those non-spherical NPs can be divided into two stages: membrane invagination and NP wrapping. Furthermore, they demonstrated that NPs having various shapes show different favorable orientations at the two stages.

Besides spherical and cylindrical NPs, graphene, which is of a two-dimensional (2D) structure, has been considered a new candidate for drug transporting in recent years, and as a consequence the interactions between graphene and lipid membranes have been studied. From experimental and theoretical study, Tu et al. [32] found that pristine graphene (PG) and graphene oxide nanosheets can induce degradation of the cell membrane of *Escherichia coli*, and reduce their viability. They also showed that the graphene could penetrate into and extract large amounts of phospholipid from the cell membrane because of the strong dispersion interaction between graphene and lipid molecule. To clarify the mechanisms of the cellular interactions of graphene nanosheets with various extents of oxidization, Mao et al. [33] found four different kinds of graphene–membrane interaction states through large-scale simulation and theoretical analysis (see Figure 8.6). Moreover, they found that higher oxidized graphene nanosheets could induce a larger scale of irregular membrane perturbation destroying the integrity of the membrane, and that the perturbation degree of the membrane increases with increasing edge length of the nanosheet.

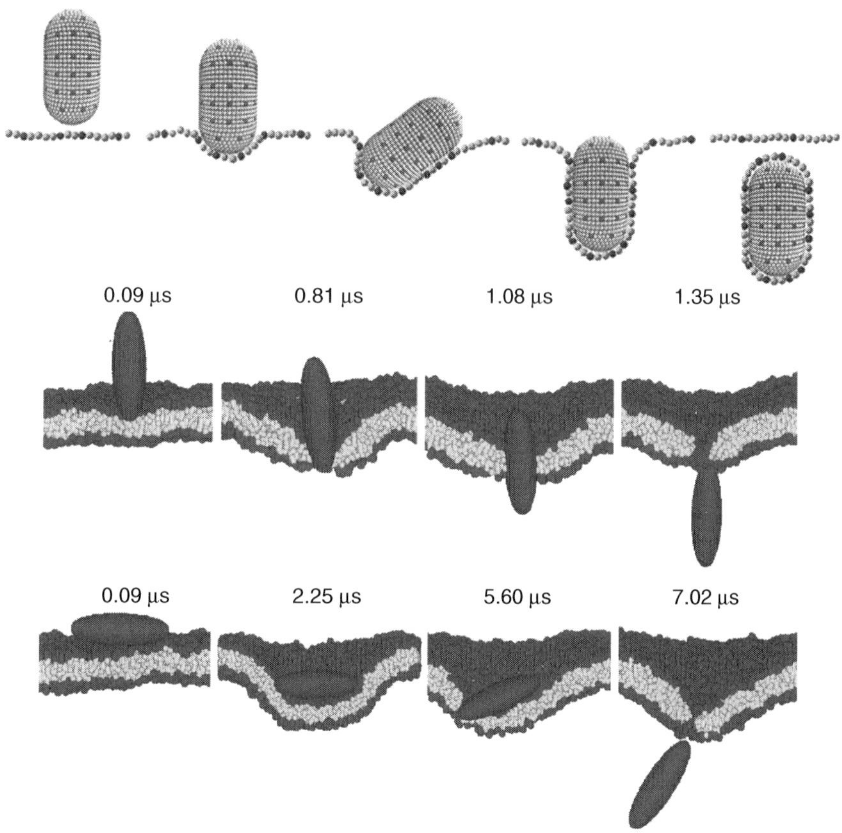

Figure 8.5 (Upper) Endocytic pathway of a spherocylindrical NP. The NP takes a general laying-down-then-standing-up sequence during endocytosis. (Reproduced from Ref. [29] with permission from the American Chemical Society.) (Lower) Computer-simulated snapshots of the translocation of ellipsoids with vertical and horizontal starting orientations. (Reproduced from Ref. [14] with permission from Nature Publication Group.)

NP Hydrophilicity and Surface Modification Numerous studies have demonstrated that adsorption of NPs often induces a change of membrane morphology and the formation of a membrane hole, strongly depending on the surface modification of the interacting NPs. Further, the cellular uptakes of hydrophilic NPs and hydrophobic NPs often adopt different mechanisms and different pathways.

To understand the effect of the surface ligands for the interaction between the surface ligand-coated NPs and the lipid bilayer, Yue and Zhang [18] used DPD simulations to show that there exist four possible kinds of membrane responses to the ligand-coated NPs. In particular, both the density and distribution of the ligands on the NP surface play a crucial role for the membrane response to the NP adsorption. They showed clearly that the NP surface hydrophobicity can induce different internalization mechanisms for NP–membrane systems. Similarly, Li et al. [34] performed CGMD simulations to investigate the effect of hydrophilic/hydrophobic properties of NPs on their interaction with dipalmitoylphosphatidylcholine (DPPC) membranes. The

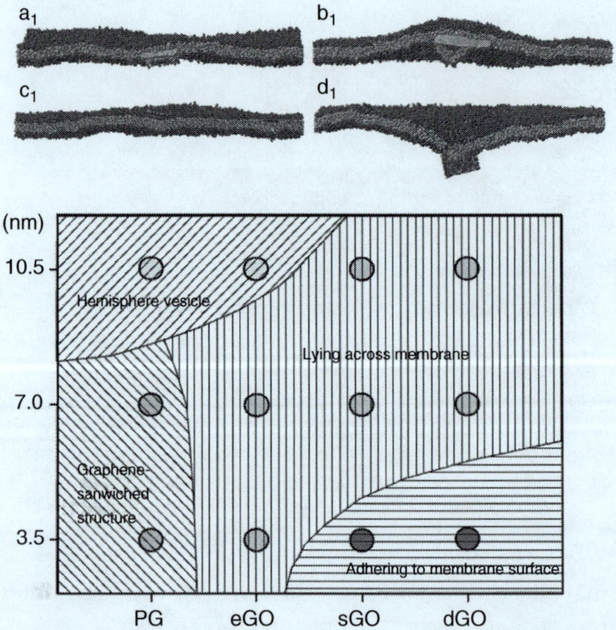

Figure 8.6 (Upper) Typical states of the interactions between a lipid bilayer membrane and a graphene nanosheet. (a_1) Graphene-sandwiched structure for PG with side length 7.0 nm. (b_1) Hemisphere vesicle structure for PG with 10.5 nm. (c_1) Adhering to the surface of the membrane for double-oxidized graphene oxide (dGO) with 3.5 nm, and (d_1) lying across the membrane for dGO with 7.0 nm. (Lower) A two-dimensional phase diagram characterizes the interrelated effects of graphene size and oxidization degree on the equilibrium states of the graphene nanosheets interacting with the lipid bilayer membrane. The various shading patterns mark the approximate regions of these states. Reproduced from Ref. [33] with permission from Elsevier.

simulation results showed that hydrophobic NPs could result in their inclusion into the membrane, while hydrophilic NPs are just absorbed on the membrane surface.

Considering the inhomogeneous surface modified NPs, Ding et al. [35] investigated the interaction between Janus NPs and the membrane, and found that there exist two different late-stage states for the Janus particle–membrane system: insertion and engulfment. For the insertion state, the hydrophobic part of the NP is inserted into the interior of the lipid bilayer, while the hydrophilic part stays outside of the membrane. For the engulfment state, the hydrophilic part is engulfed by lipid heads due to receptor–ligand interactions and the hydrophobic part is "engulfed" by lipid tails because of hydrophobic interactions. The difference between the two late-stage states is controlled by the initial orientation and the properties of the Janus NPs.

To understand the effect of the surface properties of the ligand-coated NPs, Li et al. [36] performed DPD simulations to investigate the evolution of free energy as NPs with different ligand patterns cross a lipid bilayer. In their studies, four kinds of ligand patterns were considered: NPs modified as a stripe-like structure with alternating hydrophilic and hydrophobic ligands, NPs modified with a patch-like structure, NPs with randomly distributed hydrophilic and hydrophobic ligands, as well as NPs coated with homogeneous hydrophobic and hydrophilic ligands respectively (Figure 8.7). The

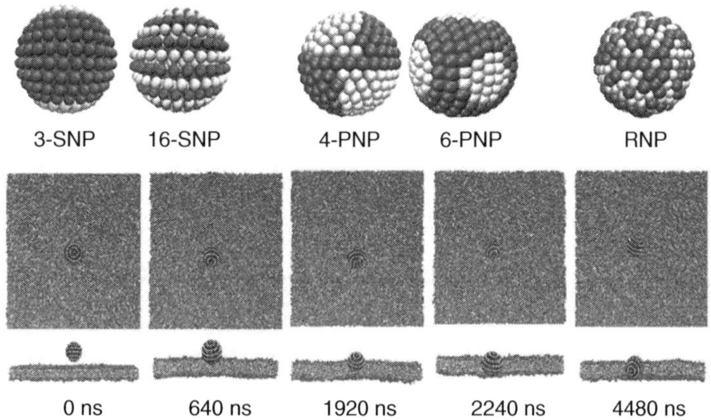

Figure 8.7 (Upper) Schematic drawings of different surface modified NPs (hydrophilic ligand on the NP surface in white, and hydrophobic ligand in purple). (Lower) Typical configurations for the penetration of a 16-SNP during its penetration processes. Reproduced from Ref. [36] with permission from the Royal Society of Chemistry.

free energy analysis indicated that NPs with small stripe/patch size normally encounter a lower energy barrier for NP translocation across the lipid membrane.

Surface Charge All cells have an electrical potential generated by the difference in cytosolic and extracellular potassium concentrations. This external potential could affect the interaction between the NPs and the membrane, and especially influence how charged NPs interact with a polar membrane. To investigate the influence of the surface charge on NPs, Shin et al. [37] combined experimental and computer simulation techniques to confirm that the decrease of membrane potential leads to decreased cellular binding of anionic NPs. They also showed that an increase in membrane potential reverses this trend and results in increased binding of anionic NPs. However, the cellular binding of cationic NPs is minimally influenced by the membrane potential. This study indicates that membrane potential is an important factor that must be considered in the design of NPs for therapeutic and sensing applications.

Similarly, surface charges on NPs also significantly affect the NP–membrane interaction. Li and Gu [38] performed CGMD simulations and showed that compared to uncharged NPs, charges on the NP surface improve the adhesion of NPs to the membrane due to the electrostatic attraction. Moreover, the binding of positively charged NPs on the phosphate terminus of the lipids increases the tilt angle of lipids and thus enlarges the area of the head group, whereas NPs with negative charges often induce the formation of the high ordered region in fluid bilayers. Therefore, the adsorption of cationic/anionic NPs on the membrane plays different roles in suppressing or promoting the membrane wrapping.

The distribution of charges on the NP surface also affects the binding of NPs to a lipid bilayer. With DPD simulations, Li et al. [39] demonstrated the importance of the pattern of charge distribution on NP surfaces, and showed that the different charge patterns may induce different morphologies of NP aggregates on a membrane. The authors also showed that the aggregation of charged NPs requires a minimum effective charged area, and the pattern of the electrostatic distribution of the NP surface determines the arrangement of self-assembled NPs. Lin et al. [40] pointed out that cationic gold NPs (AuNPs) are more disruptive to negatively charged lipid membranes, often leading to the formation of transient membrane holes. This formation of membrane pores depends on the number of positive charges on the NP surface. Similarly, positively charged NPs can also induce flipping of membrane areas leading to particle inclusion and membrane depolarization [41].

8.3.2.2 The Effect of Membrane Properties on NP–Membrane Interaction

Undoubtedly, the properties of the plasma membrane also show significant influence on the NP–membrane interaction. Among various membrane properties, the effects of membrane tension and membrane composition have been particularly studied.

Various bioprocesses taking place on the cell membrane may change the membrane tension frequently and sometimes locally, which would in turn alter the NP–membrane interaction. In this respect, Yue et al. [18] used N-varied DPD simulations to demonstrate that the membrane tension significantly affects the interaction between the membrane and the NPs. When the membrane tension is sufficiently positive, the perturbation induced by NP adsorption may rupture the membrane. At a mild membrane tension, NP adhesion or penetration is observed, depending on the strength of the NP–membrane adhesion. However, when the membrane tension decreases to a negative value, the NPs will be internalized by the cell membrane via a pathway of endocytosis.

Guo et al. [42] also confirmed through computer simulations and theoretical analysis that the membrane tension plays a significant role in the wrapping of dendrimer-like NPs. They found that when the membrane tension is larger than a critical value, the dendrimer would induce membrane disruption. Otherwise, depending on the membrane tension, three states for the NP–membrane system, including dendrimer penetration, partial wrapping, and full wrapping, are found (Figure 8.8).

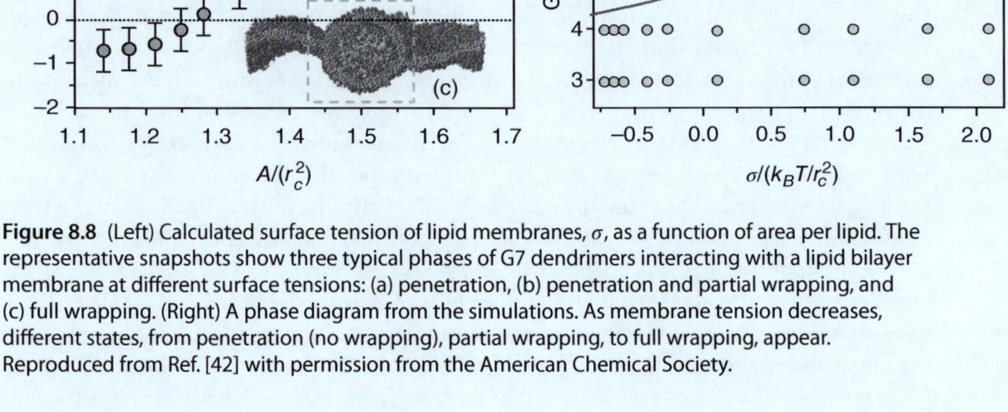

Figure 8.8 (Left) Calculated surface tension of lipid membranes, σ, as a function of area per lipid. The representative snapshots show three typical phases of G7 dendrimers interacting with a lipid bilayer membrane at different surface tensions: (a) penetration, (b) penetration and partial wrapping, and (c) full wrapping. (Right) A phase diagram from the simulations. As membrane tension decreases, different states, from penetration (no wrapping), partial wrapping, to full wrapping, appear. Reproduced from Ref. [42] with permission from the American Chemical Society.

Beside the properties of the NPs and the membrane tension, there are other factors that affect the NP interaction with the membrane. For example, it has been found that membrane proteins have a remarkable influence on the interaction between NPs and the membrane [43, 44]. Many membrane proteins have been proved to take part in the endocytosis process. It is well known that specific cell-surface receptors and Rho-family GTPases are required to achieve several types of phagocytosis processes. Note that in most situations the membrane proteins not only participate in the NP–membrane interaction, but also change the membrane properties. For instance, Yue et al. [13] and Li et al. [19] found that the self-assembly of the anchored proteins can induce membrane curvature. In fact, the membrane curvature created through the insertion and clustering of anchored proteins is shown to be influenced not only by the membrane tension but also by the hydrophobic length of the anchored proteins.

Temperature is another factor affecting NP–membrane interaction, and in general high temperature increases the fluidity of the membrane, enhancing membrane endocytosis. In addition, some environmental conditions, such as the gradient of ionic concentrations between the cytosol and extracellular medium, and additional magnetic and electric fields, also show significant influence on the interaction between NPs and membrane. For example, Tian et al. [45] and Ding et al. [46] indicated that the NP–membrane interaction is affected by the pH value for pH-sensitive dendrimers or some NPs modified by pH-sensitive polymers, as demonstrated experimentally.

8.4 Cooperative Effect between NPs during Internalization

The above-mentioned works have showed that the interaction between a single NP and a plasma membrane is controlled by NP size, NP shape, and the surface physical and chemical properties of the NP. However, in real systems the membrane often interacts with multiple NPs simultaneously. Therefore, the interaction mechanism and pathway are often cooperatively determined by multiple neighboring NPs, rather than a single one. This has been verified by experimental observation of the effects of the NP dose, which becomes increasingly important; sometimes, NP cytotoxicity is found only at high NP concentrations.

From a systematic investigation of adsorbed semi-hydrophobic NPs on supported lipid bilayers (SLBs), Jing and Zhu [47] pointed out that there exists a critical NP concentration for NPs having different degrees of surface hydrophobicity. Above this critical NP concentration the SLBs will disrupt. Consistently, Lipowsky and his coworkers [48] found that the mechanism for the interaction between multiple NPs and the membrane is different from that for single NPs interacting with the membrane. Through DPD simulations, Yue and Zhang [15] indicated that the internalization of multiple NPs was in fact a cooperative process. As shown in Figure 8.9a, small NPs with a diameter of 2.5 nm are first partially wrapped by the membrane, and at the same time adjacent NPs gradually approach each other to form a small cluster. As the simulation proceeds, the size of the cluster increases, and the cluster is encapsulated by the membrane and internalized as a whole (see Figure 8.9a). Compared to the close-packed aggregates formed by smaller NPs of 2.5 nm, the larger NPs of size 4.0 nm tend to arrange linearly on the lipid membrane and form a pearl-chain-like structure for the subsequent endocytosis (see Figure 8.9b). However, when the diameter of

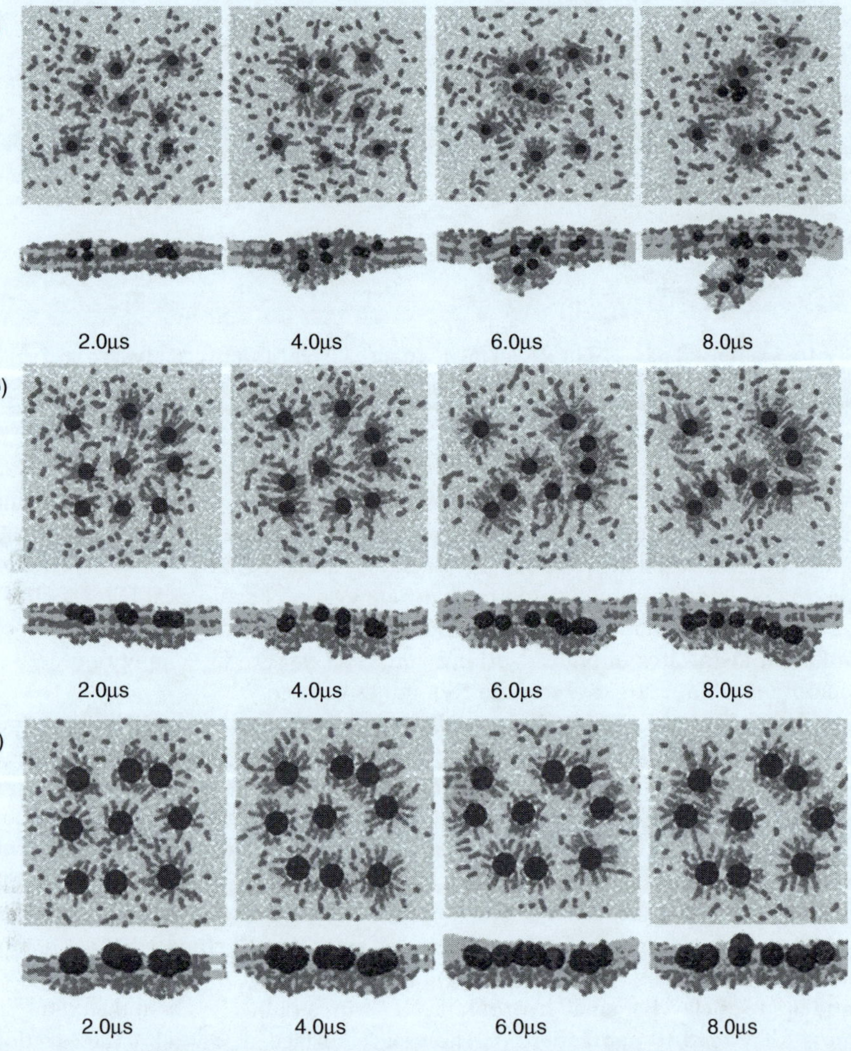

Figure 8.9 Typical snapshots during different internalization processes. (a) Nanoparticles with a diameter of 2.5 nm form close-packed aggregates before internalization. (b) Nanoparticles having a diameter of 4.0 nm aggregate into pearl-chain-like arrangements. (c) Independent endocytosis occurs for nanoparticles having a diameter of 6.0 nm. Both top view and side view are displayed. Reproduced from Ref. [15] with permission from the American Chemical Society.

the NPs increases further to 6.0 nm, no obvious cooperative endocytosis is observed (see Figure 8.9c). On the contrary, the NPs tend to be wrapped by the membrane independently. Consequently, both the internalization rate and internalization pathway of multiple NPs are strongly dependent on NP size. As mentioned above, it was rather difficult to wrap a small single NP by a cellular membrane because significant local membrane deformation was required. For this reason, the single NP with larger size can be easily wrapped because a smaller membrane curvature is needed in the wrapping process. Nevertheless, the situation for multiple NPs is quite different, because the

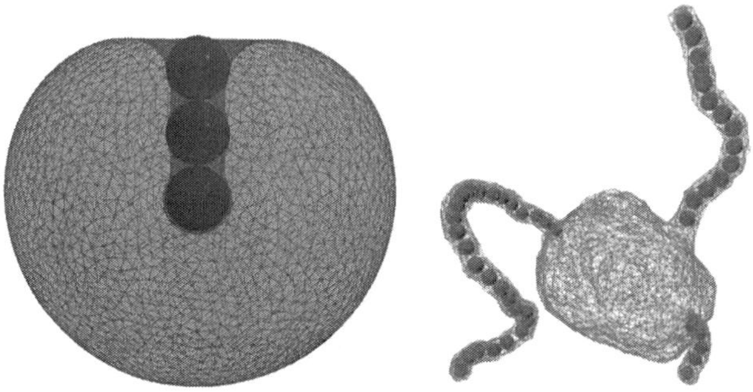

Figure 8.10 Snapshot of tube-like structures formed from multiple NPs interacting with a lipid vesicle. (Reproduced from Ref. [48] with permission from the American Physical Society.) (Left) The NPs were initially outside, and (Right) inside the vesicle. (Reproduced from Ref. [50] with permission from the American Physical Society.)

energy cost associated with the strong local membrane deformation to wrap a single small NP can be substantially reduced by NP clustering, which thus facilitates the wrapping process. Similarly, from large-scale simulations Šarić and Cacciuto [49] showed that a lipid membrane could mediate aggregation of spherical NPs binding on it for a wide range of membrane bending rigidities. The phase diagram obtained shows that different aggregates of NPs adsorbing onto a bilayer are in competition between the membrane bending rigidity and the NPs' adhesive energy.

Previous studies have indicated how NPs can affect each other on a flat bilayer. Considering that the cell membrane is a closed system with some local curvatures, it is important to understand the interaction between multiple NPs and a membrane. Recently, based on computer simulations, Bahrami et al. [48] and Šarić et al. [50] found that unexpected tube-like structures can be formed when multiple NPs interact with a lipid vesicle (Figure 8.10). Interestingly, both studies found the membrane tubules enclosing linear aggregates of NPs regardless of whether the NPs were initially placed outside of the vesicle or not. Similarly, Chen et al. [23] performed DPD simulations to show that the NP concentration has important effects on the interaction between NPs and lipid vesicles. For small hydrophilic NPs with a radius less than the membrane thickness, they tend to just adhere on the vesicle surface at low NP concentrations. With an increase in the NP concentration, the NPs would penetrate into the vesicle. Further increasing NP concentration to a threshold value induces vesicle rupture. More interestingly, three different pathways for spontaneous NP penetration, including chain-like cooperative penetration, direct penetration, and inverted micelle-like penetration were identified, depending on NP size and NP concentration.

Compared with the above spherical NP interaction with a lipid membrane, Yue et al. [51] performed DPD simulations to investigate the cooperative behavior of rod-like NPs. From the simulation they showed that not only cooperative interaction between neighboring rod-like NPs drove the NP aggregate, but also NP orientation (Figure 8.11). For NPs with a weak NP–membrane binding energy, the orientation-dependent interaction between two neighboring NPs arises purely as a result of non-homogeneous distribution of membrane curvature induced by anisotropic NP adsorption. The orientation-dependent interaction could induce the rotation of neighboring NPs from a non-equilibrium arrangement to the equilibrium one for two neighboring NPs. For

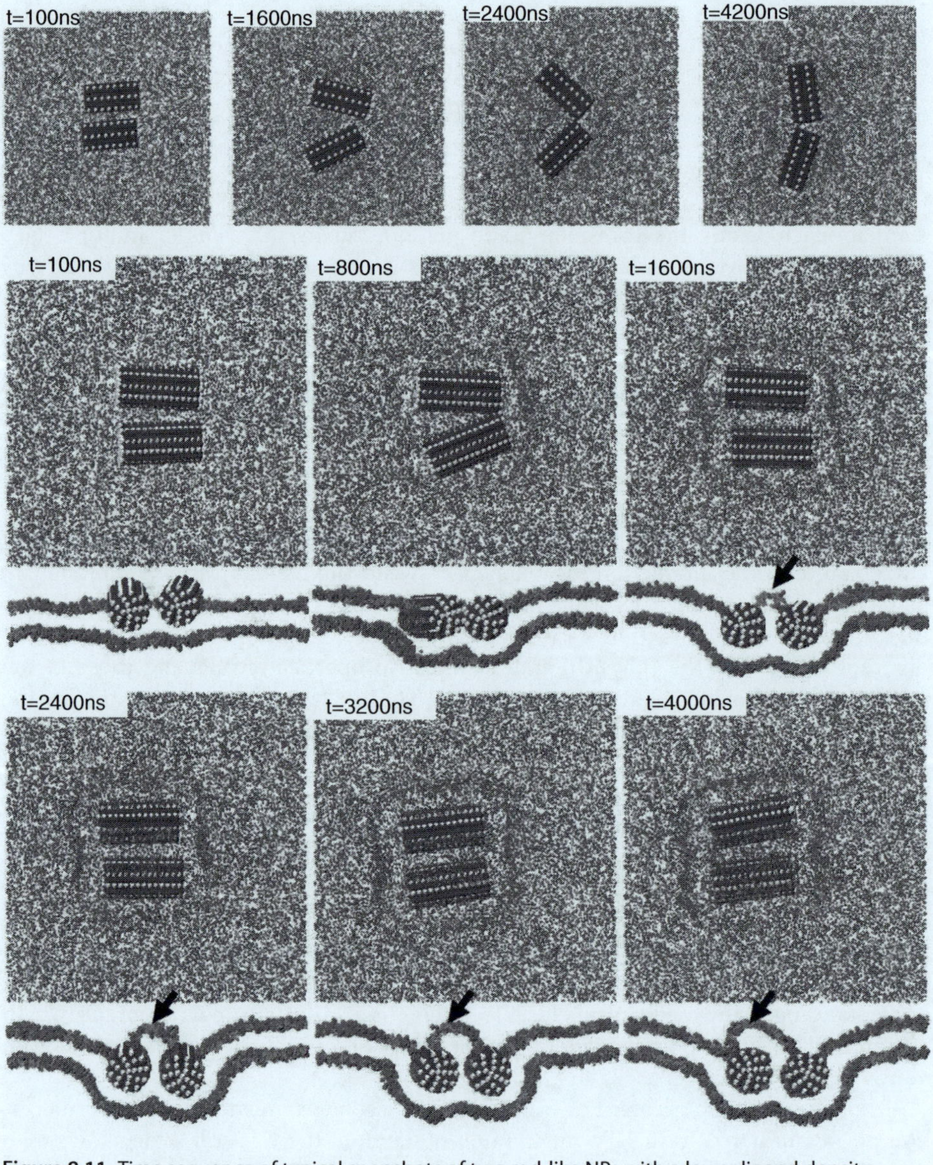

Figure 8.11 Time sequence of typical snapshots of two rod-like NPs with a lower ligand density (Upper) and higher ligand density (Lower). Reproduced from Ref. [51] with permission from the Royal Society of Chemistry.

a strong NP–membrane adhesion energy, on the other hand, an unusual wrapping pathway different from endocytosis, called asymmetrical wrapping, is found. The unusual asymmetrical wrapping also induces orientation-dependent NP interaction: the protrusion on the upper leaflet creates a short-range repulsion, the lower leaflet produces an intermediate-range attraction, and the disappearance of the asymmetrical wrapping generates a long-range repulsion. Thus, for a system with multiple rod-like

NPs, the equilibrium arrangements are not only affected by NP interaction but also by dynamic traps. Competition between the parallel and linear arrangements may transiently stabilize configurations with "incorrect" arrangements, and then rapid and extensive wrapping could cause the "incorrect" arrangements to be dynamically trapped.

Similar to NP size and topography, surface-patterned NPs could also cooperatively induce transmembrane penetration. Recently, Zhang et al. [14, 52] used DPD simulations to study the transmembrane penetration mechanisms of multiple NPs with different surface hydrophobic properties. From a series of simulations they confirmed that the quantity of NPs played a key role during the NP penetration process. Additionally, some interesting cooperative behaviors of NPs (such as aggregation, aggregation–dispersion, and aggregation–dispersion–reaggregation), which are affected by the coupling between the NP quantity and the NP surface property, were observed during the NP penetration process.

8.5 Conclusions

Understanding the interaction mechanism between the NPs and the membrane can guide us to avoid cytotoxicity while using NPs as vehicles for drug delivery. The factors that influence the interaction between NPs and the membrane can be divided into different categories: the (physical and chemical) properties of NPs and those of the membrane. Although extensive experimental studies have been carried out in the literature in order to understand NP–membrane interactions, much less is known about the molecular details of the interaction mechanisms and the kinetic pathways of these processes. In this aspect, computer simulation techniques provide a complementary tool to study NP–membrane interaction. Therefore, coarse-grained molecular dynamics (CGMD) and dissipative particle dynamics (DPD) simulations have been extensively performed investigating the interaction mechanism and corresponding interaction pathway. In this review, we have summarized the recent DPD and partial CGMD simulations on cell uptake of NPs, focusing on membrane-mediated interaction, internalization pathways, and cooperative effect. These studies undoubtly deepen our understanding of the mechanisms for NPs interacting with a plasma membrane, which can not only help us to understand the cytotoxicity of NPs, but also guide us to design new NPs for drug delivery.

Although we have a greater understanding of these mechanisms for the internalization of NPs initiated from NP–membrane interaction, there is still a long way to go to understand precisely NP uptake *in vivo*. As is well known, cell machinery is often involved *in vivo* in the internalization process, i.e., active uptake of NPs. Note that in this review, we have not discussed the role of proteins or protein complexes in the internalization pathway and how they contribute to membrane curvature, membrane pore formation, and the recruitment of other proteins.

References

1. F. Lu, S. H. Wu, Y. Hung, and C. Y. Mou. Size effect on cell uptake in well-suspended, uniform mesoporous silica nanoparticles, *Small* **5**, 1408 (2009).

2. F. Osaki, T. Kanamori, S. Sando, T. Sera, and Y. Aoyama. A quantum dot conjugated sugar ball and its cellular uptake. On the size effects of endocytosis in the subviral region, *Journal Of The American Chemical Society* **126**, 6520 (2004).
3. W. Jiang, B. Y. Kim, J. T. Rutka, and W. C. Chan. Nanoparticle-mediated cellular response is size-dependent, *Nature Nanotechnology* **3**, 145 (2008).
4. T. Mironava, M. Hadjiargyrou, M. Simon, V. Jurukovski, and M. H. Rafailovich. Gold nanoparticles cellular toxicity and recovery: Effect of size, concentration and exposure time, *Nanotoxicology* **4**, 120 (2010).
5. J. Wong-Ekkabut, S. Baoukina, W. Triampo, I.-M. Tang, D. P. Tieleman, and L. Monticelli. Computer simulation study of fullerene translocation through lipid membranes, *Nature Nanotechnology* **3**, 363 (2008).
6. K. Yang, B. Yuan, and Y.-q. Ma. Influence of geometric nanoparticle rotation on cellular internalization process, *Nanoscale* **5**, 7998 (2013).
7. L.-T. Yan and X. Yu. Charged dendrimers on lipid bilayer membranes: Insight through dissipative particle dynamics simulations, *Macromolecules* **42**, 6277 (2009).
8. H. M. Ding and Y. Q. Ma. Theoretical and computational investigations of nanoparticle–biomembrane interactions in cellular delivery, *Small* **11**, 1055 (2015).
9. A. Albanese, P. S. Tang, and W. C. Chan. The effect of nanoparticle size, shape, and surface chemistry on biological systems, *Annual Review Of Biomedical Engineering* **14**, 1 (2012).
10. Y. Li, H. Yuan, A. von dem Bussche, M. Creighton, R. H. Hurt, A. B. Kane, and H. Gao. Graphene microsheets enter cells through spontaneous membrane penetration at edge asperities and corner sites, *Proceedings of the National Academy of Sciences* **110**, 12295 (2013).
11. R. D. Groot and P. B. Warren. Dissipative particle dynamics: Bridging the gap between atomistic and mesoscopic simulation, *Journal Of Chemical Physics* **107**, 4423 (1997).
12. L. Gao, J. Shillcock, and R. Lipowsky. Improved dissipative particle dynamics simulations of lipid bilayers, *The Journal of Chemical Physics* **126**, 015101 (2007).
13. S. Li, X. Zhang, and W. Wang. Cluster formation of anchored proteins induced by membrane-mediated interaction, *Biophysical Journal* **98**, 2554 (2010).
14. K. Yang and Y.-Q. Ma. Computer simulation of the translocation of nanoparticles with different shapes across a lipid bilayer, *Nature Nanotechnology* **5**, 579 (2010).
15. T. Yue and X. Zhang. Cooperative effect in receptor-mediated endocytosis of multiple nanoparticles, *Acs Nano* **6**, 3196 (2012).
16. B. Hong, F. Qiu, H. Zhang, and Y. Yang. Budding dynamics of individual domains in multicomponent membranes simulated by N-varied dissipative particle dynamics, *The Journal of Physical Chemistry B* **111**, 5837 (2007).
17. F. Tian, X. Zhang, and W. Dong. How hydrophobic nanoparticles aggregate in the interior of membranes: A computer simulation, *Physical Review E* **90**, 052701 (2014).
18. T. Yue and X. Zhang. Molecular understanding of receptor-mediated membrane responses to ligand-coated nanoparticles, *Soft Matter* **7**, 9104 (2011).
19. T. Yue, S. Li, X. Zhang, and W. Wang. The relationship between membrane curvature generation and clustering of anchored proteins: A computer simulation study, *Soft Matter* **6**, 6109 (2010).

20. T. Yue and X. Zhang. Signal transduction across cellular membranes can be mediated by coupling of the clustering of anchored proteins in both leaflets, *Physical Review E* **85**, 011917 (2012).

21. B. J. Reynwar, G. Illya, V. A. Harmandaris, M. M. Müller, K. Kremer, and M. Deserno. Aggregation and vesiculation of membrane proteins by curvature-mediated interactions, *Nature* **447**, 461 (2007).

22. Q. Liang, Q.-H. Chen, and Y.-Q. Ma. Membrane-mediated interactions between nanoparticles on a substrate, *The Journal of Physical Chemistry B* **114**, 5359 (2010).

23. X. Chen, F. Tian, X. Zhang, and W. Wang. Internalization pathways of nanoparticles and their interaction with a vesicle, *Soft Matter* **9**, 7592 (2013).

24. T. Yue, X. Zhang, and F. Huang. Molecular modeling of membrane tube pearling and the effect of nanoparticle adsorption, *Physical Chemistry Chemical Physics* **16**, 10799 (2014).

25. X. Lin, Y. Li, and N. Gu. Nanoparticle's size effect on its translocation across a lipid bilayer: A molecular dynamics simulation, *Journal of Computational and Theoretical Nanoscience* **7**, 269 (2010).

26. Y. Roiter, M. Ornatska, A. R. Rammohan, J. Balakrishnan, D. R. Heine, and S. Minko. Interaction of nanoparticles with lipid membrane, *Nano Letters* **8**, 941 (2008).

27. W. Helfrich. Elastic properties of lipid bilayers: Theory and possible experiments, *Zeitschrift für Naturforschung C* **28**, 693 (1973).

28. X. Shi, A. von Dem Bussche, R. H. Hurt, A. B. Kane, and H. Gao. Cell entry of one-dimensional nanomaterials occurs by tip recognition and rotation, *Nature Nanotechnology* **6**, 714 (2011).

29. C. Huang, Y. Zhang, H. Yuan, H. Gao, and S. Zhang. Role of nanoparticle geometry in endocytosis: Laying down to stand up, *Nano Letters* **13**, 4546 (2013).

30. R. Vácha, F. J. Martinez-Veracoechea, and D. Frenkel. Receptor-mediated endocytosis of nanoparticles of various shapes, *Nano Letters* **11**, 5391 (2011).

31. Y. Li, T. Yue, K. Yang, and X. Zhang. Molecular modeling of the relationship between nanoparticle shape anisotropy and endocytosis kinetics, *Biomaterials* **33**, 4965 (2012).

32. Y. Tu et al. Destructive extraction of phospholipids from *Escherichia coli* membranes by graphene nanosheets, *Nature Nanotechnology* **8**, 594 (2013).

33. J. Mao, R. Guo, and L.-T. Yan. Simulation and analysis of cellular internalization pathways and membrane perturbation for graphene nanosheets, *Biomaterials* **35**, 6069 (2014).

34. Y. Li, X. Chen, and N. Gu. Computational investigation of interaction between nanoparticles and membranes: Hydrophobic/hydrophilic effect, *The Journal of Physical Chemistry B* **112**, 16647 (2008).

35. H.-M. Ding and Y.-Q. Ma. Interactions between Janus particles and membranes, *Nanoscale* **4**, 1116 (2012).

36. Y. Li, X. Zhang, and D. Cao. A spontaneous penetration mechanism of patterned nanoparticles across a biomembrane, *Soft Matter* **10**, 6844 (2014).

37. E. H. Shin, Y. Li, U. Kumar, H. V. Sureka, X. Zhang, and C. K. Payne. Membrane potential mediates the cellular binding of nanoparticles, *Nanoscale* **5**, 5879 (2013).

38. Y. Li and N. Gu. Thermodynamics of charged nanoparticle adsorption on charge-neutral membranes: A simulation study, *The Journal of Physical Chemistry B* **114**, 2749 (2010).
39. Y. Li, X. Zhang, and D. Cao. Self-assembly of patterned nanoparticles on cellular membranes: Effect of charge distribution, *The Journal of Physical Chemistry B* **117**, 6733 (2013).
40. J. Lin, H. Zhang, Z. Chen, and Y. Zheng. Penetration of lipid membranes by gold nanoparticles: Insights into cellular uptake, cytotoxicity, and their relationship, *Acs Nano* **4**, 5421 (2010).
41. R. R. Arvizo, O. R. Miranda, M. A. Thompson, C. M. Pabelick, R. Bhattacharya, J. D. Robertson, V. M. Rotello, Y. Prakash, and P. Mukherjee. Effect of nanoparticle surface charge at the plasma membrane and beyond, *Nano Letters* **10**, 2543 (2010).
42. R. Guo, J. Mao, and L.-T. Yan. Unique dynamical approach of fully wrapping dendrimer-like soft nanoparticles by lipid bilayer membrane, *Acs Nano* **7**, 10646 (2013).
43. A. Grafmüller, J. Shillcock, and R. Lipowsky. Dissipative particle dynamics of tension-induced membrane fusion, *Molecular Simulation* **35**, 554 (2009).
44. A. Šarić and A. Cacciuto. Self-assembly of nanoparticles adsorbed on fluid and elastic membranes, *Soft Matter* **9**, 6677 (2013).
45. W.-D. Tian and Y.-Q. Ma. Molecular dynamics simulations of a charged dendrimer in multivalent salt solution, *The Journal of Physical Chemistry B* **113**, 13161 (2009).
46. H.-M. Ding and Y.-Q. Ma. Controlling cellular uptake of nanoparticles with pH-sensitive polymers, *Scientific Reports* **3** (2013).
47. B. Jing and Y. Zhu. Disruption of supported lipid bilayers by semihydrophobic nanoparticles, *Journal of the American Chemical Society* **133**, 10983 (2011).
48. A. H. Bahrami, R. Lipowsky, and T. R. Weikl. Tubulation and aggregation of spherical nanoparticles adsorbed on vesicles, *Physical Review Letters* **109**, 188102 (2012).
49. A. Šarić and A. Cacciuto. Fluid membranes can drive linear aggregation of adsorbed spherical nanoparticles, *Physical Review Letters* **108**, 118101 (2012).
50. A. Šarić and A. Cacciuto. Mechanism of membrane tube formation induced by adhesive nanocomponents, *Physical Review Letters* **109**, 188101 (2012).
51. T. Yue, X. Wang, F. Huang, and X. Zhang. An unusual pathway for the membrane wrapping of rodlike nanoparticles and the orientation-and membrane wrapping-dependent nanoparticle interaction, *Nanoscale* **5**, 9888 (2013).
52. H. Zhang, Q. Ji, C. Huang, S. Zhang, B. Yuan, K. Yang, and Y.-Q. Ma. Cooperative transmembrane penetration of nanoparticles, *Scientific Reports* **5** (2015).

9

Theories for Polymer Melts Consisting of Rod–Coil Polymers

Ying Jiang[1] *and Jeff Z. Y. Chen*[2]

[1] *School of Chemistry and Environment, Beihang University, Beijing, China*
[2] *Department of Physics and Astronomy, University of Waterloo, Canada*

9.1 Introduction

9.1.1 Rod–Coil Polymers and Recent Theoretical Progress

Block copolymers consisting of rigid and flexible blocks are commonly known as rod–coil block copolymers. The rigid block can have interesting electrical and photoactive properties that are exploited in device making. For example, the practical application of organic optoelectronic devices requires self-assembly of such copolymers [1–3]. The rod block can be made from conjugated backbone, aromatic groups, or even the helical secondary structure of a peptide [4]. Experimental studies on self-assembly of rod–coil copolymers have suggested remarkably rich structural morphologies arising from microphase separation driven by immiscibility and orientation-dependent interaction. So far, morphologies such as smectic-A, smectic-C, lamellar, hexagonally perforated lamellar, hexagonal cylinder, gyroid, body-centered cubic, and zigzag structures have been observed in a variety of rod–coil diblock copolymer melts [5–11]. This has provided a wide playground for manipulating microstructure by alternating molecular-level architectures.

Past decades have witnessed surging interest in the theoretical understanding of flexible–flexible block copolymers, particularly due to the success of self-consistent field theory (SCFT), which incorporates the Gaussian chain (GSC) model to describe both flexible polymer blocks [12–14]. The SCFT formalism can also be developed for rod–coil copolymers following the same idea, where the flexible block is treated as GSC [15–17]. The rigid block is modeled by a rod, which typically has an orientation. The rigid segment behaves differently from the flexible block, because of its lack of conformational entropy; in replacement, the orientational entropy becomes more important. Depending on the physical system, the rod–rod interaction arising from different polymer chains can also produce orientationally dependent interactions, typically seen in liquid crystals. Self-assembly of rod–coil block copolymers forms a distinct class of problems in polymer physics.

Self-Assembling Systems: Theory and Simulation, First Edition. Edited by Li-Tang Yan.

Table 9.1 summarizes the main theoretical efforts made in understanding the problem of rod–coil polymer melts and microphase structures determined from these models. There are basically three approaches: ideal rod–coil polymer model, lattice model, and semiflexible–semiflexible polymer model; each is reviewed in this chapter.

The first theoretical model for rod–coil diblock copolymers was introduced by Semenov and Vasilenko [18, 19], who focused on studying the nematic and smectic phases. In this study, they considered an ideal rod–coil diblock copolymer chain where one end of a rigid rod is connected to a flexible chain. Later, performing a scaling analysis, Williams and Fredrickson [20] extended this model to study more complicated "hockey puck" micelles where the tightly packed rods form cylinders at a high volume fraction of coil blocks. Taking the isotropic Flory–Huggins interaction and the anisotropic Maier–Saupe interaction, Holyst and Schick [21] explored the rod–coil diblock phase diagram in the regime of weak segregation by the Landau–Ginzburg free-energy expansion with respect to the compositional order parameter and the orientational order parameter. Reenders and ten Brinke [22] adopted a similar method and addressed the role played by the nematic ordering in microphase separation. In order to explicitly investigate the effects of chain flexibility, Singh et al. [23] studied the structural properties of semiflexible rod–coil diblock copolymers in the same spirit [21].

The extraordinary success of the self-consistent field theory (SCFT) on the prediction of the equilibrium structures self-assembled by block copolymers allows us to determine the rod–coil morphologies with more precision. Müller and Schick [15] calculated the two-dimensional (2D) phase diagram of rod–coil diblock copolymers explicitly involving the hexagonal cylinder phase. Adopting a 2D lattice model accounting for chain rigidity in terms of the rotational isomeric state, Li and Gersappe [24] studied more complicated 2D morphologies observed in experiments, such as zigzag lamellar and elliptical cross-sectional cylinders. Chen et al. [25] developed a self-consistent-field lattice model and calculated the rod–coil phase diagram in which novel three-dimensional (3D) structures are included. However, the orientational interactions arising from the steric effects between rods are all ignored in these studies. On the basis of the Semenov–Vasilenko model [18], where the orientational interaction is strongly imposed, Matsen and Barrett [16] concentrated on an analysis of one-dimensional (1D) monolayer and bilayer smectic phases. In order to explore the high-dimensional morphologies, Pryamitsyn and Ganesan [26] proposed a general SCFT scheme in which the effects of the orientational alignment of rods were taken in the form of the Maier–Suape interaction. They mainly focused on calculating structures that are 2D in nature. Recently, Kriksin and Khalatur [27] extended this model to the study of 3D morphologies by means of a parallelized numerical algorithm.

The wormlike model describes the crossover of a polymer chain from the rod limit to the flexible limit. The local chain flexibility can be specified in a typical model. Sullivan et al. [28, 29] used the wormlike chain model to study the 1D smectic structures formed by rod–coil copolymers, where the Onsager steric interaction between polymer segments was used. Implementing a finite volume algorithm, Song et al. [30] extended the model to study the smectic-C phase for wormlike rod–coil copolymers. Following a similar approach, Gao et al. [31] explored the 2D phases beyond smectics by high-performing GPU techniques. The wormlike chain model incorporating the

Table 9.1 Theoretical studies of rod–coil (RC) diblock copolymers.

Model	Microstructures studied	Description of approaches	References (year)
Ideal RC diblock copolymers	Nematic and smectic-A phases	Approximate mean-field calculations in terms of a combinational free-energy accounting for interactions between unlike segments, chain conformation entropy for both blocks, and steric interactions between rods.	Semenov and Vasilenko [18] (1986)
Ideal RC diblock copolymers	Smectic-A and smectic-C phases	Scaling analysis on the free-energy composed of stretching entropy of coils and interfacial energy between RC diblocks in a selective solvent.	Halperin [36] (1990)
Ideal RC diblock copolymers	Nematic, smectic-A, and smectic-C phases	The same theoretical framework as in Ref. [18]; a special consideration on tilting rods from the lamellar normal.	Semenov [19] (1991)
Ideal RC diblock copolymers	Smectic-A/C phases and "hockey puck" cylindrical micelles	Scaling analysis of the Semenov–Vasilenko model [18].	Williams and Fredrickson [20] (1992)
Ideal RC diblock copolymers	Isotropic lamellae and nematic phase	Landau expansion of the free-energy functional up to second order using the random phase approximation. The interaction potentials consist of the isotropic part described by the Flory interaction and the anisotropic part described by the Maier–Saupe interaction.	Holyst and Schick [21] (1992)
Semiflexible–semiflexible diblock copolymers	Nematic and smectic-A phases	Similar to Ref. [21]. Block flexibility is modeled by the freely jointed chain model.	Singh, Goulian, Liu, and Fredrickson [23] (1994)
Ideal RC diblock copolymers	Lamellae and hexagonal cylinders	Solving the self-consistent field equations by numerically evaluating the single chain partition function and scaling analysis in terms of the brushlike approximation in the strong segregation limit.	Müller and Schick [15] (1996)
Ideal RC diblock copolymers	Nematic and smectic-A/C phases	Mean-field calculations of the Semenov–Vasilenko model [18].	Matsen and Barrett [16] (1998)

RC diblock copolymers based on rotational isomeric state (RIS) model	Lamella, zigzag lamella, elliptical cylinders, and hexagonal cylinders	Two-dimensional self-consistent field lattice model incorporating the RIS scheme to account for chain rigidity.	Li and Gersappe [24] (2001)
Wormlike–wormlike RC diblock copolymers	Nematic and smectic-A phases	Wormlike SCFT calculations accounting for the Onsager-type interaction.	Düchs and Sullivan [28] (2002)
Ideal RC diblock copolymers	Lamellae, hexagonal cylinders, spheres (BCC), nematic, and smectic-A/C phases	Extending the model in Ref. [21]; Landau expansion up to the fourth order in composition and orientation order parameters, in the weak segregation limit.	Reenders and ten Brinke [22] (2002)
RC diblock copolymers based on the discrete wormlike chain model	Smectic structures and "hockey puck" cylindrical micelles	Dynamic SCFT simulations via the introduction of the diffusion equations individually for the compositional and orientational order parameters.	Hamm, Goldbeck-Wood, Zvelindovsky, and Fraaije [37] (2003)
Ideal RC diblock copolymers	Smectic-A/C phases, broken lamellas, pucks, arrowheads, and zigzag phase	An exact 2D SCFT calculation based on the ideal RC diblock copolymer model accounting for the Flory interaction and Maier–Saupe interaction.	Pryamitsyn and Ganesan [26] (2004)
Ideal RC diblock copolymers	Smectic lamellae, ellipsoidal micelles	The dynamic evolution of the composition and orientation order parameters via a set of diffusion equations associated with the Landau-type phenomenological free-energy.	Yamazaki, Motoyama, Nonomura, and Ohta [38] (2004)
Ideal RC diblock copolymers based on the lattice model	Lamellaes, zigzag, perforated lamella, strip, gyroid, and spherical micelle	Self-consistent field lattice model employing a two-stage relaxation scheme.	Chen, Zhang, Sun, Zheng, and An [25] (2006)
Coil–wormlike diblock copolymers	Smectic-A structures	SCFT calculation implemented for multiblock copolymers accounting for the Flory-Huggins interaction and Maier–Saupe interaction.	Shah and Ganesan [32] (2009)
RC wormlike diblock copolymers	Nematic and smectic-A/C structures	1D SCFT calculation accounting for the Onsager potential to describe steric interactions between segments.	Song, Tang, Zhang, Yang, and Shi [30] (2009)
Semiflexible-coil diblock copolymers	Lamellae, nematic, and smectic-A/C structures	1D SCFT calculation accounting for the Flory-Huggins and Maier–Saupe interactions.	Song, Tang, Qiu, Yang, and Shi [33] (2011)

(Continued)

Table 9.1 (Continued)

Model	Microstructures studied	Description of approaches	References (year)
RC wormlike diblock copolymers	Nematic, smectic-A/C structures, cylindrical micelles, and zigzag phase	2D SCFT calculation. GPU (graphics processing units) technique.	Gao, Song, Tang, and Yang [31] (2011)
Ideal RC diblock copolymers	Cylinders, spheres, zigzags, pucks, polyhedra, and hexagonal chiral columnar phase	Parallel numerical algorithm. 3D calculation of the same model as in Ref. [26]	Kriksin and Khalatur [27] (2012)
Semiflexible-coil diblock copolymers	Lamellae, cylinders, nematic, smectic-A/C, and pucks	2D self-consistent Brownian dynamics simulations accounting for the nematic ordering of self-assembled morphologies.	Kumar and Ganesan [39] (2012)
Coil–wormlike diblock copolymers	Lamellae, cylinders, nematic, smectic-A/C, hexagonal pucks, and rhombic pucks	2D SCFT calculation accounting for the Flory-Huggins interaction and Maier–Saupe interaction on the study of the chain conformation combining Gaussian chain and wormlike chain models.	Gao, Tang, and Yang [34] (2013)
Wormlike-rod diblock copolymers	Smectic-A/C structures, hexagonal cylinders, elliptical cylinders	2D SCFT calculation accounting for the Flory-Huggins interaction and Maier–Saupe interaction on the study of a wormlike chain theory.	Li, Jiang, and Chen [35] (2014)
RC diblock copolymers	Spheres (BCC, A15), hexagonal cylinders, gyroids, and lamellae.	3D SCFT calculation based on the wormlike chain theory.	Tang, Jiang, Zhang, Yan, and Chen [40] (2015)

Maier–Saupe interaction in the framework of SCFT has been extensively implemented in a variety of copolymer systems, such as semicrystalline multiblock copolymers [32], semiflexible-coil diblock copolymers [33, 34], and semiflexible-rod diblock copolymers [35].

9.1.2 Basic Parameters

In this section, we review the basic parameters used in a typical theoretical model to represent the properties of a rod–coil system. Depending on the theoretical approach, not all parameters are present in a model.

9.1.2.1 Molecular Parameters

An ideal model simply treats the block copolymer as a polymer chain made of two blocks: a coil block that has one end jointing with a rod block. The coil and rodlike blocks are modeled as a Gaussian coil of length L_C and a rigid rod of length L_R. To denote the position of a segment, a contour variable s can be used, which has the value $s = 0$ starting from the coil terminal end, $s = L_C$ at the joint point, and $s = L = L_C + L_R$ at the rod terminal end (Figure 9.1). A basic segmental length a_C is defined on the coil block, which is usually identified with the Kuhn length of a coil polymer. The volume of such a segment, $1/\rho_0$, defines the volume occupied by one basic segment. Correspondingly, we define a basic statistical unit a_R on the rod block, which also has a volume $1/\rho_0$. In general $a_R \neq a_C$. The number of these segments can then be identified. In total, a single polymer consists of N segments among which there are N_C segments on the coil block and the remaining $N_R = N - N_C$ on the rodlike block. Below we use the volume fraction for coil segments,

$$f \equiv f_C = N_C/N, \tag{9.1}$$

which turns out to be an essential parameter. Here, following the same notation as in Refs. [16, 26], the symbol f is used for the coil volume fraction f_C where $f_R = 1 - f_C$.

Within a typical theoretical framework for the rod–coil problem, another essential parameter that measures the inherited conformational asymmetries of the two blocks, ν, is needed [16, 26]. Taking the ratio between the mean square end-to-end distances of an ideal coil and an ideal rod we have

$$\nu^2 = (a_C\sqrt{N})^2/(a_R N)^2 = a_C^2(a_R^2 N)^{-1}. \tag{9.2}$$

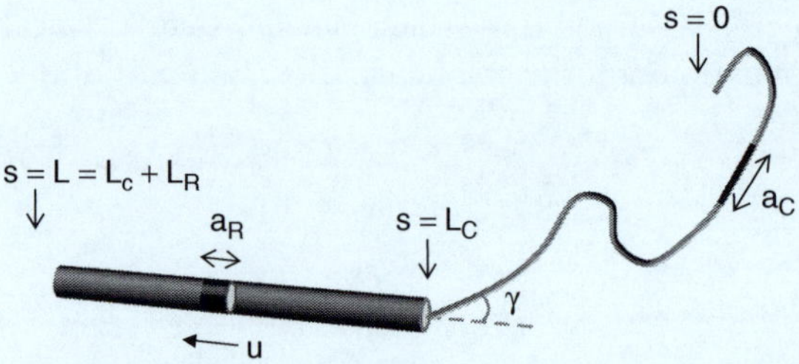

Figure 9.1 A rod–coil block copolymer of total contour length L. Basic segmental volumes are considered for lengths a_C and a_R on coil and rod blocks, respectively. *(See insert for color representation of the figure.).*

In some references [26, 33], a prefactor of 1/6 is used in the definition. A similar parameter is also needed for describing other conformationally asymmetric systems such as coil–coil block copolymers with different backbone Kuhn length [41, 42].

9.1.2.2 Polymer-Melt Parameters

A typical polymer melt is composed of n rod–coil polymers. The interaction between diblock copolymers is commonly modeled by the Flory–Huggins parameter χ, to be defined more precisely below, which shows up in a typical theory as the combination χN.

To complicate the matter further, in a theory that includes the orientational interaction, an additional parameter, for example the Maier–Saupe coefficient μ, needs to be considered [43]. Depending on the magnitude of μ, in general a strong-μ system prefers orientationally ordered states and a weak-μ system prefers orientationally isotropic states [26]. A weak-μ system, however, can already contain an orientationally ordered state—for example, rod blocks are directionally ordered in a lamellar state, driven from microphase separation caused by χN alone. In a different approach, the steric interaction avoiding the overlapping of rods can be simulated by Flory's lattice theory [18, 16].

For a polymer melt, in a self-consistent field theory the incompressibility condition

$$\phi_{\mathrm{R}}(\mathbf{r}) + \phi_{\mathrm{C}}(\mathbf{r}) = 1 \tag{9.3}$$

is generally assumed, where ϕ_{R} and ϕ_{C} are the local volume fractions for of the rod segments and the coil segments, respectively. This implies that, on average, space is filled by either segments a_{R} or a_{C}, uniformly, regardless of the nonuniform distribution of one type of monomer. Thus, within such an assumption, a total of nN segments occupy the system volume V, $nN/V = \rho_0$. In a typical SCFT, ρ_0 is scaled out at the free-energy level and becomes a nonessential parameter.

9.1.2.3 Other Parameters

Strictly speaking, a conformationally asymmetric block copolymer with a small N cannot be called a rod–coil copolymer. Instead, one must consider a model for rod–semiflexible block copolymers. Two more parameters enter into play in this case, one of which is the finite N that is a separate parameter beyond ν and χN. Because $a_{\mathrm{C}}^2 N$ is no longer the mean square end-to-end distance of an ideal coil, the parameter pair N, ν should now be replaced by N, $a_{\mathrm{C}}/a_{\mathrm{R}}$.

Another parameter becomes important for a rod–semiflexible copolymer system. The bond angle or the lack of a fixed bond angle γ determines the *local* conformation and *global* microphase-separated structures yielded from the chain persistency. In a large-N system, however, the influence of γ is only *local*, within the length scale of the persistence length of the coil segment, which is assumed to be much smaller than a typical microphase domain size, the radius of gyration of the coil block. The main part of this chapter concerns the $N \gg 1$ limit.

9.2 Theoretical Models

9.2.1 The Ideal Rod–Coil Diblock Model

9.2.1.1 Comments

The ideal rod–coil diblock model, which simply treats a block copolymer as a polymer chain where one end of the rodlike block is jointed to that of the coil block, was originally proposed by Semenov and Vasilenko [18] to theoretically study the smectic-A lamellar structure where the rods are aligned perpendicular to the periodic interfaces. Following the same idea, Semenov extended the study to the smectic-C phase [19] where the tilt angle relative to the lamellar normal was allowed to vary in the model. Implementing a scaling analysis, Williams and Fredrickson [20] predicted the existence of a hockey puck state which was shown to have higher stability than monolayer lamellae. Further developing SCFT based on the Semenov–Vasilenko model, Matsen and Barrett [16] eliminated some approximations used in previous studies [18–20] and generated a phase diagram which coincides reasonably well with experiments [5, 6]. In order to describe the steric interactions between rodlike segments, they used a strong orientational interaction based on Flory's lattice theory [44]. In the following, we review a commonly used model, basically following Pryamitsyn and Ganesan [26]. This model, when applied to systems that display both positional and orientational inhomogeneities, has yielded a variety of morphologies for rod–coil copolymers [22, 26, 27, 34, 35, 37, 45].

9.2.1.2 Formalism

Associated with the contour length s illustrated in Figure 9.1, which has the range $[0, L]$ over the entire chain, we now define a reduced variable t such that

$$t = \begin{cases} s/Na_C & \text{for } s = [0, L_C], \\ s/Na_R & \text{for } s = [L_C, L]. \end{cases} \tag{9.4}$$

Note that the coil and rodlike blocks can be described by t in the ranges $[0, f_C]$ and $[f_C, 1]$, respectively. The relation between t and s is nonlinear if $a_R \neq a_C$.

In a coarse-grained formalism, the configuration of the coil block of the kth chain is denoted by a space curve $\mathbf{R}_k(t)$ where the contour variable t has the range $[0, f_C]$. The reduced Hamiltonian is given by

$$\mathcal{H}_0 = \frac{3}{2Na_C^2} \sum_{k=1}^{n} \int_0^{f_C} \left| \frac{d\mathbf{R}_k(t)}{dt} \right|^2 dt, \tag{9.5}$$

which gives rise to a Gaussian statistical weight on the chain configuration. One defines a configurational-dependent coil volume fraction by

$$\hat{\phi}_C(\mathbf{r}) = \frac{N}{\rho_0} \sum_{k=1}^{n} \int_0^{f_C} \delta[\mathbf{r} - \mathbf{R}_k(t)] dt, \tag{9.6}$$

and rod volume fraction by

$$\hat{\phi}_{\mathrm{R}}(\mathbf{r}) = \frac{N}{\rho_0} \sum_{k=1}^{n} \int_{f_{\mathrm{C}}}^{1} \delta\left\{\mathbf{r} - [\mathbf{R}_k(f_{\mathrm{C}}) + a_{\mathrm{R}}(t - f_{\mathrm{C}})N\mathbf{u}_k]\right\} \mathrm{d}t. \tag{9.7}$$

The Flory–Huggins interaction energy, which gives an enthalpic penalty between dissimilar segments, is then

$$\mathcal{H}_1 = \chi\rho_0 \int \mathrm{d}\mathbf{r}\hat{\phi}_{\mathrm{R}}(\mathbf{r})\hat{\phi}_{\mathrm{C}}(\mathbf{r}), \tag{9.8}$$

where χ is the well-known Flory–Huggins parameter.

For any given system configuration, the microscopic state of the kth rodlike block can be specified by a unit vector $\mathbf{u}_k$ (see Figure 9.1). Depending on the physical system, the rod–rod interaction can give rise to an orientational-dependent interaction, not accounted for in $\mathcal{H}_1$. This interaction could be of the form of an Onsager interaction [46], or, more generally, phenomenologically represented by a Maier–Saupe interaction term. According to the liquid crystal theory [47], the configurational-dependent orientational order-parameter tensor, which measures orientation of the rodlike blocks, is defined by

$$\hat{\mathbf{S}}_{\mathrm{R}}(\mathbf{r}) = \frac{N}{\rho_0} \sum_{k=1}^{n} \int_{f_{\mathrm{C}}}^{1} \delta\left\{\mathbf{r} - [\mathbf{R}_k(f_{\mathrm{C}}) + a_{\mathrm{R}}(t - f_{\mathrm{C}})N\mathbf{u}_k]\right\} \left[\mathbf{u}_k\mathbf{u}_k - \frac{\mathbf{I}}{3}\right] \mathrm{d}t, \tag{9.9}$$

where $\mathbf{I}$ is a 3×3 unit tensor. The additional Maier–Saupe energy then has the form

$$\mathcal{H}_2 = -\frac{\mu\rho_0}{2} \int \mathrm{d}\mathbf{r}\hat{\mathbf{S}}_{\mathrm{R}}(\mathbf{r}) : \hat{\mathbf{S}}_{\mathrm{R}}(\mathbf{r}), \tag{9.10}$$

where μ is the Maier–Saupe parameter. Usually, in an ideal chain model, the orientational ordering associated with the flexible block is ignored.

For rod–coil block copolymers, the above basic energetics were first suggested by Pryamitsyn and Ganesan [26]. In the SCFT framework, the Helmholtz free energy of the present system can be straightforwardly expressed as a functional of $\phi_{\mathrm{C}}(\mathbf{r})$, $\phi_{\mathrm{R}}(\mathbf{r})$, $\mathbf{S}_{\mathrm{R}}(\mathbf{r})$ (now averaged over the configurations) and its conjugating fields $w_{\mathrm{C}}(\mathbf{r})$, $w_{\mathrm{R}}(\mathbf{r})$, $\mathbf{M}_{\mathrm{R}}(\mathbf{r})$. Taking the saddle-point approximation we obtain the reduced free energy per chain

$$\begin{aligned}\frac{F}{nk_{\mathrm{B}}T} = -\ln\ Q + \frac{1}{V}\int \mathrm{d}\mathbf{r}\Big\{ & \chi N\phi_{\mathrm{C}}(\mathbf{r})\phi_{\mathrm{R}}(\mathbf{r}) \\ & - \frac{\mu N}{2}\mathbf{S}_{\mathrm{R}}(\mathbf{r}) : \mathbf{S}_{\mathrm{R}}(\mathbf{r}) \\ & - w_{\mathrm{C}}(\mathbf{r})\phi_{\mathrm{C}}(\mathbf{r}) - w_{\mathrm{R}}(\mathbf{r})\phi_{\mathrm{R}}(\mathbf{r}) \\ & + \mathbf{M}_{\mathrm{R}}(\mathbf{r}) : \mathbf{S}_{\mathrm{R}}(\mathbf{r}) + \xi(\mathbf{r})[\phi_{\mathrm{C}}(\mathbf{r}) + \phi_{\mathrm{R}}(\mathbf{r}) - 1]\Big\}.\end{aligned} \tag{9.11}$$

The last term on the right-hand side incorporates the incompressibility condition (9.3) imposed on the component volume fractions by means of the Lagrangian multiplier $\xi(\mathbf{r})$. The single-chain partition function can be expressed by

$$Q = \frac{1}{4\pi V}\int \mathrm{d}\mathbf{r} \int \mathrm{d}\mathbf{u}\ \exp\left[-\int_{f_{\mathrm{C}}}^{1} \mathrm{d}t\Gamma\left(\mathbf{r} + (t - f_{\mathrm{C}})Na_{\mathrm{R}}\mathbf{u}, \mathbf{u}\right)\right] q(\mathbf{r}, f_{\mathrm{C}}), \tag{9.12}$$

where

$$\Gamma(\mathbf{r},\mathbf{u}) \equiv w_{\mathrm{R}}(\mathbf{r}) - \mathbf{M}(\mathbf{r}) : \left[\mathbf{uu} - \frac{\mathbf{I}}{3}\right]. \tag{9.13}$$

The function $q(\mathbf{r}, f_{\mathrm{C}})$ represents the probability of finding the rod–coil junction in the vicinity of $\mathbf{r}$. The so-called propagator satisfies a modified diffusion equation

$$\frac{\partial}{\partial t} q(\mathbf{r},t) = \frac{Na_{\mathrm{C}}^2}{6}\nabla^2 q(\mathbf{r},t) - w_{\mathrm{C}}(\mathbf{r})q(\mathbf{r},t) \tag{9.14}$$

for $t = [0, f_{\mathrm{C}}]$, where an initial condition $q(\mathbf{r}, 0) = 1$ is assumed.

Minimizing the free energy in (9.11) with respect to functions $\phi_{\mathrm{C}}(\mathbf{r})$, $\phi_{\mathrm{R}}(\mathbf{r})$, $\xi(\mathbf{r})$, $\mathbf{S}(\mathbf{r})$, $w_{\mathrm{C}}(\mathbf{r})$, $w_{\mathrm{R}}(\mathbf{r})$, and $\mathbf{M}(\mathbf{r})$, we obtain a set of expressions:

$$w_{\mathrm{C}}(\mathbf{r}) = \chi N \phi_{\mathrm{R}}(\mathbf{r}) + \xi(\mathbf{r}), \tag{9.15}$$

$$w_{\mathrm{R}}(\mathbf{r}) = \chi N \phi_{\mathrm{C}}(\mathbf{r}) + \xi(\mathbf{r}), \tag{9.16}$$

$$\phi_{\mathrm{C}}(\mathbf{r}) + \phi_{\mathrm{R}}(\mathbf{r}) = 1, \tag{9.17}$$

$$\mu N \mathbf{S}(\mathbf{r}) = \mathbf{M}(\mathbf{r}), \tag{9.18}$$

$$\phi_{\mathrm{C}}(\mathbf{r}) = \frac{1}{4\pi Q}\int_0^{f_{\mathrm{C}}} \mathrm{d}t q(\mathbf{r},t) q^*(\mathbf{r}, f_{\mathrm{C}} - t), \tag{9.19}$$

$$\phi_{\mathrm{R}}(\mathbf{r}) = \frac{1}{4\pi Q}\int \mathrm{d}\mathbf{u} \int_{f_{\mathrm{C}}}^1 \mathrm{d}t \ \exp\left[-\int_{f_{\mathrm{C}}}^1 \mathrm{d}t'\Gamma(\mathbf{r} - tNa_{\mathrm{R}}\mathbf{u} + t'Na_{\mathrm{R}}\mathbf{u}, \mathbf{u})\right] q(\mathbf{r} - (t - f_{\mathrm{C}})Na_{\mathrm{R}}\mathbf{u}, f_{\mathrm{C}}), \tag{9.20}$$

$$\mathbf{S}(\mathbf{r}) = \frac{1}{4\pi Q}\int \mathrm{d}\mathbf{u}\left[\mathbf{uu} - \frac{\mathbf{I}}{3}\right] \int_{f_{\mathrm{C}}}^1 \mathrm{d}t \ \exp\left[-\int_{f_{\mathrm{C}}}^1 \mathrm{d}t'\Gamma(\mathbf{r} - tNa_{\mathrm{R}}\mathbf{u} + t'Na_{\mathrm{R}}\mathbf{u}, \mathbf{u})\right] q(\mathbf{r} - (t - f_{\mathrm{C}})Na_{\mathrm{R}}\mathbf{u}, f_{\mathrm{C}}). \tag{9.21}$$

Here we have introduced a complementary propagator $q^*(\mathbf{r}, t)$, which contains an arc variable t starting from the unconnnected terminal end of the rod block. For the coil block, this propagator satisfies a similar modified diffusion equation,

$$\frac{\partial}{\partial t} q^*(\mathbf{r},t) = -\frac{Na_{\mathrm{C}}^2}{6}\nabla^2 q^*(\mathbf{r},t) + w_{\mathrm{C}}(\mathbf{r})q^*(\mathbf{r},t), \tag{9.22}$$

with the initial condition

$$q^*(\mathbf{r}, f_{\mathrm{C}}) = \int \mathrm{d}\mathbf{u} \ \exp\left[-\int_{f_{\mathrm{C}}}^1 \mathrm{d}t\Gamma(\mathbf{r} + (t - f_{\mathrm{C}})Na_{\mathrm{R}}\mathbf{u}, \mathbf{u})\right]. \tag{9.23}$$

The expressions and equations in Equations (9.12)–(9.22) constitute the basic self-consistent set of equations. This theoretical framework contains a number of essential parameters. The overall volume fraction f_{C} appears in Equations (9.12),

(9.14), (9.19)–(9.21), and (9.22); the combination χN and μN appear in Equations (9.11), (9.15), (9.16), and (9.18). Upon rescaling of all $\mathbf{r}$ dependent variables by $\sqrt{Na_C^2}$, Equations (9.14) and (9.22) become independent of this factor. The places where they show up are now in Equations (9.12), (9.20), and (9.21), as the combination ν defined in Equation (9.2). Hence there are four parameters in this formalism: f_C, χN, μN, and ν.

9.2.2 The Lattice Model

9.2.2.1 Comments

The lattice model is based on the assumption that every polymer segment occupies one lattice, the most primitive unit of the calculated system. The treatment substantially simplifies the calculation of chain conformations because the propagating probability of segments can be conveniently determined according to the finite coordinate number of lattices. For rod–coil diblock copolymer melts, this scheme was first adopted by Li and Gersappe [24] in a 2D study and then extended by An et al. [25, 48, 49] to study 3D microphase separation. Here, we outline the main steps.

9.2.2.2 Formalism

We consider n diblock copolymers with the total polymerization degree N distributed on N_L lattice sites, where the rod block and the coil block occupy N_R and $N_C = N - N_R$, respectively. The same volume for each type of segment is assumed and thus $N_L = nN$. The mean-field free energy functional per chain can readily be written as

$$\frac{F}{nk_BT} = -\ln Q + \frac{1}{N_L}\sum_{\mathbf{r}}\left\{\frac{1}{z}\sum_{\mathbf{r}'}\chi N\phi_C(\mathbf{r})\phi_R(\mathbf{r}') - w_C(\mathbf{r})\phi_C(\mathbf{r}) - w_R(\mathbf{r})\phi_R(\mathbf{r}) + \xi(\mathbf{r})[\phi_C(\mathbf{r}) + \phi_R(\mathbf{r}) - 1]\right\}. \tag{9.24}$$

The summation $\sum_{\mathbf{r}'}$ is taken for the nearest-neighbor z sites around the site at $\mathbf{r}$. The single chain partition function is calculated from

$$Q = \frac{1}{zN_L}\sum_{\mathbf{r}}\sum_{\mathbf{u}} G(\mathbf{r},\mathbf{u},s)G^*(\mathbf{r},\mathbf{u},s). \tag{9.25}$$

The distribution function $G(\mathbf{r},\mathbf{u},s)$ denotes the probability of finding the end segment of a polymer chain of length s at site $\mathbf{r}$ with an orientation $\mathbf{u}$. It is calculated from [50]

$$G(\mathbf{r},\mathbf{u},s) = G(\mathbf{r},s)\sum_{\mathbf{r}'}\sum_{\mathbf{u}'}\lambda_{\mathbf{r}-\mathbf{r}'}^{\mathbf{u}-\mathbf{u}'}G(\mathbf{r}',\mathbf{u}',s-1). \tag{9.26}$$

The initial condition for the end-segment distribution function reads

$$G(\mathbf{r},\mathbf{u},1) = G(\mathbf{r},1), \tag{9.27}$$

which is independent of the orientation of the first segment. Owing to the nonreversal symmetry of the RC diblock copolymer, a second end-segment probability function starting from Nth segment is needed,

$$G^*(\mathbf{r},\mathbf{u},s) = G(\mathbf{r},s)\sum_{\mathbf{r}'}\sum_{\mathbf{u}'}\lambda_{\mathbf{r}-\mathbf{r}'}^{\mathbf{u}-\mathbf{u}'}G^*(\mathbf{r}',\mathbf{u}',s+1), \tag{9.28}$$

which is subject to the initial condition $G^*(\mathbf{r}, \mathbf{u}, N) = G(\mathbf{r}, N)$. In all expressions, the function $G(\mathbf{r}, s)$ represents

$$G(\mathbf{r}, s) = \begin{cases} \exp\,[-w_\mathrm{C}(\mathbf{r}_s)] & 1 \le s \le N_\mathrm{C}, \\ \exp\,[-w_\mathrm{R}(\mathbf{r}_s)] & N_\mathrm{C} < s \le N. \end{cases}$$

The chain statistics in space is manipulated by the propagating probability factor λ. For the random-walk block, it takes the form [25]

$$\lambda_{\mathbf{r}-\mathbf{r}'}^{\mathbf{u}-\mathbf{u}'} = \begin{cases} 0 & \mathbf{u} = -\mathbf{u}', \\ \dfrac{1}{z-1} & \text{otherwise,} \end{cases}$$

which excludes direct back folding. For the rod block it takes the form

$$\lambda_{\mathbf{r}-\mathbf{r}'}^{\mathbf{u}-\mathbf{u}'} = \begin{cases} 1 & \mathbf{u} = \mathbf{u}', \\ 0 & \text{otherwise.} \end{cases}$$

Li and Gersappe considered more complicated local chain configurations that account for the rotational isomeric state. The λ factor was taken to describe details of the trans and gauche states, [24]

$$\lambda_{\mathbf{r}-\mathbf{r}'}^{\mathbf{u}-\mathbf{u}'} = \begin{cases} \lambda^g & \text{gauche-state connection between } s \text{ and } s-1, \\ 1-2\lambda^g & \text{otherwise.} \end{cases}$$

The parameter $\lambda^g = 1/[2 + \exp\,(U_g/k_\mathrm{B}T)]$, where U_g determines the flexibility of the polymer chain.

Minimizing the free energy functional Equation (9.24) with respect to ϕ_C, ϕ_R, ξ, w_C, and w_R, one arrives at a set of self-consistent equations:

$$w_\mathrm{C}(\mathbf{r}) = \frac{\chi N}{z} \sum_{\mathbf{r}'} \phi_\mathrm{R}(\mathbf{r}') + \xi(\mathbf{r}), \tag{9.29}$$

$$w_\mathrm{R}(\mathbf{r}) = \frac{\chi N}{z} \sum_{\mathbf{r}'} \phi_\mathrm{C}(\mathbf{r}') + \xi(\mathbf{r}), \tag{9.30}$$

$$\phi_\mathrm{C}(\mathbf{r}) + \phi_\mathrm{R}(\mathbf{r}) = 1, \tag{9.31}$$

$$\phi_\mathrm{C}(\mathbf{r}) = \frac{1}{zN_\mathrm{L}Q} \sum_{s=1}^{N_\mathrm{C}} \sum_{\mathbf{u}} \frac{G(\mathbf{r}, \mathbf{u}, s) G^*(\mathbf{r}, \mathbf{u}, s)}{G(\mathbf{r}, s)}, \tag{9.32}$$

$$\phi_\mathrm{R}(\mathbf{r}) = \frac{1}{zN_\mathrm{L}Q} \sum_{s=N_\mathrm{C}+1}^{N} \sum_{\mathbf{u}} \frac{G(\mathbf{r}, \mathbf{u}, s) G^*(\mathbf{r}, \mathbf{u}, s)}{G(\mathbf{r}, s)}. \tag{9.33}$$

By solving these self-consistent equations for a given parameter set, a particular morphology can then be established. Three parameters, $f_\mathrm{C} = N_\mathrm{C}/N$, χN, and z, are present in the theory. The lattice coordination number z is determined by the selection of the specific lattice, hence only two system parameters exist. In comparison with the ideal chain model described above, the models in Refs. [24, 25, 48, 49] effectively take

$$\mu N = 0, \quad \text{and} \quad v^2 = 1/N. \tag{9.34}$$

The orientational interaction is not considered and the asymmetry parameter takes a specific form.

9.2.3 The Wormlike–Wormlike Diblock Model

9.2.3.1 Comments

In this section, the model for a diblock-copolymer melt consisting of semiflexible–semiflexible polymer chains in the SCFT framework is discussed. The model was first introduced by Matsen [51] and Jiang and Chen [52–54] to study microphase separation of semiflexible–semiflexible polymer melts where, in these studies, the two polymer blocks have the same persistence lengths. Here we generalize the formalism to deal with the current system. Such a treatment accounts for the stiffness disparity between individual blocks as well as the steric excluded-volume effects between rods. The model allows us to adjust chain rigidity from flexible to rodlike through specifying the ratio between block length and persistence length, and hence can be used for the rod–coil molecular architecture [35]. In particular, by taking appropriate limits of the persistent lengths, the model can exactly reduce to the ideal rod–coil diblock model.

9.2.3.2 Formalism

To describe the configuration of the kth wormlike–wormlike copolymer, we use the position vector of the sth monomer, $\mathbf{R}_k(s)$, defined for the entire chain, $s = [0, L]$. According to the Saito–Takahashi–Yunoki (STY) model [55], a continuum version of the discrete Kratky and Porod model [56], the polymer chain is an inextensible thread with a linear elastic bending energy penalty

$$\mathcal{H}_0 = \frac{\lambda_C}{2}\int_0^{L_C} ds \left|\frac{d\mathbf{u}_k(s)}{ds}\right|^2 + \frac{\lambda_R}{2}\int_{L_C}^{L} ds \left|\frac{d\mathbf{u}_k(s)}{ds}\right|^2. \tag{9.35}$$

The tangent vector $\mathbf{u}_k(s) = d\mathbf{R}_k(s)/ds$, which specifies the local orientation at location s of the kth polymer chain, is a unit vector, $|\mathbf{u}_k(s)| = 1$. Two separate bare persistence lengths λ_C and λ_R need to be introduced for the coil and rodlike blocks. Here we maintain the more general case of $a_R \neq a_C$.

Introducing the t variable defined in (9.4), here we can use the same configurational-dependent volume fractions defined in (9.6) and (9.7). The Flory–Huggins interaction energy $\mathcal{H}_1$ in (9.8) remains in the same form. In the most general case we introduce two orientational order parameter tensors, one for the rod blocks as before in (9.9) and another for the coil blocks,

$$\hat{\mathbf{S}}_C(\mathbf{r}) \equiv \frac{N}{\rho_0}\sum_{k=1}^{n}\int_0^{f_C}\left[\mathbf{u}_k(t)\mathbf{u}_k(t) - \frac{\mathbf{I}}{3}\right]\delta[\mathbf{r} - \mathbf{R}_k(t)]dt, \tag{9.36}$$

$$\hat{\mathbf{S}}_R(\mathbf{r}) \equiv \frac{N}{\rho_0}\sum_{k=1}^{n}\int_{f_C}^{1}\left[\mathbf{u}_k(t)\mathbf{u}_k(t) - \frac{\mathbf{I}}{3}\right]\delta[\mathbf{r} - \mathbf{R}_k(t)]dt. \tag{9.37}$$

We generalize the Maier–Saupe energy by the form

$$\begin{aligned}\mathcal{H}_2^{\text{worm}} &= -\frac{1}{2}\mu_{RR}\rho_0 \int d\mathbf{r}\hat{\mathbf{S}}_R(\mathbf{r}) : \hat{\mathbf{S}}_R(\mathbf{r}) \\ &- \mu_{RC}\rho_0 \int d\mathbf{r}\hat{\mathbf{S}}_R(\mathbf{r}) : \hat{\mathbf{S}}_C(\mathbf{r}) \\ &- \frac{1}{2}\mu_{CC}\rho_0 \int d\mathbf{r}\hat{\mathbf{S}}_C(\mathbf{r}) : \hat{\mathbf{S}}_C(\mathbf{r}),\end{aligned} \tag{9.38}$$

where μ_{RR}, μ_{RC}, and μ_{CC} represent the magnitudes of the anisotropic interactions between rod–rod, rod–coil, and coil–coil segments, which need to be considered if all segments are wormlike.

We can then proceed to establishing the SCFT free energy per chain for the system, which takes the reduced form

$$\begin{aligned}\frac{F}{nk_BT} &= -\ln Q + \frac{1}{V}\int d\mathbf{r}\Big\{ \chi N\phi_C(\mathbf{r})\phi_R(\mathbf{r}) \\ &- \frac{\mu_{RR}N}{2}\mathbf{S}_R(\mathbf{r}) : \mathbf{S}_R(\mathbf{r}) - \mu_{RC}N\mathbf{S}_R(\mathbf{r}) : \mathbf{S}_C(\mathbf{r}) - \frac{\mu_{CC}N}{2}\mathbf{S}_C(\mathbf{r}) : \mathbf{S}_C(\mathbf{r}) \\ &- w_C(\mathbf{r})\phi_C(\mathbf{r}) - w_R(\mathbf{r})\phi_R(\mathbf{r}) \\ &+ \mathbf{M}_R(\mathbf{r}) : \mathbf{S}_R(\mathbf{r}) + \mathbf{M}_C(\mathbf{r}) : \mathbf{S}_C(\mathbf{r}) \\ &+ \xi(\mathbf{r})[\phi_C(\mathbf{r}) + \phi_R(\mathbf{r}) - 1]\Big\}.\end{aligned} \tag{9.39}$$

This saddle-point approximation for F is the functional of the auxiliary fields $w_C(\mathbf{r})$, $w_R(\mathbf{r})$, $\mathbf{M}_R(\mathbf{r})$, and $\mathbf{M}_C(\mathbf{r})$ that the segments experience, the mean volume fractions $\phi_C(\mathbf{r})$, $\phi_R(\mathbf{r})$, and orientational order-parameters $\mathbf{S}_R(\mathbf{r})$, $\mathbf{S}_C(\mathbf{r})$ at coordinate $\mathbf{r}$, and a Lagrangian multiplier $\xi(\mathbf{r})$ that enforces the incompressibility constraint on the system. In total, nine functions need to be dealt with.

The single-chain partition function can be obtained from

$$Q = \frac{1}{4\pi V}\int d\mathbf{r}d\mathbf{u}q(\mathbf{r}, \mathbf{u}, 1), \tag{9.40}$$

where $q(\mathbf{r}, \mathbf{u}, t)$ represents the probability of finding the tth segment of a polymer chain starting from the $t = 0$ end, which is located at a spatial position $\mathbf{r}$ and points in a direction $\mathbf{u}$. The propagator satisfies the modified diffusion equation (MDE) [57, 58]

$$\frac{\partial}{\partial t}q(\mathbf{r}, \mathbf{u}, t) = \left[\frac{Na_C}{2\lambda_C}\nabla_{\mathbf{u}}^2 - Na_C\mathbf{u}\cdot\nabla_{\mathbf{r}} - \Gamma_C(\mathbf{r}, \mathbf{u})\right] q(\mathbf{r}, \mathbf{u}, t) \qquad t \in [0, f_C], \tag{9.41}$$

$$\frac{\partial}{\partial t}q(\mathbf{r}, \mathbf{u}, t) = \left[\frac{Na_R}{2\lambda_R}\nabla_{\mathbf{u}}^2 - Na_R\mathbf{u}\cdot\nabla_{\mathbf{r}} - \Gamma_R(\mathbf{r}, \mathbf{u})\right] q(\mathbf{r}, \mathbf{u}, t) \qquad t \in [f_C, 1], \tag{9.42}$$

subject to the initial condition $q(\mathbf{r}, \mathbf{u}, 0) = 1$.

The free energy contains four interaction parameters: χN, $\mu_{RR}N$, $\mu_{RC}N$, $\mu_{CC}N$. Taking $a_C = 2\lambda_C$ as the basic unit length and using it to scale all length-dependent variables, from Equations (9.41) and (9.42) we see that the model contains parameters N, a_R/λ_R, and a_R/a_C. In total, seven parameters exist in a complete wormlike–wormlike diblock copolymer theory.

Minimization of the free energy functional in Equation (9.39) with respect to the functions ϕ_C, ϕ_R, $\mathbf{S}_C$, $\mathbf{S}_R$, ξ, w_C, w_R, $\mathbf{M}_C$, and $\mathbf{M}_R$ leads to the following self-consistent equations:

$$w_C(\mathbf{r}) = \chi N \phi_R(\mathbf{r}) + \xi(\mathbf{r}), \tag{9.43}$$

$$w_R(\mathbf{r}) = \chi N \phi_C(\mathbf{r}) + \xi(\mathbf{r}), \tag{9.44}$$

$$\mathbf{M}_C(\mathbf{r}) = \mu_{CC} N \mathbf{S}_C(\mathbf{r}) + \mu_{RC} N \mathbf{S}_R(\mathbf{r}), \tag{9.45}$$

$$\mathbf{M}_R(\mathbf{r}) = \mu_{RR} N \mathbf{S}_R(\mathbf{r}) + \mu_{RC} N \mathbf{S}_C(\mathbf{r}), \tag{9.46}$$

$$\phi_C(\mathbf{r}) + \phi_R(\mathbf{r}) = 1, \tag{9.47}$$

$$\phi_C(\mathbf{r}) = \frac{1}{4\pi Q} \int d\mathbf{u} \int_0^{f_C} dt q(\mathbf{r}, \mathbf{u}, t) q^*(\mathbf{r}, \mathbf{u}, t), \tag{9.48}$$

$$\phi_R(\mathbf{r}) = \frac{1}{4\pi Q} \int d\mathbf{u} \int_{f_C}^{1} dt q(\mathbf{r}, \mathbf{u}, t) q^*(\mathbf{r}, \mathbf{u}, t), \tag{9.49}$$

$$\mathbf{S}_C(\mathbf{r}) = \frac{1}{4\pi Q} \int d\mathbf{u} \left[\mathbf{uu} - \frac{\mathbf{I}}{3}\right] \int_0^{f_C} dt q(\mathbf{r}, \mathbf{u}, t) q^*(\mathbf{r}, \mathbf{u}, t), \tag{9.50}$$

$$\mathbf{S}_R(\mathbf{r}) = \frac{1}{4\pi Q} \int d\mathbf{u} \left[\mathbf{uu} - \frac{\mathbf{I}}{3}\right] \int_{f_C}^{1} dt q(\mathbf{r}, \mathbf{u}, t) q^*(\mathbf{r}, \mathbf{u}, t). \tag{9.51}$$

By adjusting the coupled parameter $Na_\alpha/2\lambda_\alpha$ (here α = R, C) on the individual block, the current model enables us to investigate self-assembly of block copolymers ranging from ideal rod–coil copolymers, that were previously modeled by rigid-rod blocks connected to Gaussian blocks as discussed above [15, 16, 26], to semiflexible–coil copolymers, that were modeled by wormlike blocks connected to Gaussian blocks [30, 33, 39, 59], to semiflexible–semiflexible copolymers [35, 51–54].

9.2.3.3 Reduction to the Rod–Coil Problem

We now examine the rod–coil problem by taking the asymptotic limit,

$$a_R/\lambda_R = 0, \tag{9.52}$$

and remove the anisotropic interaction associated with the coil block,

$$\mu_{RC} = \mu_{CC} = 0, \quad \mu_{RR} = \mu. \tag{9.53}$$

From the expression in Equations (9.39), (9.41), and (9.42) we can see that the system now contains five parameters, χN, μN, f_C, N, and a_R/a_C, if we use a_C as the basic length unit.

Within the coil limit

$$N \gg 1, \tag{9.54}$$

in Ref. [54] we demonstrated that Equation (9.41) is equivalent to Equation (9.14), provided that $q(\mathbf{r},t)$ in that equation is understood as the reduced $\int d\mathbf{u}q(\mathbf{r},\mathbf{u},t)$. Following the same procedure as in Section 9.2.1, and using $\sqrt{Na_C^2}$ as the scaling unit for all $\mathbf{r}$ variables instead of using a_C, we can then show that the only place where this factor appears is the prefactor of the $\mathbf{u}\cdot\nabla_\mathbf{r}$ term, as the combination $\sqrt{N}a_R/a_C$. Hence, we completely return to the four-parameter model, χN, μN, f_C, ν, used in Section 9.2.1.

9.2.4 Numerical Algorithms

9.2.4.1 Comments

All SCFT models presented in this chapter require self-consistent determination of functional sets, and this is usually achieved by designing a numerical algorithm. The most challenging task is to numerically solve the partial differential equations, such as those in Equations (9.14) or (9.41). After obtaining the propagator of polymer chains from solving these equations, we can readily calculate the free energy arising from the saddle-point approximation. Through a comparison of the free energies of different structures, the most stable structure is then determined.

For polymer models described by the Gaussian chain, many mature numerical schemes are available in the literature to solve MDEs, made possible by recent progress in studying the properties of block copolymers [14]. In contrast, designing an efficient numerical algorithm for solving the problem of rod–coil block copolymers is more difficult, because the simultaneous treatment of both $\mathbf{r}$ and $\mathbf{u}$ requires the use of three spatial variables and two angular variables, which is computationally much more involved. The full-space computation requires sufficient resolution in six-dimensional coordinates, including three Cartesian spatial variables, x, y, and z, generally represented by $\mathbf{r}$, two angular variables, θ and ϕ, generally represented by $\mathbf{u}$, and one time-like variable t. Here we review a number of numerical algorithms used to solve a typical MDE such as Equations (9.14) or (9.41); they are classified roughly into categories according to numerical methodologies.

9.2.4.2 Lattice Sampling

As an alternative strategy to consideration of the full orientational configurations, one can resort to the sampling technique that is based on a discrete polymer chain model. A representative method is the lattice model used by Li and Gersappe [24] and by Chen et al. [25]. The propagation of the end-segment statistics is calculated based on several fixed lattice directions. According to the different chain models simulating the individual blocks, the propagator in the external fields can be conveniently computed by the transfer matrix method based on step-weighted walks [50]. This can be done with or without incorporating the rotational isomeric states. The method is questionable when an accurate computation of the free energy for a directionally ordered state is required.

9.2.4.3 Spectral Method

In order to avoid the mathematical complexity of calculating (9.41) analytically, Müller and Schick [15] adopted a "single-chain in the mean-field" approach to numerically evaluate the single-chain partition function. To speed up the iterative computation of the free energy on microstructures with a given symmetry, Müller and Schick [15] adopted a spectral method to deal with both compositional densities and external fields by a complete set of orthonormal functions satisfying the specified symmetry. Their approach in principle can be implemented in exploration of stability of any microstructure without the constraint of chain architectures. The requirement of a priori determination of the resulting structure, however, limits this method to only known microphase-separated cases.

9.2.4.4 Pseudo-Spectral Method for GSC Propagator and Finite Difference for Rod Probability

Pryamitsyn and Ganesan [26] employed the pseudospectral operator-splitting scheme proposed by Rasmussen and Kalosakas [60] to solve the MDE in Equation (9.14). The propagator is calculated step by step from

$$q(\mathbf{r}, t + \Delta t) = e^{-w_{\mathrm{C}}(\mathbf{r})\Delta t/2} e^{(Na_{\mathrm{C}}^2/6)\Delta t \nabla^2} e^{-w_{\mathrm{C}}(\mathbf{r})\Delta t/2} q(\mathbf{r}, t) + O(\Delta t^3), \tag{9.55}$$

where Δt is a small incremental step for the arc variable t. The method separately treats the diffusive and potential contributions within a second-order accuracy in Δt. Applying the eigen-relation between the operator ∇^2 and the plane-wave function, one can conveniently obtain an update of function q. The method is typical of SCFT calculations for Gaussian chain problems of other systems.

In Ref. [26], the angular dependence for the rod block is handled by a Gaussian quadrature technique. In Ref. [45], Yang et al. applied a discretization approach where the ends of all possible unit vectors $\mathbf{u}$ are assumed to form a unit sphere surface, which is then discretized by a nearly uniform icosahedron triangular lattice to improve the angular resolution of rod orientation. In a 2D study of rod–coil diblock copolymers confined between two plates, their approach demonstrated better numerical accuracy and stability than the Gaussian quadrature approach.

Taking advantage of a pseudospectral scheme in a parallel computation environment, Kriksin and Khalatur [27] developed an algorithm which can be implemented in large-scale 3D SCFT calculation, using open-source packages which employ fast Fourier transforms [61]. According to Ref. [27], the algorithm scales well up to 1024 processors and achieves more than 200-fold speedups over a single-processor calculation. The orientation-dependent calculations in Equation (9.20) and (9.21) are taken by the method of trigonometric interpolation of integrands.

9.2.4.5 Single-Chain Mean-Field Calculation

To avoid the direct calculation of the propagator which requires the solution of the MDE for a wormlike chain (WLC), an alternative strategy is the adoption of single-chain computer simulation in an external potential field [62]. In this technique, a large ensemble of noninteracting molecules subjected to the self-consistent fields is investigated by particle-based simulations such as the Monte Carlo method. Using this technique, Ganesan and coworkers [39, 63] obtained theoretical predictions comparable to experimental observations in exploration of the phase diagram of wormlike diblock copolymers in two dimensions. This method can be very efficient in low-dimensional space.

9.2.4.6 Finite Difference Method for a WLC Problem

The most naive numerical treatment of a wormlike MDE is discretizing both positional and orientational dependencies of the propagator. This can be performed in low dimensions where the number of independent variables is reduced. In real space, the dependence on a particular variable is divided into a finite number of representative nodes. This technique can be traced back to those used in earlier studies of the isotropic nematic interface of rod molecules which require one spatial variation and two orientational variations [64, 65]. Instead of dividing the polar and azimuthal angles, Song et al. [30] employed a finite volume algorithm where a unit spherical surface is discretized with an icosahedron triangular mesh for the treatment of the **u** dependence. Resorting to graphics processing unit (GPU) technology, Gao et al. [31] further extended this real-space scheme to study a problem with 2D spatial variation.

9.2.4.7 Combined Finite Difference and Spherical Harmonics Expansion

The fact that the spherical harmonics are the eigenfunctions of the operator $\nabla^2_{\mathbf{u}}$ enables expansion of the propagator in terms of spherical harmonics. Writing

$$q(\mathbf{r}, \mathbf{u}, t) = \sum_{l,m} q_{j,l,m}(\mathbf{r}, t) Y_l^m(\mathbf{u}), \tag{9.56}$$

one can then treat the **r** dependence by a finite difference method. This method was adopted in earlier studies of surfaces and interfaces of semiflexible liquid crystal polymers [66–68] and smectic structures of rod–coil block copolymers [28], all with 1D spatial variation. The application of SPHEREPACK provides an efficient spherical harmonic analysis and synthesis transformation [69], which enabled Jiang and Chen to calculate the liquid crystal interface formed by semiflexible polymers [70].

9.2.4.8 Full Spectral Method for a WLC Problem

Formally, an exact expression for the MDE of a wormlike chain, Equation (9.41), can be written as

$$\frac{\partial}{\partial t} q(\mathbf{r}, \mathbf{u}, t) = \left(\mathcal{L}^{\nabla^2_{\mathbf{u}}} + \mathcal{L}^{\nabla_{\mathbf{r}}} + \mathcal{L}^{w}\right) q(\mathbf{r}, \mathbf{u}, t), \tag{9.57}$$

where the three operators appearing in the above expression are $\mathcal{L}^{\nabla^2_{\mathbf{u}}} \equiv (Na_\alpha/2\lambda_\alpha)\nabla^2_{\mathbf{u}}$, $\mathcal{L}^{\nabla_{\mathbf{r}}} \equiv -Na_\alpha \mathbf{u} \cdot \nabla_{\mathbf{r}}$, and $\mathcal{L}^{w} \equiv -\Gamma_\alpha(\mathbf{r}, \mathbf{u})$. Replacing Equation (9.57) by its discrete formula,

$$q(\mathbf{r}, \mathbf{u}, t + \Delta t) = \exp[\Delta t \left(\mathcal{L}^{\nabla^2_{\mathbf{u}}} + \mathcal{L}^{\nabla_{\mathbf{r}}} + \mathcal{L}^{w}\right)] q(\mathbf{r}, \mathbf{u}, t), \tag{9.58}$$

one can obtain an accurate solution of the propagators as long as the number of bases is adequate. Matsen [51] suggested a full spectral approach where plane waves and spherical harmonics are used to describe the **r** dependence and the **u** dependence, respectively. With this choice of eigenfunctions, the objective function can be expanded by a suitable spectral approximation:

$$q(\mathbf{r}, \mathbf{u}, t) = \sum_{k} \sum_{j=0}^{M-1} q_{kj}(t) \exp(i\mathbf{k} \cdot \mathbf{r}) Y_l^m(\mathbf{u}), \tag{9.59}$$

where $Y_l^m(\mathbf{u})$ denotes the standard spherical harmonic functions [71]. The abbreviation index j covers the integers l and m in a spherical harmonic. The summation in Equation

(9.59) is truncated at M, where $M = (\overline{L} + 1)^2$ represents the total number of spherical harmonic functions used in the expansion and $\overline{L}$ is the highest order of the truncated associated Legendre function. A 1D study on the lamellar morphology formed by wormlike diblocks was demonstrated in this way. This numerical scheme takes advantage of the reality that the the plane wave/spherical harmonic bases are the eigenfunctions of the operators $\mathcal{L}^{\nabla^2_{\mathbf{u}}}$ and $\mathcal{L}^{\nabla_{\mathbf{r}}}$. However, the basis functions are not diagonalizable for the operator $\mathcal{L}^{w}$. This fact leads to an expensive cost in 3D computation [14], even if the basis functions accounting for the specific symmetry are implemented.

9.2.4.9 Pseudospectral Method for a WLC Problem

A pseudospectral operator splitting scheme for the wormlike chain model was suggested by Fredrickson in his book [14],

$$q(\mathbf{r}, \mathbf{u}, t + \Delta t) = e^{\Delta t \mathcal{L}^{w}/2} e^{\Delta t \mathcal{L}^{\nabla_{\mathbf{r}}}/2} e^{\Delta t \mathcal{L}^{\nabla^2_{\mathbf{u}}}} e^{\Delta t \mathcal{L}^{\nabla_{\mathbf{r}}}/2} e^{\Delta t \mathcal{L}^{w}/2} q(\mathbf{r}, \mathbf{u}, t) + \mathcal{O}(\Delta t^3), \tag{9.60}$$

after the Baker–Campbell–Hausdorff operator identity is performed twice [52, 72]. One advantage of using the above is that for a small Δt, terms of order Δt^2 exactly cancel out and the error only amounts to an order of Δt^3 and higher. The spherical harmonics and plane waves are the eigenfunctions of the operators $\mathcal{L}^{\nabla^2_{\mathbf{u}}}$ and $\mathcal{L}^{\nabla_{\mathbf{r}}}$, respectively. Existing computation packages can be used to deal with transformations connecting the variables **u** and **r** to the expansion coefficients of spherical harmonics and Fourier functions [73, 74]; the method is known to be unconditionally stable and highly accurate, which means that fewer expansion coefficients and a relatively large step length Δt can be taken in the actual calculation. The algorithm has been employed in one [52] and two dimensions [34] for wormlike diblock copolymers.

9.2.4.10 Pseudospectral Backward Differentiation Formula Method for a WLC Problem

The numerical procedure introduced in the following was recently developed by Jiang and Chen for successfully calculating microphase structures self-assembled from WLC diblock copolymers in the full 3D space [53, 54]. In this algorithm, the propagator as a function of six variables was treated by a hybrid numerical algorithm, composed of three main ingredients: fast Fourier transformation to handle the spatial dependence [75], spherical harmonic expansion to handle the orientational dependence [71], and the backward differentiation formula (BDF) to handle the t dependence [76].

The treatment starts from the usual expansion of the propagator in terms of the spherical harmonics in the real-space version. We consider

$$q(\mathbf{r}, \mathbf{u}, t) = \sum_{j=0}^{M-1} q_j(\mathbf{r}, t)\, \tilde{Y}_l^m(\mathbf{u}), \tag{9.61}$$

where the real spherical harmonics are defined by

$$\tilde{Y}_l^m(\mathbf{u}) = \begin{cases} \frac{1}{\sqrt{2}} \left[Y_l^m(\mathbf{u}) + (-1)^m Y_l^{-m}(\mathbf{u}) \right] & \text{if } m > 0 \\ Y_l^0(\mathbf{u}) & \text{if } m = 0 \\ \frac{1}{i\sqrt{2}} [Y_l^{-m}(\mathbf{u}) - (-1)^m Y_l^m(\mathbf{u})] & \text{if } m < 0. \end{cases}$$

Substituting the expansion in Equation (9.61) into Equation (9.41) and using the orthonormal condition, we arrive at

$$\frac{\partial}{\partial t}q_j(\mathbf{r},t) = \left[-l(l+1)\frac{Na_\alpha}{2\lambda_\alpha} - w_\alpha(\mathbf{r})\right] q_j(\mathbf{r},t) - Na_\alpha \sum_{j'=0}^{M-1} \mathbf{R}_{jj'} \cdot \nabla q_{j'}(\mathbf{r},t) + \sum_{j'=0}^{M-1} \Theta_{jj'}(\mathbf{r}) q_{j'}(\mathbf{r},t), \tag{9.62}$$

where two vector matrices have been defined:

$$\mathbf{R}_{jj'} \equiv \frac{1}{4\pi}\int d\mathbf{u}\, \mathbf{u} \tilde{Y}_l^m(\mathbf{u}) \tilde{Y}_{l'}^{m'}(\mathbf{u}), \tag{9.63}$$

$$\Theta_{jj'}(\mathbf{r}) \equiv \frac{1}{4\pi}\int d\mathbf{u}\, \mathbf{M}(\mathbf{r}) : \left(\mathbf{u}\mathbf{u} - \frac{\mathbf{I}}{3}\right) \tilde{Y}_l^m(\mathbf{u}) \tilde{Y}_{l'}^{m'}(\mathbf{u}), \tag{9.64}$$

where the index j' abbreviates both l' and m'. The t dependence in Equation (9.62) is dealt with by a multistep implicit–explicit scheme. Under a third-order BDF (BDF3) scheme, Equation (9.62) is transformed into

$$\begin{aligned}
&\frac{11}{6}q_j(\mathbf{r},t) - 3q_j(\mathbf{r},t-\Delta t) + \frac{3}{2}q_j(\mathbf{r},t-2\Delta t) - \frac{1}{3}q_j(\mathbf{r},t-3\Delta t) \\
&= -\Delta t l(l+1)\frac{Na_\alpha}{2\lambda_\alpha} q_j(\mathbf{r},t) - \Delta t Na_\alpha \sum_{j'=0}^{M-1} \mathbf{R}_{jj'} \cdot \nabla q_{j'}(\mathbf{r},t) - \Delta t \sum_{j'=0}^{M-1} (w_\alpha(\mathbf{r})\delta_{jj'} - \Theta_{jj'}(\mathbf{r})) \\
&\times [3q_{j'}(\mathbf{r},t-\Delta t) - 3q_{j'}(\mathbf{r},t-2\Delta t) + q_{j'}(\mathbf{r},t-3\Delta t)] + O(\Delta t^4),
\end{aligned} \tag{9.65}$$

where Δt is the step size taken in the computation for t. The method depends on multiple time steps for the calculation of the propagator at the current t. Initially, Euler's method and Richardson extrapolation [77] are used to produce the function in the first few time steps.

The $\mathbf{r}$ dependence in the above equation is then treated by a spectral collocation method which uses the plane wave basis. Denoting $\hat{\mathcal{F}}$ as the standard Fourier transformation which manipulates a function from its $\mathbf{r}$-space to $\mathbf{k}$-space representations, we can simplify Equation (9.65) by

$$\sum_{j'=0}^{M-1} G_{jj'}(\mathbf{k}) q_{j'}(\mathbf{k},t) = P_j(\mathbf{k}, t-\Delta t, t-2\Delta t, t-3\Delta t), \tag{9.66}$$

where

$$G_{jj'}(\mathbf{k}) = \frac{11}{6}\delta_{jj'} - \Delta t\left(-\frac{Na_\alpha}{2\lambda_\alpha} l(l+1)\delta_{jj'} - Na_\alpha \mathbf{R}_{jj'} \cdot (i\mathbf{k})\right)$$

and

$$\begin{aligned}
P_j(\mathbf{k}, t-\Delta t, t-2\Delta t, t-3\Delta t) = \hat{\mathcal{F}}[&3q_j(\mathbf{r},t-\Delta t) - \frac{3}{2}q_j(\mathbf{r},t-2\Delta t) + \frac{1}{3}q_j(\mathbf{r},t-3\Delta t) \\
&- \Delta t \sum_{j'=0}^{M-1} (w_\alpha(\mathbf{r})\delta_{jj'} - \Theta_{jj'}(\mathbf{r}))(3q_{j'}(\mathbf{r},t-\Delta t) \\
&- 3q_{j'}(\mathbf{r},t-2\Delta t) + q_{j'}(\mathbf{r},t-3\Delta t))].
\end{aligned}$$

One important advantage of this formalism is that the matrix element $G_{jj'}(\mathbf{k})$ is independent of t and the iterative field $w_\alpha(\mathbf{r})$; we only need to calculate the inversion matrix $G(\mathbf{k})$ once, in the initial setup of the computation. Finally, the $\mathbf{k}$-space propagator can be calculated from

$$q_j(\mathbf{k}, t) = \sum_{j'=0}^{M-1} G_{jj'}^{-1}(\mathbf{k}) P_{j'}(\mathbf{k}, t - \Delta t, t - 2\Delta t, t - 3\Delta t). \tag{9.67}$$

Eventually, an inverse Fourier transformation yields the real-space propagator $q_j(\mathbf{r}, t)$.

The total cost to propagate a solution forward from the initial condition to the end of the chain using N steps is estimated to be $O(NKM)$ [53, 54], which can be contrasted with the computational cost $O(NKM^{3/2}\log_2 M)$ estimated for the pseudospectral operator-splitting scheme outlined above [14]. The current numerical algorithm has a clear advantage. The fact that BDF3 has a truncation error of order Δt^4 also enables a larger step size in t.

9.3 Concluding Remarks

Despite recent efforts in calculating the phase diagram for the problem of rod–coil polymer melts, the calculation of the entire phase diagram is far from complete for a couple of reasons.

First, a typical formalism involves spatial inhomogeneity coupled with orientational ordering of polymer segments. The coupling requires specification of both spatial and orientational variables mathematically, and hence creates the computational challenge of handling multidimensional variables; this can be contrasted with SCFT for flexible–flexible block copolymer melts which normally involves spatial variation only. The SCFT formalism for semiflexible–semiflexible diblock copolymer melts is now generally possible by incorporating the wormlike chain model. With appropriate identification of the persistence lengths, the formalism can be used to study rod–coil systems. An algorithm that exploits a simple parallel computational environment is now possible to tackle a problem that consists of both positional and orientational dependencies [53, 54].

Second, there are a number of essential physical parameters in the system, yielding a multidimensional parameter space. In the above, we demonstrated that a complete model contains four parameters for a rod–coil melt: $\chi N, f, \nu$, and μN. Investigating the complete phase behavior in the entire $\chi N - f - \nu - \mu$ space is a daunting task and previous efforts (such as those listed in Table 9.1) usually focus on a specific area of this four-dimensional parameter space [16, 22, 25–27, 34, 35]. A complete description of the phase diagram for the isotropic phases where $\mu N = 0$ requires the consideration of the three-parameter $\chi N - f - \nu$ subspace; such a calculation has just become available recently [40]. Given the fact that liquid crystal phases exist for a nonzero μN system [16, 26], completion of the calculation for the phase diagram based on the full $\chi N - f - \nu - \mu$ space becomes a desirable future task.

References

1. K. M. Coakley and M. D. McGehee, *Chemistry of Materials* **16**, 4533 (2004).
2. S. Günes, H. Neugebauer, and N. S. Sariciftci, *Chemical Reviews* **107**, 1324 (2007).
3. R. A. Segalman, B. McCulloch, S. Kirmayer, and J. J. Urban, *Macromolecules* **42**, 9205 (2009).
4. B. D. Olsen and R. A. Segalman, *Materials Science and Engineering: R: Reports* **62**,37 (2008).
5. J. T. Chen, E. L. Thomas, C. K. Ober, and G.-p. Mao, *Science* **273**, 343 (1996).
6. J. T. Chen, E. L. Thomas, C. K. Ober, and S. S. Hwang, *Macromolecules* **28**, 1688 (1995).
7. S. A. Jenekhe and X. L. Chen, *Science* **283**, 372 (1999).
8. J.-H. Ryu, N.-K. Oh, W.-C. Zin, and M. Lee, *Journal of the American Chemical Society* **126** ,3551 (2004).
9. Y. B. Lim, K. S. Moon, and M. Lee, *J. Mater. Chem.* **18**, 2909 (2008).
10. F. J. M. Hoeben, P. Jonkheijm, E. W. Meijer, and A. P. H. J. Schenning, *Chemical Reviews* **105**, 1491 (2005).
11. L.-Y. Shi, Y. Zhou, X.-H. Fan, and Z. Shen, *Macromolecules* **46**, 5308 (2013).
12. M. W. Matsen and F. S. Bates, *Macromolecules* **29**, 1091 (1996).
13. F. S. Bates, M. A. Hillmyer, T. P. Lodge, C. M. Bates, K. T. Delaney, and G. H. Fredrickson, *Science* **336**, 434 (2012).
14. G. H. Fredrickson, *The Equilibrium Theory of Inhomogeneous Polymers* (Clarendon, Oxford, 2006).
15. M. Müller and M. Schick, *Macromolecules* **29**, 8900 (1996).
16. M. W. Matsen and C. Barrett, *The Journal of Chemical Physics* **109**, 4108 (1998).
17. M. Shah, V. Pryamitsyn, and V. Ganesan, *Macromolecules* **41**, 218 (2008).
18. A. N. Semenov and S. V. Vasilenko, *Sov. Phys. JETP* **63**, 70 (1986).
19. A. N. Semenov, *Molecular Crystals and Liquid Crystals* **209**, 191 (1991).
20. D. R. M. Williams and G. H. Fredrickson, *Macromolecules* **25**, 3561 (1992).
21. R. Holyst and M. Schick, *The Journal of Chemical Physics* **96**, 730 (1992).
22. M. Reenders and G. ten Brinke, *Macromolecules* **35**, 3266 (2002).
23. C. Singh, M. Goulian, A. J. Liu, and G. H. Fredrickson, *Macromolecules* **27**, 2974 (1994).
24. W. Li and D. Gersappe, *Macromolecules* **34**, 6783 (2001).
25. J.-Z. Chen, C.-X. Zhang, Z.-Y. Sun, Y.-S. Zheng, and L.-J. An, *The Journal of Chemical Physics* **124**,104907 (2006).
26. V. Pryamitsyn and V. Ganesan, *The Journal of Chemical Physics* **120**, 5824 (2004).
27. Y. A. Kriksin and P. G. Khalatur, *Macromolecular Theory and Simulations* **21**, 382 (2012).
28. D. Düchs and D. E. Sullivan, *Journal of Physics: Condensed Matter* **14**, 12189 (2002).
29. R. C. Hidalgo, D. E. Sullivan, and J. Z. Y. Chen, *Journal of Physics: Condensed Matter* **19**, 376107 (2007).
30. W. Song, P. Tang, H. Zhang, Y. Yang, and A.-C. Shi, *Macromolecules* **42**, 6300 (2009).
31. J. Gao, W. Song, P. Tang, and Y. Yang, *Soft Matter* **7**, 5208 (2011).
32. M. Shah and V. Ganesan, *The Journal of Chemical Physics* **130**, 054904 (2009).
33. W. Song, P. Tang, F. Qiu, Y. Yang, and A.-C. Shi, *Soft Matter* **7**, 929 (2011).

34. J. Gao, P. Tang, and Y. Yang, *Soft Matter* **9**, 69 (2013).
35. S. Li, Y. Jiang, and J. Z. Y. Chen, *Soft Matter* **10**, 8932 (2014).
36. A. Halperin, *Macromolecules* **23**, 2724 (1990).
37. M. Hamm, G. Goldbeck-Wood, A. V. Zvelindovsky, and J. G. E. M. Fraaije, *The Journal of Chemical Physics* **118**, 9401 (2003).
38. N. Yamazaki, M. Motoyama, M. Nonomura, and T. Ohta, *The Journal of Chemical Physics* **120**, 3949 (2004).
39. N. A. Kumar and V. Ganesan, *The Journal of Chemical Physics* **136**, 101101 (2012).
40. J. Tang, Y. Jiang, X. Zhang, D. Yan, and J. Z. Y. Chen, submitted (2015).
41. J. D. Vavasour and M. D. Whitmore, *Macromolecules* **26**, 7070 (1993).
42. M. W. Matsen and F. S. Bates, *Journal of Polymer Science Part B: Polymer Physics* **35**, 945 (1997).
43. W. Maier and A. Saupe, *Zeitschrift Fur Naturforsch. Part a-Astrophysik Phys. Und Phys. Chemie* **14**, 882 (1959).
44. P. J. Flory, in *Liquid Crystal Polymers I*, Advances in Polymer Science, Vol. **59**, edited by N. Plat (Springer, Berlin, Heidelberg, 1984), pp. 1–36.
45. G. Yang, P. Tang, Y. Yang, and Q. Wang, *The Journal of Physical Chemistry B* **114**, 14897 (2010).
46. L. Onsager, *Ann. N. Y. Acad. Sci.* **51**, 627 (1949).
47. A. Y. Grosberg and D. V. Pachomov, *Liquid Crystals* **10**, 539 (1991).
48. J.-Z. Chen, C.-X. Zhang, Z.-Y. Sun, L.-J. An, and Z. Tong, *The Journal of Chemical Physics* **127**, 024105 (2007).
49. J.-Z. Chen, Z.-Y. Sun, C.-X. Zhang, L.-J. An, and Z. Tong, *The Journal of Chemical Physics* **128**, 074904 (2008).
50. F. A. M. Leermakers and J. M. H. M. Scheutjens, *The Journal of Chemical Physics* **89**,3264 (1988).
51. M. W. Matsen, *The Journal of Chemical Physics* **104**, 7758 (1996).
52. Y. Jiang, W.-Y. Zhang, and J. Z. Y. Chen, *Phys. Rev. E* **84**, 041803 (2011).
53. Y. Jiang and J. Z. Y. Chen, *Phys. Rev. Lett.* **110**, 138305 (2013).
54. Y. Jiang and J. Z. Y. Chen, *Phys. Rev. E* **88**, 042603 (2013).
55. N. Saitô, K. Takahashi, and Y. Yunoki, *J. Phys. Soc. Jpn.* **22**, 219 (1967).
56. O. Kratky and G. Porod, *Recl. des Trav. Chim. des Pays-Bas* **68**, 1106 (1949).
57. K. F. Freed, *Adv. Chem. Phys.* **22**, 1 (1972).
58. Q. Liang, J. Li, P. Zhang, and J. Z. Y. Chen, *The Journal of Chemical Physics* **138**, 244910 (2013).
59. W. Song, P. Tang, F. Qiu, Y. Yang, and A.-C. Shi, *The Journal of Physical Chemistry B* **115**, 8390 (2011).
60. K. Ø. Rasmussen and G. Kalosakas, *Journal of Polymer Science Part B: Polymer Physics* **40**, 1777 (2002).
61. N. Li and S. Laizet, *2DECOMPFFT: A Highly Scalable 2d Decomposition Library and FFT Interface* (Cray User Group, Edinburgh, 2010).
62. K. C. Daoulas and M. Müller, *The Journal of Chemical Physics* **125**, 184904 (2006).
63. D. Kipp and V. Ganesan, *The Journal of Physical Chemistry B* **118**, 4425 (2014).
64. Z. Y. Chen and J. Noolandi, *Phys. Rev. A* **45**, 2389 (1992).
65. Z. Y. Chen, *Macromolecules* **26**, 3419 (1993).
66. S.-M. Cui, O. Akcakir, and Z. Y. Chen, *Phys. Rev. E* **51**,4548 (1995).
67. Z. Y. Chen and S.-M. Cui, *Phys. Rev. E* **52**, 3876 (1995).

68. J. Z. Y. Chen, D. E. Sullivan, and X. Yuan, *Macromolecules* **40**, 1187 (2007).
69. P. N. Swarztrauber, *Mon. Weather Rev.* **121**, 3415 (1993).
70. Y. Jiang and J. Z. Y. Chen, *Macromolecules* **43**,10668 (2010).
71. G. B. Arfken and H. J. Weber, *Mathematical Methods for Physicists* (Academic Press, Waltham, MA, 2000), 5th edn.
72. M. W. Reinsch, *Journal of Mathematical Physics* **41**, 2434 (2000).
73. R. A. Fisher and W. Bischel, *Applied Physics Letters* **23**, 661 (1973).
74. R. A. Fisher and W. K. Bischel, *Journal of Applied Physics* **46**, 4921 (1975).
75. M. Frigo and S. G. Johnson, *Proc. ICASSP*, 1381 (1998).
76. J. D. Lambert, The initial value problem.;In *Numerical Methods for Ordinary Differential Systems* (John Wiley & Sons, Chichester, 1991).
77. W. H. Press, S. A. Teukolsky, W. T. Vetterling, and B. P. Flannary, *Numerical Recipes in C* (Cambridge University Press, Cambridge, 1992), 2nd edn.

10

Theoretical and Simulation Studies of Hierarchical Nanostructures Self-Assembled from Soft Matter Systems

Liangshun Zhang and Jiaping Lin

Shanghai Key Laboratory of Advanced Polymeric Materials, State Key Laboratory of Bioreactor Engineering, Key Laboratory for Ultrafine Materials of Ministry of Education, School of Materials Science and Engineering, East China University of Science and Technology, Shanghai, China

10.1 Introduction

Starting from elemental building units at microscopic scales, it is a scientist's dream to fabricate synthetic functional materials at macroscopic scales, whose self-assembled and responsive characteristics mimic those of living creatures. This grand dream has stimulated a large number of scientific endeavors, originating from the design philosophy of nanomaterials in supramolecular chemistry of small molecules [1], but now encompassing the self-assembly of copolymers, biological molecules and colloidal particles [2–9]. In comparison with the simple patterns of molecular self-organization, complicated soft matter systems self-assemble into well-defined multiple-length-scale superstructures without synthesizing the whole structures.

In experiments, general approaches to construct such hierarchical structures include multilevel driving forces and multistep processes [10–12]. The approach of multilevel forces is initiated by assembly of elemental building units into pre-assembled structural motifs via non-covalent interactions, which act as the secondary building blocks to generate more sophisticated architectures. In the case of the multistep process usually emerging in block copolymer systems, the stepwise phase separation of distinct species can be finely regulated, resulting in a structural hierarchy of self-assembled patterns. Benefiting from the progresses of synthetic techniques and structural characterizations, the fields of hierarchical structures from polymer systems have undergone rapid growth. A diversity of synthetic polymers on the basis of molecular design have been proposed to target and manipulate hierarchical nanostructures [13–16], such as the structures-within-structures self-assembled from supramolecular polymers, the multicompartment micelles from multicomponent copolymer solutions and the hierarchically ordered nanocomposites from block copolymer/nanoparticle systems.

The rational design of the novel superstructures demands comprehensive understanding of the self-assembly behaviors of tailor-made polymers at the microscopic level. Over the last decade, numerous advancements have been made in theoretical/computational modeling and methods to reproduce and predict the hierarchical

Self-Assembling Systems: Theory and Simulation, First Edition. Edited by Li-Tang Yan.

superstructures of complex polymeric fluids. More importantly, the computer simulations enable researchers to gain detailed thermodynamic and dynamic information on self-assembled nanostructures, which is difficult to deduce from experimental measurements.

The purpose of this chapter is to focus on recent progress in design strategies of structural hierarchy from the point of view of computational modeling. The simulation approaches for the self-assembly of soft matter are briefly summarized in the next section. Then, we concentrate on the superstructures of polymer-based systems via the multistep process of self-assembly. The multilevel approach to building the hierarchical structures is not included. The following topics are involved: (1) hierarchical nanostructures self-assembled from block copolymer melts, (2) multicompartment aggregates from block copolymer solutions, and (3) hierarchically ordered nanocomposites formed by organic–inorganic systems. Finally, we conclude with a brief outlook on challenges and perspectives.

10.2 Computational Modeling and Methods

On the basis of the magnitude of characteristic relaxation time and length, modern computer simulation methods for polymers and other soft materials are classified into two major categories: atomistic/molecular and coarse-grained approaches [17–19]. Although the atomistic/molecular simulations provide detailed and valuable information about the equilibrium or non-equilibrium properties of inhomogeneous polymeric fluids, they cannot access the millisecond and micrometer scales involved in the self-assembly due to the restriction of computational resources. A reasonable alternative approach is coarse-grained simulations, which attempt to describe the large-scale phenomena via lumping a small number of atoms or a group of atoms into an effective particle. The reduced freedom degree and the simplified soft interactions in the coarse-grained model lead to significant computational acceleration for simulations of large systems, which makes possible full-scale investigations of self-assembly phenomena. Within the framework of the statistical mechanism, there exist particle- and field-based coarse-grained approaches [20, 21].

10.2.1 Particle-Based Methods

Basically, there are three classic particle-based methods particularly suited to studying the self-assembly behaviors of soft matter at mesoscopic levels: Monte Carlo (MC), Brownian dynamics (BD) and dissipative particle dynamics (DPD) [17]. MC is a stochastic computer simulation technique frequently used in the study of soft matter [22]. The MC method has been widely applied in the fields of simple physical systems based on the Metropolis algorithm. For the many-chain systems of polymers, advanced MC algorithms have been developed to facilitate the sampling, which are reviewed by Binder and Paul [23]. More significantly, the MC method is conducted in terms of a multiscale modeling framework to interpret the relationships between the molecular-level information and the mesoscopic or macroscopic properties of simulated systems.

Coarse-grained BD is a mesoscopic method that uses an implicit solvent description. The loss of freedom degree is compensated by adding stochastic events [24], i.e., independent white noise terms are incorporated into the motion equations of particles. This allows one to simulate systems with large spatial and long temporal scales. In the traditional BD method, the momentum and energy of molecular systems are not conserved.

DPD is also a stochastic simulation technique, which was first developed by Hoogerbrugge and Koelman and subsequently reformulated to ensure the thermal equilibrium states [25, 26]. Similarly, the beads in the model represent whole molecules or fluid regions. Simplified pairwise dissipative and random forces are incorporated into the motion equations. As a result, DPD simulations locally conserve momentum and guarantee true hydrodynamic behavior. Since DPD simulations give access to long time and large length scales, they have been more popular for studying the self-assembly behaviors of soft matter.

10.2.2 Field-Based Methods

In particle-based methods, the freedom degree is the generalized coordinates associated with the coarse-grained particles. An alternative approach to represent the particles is on the basis of several statistical density fields. The Hamiltonian or energy of multicomponent systems is described by the functional of the density fields rather than as functions of the coordinates. Such ideas enter into polymer physics in the form of self-consistent (mean) field theory (SCFT) [27], which derives from the field-theoretic approach of Edwards in the 1960s [28]. Recent developments in numerical methods have accelerated applications of SCFT to studying the statistical properties of self-assembled nanostructures of multicomponent polymers [29–31].

The standard field-based methods are extended to describe non-equilibrium states. The dynamic version of mean-field theory for polymeric fluids was originally proposed by Fraaije et al. [32, 33], and is called dynamic density functional theory or dynamic SCFT. The dynamic method shares the thermodynamic basis of static SCFT, i.e., the free energy of polymer systems is described by the functional of the density fields in the local equilibrium states. The kinetics of systems are assumed to obey the diffusion equations, which push the systems toward states with lower energies. The driving force for the diffusion is the intrinsic chemical potential obtained from a trick using hypothetical external fields. Dynamic SCFT has successfully been applied to explore the morphological evolution and phase separation of block copolymers in the nonlinear regime [34, 35].

10.3 Hierarchical Nanostructures of Block Copolymer Melts

Block copolymers composed of two or more chemically different species constitute an important class of soft materials, which are capable of spontaneously assembling into hierarchical structures. The first observations of such superstructures with two-length scales were reported by ten Brinkes and co-workers using comb-shaped supramolecules [36–39], where low-molecular-weight compounds are weakly connected to one block of copolymers via hydrogen bonds. With the development of synthesizing methods, it is possible to make more complicated block copolymers, which self-assemble into novel structures involving multiple-length scales. One way is to introduce more chemical components into the copolymers (such as ABC linear and star-like terpolymers) [40–44], and another way is to add more blocks with various lengths into the chains (such as $A(BA)_n$ and $A(BC)_nA'$ multiblock polymers) [43, 44]. Theoretical and computational efforts have been made to achieve a fundamental understanding of the hierarchical nanostructures of block copolymers with various molecular architectures. In particular, we focus on three issues: (1) hierarchical structures self-assembled from ABC terpolymers, (2) hierarchical patterns from multiblock polymers, and (3) hierarchical structures formed by supramolecular polymers.

10.3.1 Hierarchical Structures Self-Assembled from ABC Terpolymers

Nanostructures formed by the simplest AB linear diblock copolymers involve one characteristic length scale. Introducing chemically distinct C blocks into AB diblock copolymers leads to ABC triblock terpolymers with linear or star-like topological architectures, which self-assemble into abundant ordered nanophases beyond the ordinary structures of AB diblock copolymers. The rich nanostructures provide a platform to screen the novel superstructures involving multiple-length scales, which are impacted by the interaction parameters between chemically distinct blocks, the volume fractions of blocks, the topologies of chain architectures and so forth.

As the interaction parameters between the two end blocks are comparable to or larger than those of the neighboring blocks, non-frustrated ABC linear triblock terpolymers self-assemble into the core-shell or alternating nanostructures [45], where the arrangement of domains coincides with the block sequence of polymers. In frustrated block copolymer systems, the repulsions between the two end blocks are much weaker than those of other blocks. To balance the interface energy of incompatible blocks and the conformational entropy of polymer chains, frustrated ABC linear triblock terpolymers self-assemble into a large number of structures-within-structure nanophases [46, 47], including cylinders-within-lamella, spheres-within-lamella, cylinders-on-cylinder, spheres-on-cylinder, spheres-on-sphere and knitting pattern. In particular, Li and co-workers utilized three-dimensional SCFT to investigate the self-assembly behaviors of cylinder-forming ABC linear triblock terpolymers with frustrated feature of interactions [48, 49]. The self-assembled superstructures are composed of hexagonally arrayed cylinders in large-length scales. Each of these structure units also contains substructures, including triple/quadruple cylinders and single/double/triple helices (Figure 10.1a). Phase diagrams in terms of the copolymer compositions and the interactions are constructed via comparing the free energies of candidate phases. It is predicted that the substructures of hierarchical patterns transit from disorder states to spheres, then to helices and finally to lamellae as the length of the middle blocks is tuned. In addition, the geometrical properties of substructures are also modulated through the interaction parameters between the neighboring blocks. These theoretical predictions are generally consistent with experimental observations [50–52].

Unlike the linear triblock terpolymers, the ends of arms are jointed together in ABC star terpolymers. A striking feature of self-assembled structures from star-shaped terpolymers is that the topology of polymer chains forces junctions of three arms to be in one-dimensional lines [44]. Gemma et al. used MC simulations to explore the self-assembly behaviors of a series of $A_{1.0}B_{1.0}C_x$ star terpolymers with a wide range

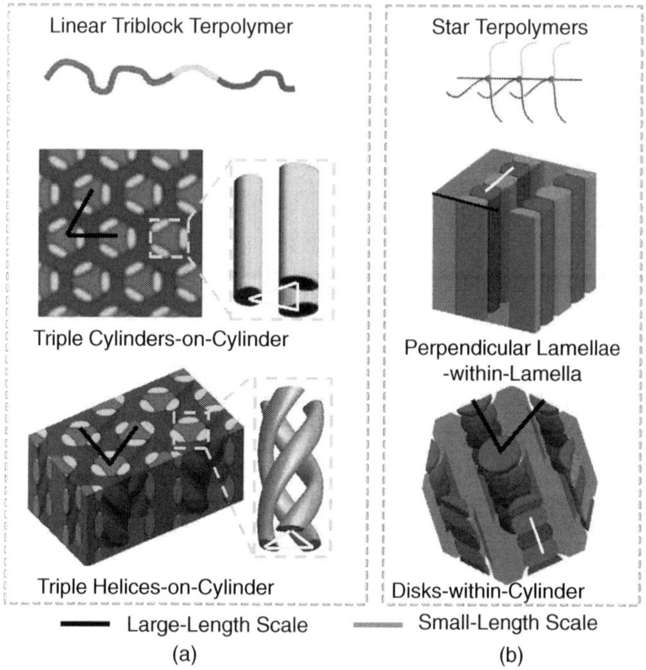

Figure 10.1 (a) Representative hierarchical structures self-assembled from ABC linear triblock terpolymers. Adapted with permission from Ref. [49]. Copyright 2012 American Chemical Society. (b) Representative hierarchical structures self-assembled from ABC star terpolymers. Adapted with permission from Ref. [60]. Copyright 2013 American Chemical Society. *(See insert for color representation of the figure.)*.

of C arm length x from 0.2 to 25, and discovered polygonal-tiling patterns with the three-arm junctions aligning along straight lines [53], which were ascertained by the experiments of Matsushita's group [54, 55]. DPD simulations reproduced the above novel patterns, and further predicted sophisticated structures containing a mixture of lamellae and cylinders [56]. Such structures can be regarded as hierarchical lamella phases consisting of C layers with large repeat spacing and A/B layers with small repeat spacing. Besides the tiling patterns, SCFT simulations systemically uncovered lamella-, cylinder-, sphere-, gyroid- and helix-based hierarchical structures [57–60]. It was also discovered that lamellae containing vertically oriented lamellae and cylinders containing perpendicular disks may be stable in star terpolymer systems (Figure 10.1b). The perpendicular arrangement of large- and small-length-scale domains is not currently observed in the self-assembled patterns of linear ABC molecules.

The computer simulations not only show the equilibrium phase behaviors of complex polymers, but also capture the formation mechanism of sophisticated structures. In a recent study, a dynamic extension SCFT coupling with the variable cell shape method was developed to probe the ordering mechanism of ABC star terpolymers from homogenous initial configurations [61]. The integrated method seeks to achieve the stress-free states of phase-separated nanostructures. Importantly, according to the incubation times and the rates of phase separation, three types of ordering processes (one-step, quick–slow and stepwise mechanisms) are distinguished during

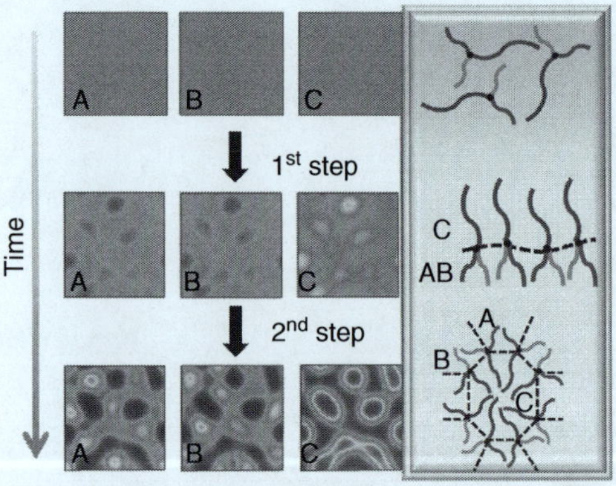

Figure 10.2 Stepwise mechanism of microphase separation of ABC star terpolymers. The left panel represents the density profiles of A, B and C components; the right panel schematically illustrates the disorder-to-order transition of ABC star terpolymers. Adapted with permission from Ref. [61]. Copyright 2014 Royal Society of Chemistry.

the structural formation of ABC star-shaped terpolymers. In particular, the stepwise mechanism corresponds to the procedure of phase separation, where one component is first separated from the mixed domains and the remaining mixed components are subsequently segregated from each other (Figure 10.2). This type of ordering mechanism is also found in the disorder-to-order transition of ABC linear terpolymers [62].

While the complex nanopatterns of ABC terpolymers have been thoroughly explored, further computer simulations for such systems lie in long-range ordered hierarchical structures and device-oriented patterns via directed self-assembly. For instance, Ross and colleagues pointed out that ABC star terpolymers on template substrates can produce well-aligned hierarchical nanostructures with tetragonal symmetry [63], which offer an attractive route to fabricating lithographic elements. The simulation techniques can be applied to predict how the chemical or topographical templates affect the well-ordered nanostructures, and thereby propose promising strategies for engineering arbitrary lattices of nanodomains.

10.3.2 Hierarchical Patterns Self-Assembled from Multiblock Copolymers

Multiblock copolymers composed of one or two long tails and many short middle blocks provide another possibility to engineer hierarchical structures with double periodicity. In these systems, the microphase separation between the tails and the middle blocks leads to large-length-scale domains, and the segregation within the middle blocks generates nanodomains possessing small-length scales. The structures-within-structure nanophases are corroborated by the experimental observations in Matsuhita's group [64, 65]. To understand the stable states of hierarchical structures for given multiblock copolymers, they also developed a simple model counting the possible loop and bridge conformations [64]. Although the predictions of the model coincide with the experimental findings, the model ignores the contributions of interfacial tensions and chain stretching.

Apart from the experimental and theoretical investigations, computational modeling that provides molecular level insights has been applied to advance understanding of the hierarchical nanostructures of $A_m(BA)_n$ multiblock copolymers [66, 67], comprising long A_m tails and short $(BA)_n$ multiblocks (n is the repeating number). Above the disorder-to-order transition, the phase separation between the long A_m tails and the short $(BA)_n$ multiblocks forms the large-length-scale lamella structures,

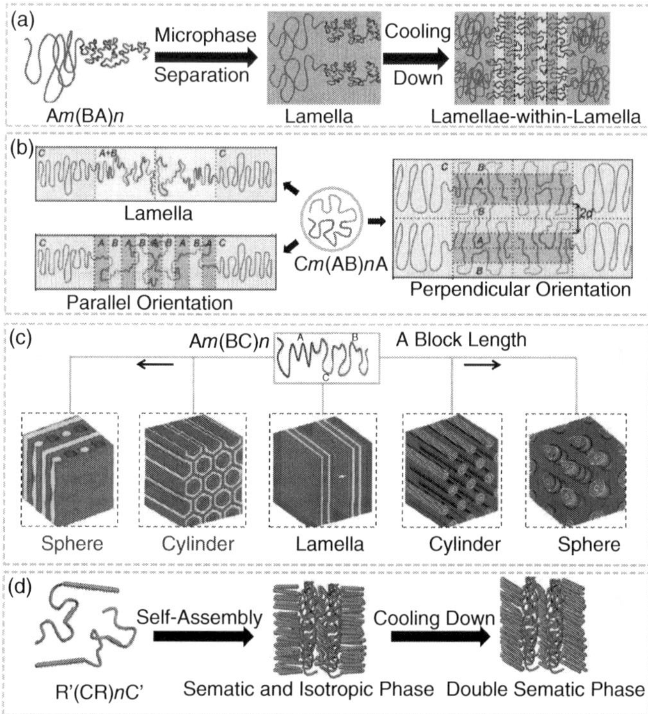

Figure 10.3 (a) Hierarchical patterns of $A_m(BA)_n$ linear-alternating multiblock copolymers. Adapted with permission from Ref. [67]. Copyright 2005 American Chemical Society. (b) Parallel and perpendicular lamellae-within-lamella structures of ternary multiblock terpolymers. Adapted with permission from Ref. [71]. Copyright 2010 American Chemical Society. (c) Lamella-, cylinder- and sphere-based hierarchical nanostructures of $A_m(BC)_n$ multiblock terpolymers. Adapted with permission from refs. [75]. Copyrights 2010 American Chemical Society. (d) Nanostructures with liquid crystalline phases from rod–coil multiblock copolymers. Adapted with permission from refs. [77]. Copyrights 2013 American Chemical Society.

and the segregation within the middle A and B blocks subsequently generates the small-length-scale layers as the repulsions between the A and B species are strong enough (Figure 10.3a). It has been demonstrated that the periodicity of nanophases displays a local maximum as the polymers begin to form the hierarchical structures. The phenomena are markedly different from the monotonic behaviors of periodicity for diblock copolymers.

Motivated by the experimental observations, the subsequent studies of the same group focused on the hierarchical structures of multiblock terpolymers via a theoretical approach in the strong segregation limit and the SCFT [68–71]. Unlike the simple counting model, the free energy of lamella-based superstructures explicitly includes the contributions of the stretching energy of blocks, the energy of loop/bridge conformations and the interfacial energy of incompatible blocks. The stability of lamella, parallel and perpendicular lamellae-within-lamella structures is affected by the number of multiblocks and the Flory–Huggins interaction parameters (Figure 10.3b). In particular, the orientation of thin layers relative to the thick layers is sensitive to the interaction parameters between the tails and the middle blocks. The theoretically predicted tendencies are corroborated by the results of SCFT simulations [69–71].

Besides the orientations, the number of internal layers is also regulated by the competition between the chain entropy and the internal energy. Li and colleagues utilized real-space SCFT to investigate the phase behaviors of $A(BC)_nBA$ multiblock terpolymers [72], and obtained lamella-based hierarchical structures with various numbers of internal layers. The phase diagrams are mapped via comparing the free-energy density of candidate structures. For given ternary multiblock polymers, as the repulsions between the tails and the middle blocks become strong, the number of internal layers increases. These findings are further demonstrated by the subsequent works of Li et al. [73, 74].

The above studies concentrate on the lamellae-within-lamella patterns of multiblock polymers. As elaborated by the phase diagram of diblock copolymers, change of copolymer composition is an effective way to modulate the morphologies of self-assembled nanostructures. With the help of the high accuracy and stability of numerical algorithms, Wang et al. tackled the diversity and complexity of structures-within-structure patterns self-assembled from the $A(BC)_n$ multiblock terpolymers [75]. In the parameter settings, the volume fractions of the B and C blocks were equal, and only the volume fraction of the A tails was tuned. As the A tails of polymers were short, the A-rich spheres or cylinders were embodied in the alternating B and C layers as a matrix, as depicted in Figure 10.3c. When the A tails formed the matrix, the B and C blocks self-assembled into concentrically cylindrical or spherical subdomains, i.e., the cylinders-within-cylinder or the spheres-within-sphere patterns. It was also corroborated that the orientation of substructures relative to the large-length-scale domains could be regulated by the interfacial tensions.

Nanostructures with liquid crystalline phases from rod–coil copolymers exhibit more advanced microscopic and macroscopic properties than the corresponding structures formed by coil–coil copolymers. Combining the concepts of hierarchical self-assembly with the rigid features of polymer chains opens the door to building a novel class of superstructures simultaneously possessing double periodicity and orientation. Referring to the model of Pryamitsyn and Ganesan [76], Zhu et al. incorporated Maier–Saupe orientation interactions into field-theoretic simulations of polymers containing various lengths of rod blocks [77]. Novel liquid crystalline lamellae-within-lamella structures with double orientation (sematic-within-sematic and isotropic-within-sematic) are achieved in rod–coil multiblock copolymer melts (Figure 10.3d). The transition between these phases is triggered by the change of rod block length and interaction strength. In comparison with the flexible multiblock copolymers, one important difference is that the rod–coil multiblock copolymers at small values of block numbers and interaction strength self-assemble into structures with multiple-length scales.

The formation process of hierarchical nanostructures has been examined by DPD simulations [78]. The ordering procedure of nanostructures with double periodicity obeys the stepwise mechanism of microphase separation. Specifically, the long tails and the multiblocks first form ordered domains with large-length scales, and then the multiblocks are gradually guided to self-assemble into small-length-scale structures with tunable relative orientations. In particular, the second step of the ordering process is very slow.

As demonstrated in experiments [79], superstructures with multiple-length scales from linear multiblock terpolymers significantly enhance the mechanical properties of polymer-based materials in comparison with the simple architectures like the AB and ABA topologies. To elucidate the excellent mechanical properties of the hierarchical structures, Zhu et al. combined the linear elasticity model and the SCFT to investigate the elastic properties of $A(BC)_n$ multiblock terpolymers in the hierarchical lamellar states [80]. As the structures of terpolymers transit from conventional lamellae to hierarchical lamellae, the elastic moduli of phase-separated materials are remarkably strengthened. The increase of extension and shear moduli in the lamellae-within-lamella states essentially originates from the contribution of internal energy in the small-length-scale structures. On the basis of these findings, the authors increased the number of small-length-scale layers to further raise the Young's modulus of multiblock terpolymer materials. These computational results provide useful guidelines for designing the molecular architectures of block copolymers and constructing complex structures with hierarchical features.

Computer simulations significantly enhance our understanding of the mechanical properties of hierarchical nanostructures under small deformations. However, there are many interesting problems or challenges waiting to be addressed. For example, one long-standing problem is the origin of unique properties of hierarchical structures in the nonlinear regime. Another important issue is that simulation results for electric, magnetic and photonic properties seem to be relatively scarce, although the fields are very important in the applications of hierarchical nanostructures such as quantum dots, magnetic storage media and next-generation lithography.

10.3.3 Hierarchical Structures Self-Assembled from Supramolecular Polymers

The structural hierarchy of polymer melts was first uncovered by ten Brinke and co-workers using comb-shaped supramolecular systems [36, 37], where low-molecular-weight molecules are attached to a block via hydrogen bonds (Figure 10.4a). Since the hydrogen bonds are sensitive to the temperature, the environmental alteration triggers transitions between the conventional and hierarchical structures [36]. Due to the complexity of such systems, the related theoretical and computational studies are very limited. To grasp the universal features of structural hierarchy, the computer simulations focus on the simplified model of comb-shaped supramolecular systems.

The simplest system consists of A-*block*-(A-*graft*-B) coil–comb molecules [81–84], where the B blocks are covalently bonded with the A blocks. This model seems to be a drastic simplification of the comb-shaped supramolecular systems in the experiments. However, it can reproduce the essential features of hierarchical patterns of supramolecular self-assembly. In the investigation of Wang et al. [85], parallel and perpendicular lamellae-within-lamella as well as cylinders-within-lamella patterns were reproduced (Figure 10.4b). The segregation between the coil and comb blocks produces the large-length-scale domains, and the comb blocks further self-organize into the small-length-scale subdomains. Another finding is that the small-length-scale structures remain whereas the morphologies of the large-length-scale structures are changed by the length of tails.

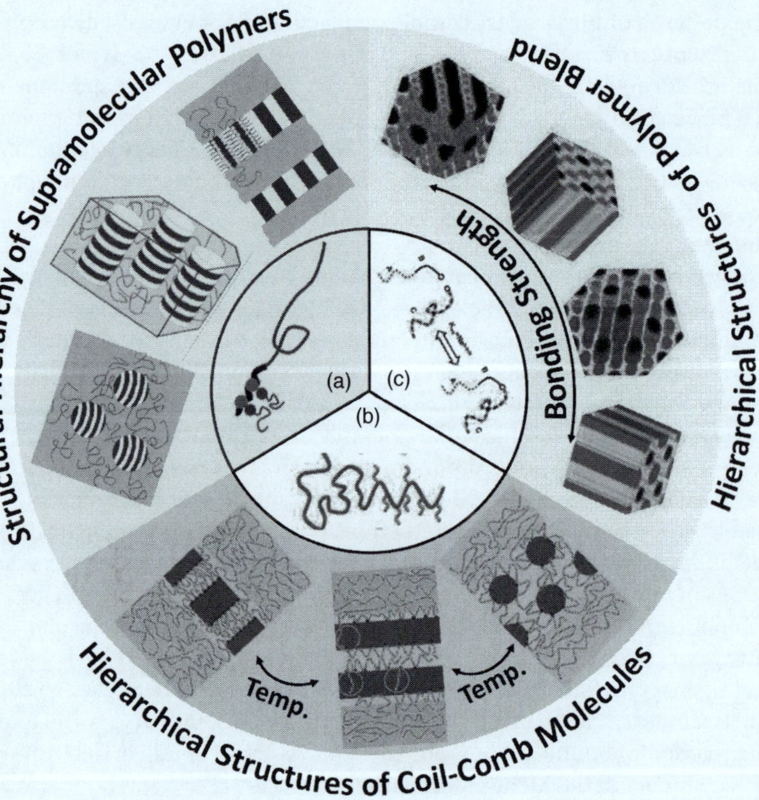

Figure 10.4 (a) Structural hierarchy of comb-shaped supramolecular polymers in experiments. Adapted with permission from Ref. [37]. Copyright 1999 Wiley-VCH. (b) Hierarchical structures self-assembled from simplified coil–comb molecules. Reproduced with permission from Ref. [85]. Copyright 2009 American Chemical Society. (c) Hierarchical self-assembly of AB diblock copolymer/C homopolymer blends via reversibly interactions. Adapted with permission from Ref. [93]. Copyright 2011 American Chemical Society.

Indeed, the A-*block*-(B-*graft*-C) coil–comb terpolymers may be more suited to model the hierarchical self-assembly of supramolecular systems [36]. Using DPD simulations, Huang et al. found that coil–comb terpolymers are easy to self-assemble into multiple-length-scale superstructures [86], which are strongly dependent on the composition of A tails, the polymerization degree and the incompatibility of different blocks. In particular, as the polymer chains are short, the comb blocks cannot form highly ordered lamellae; increasing the chain length leads to the experimentally observed structures such as spheres-within-lamella, cylinders-within-lamella and lamellae-within-lamella.

From the model point of view, another route to comprehend supramolecular self-assembly involves the polymer blend via a simple interaction. Linking experiment and theory, Tang et al. modeled the hydrogen bond as a simple contact attraction, and explored the square arrays of cylinders formed by AB/B′C copolymer blends (attraction between the B and B′ segments) [87]. Although the study does not concern the structural hierarchy of supramolecular assembly, it offers a hint to investigating the hierarchical patterns of host–guest complex molecules. Park et al. used computer simulations to explore the self-assembly mechanism of AB/BC copolymer blends [88]. The hosts are formed by hexagonally perforated lamellae via the spontaneous assembly of AB block copolymers in the films. The guests are made from the spherical morphologies of BC block copolymers. By comparing the free energy of candidate configurations, it is demonstrated that the hierarchical structures with the sphere position at the center of the rings are stable.

Although the above models have the ability to reproduce the complex superstructures of supramolecular assembly, they cannot explicitly treat the reversible features of dynamic self-assembly resulted from the existence of hydrogen bonds or metal–ligand interactions. Numerical SCFT for fixed architectures was recently extended to study the supramolecular assembly of block copolymers including dynamically bonded polymeric mixtures and reversibly bonded polymeric networks [89–92]. Unfortunately, applications of such an extension to hierarchical self-assembly of supramolecular polymers are extremely rare. Zhuang et al. introduced bonding reaction into the canonical SCFT to investigate the structural hierarchy of AB diblock copolymer/C homopolymer blends through reversible B–C supramolecular interactions [93], as depicted in Figure 10.4c. Via raising the bonding strength, the content of bound AB-C terpolymers increases and the concentrations of unbound AB diblock copolymers and C homopolymers decrease. These give the opportunity to tune the hierarchical structures of supramolecular systems. The nanostructures change from tetragonal C cylinders surrounded by octagonal A spheres, to tetragonal C cylinders surrounded by hexagonal A cylinders, then to alternating hexagonal cylinders, and finally to alternating tetragonal cylinders. It is also found that the volume fraction of bound C homopolymers plays a critical role in determining the packing of A cylinders in the hierarchical patterns.

Most of the simulations for the hierarchical structures in the melts are restricted to small cells containing one or two periodicities, leading to configurations with high internal stress and unrealistic dynamics of phase separations. In order to properly probe the self-assembly behaviors of structured fluids, large-cell simulations should be conducted to reduce the internal stress effects on the morphologies and the ordering kinetics. Unfortunately, the computational intensity required by large-cell simulations is enlarged greatly. Acceleration techniques via graphics processing units are addressing the challenges [94–96], and significantly facilitate the development and application of new simulation methods for comprehensively grasping the fundamental principles of hierarchical self-assembly.

10.4 Hierarchical Aggregates of Block Copolymer Solutions

In solutions of polymers with multiblocks or multicomponents, the phase separation between the polymers and the solvent molecules generates the large-length-scale structures such as spheres, cylinders and platelets. These structures provide soft confinement environments to program the small-length-scale substructures of blocks or solvent molecules. In the cases of (semi-)dilute solutions, aggregates with structural hierarchy are termed multicompartment micelles, which have been widely investigated in the experimental community [13, 14, 97–99]. Various self-assembly strategies are applied to prepare the multicompartment nanostructures, including AB block copolymers, ABC triblock terpolymers and copolymer blends. Meanwhile, theory and computational modeling for polymer melts have been extended to reproduce the observed morphologies and predict the novel sophisticated nanostructures of polymers in solutions. These studies include hierarchical aggregates from AB block copolymer solutions, multicompartment nanostructures from ABC triblock terpolymer solution, and multicompartment nanostructures from amphiphilic block copolymer blends.

10.4.1 Hierarchical Aggregates Self-Assembled from Block Copolymer Solutions

For block copolymers in poor solvent, macrophase separation between the polymers and the solvent molecules leads to large-length-scale aggregates, and microphase separation of block copolymers inside the aggregates determines the morphologies of subdomains under soft-confinement environments. Using dynamic SCFT simulations of discrete Gaussian chains, Sevink et al. placed the block copolymers into droplets, and investigated the morphologies of self-assembled nanostructures [100–102]. A rich variety of nanostructures with complex internal subdomains appeared in the droplets, including concentric lamellae, stacked lamellae, perforated lamellae, interconnected cylinders and spheres, as illustrated in Figure 10.5a. These hierarchical structures are related to the copolymer composition, the droplet radius, the segregation strength of the block copolymers, and the solvent properties. In particular, the formation process of multilayer superstructures from the small aggregates follows the sphere–rod–platelet pathway, and further experiences the bend step of platelets from the outside to the inside. Chi et al. used the simulated annealing MC technique to systematically explore the morphologies of block copolymers under the soft confinements realized by a poor solvent [103]. The similar internal morphologies of subdomains in the droplets were confirmed. As cylinder-forming block copolymers were placed inside the droplets, stacked toroids and helices were predicted. Moreover, the shape of the structured droplets became non-spherical due to the existence of anisotropic internal substructures.

The block copolymer/poor solvent mixtures generally aggregate into compact internal structures with low solvent content, which are difficult to use in the field of nanomedicine. The vesicles formed by amphiphilic block copolymers are the simplest multicompartment nanostructures, with hollow, solvent-filled interiors and solvophobic walls. Besides the simple vesicles, several classes of multicompartment vesicles were also prepared by the self-assembly of diblock copolymer/good solvent mixtures. On the basis of the new mechanism of vesicle formation from amphiphilic block copolymers [104], He et al. exploited computational modeling to devise a prenucleation strategy for control over the size and shape of vesicles with internal multicompartments [105, 106], as shown in Figure 10.5b. As the number density of nucleation seeds decreases, the unilamellar vesicles change to vesicles with multicompartments arranged in a rosette-like fashion. Furthermore, the variation of solvent properties triggers the morphological transition from multicompartment vesicles to multilamellar vesicles. With further optimization of the parameter settings of block copolymer/good solvent systems, complex cage aggregates are formed through the mechanism of anisotropic micellar growth [106].

Although the prenucleation strategy can prepare metastable or stable hierarchical structures, the experimental implementation of such kinetic control is extraordinarily difficult. More simple and economic approaches on the basis of molecular

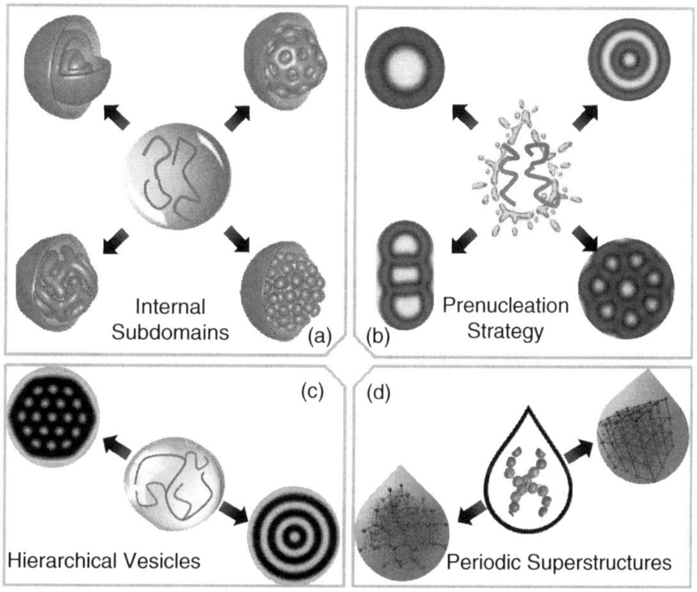

Figure 10.5 (a) Self-assembled structures of block copolymers in poor solvent. Adapted with permission from Ref. [100]. Copyright 2003 American Chemical Society. (b) Prenucleation strategy to achieve hierarchical vesicles formed by amphiphilic diblock copolymers. Adapted with permission from Ref. [105]. Copyright 2006 American Chemical Society. (c) Hierarchical vesicles self-assembled from graft copolymers in backbone-selective solvent. Adapted with permission from Ref. [107]. Copyright 2013 Royal Society of Chemistry. (d) Periodic ordered structures hierarchically self-assembled from diblock copolymer stars. Adapted with permission from Ref. [108]. Copyright 2007 American Physical Society.

design are required to prepare hierarchical vesicles. Employing SCFT and DPD simulations, Wang et al. demonstrated that the self-assembly of graft copolymers in backbone-selective solvent opens a route to spontaneous assembly into hierarchical aggregates such as multilamellar vesicles and multicompartment vesicles [107], as depicted in Figure 10.5c. Examining the excess grand potential of hierarchical vesicles suggests that the nanocontainers are thermodynamically stable due to the topological features of amphiphilic graft copolymers. Moreover, the DPD simulations reveal that the formation of hierarchical vesicles obeys the stepwise mechanism. Specifically, the first stage of hierarchical self-assembly is the formation of unilamellar vesicles via the standard pathway, and the second stage is the coalescence of unilamellar vesicles, leading to the production of multicompartment or multilamellar vesicles.

Star polymers are another molecular topology to hierarchically self-assemble into intermediate soft-patchy particles and stabilized periodic structures in solutions. Capone et al. theoretically devised a type of diblock copolymer stars [108, 109] where the solvophilic A blocks are jointed at the center of stars and followed by solvophobic B blocks. In the first stage of hierarchical self-assembly, the macromolecules spontaneously assemble into soft-patchy particles, where the number of patches is influenced by the number of arms and the composition of the stars. Unlike hard-patchy colloids, the patches formed by the copolymer stars can recombine and rearrange in the second stage of hierarchical self-assembly. In particular, diamond and cubically arrayed structures are produced over a wide range of polymer concentrations (Figure 10.5d). Additionally, wormlike super-micelles are generated by the hierarchical self-assembly of copolymer stars via intermediate pearl-necklace micelles [110].

10.4.2 Multicompartment Aggregates Self-Assembled from Triblock Terpolymer Solutions

AB diblock copolymers in selective solvent self-assemble into limited morphologies of nanostructures such as spheres and rods, where the space is divided into two subdomains (inside and outside). Polymers with three or more types of species provide the feasibility for constructing microphase-separated subdomains inside the large aggregates, leading to the formation of multicompartment micelles. To build multi-length-scale aggregates, researchers have expanded on the strategies of ABC triblock terpolymers with various molecular architectures such as linear and star-shaped topologies of polymers [111–115], which strongly depend on the compatibility of different blocks, the solvent properties, the block sequence and the spatial restriction imparted by the chain architectures.

The resulting morphologies of ABC linear triblock terpolymers are related to the relative lengths of the three blocks and the strength of the interfacial energies. Jiang et al. utilized DPD simulations to investigate the morphological transition of self-assembled structures formed by ABC linear terpolymers in an A-selective solvent [116]. It is assumed that the terpolymers with solvated A blocks self-assemble into concentric core-shell–corona structures with only two solvophobic compartments, where the B blocks enwrap all the C-rich inner domains (Figure 10.6a). As the B blocks become short, the concentric structures change to raspberry-like aggregates, where the bumps formed by the B blocks are separated by C blocks. When the solvent properties are tuned, the C blocks come out from the inner cores and aggregate into C bumps around the B surfaces to reduce the interfacial energy.

The solvophilic-block sequence in linear terpolymers plays a crucial role in determining the placement of solvophobic cores and the self-assembled morphologies. In the solvophilic–solvophobic–solvophobic sequence of ABC terpolymers, the self-assembled nanostructures are composed of solvophilic coronae and microphase-separated solvophobic cores with common interfaces between the distinct domains. In the solvophobic–solvophilic–solvophobic sequence, the solvophobic cores are partitioned by the solvophilic shells, i.e., the inner solvophobic domains have no common interfaces with the distinct solvophobic subdomains. The complicated aggregates are named multicore micelles. Wang et al. employed three-dimensional SCFT to investigate the multicore aggregates from ABC linear triblock terpolymers in a B-selective solvent [117]. A variety of micelles with disk, sphere and helix cores were observed (Figure 10.6b). The formation range of such multicore micelles is related to the composition of the triblock terpolymers and the interaction parameters between the middle blocks and the solvent molecules. In particular, cores with double-stranded helix features were identified in the systems, and were only stable at the intermediate solubility of solvophilic blocks.

The self-assembly of ABC star terpolymers provides another versatile and powerful route toward preparing nanostructures with hierarchical features. One distinct feature

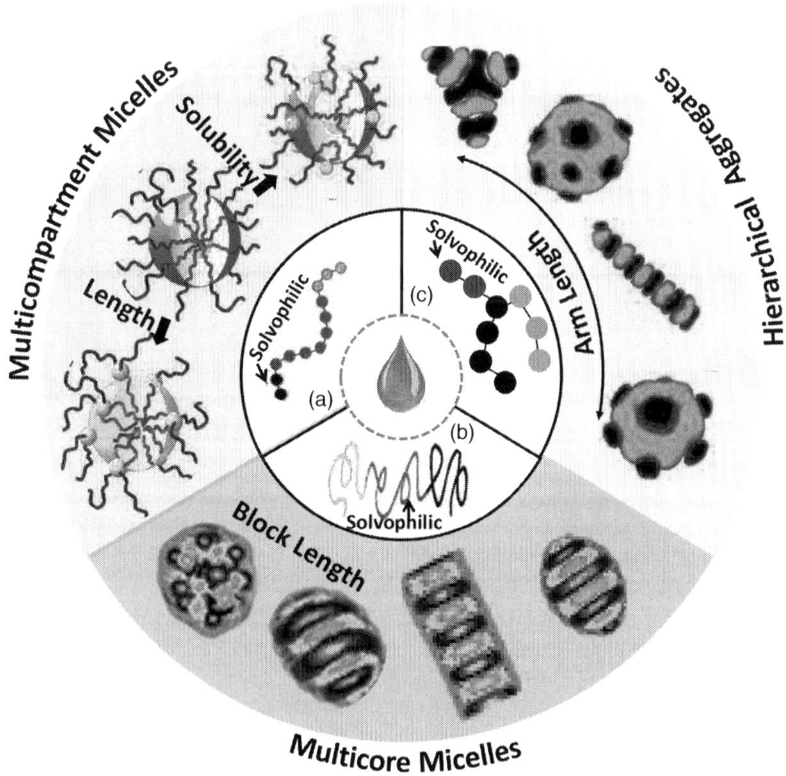

Figure 10.6 (a) Multicompartment micelles from ABC linear triblock terpolymers in A-selective solvent. Reproduced with permission from Ref. [116]. Copyright 2011 American Chemical Society. (b) Multicore micelles self-assembled from ABC linear triblock terpolymers in B-selective solvent. Adapted with permission from Ref. [117]. Copyright 2011 Royal Society of Chemistry. (c) Self-assembled nanostructures of ABC star terpolymers in A-selective solvent. Adapted with permission from Ref. [121]. Copyright 2009 American Chemical Society.

of the star-shaped architectures is that the different arms are linked at the junction points, thereby constraining the various subdomains to meet along a common curve space [44]. The topological requirement effectively suppresses the formation of concentric core-shell–corona structures, leading to the production of complex multicompartment aggregates.

Zhulina and Borisov carried out a scaling analysis for the self-assembled multicompartment structures of ABC star terpolymers in an A-selective solvent [118]. The free energy contribution from the topological constraints was incorporated into the total free energy per macromolecule in a micelle, and the spherical and cylindrical micelles with microphase-separated subdomains were considered as candidate structures. It was proved that the morphologies of large-length-scale structures are determined by the relative lengths of the solvophilic arms, and the number of subdomains is related to the polymer–polymer and polymer–solvent interface tensions.

DPD simulations have also been utilized to examine the multicompartment micelles formed by ABC star terpolymers in an A-selective solvent. For example, Xia et al. thoroughly explored the effects of arm length on the morphologies of hierarchical self-assembled nanostructures [119, 120]. As the solvophilic arms are long, the terpolymers self-assemble into "hamburger" micelles with three layers of solvophobic disks. When the volume fraction of solvophobic arms is dominant, the ABC star terpolymers first self-organize into the discrete "hamburger" micelles. Subsequently, the solvophobic cores of micelles contact each other, leading to the production of wormlike micelles with multilayers of segregated solvophobic disks. The simulation results agree well with the predictions of the scaling theory of Zhulila and Borisov [118].

Comprehensive understanding of the thermodynamic behaviors of ABC star terpolymers is presented by Li and co-workers [121]. The annealing technique of MC simulations was applied to map the generic phase diagrams in terms of the composition of terpolymers, the solvent quality and the repulsions between the incompatibility arms. The simulations reproduce both the morphologies and the corresponding transition sequences observed in the experiments (Figure 10.6c). Bilayer membrane structures with complex inner subdomains are predicted. Moreover, two basic thermodynamic features of nanostructures were discovered: The morphologies of large-length-scale structures are mainly governed by the volume fractions of the solvophilic arms, and the morphologies of small-length-scale subdomains are determined by the relative lengths of the solvophobic arms and the confinement environments.

The above investigations concentrate on the structural hierarchy of ABC terpolymers in solution. Recently, reports have been produced about the formation pathway of multicompartment micelles. Wang et al. extended polymeric field theory to examine the hierarchical self-assembly behaviors of ABC star terpolymers in dilute solutions [122]. Apart from the multicompartment micelles reported above, segmented cage-like micelles and spotted vesicles were observed. More importantly, two different pathways were identified via analyzing snapshots of nanostructure formation, as depicted in Figure 10.7. In the cases of short solvophilic arms, the formation of hierarchical structures obeys the nucleation–coalescence–growth mechanism. When the solvophilic

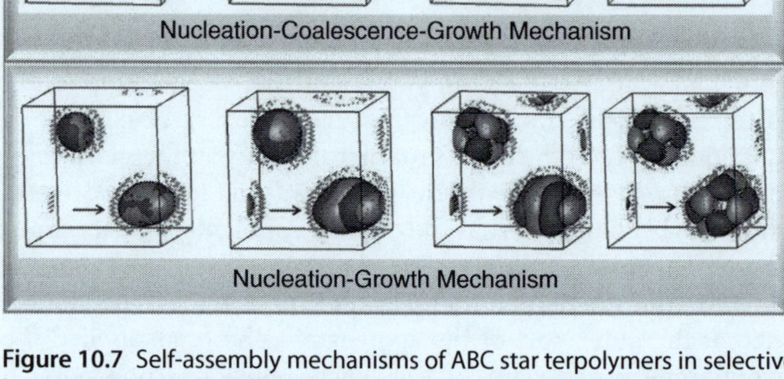

Figure 10.7 Self-assembly mechanisms of ABC star terpolymers in selective solvent. The upper and lower images correspond to the cases of polymers with short and long lengths of solvophilic arms, respectively. Adapted with permission from Ref. [122]. Copyright 2012 Royal Society of Chemistry.

arms become long, the formation process of self-assembled structures only contains the nucleation and growth steps.

10.4.3 Multicompartment Aggregates Self-Assembled from Amphiphilic Copolymer Blends

To further expand on the design strategies of inner subdomains in the superstructures, polymer blends are a prominent pursuit for multicompartment nanostructures with polymer/polymer subdomains in the cores or coronae. In the computer simulations, several blending strategies emerge to achieve the structural hierarchy of multicomponent systems, such as mixtures of AB and BC diblock copolymers in A- and C-selective solvent, blends of ABC terpolymers and AB copolymers, blends of distinct ABC terpolymers, as well as multicomponent rigid homopolymer/rod–coil diblock copolymer systems in selective solvent.

10.4.3.1 Mixtures of Diblock Copolymers

The simplest example of polymer blend is the co-assembly of AB and BC diblock copolymer mixtures in an A- and C-selective solvent. The solvophobic B blocks produce homogenous B cores stabilized by solvophilic A/C coronae, and the incompatibility between the A and C blocks drives the de-mixing of the coronae to form patchy and Janus subdomains, depending on the repulsions between the A and C blocks. As shown in Figure 10.8a, Srinivas and Pitera used the coarse-grained molecular dynamics approach to simulate the binary mixtures of two different copolymers with common solvophobic blocks (B) but incompatible solvophilic blocks (A and C) [123]. The patches formed by the C blocks are distributed at the surfaces of the B-rich cores. The number and size of the patches are dependent on the copolymer composition as well as the surface tension between the B and C blocks. Altering the chemical properties of copolymers allows the transition between spherical and cylindrical patchy micelles. Zhu and co-workers applied the MC simulations to screen Janus-like aggregates from AB/BC amphiphilic block copolymers [124]. Lots of control parameters were examined and a generic phase diagram was mapped, including Janus-like cylinders, lamellae and rings. In particular, the formation of Janus-like aggregates is sensitive to the repulsions between the A and C solvophilic blocks. As the repulsions are increased, the morphological transition follows the sequence of mixed micelles↔Janus micelles↔ mixture of pure micelles.

10.4.3.2 Blends of Terpolymers and Copolymers

The blending of ABC terpolymers with AB linear diblock copolymers in solutions is also a promising approach to tune the hierarchical nanostructures. For instance, Ma et al. implemented SCFT simulations to study the effects of mixture ratio on the morphologies of ABC star terpolymers in an A-selective solvent [125]. It was found that the length of segmented wormlike micelles becomes short as the diblock copolymers are incorporated into the systems (Figure 10.8b). The phenomena observed in the SCFT simulations are demonstrated by the experimental findings [112, 113]. Moreover, the diblock copolymers are distributed at the outer part of the aggregates. Zhu et al. applied the MC method to study the hierarchical aggregates of ABC linear triblock terpolymer/AB diblock copolymer mixtures in solutions [126]. The pure ABC linear terpolymers are

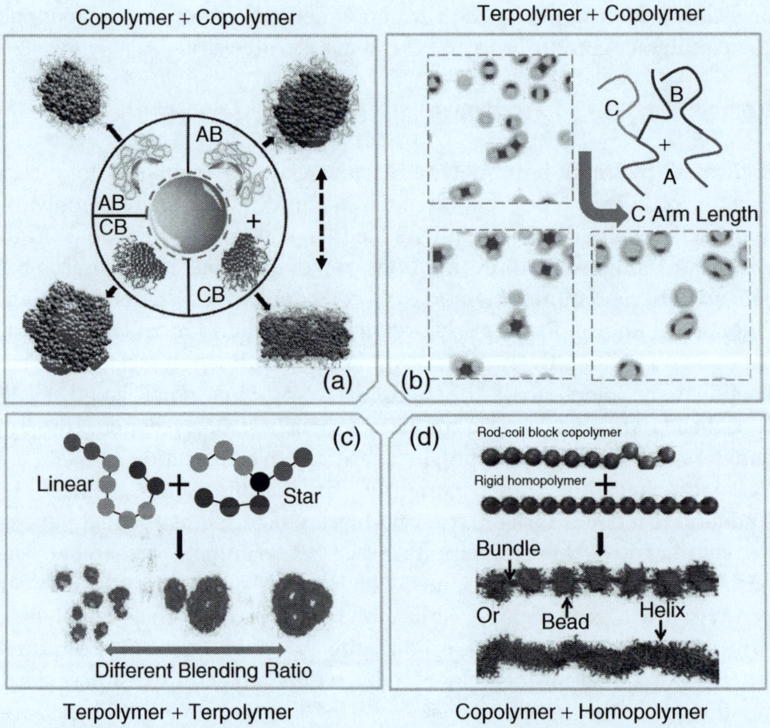

Figure 10.8 (a) Co-assembly behaviors of AB and BC block copolymer blends in A- and C-selective solvent. Adapted with permission from Ref. [123]. Copyright 2008 American Chemical Society. (b) Aggregate morphologies of ABC star terpolymer/AB diblock copolymer mixtures in dilute solutions. Reproduced with permission from Ref. [125]. Copyright 2007 American Chemical Society. (c) Novel multicompartment micelles self-assembled from blends of ABC linear triblock terpolymers and ABC star terpolymers. Adapted with permission from Ref. [127]. Copyright 2007 American Chemical Society. (D) Novel hierarchical nanostructures from cooperative self-assembly of rigid homopolymer/rod–coil diblock copolymer mixtures. Adapted with permission from Ref. [130]. Copyright 2015 Nature Publishing Group.

assumed to produce core-shell–corona structures. Incorporating diblock copolymers into the solutions leads to the formation of segmented and double-helix multicompartment micelles, which are affected by the solvent quantity, the incompatibility of the different blocks and the mixture ratio. It was also demonstrated that the shape of the structured aggregates is also influenced by these factors.

10.4.3.3 Blends of Distinct Terpolymers

Blends of two unlike terpolymers respectively forming various aggregates require modulations of self-assembled structures, which results in novel hierarchical morphologies. For example, pure star-shaped and linear terpolymers self-assemble into "hamburger" and toroid micelles, respectively. Mixtures of terpolymers produce new multicompartment micelles with cogwheel-like cores [127, 128], as displayed in Figure 10.8c. The evolution behaviors of the novel hierarchical structures were also studied via preparing different initial configurations. The snapshots demonstrate that the initial states only influence the formation pathway of self-assembled micelles, but the final morphologies of the novel micelles are determined by the parameter settings of systems such as the mixture ratio, the copolymer composition and the chain architectures.

10.4.3.4 Multicomponent Rigid Homopolymer/Rod–Coil Diblock Copolymer Systems

The structures formed by rigid homopolymers can serve as templates to direct the self-assembly of rod–coil block copolymers. Such directed self-assembly of non-Gaussian polymers results in novel sophisticated aggregates. Lin's group utilized the coarse-grained molecular dynamics method to comprehend the hierarchical self-assembly behaviors of rigid homopolymer/rod–coil copolymer systems [129, 130]. Mixtures with short rigid homopolymers self-organize into spheres or rods. However, long rigid homopolymers cooperatively self-assemble into large-length-scale bundles, and rod–coil copolymers align along the bundles to produce small-length-scale structures such as beads and helices (Figure 10.8d), which strongly depend on the interaction parameters, the rigidity of the polymers and the mixture ratio.

As demonstrated by the above cases, computer simulations can be utilized to understand the general principles for constructing hierarchical self-assembled nanostructures. More significantly, the simulations assist experimental researchers in rationally synthesizing a number of compounds, which self-assemble into hierarchical on-target patterns. In the cases of rigid homopolymer/rod–coil block copolymer systems [129], the molecular simulations predict the novel hierarchical structures as well as the structural transition sequences (abacus-like structures↔superhelical structures↔plain fibers) as the interaction strength of the hydrophobic blocks are changed. To match the simulations, peptide amphiphiles and the corresponding homopolymers are designed in experiments. The blends can self-assemble into the novel morphologies predicted by the simulations. As the environmental temperature is tuned, the transition sequences of morphologies are confirmed. The validation between the experimental findings and the simulation predictions manifests the fact that with the assistance of computer simulations, researchers accelerate the exploration of novel nanostructures from the self-assembly of block copolymers.

We note that almost all the simulations presented above encompass certain limitations. For instance, the final morphologies of block copolymer solutions may trap at the metastable states. In a dilute solution, fluctuation effects are important in the self-assembly of block copolymers. However, the formalism of SCFT neglects the influence of thermal fluctuations. This results in the alternation of mean-field theoretical predictions. Recent development in field-theoretic simulation makes it possible to numerically sample the fluctuations in a complete model and accurately capture the thermodynamic properties [27]. Unfortunately, the progress along this line is very slow due to implementation difficulties of the sampling strategies.

10.5 Hierarchically Ordered Nanocomposites Self-Assembled from Organic–Inorganic Systems

Hierarchical control over the placement of inorganic nanoparticles greatly facilitates the fabrication of nanodevices with enhanced functions. The ordered microphase-separated nanostructures of organic block copolymers serve as templates to guide the assembly of nanoparticles [131–133]. Organic–inorganic hybrid systems provide the possibility of achieving hierarchically ordered nanocomposites via the rational design of the molecular architectures and the geometry parameters and wetting properties of the nanoparticles. Meanwhile, the incorporation of nanoparticle guests also acts on the thermodynamic behaviors of block copolymer hosts [134]. Therefore, the final self-assembled nanostructures are elegantly determined by the balance between the entropy and the enthalpy within the polymer/nanoparticle systems. Computational modeling for such systems advances our understanding of the cooperative self-assembly behaviors of nanocomposites at microscopic levels. There have been several reviews of nanoparticle assembly in a polymer matrix [135, 136]. Here, only the significant advancements in the subject of hierarchically ordered nanocomposites in bulks and solutions are brought together and highlighted.

10.5.1 Hierarchical Self-Assembly of Block Copolymer/Nanoparticle Mixtures

An experimental method to control nanoparticle arrangements at nanometer scales is the hierarchical approach from the assembly of nanoparticles within the microphase-separated structures of multicomponent polymers. The polymers first self-assemble into ordered nanostructures, and in the second step the selective nanoparticles are located in the formed nanostructures. Besides the experimental studies on such unique systems, computational modeling is also exploited to design various strategies for preparing hierarchically ordered nanocomposites from the polymer and nanoparticle blends. Furthermore, it enables researchers to understand the formation mechanism of hybrid structures and predict novel superstructures that are not observed in the current experiments. Recent works on these aspects are summarized below according to the nanocomposite design strategies.

The first method to construct hierarchically ordered nanocomposites is the blending of bidisperse particles and block copolymers. In block polymer/monodisperse particle systems, the particles are placed at the center or the interface of phase-separated structures, depending on the sphere radius and the wetting properties of the particles [133]. Bidisperse nanoparticles display a distinct feature of particle arrangement in the host structures. Lee et al. modified the method [133], which combines the SCFT describing the polymeric species and statistical mechanical density functional theory (DFT) characterizing the solid particles, to predict the self-assembled structures of blending systems [137]. The block copolymers guide the particles to distribute in hexagonally arrayed cylinders, which lead to the formation of large-length-scale structures (Figure 10.9a). In the cylinders, the large particles are concentrated at the center, but the small particles are placed near the edge of the domains. The inhomogeneous distribution of nanoparticles in the polymeric cylinders results in small-length-scale structures. From decomposing the contributions of free energy, it is found that the entropy-driven mechanism results in hierarchical distributions of nanoparticles in microphase-separated nanostructures. The novel hybrid structures of binary nanoparticles open enormous opportunities for preparing three-dimensional materials. Wiesner and co-workers linked experiment and theory to comprehend the self-assembled superstructures from the mixture of ABC triblock terpolymers and binary particles [138–140]. The alternating gyroids of terpolymers offer templates to register the hierarchical distributions of different particles.

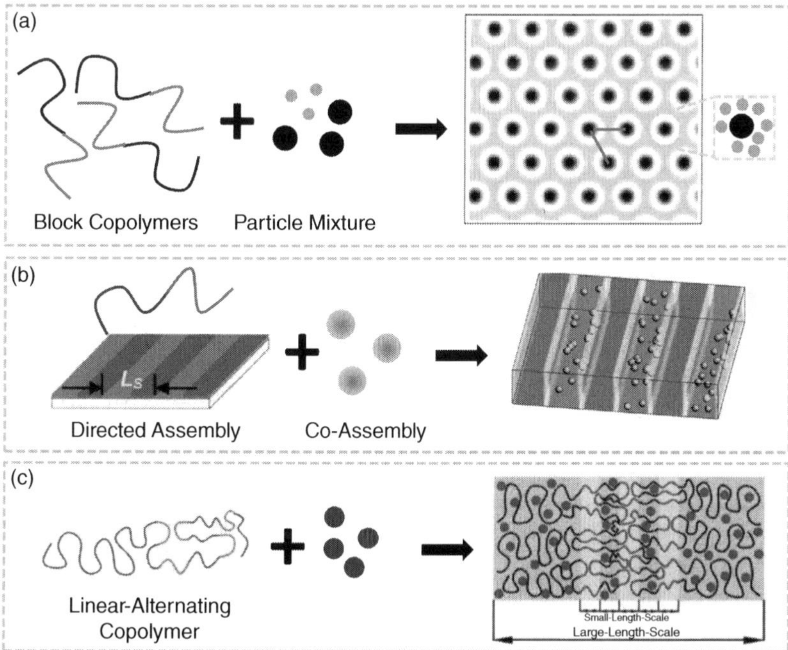

Figure 10.9 (a) Hierarchically ordered structures formed by block copolymer/bidisperse nanoparticle mixtures. Adapted with permission from Ref. [137]. Copyright 2002 American Physical Society. (b) Hierarchical nanocomposites formed by directed co-assembly of block copolymer/nanoparticle blends in chemical templates. Adapted with permission from Ref. [141]. Copyright 2008 American Physical Society. (c) Lamellae-within-lamella structures obtained from linear-alternating multiblock copolymer/nanoparticle mixtures. Reproduced with permission from Ref. [142]. Copyright 2009 American Chemical Society.

In particular, various repulsions between the particles and the blocks control the programmable arrangement of nanoparticles in the gyroids of block copolymers.

The second approach for hierarchical distributions of nanoparticles is the directed self-assembly of block copolymers on chemically patterned substrates. MC simulations of the coarse-grained model are extended to understand the self-assembly behaviors of block copolymer–nanoparticle hybrid systems in the chemical templates [141]. Strong and specific interactions between the surfaces and the blocks force the copolymers to form well-ordered lamellae. The tailored interactions between the nanoparticles and the blocks drive the nanoparticles to locate at the center domains of the lamellae. As homopolymers are introduced into the hybrid systems, two arrays of particles are accommodated in each lamella of the block copolymers (Figure 10.9b). Meanwhile, the hierarchical distributions of particles are finely tuned through the commensurability conditions between the dimensions of the lamellae and the periodicity of the chemical stripes. The above design rules are also suited to hierarchically placing the nanoparticles into device-oriented structures of block copolymers.

The third approach is inspired by the hierarchical structures of block copolymers. When the interaction parameters are large enough, linear-alternating copolymers self-assemble into structures-within-structure nanophases, which provide guided templates to program nanoparticle placements. Zhang et al. applied SCFT/DFT to investigate the cooperative self-assembly behaviors and novel superstructures of linear-alternating copolymer/nanoparticle mixtures [142]. The nanoparticles are hierarchically located in the lamellae-within-lamellae nanostructures of linear-alternating copolymers (Figure 10.9c). Meanwhile, the incorporation of nanoparticles also triggers a change of substructures due to the entropy contributions. Subsequently, the same group replaced one block of the ABC terpolymers by the nanoparticles, termed block copolymers tethered by nanoparticles (ABP molecules) [143]. Like the hierarchical nanostructures of ABC terpolymers, the nanoparticles act as an element to achieve the hybrid superstructures with hierarchical arrangements of particles. The large-length-scale distributions of particles are generated by the microphase separation between the A blocks and the BP blocks, and the segregation between the B blocks and the particles further form the small-length-scale distributions. It is demonstrated that the morphologies of superstructures are related to the composition of the molecules and the radius of the particles.

The formation kinetic of hybrid structures in polymer/nanoparticle systems obeys the stepwise mechanism. Sevink and co-workers developed a particle–field hybrid method to explore the phase separation dynamics of polymer-based nanocomposites [144]. The dynamic extension of SCFT is used to capture the evolution of block copolymer fluids, and Brownian dynamics is applied to describe the motion of the nanoparticles. The nanoparticles are assumed to be incompatible with both blocks, but to different extents. During the first stage of phase separation, the nanoparticles are placed at the interfaces of self-assembled structures. The subsequent microphase separation of block copolymers guides the nanoparticles to depart the interfaces and to form clusters at the center of the block copolymer domains.

The simulation results demonstrate that various design strategies based on polymer/particle mixtures can be proposed to construct hybrid structures with hierarchical distributions of nanoparticles, such as the blend of bidisperse particles, the directed self-assembly of block copolymers and the hierarchical structures of multiblock copolymers. These data advance our understanding of the structural formation mechanism of nanocomposites and provide useful guidelines for experimental design. For example, Xu's group carried out experimental studies of the co-assembly of supramolecular block copolymers and nanoparticles [15]. The distributions of nanoparticles in the polymer matrix were rationalized by the simulation results of Balazs and co-workers [133]. Furthermore, computational modeling has also predicted novel hybrid superstructures of polymer/nanoparticle mixtures via the molecular design of organic compounds [142], which provide new opportunities for the development of functional materials.

10.5.2 Hierarchical Self-Assembly of Polymer/Nanoparticle/Solvent Systems

Nanoparticles tethered by flexible polymer tails self-assemble into highly diverse, thermodynamically stable hierarchical superstructures in solutions. Experimentally, the class of organic–inorganic tethered materials has attracted extensive research interest [145, 146]. In computational modeling, Ma et al. conceived a model of solvophobic homopolymer tails tethered by a solvophilic nanoparticle head [147]. The DPD simulations show that molecules with a small diameter of solvophilic heads self-assembled into bilayer vesicles, which are similar to those of amphiphilic block copolymers. As the radius of the solvophilic heads becomes large, the tethered molecules self-organize into

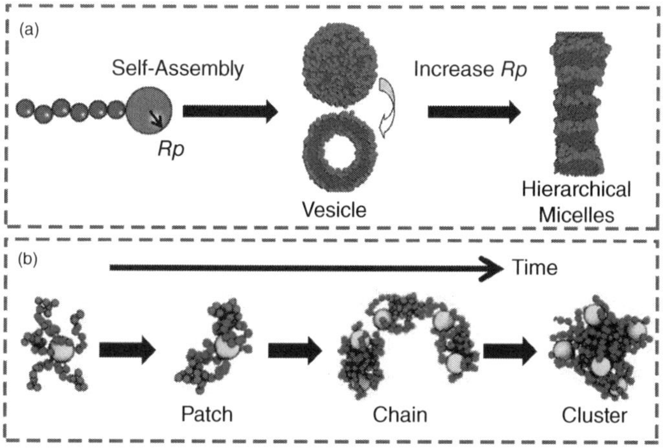

Figure 10.10 (a) Self-assembled structures of nanoparticles tethered by flexible polymer tails in dilute solution. Adapted with permission from Ref. [147]. Copyright 2015 American Chemical Society. (b) Hierarchical self-assembly of diblock copolymer functionalized nanoparticles in selective solvent. Reproduced with permission from Ref. [148]. Copyright 2014 American Institute of Physics.

wormlike micelles with multicompartment features (Figure 10.10a). It is also demonstrated that the formation mechanism of such superstructures obeys the conventional aggregation–coalescence–growth pathway.

In comparison with the cases of grafted homopolymers, the self-assembly behaviors of diblock copolymers tethered by nanoparticles display a distinct difference. In particular, Estridge and Jayaraman used coarse-grained molecular dynamics simulations to probe the hierarchical self-assembly of functionalized spherical nanoparticles at low grafting density of block copolymers [148]. At the optimal design parameters, the hybrid systems first self-assemble into anisotropic soft particles with various numbers of patches due to the amphiphilic characteristics of the block copolymers. To lower the free energy, such patchy particles are further driven to form chain-like intermediate states, and finally achieve large clusters with solvophilic coronae and solvophobic cores (Figure 10.10b). The above self-assembly pathway of nanoparticle/block copolymer systems is termed the stepwise "polymerization" process, which is different from the aggregation–coalescence–growth mechanism for nanoparticle–homopolymer tethered molecules.

Amphiphilic block copolymers can spontaneously organize into aggregates with controllable size and shape in a selective solvent, which serve as templates to program nanoparticle arrangements. An SCFT-based theoretical approach was extended to investigate the co-assembly behaviors of amphiphilic block copolymer/nanoparticle mixtures in dilute solutions [149]. It was found that the solvophobic nanoparticles are distributed at the cores or walls of aggregates, but the arrangements of nanoparticles are not hierarchical. As demonstrated in experimental studies [13], triblock terpolymers in selective solvent self-assemble into multicompartment aggregates, which could potentially be exploited to hierarchically place nanoparticles. Therefore, one important further study topic is that the aggregates formed by ABC triblock terpolymers are utilized to guide the hierarchical distributions of particles.

Another important further topic is new modeling and simulation methods for the structural hierarchy of nanocomposites. Recently, hybrid particle–field methods have been developed to study the distributions of nanoparticles in self-assembled structures [134, 150]. The shape and size of inorganic particles are represented by cavity functions. Coupling terms are introduced to model the wetting properties of particles. In the hybrid methods, the SCFT is used to obtain the local equilibrium states of polymeric fluids. In the spirit of Sides et al. [134], the force acting on a single particle follows the Hellmann–Feynman theorem. The scheme is seemingly suited to spherical nanoparticles, and it was modified to adapt for rod-shaped anisotropic particles [151]. In the treatment of Hall et al. [150], the force is calculated by the gradient of the chemical potential. The advantage of the method is that the translation and rotation of particles with various shapes are described in the unified framework of rigid motion. The above modeling and methods may offer a novel avenue to screen the hierarchical structures of nanocomposites. For instance, nanorod/nanotriangle mixtures are guided to produce hierarchical distributions of nanoparticles in the copolymer matrix, as well as nanorods tethered by polymer tails self-assembling into hierarchical structures with the helix arrangement of anisotropic particles.

10.6 Conclusions and Perspectives

The hierarchical nanostructures formed by polymer-based systems provide a powerful platform for mankind to understand and mimic nature. The rational design of hierarchical structures on demand is a critical step that promotes researchers to achieve high performance nanomaterials. Computational modeling and simulations are exploited to grasp the universal thermodynamic and dynamic principles of multiple-length-scale superstructures self-assembled from block copolymers. The examples and design strategies summarized in this chapter involve multicomponent polymers with complex topologies in bulk and solution, as well as hybrid nanocomposites. In principle, the phase separation of unlike molecules leads to the emergence of large-length-scale structures, which guide the production of small-length-scale subdomains from the microphase separation of incompatible components. The formation mechanism of structural hierarchy generally obeys the stepwise pathway of phase separation.

While the reported results are encouraging, there remain lots of challenges and opportunities for further understanding the hierarchical self-assembly of complicated molecules and rationally designing the structural hierarchy formed by soft matter.

10.6.1 New Theoretical Insights

The hierarchical nanostructures of block copolymers may be formed via various dynamic pathways of self-assembly, including non-hierarchical and hierarchical assembly procedures. These pathways involve intermediates with counterparts on the equilibrium structures (for example, lamellae-with-lamella structures via multistep phase separation) or intermediates with no such counterparts (for instance, cage aggregates via the nucleation–growth process). Classical theories on the basis of near-equilibrium assumptions are not sufficient to physically explain the self-assembly phenomena of such complex systems. New theoretical insights are required to further understand the self-assembly mechanisms and finely control the hierarchical structures of soft matter. For example, the far-from-equilibrium theoretical framework is proving important, and it is likely to connect naturally with the hierarchical self-assembly of biological systems [152].

10.6.2 Horizontal Multiscale Modeling

To build realistic models for hierarchical nanostructures, the mesoscopic information for the structural hierarchy and certain nanoscale features (such as hydrogen bonds, metal–ligand interactions and host–guest interactions) should be retained. Currently, the most available methodology is hierarchical multiscale modeling, where simulations at the lower level pass information to higher-level simulations ("vertical" multiscale modeling). The scheme strongly depends on systematic coarse-graining and mapping. Another alternative route is "horizontal" multiscale modeling, where fine-grained details are incorporated into the existing methods from different spatial levels [153]. The integrated approach is especially appropriate for hierarchical self-assembly involving multilevel driving forces.

10.6.3 Inverse Design Strategy

As the self-assembly systems become more complex, the computer simulations face greater challenges in screening the hierarchical structures via a "forward" strategy (from primary building elements to superstructures). An alternative and more promising strategy is to begin with the superstructures and then to determine the composition, interaction parameters and sequences of polymers via "inverse" optimization methods [154]. This design philosophy was recently implemented to develop guided templates for registering device-oriented structures of block copolymers [155, 156]. Incorporating the inverse design strategy into existing simulation methods will significantly accelerate the construction of hierarchical nanostructures.

10.6.4 Element–Structure–Property Relationships

A critical challenge is the relationships between hierarchical nanostructures and their macroscopic properties. Although the enhanced mechanical properties of multiple-length-scale structures were confirmed by computer simulations [80], relatively few studies focused on the relationships between the self-assembled superstructures and other excellent properties, such as switchable optical property and protonic conductivity. An integrated method, which combines the mesoscopic approach for determining the morphologies and existing theories (Maxwell's or drift–diffusion equations) for predicting the collective properties, may provide an effective route to realize the accurate prediction of element–structure–property relationships for novel materials.

Acknowledgments

This work is supported by the National Natural Science Foundation of China (51203049, 21574040, 21234002). Support from the 111 project (B14018) is also appreciated.

References

1. Lehn, J.-M. (1993) Supramolecular chemistry. *Science*, **260**, 1762–1763.
2. Qiu, H., Hudson, Z. M., Winnik, M. A., Manners, I. (2015) Multidimensional hierarchical self-assembly of amphiphilic cylindrical block comicelles. *Science*, **347**, 1329–1332.
3. Liu, K., Nie, Z., Zhao, N., Li, W., Rubinstein, M., Kumacheva, E. (2010) Step-growth polymerization of inorganic nanoparticles. *Science*, **329**, 197–200.
4. Douglas, S. M., Dietz, H., Liedl, T., Högberg, B., Graf, F., Shih, W. M. (2009) Self-assembly of DNA into nanoscale three-dimensional shapes. *Nature*, **459**, 414–418.
5. Ke, Y., Ong, L. L., Shih, W. M., Yin, P. (2012) Three-dimensional structures self-assembled from DNA bricks. *Science*, **338**, 1177–1183.
6. Chen, Q., Bae, S. C., Granick, S. (2011) Directed self-assembly of a colloidal kagome lattice. *Nature*, **469**, 381–384.
7. Chen, Q., Whitmer, J. K., Jiang, S., Bae, S. C., Luijten, E., Granick, S. (2011) Supracolloidal reaction kinetics of Janus spheres. *Science*, **331**,199–202.
8. Glotzer, S. C., Solomon, M. J. (2007) Anisotropy of building blocks and their assembly into complex structures. *Nature Mater.*, **6**, 557–562.
9. Damasceno, P. F., Engel, M., Glotzer, S. C. (2012) Predictive self-assembly of polyhedra into complex structures. *Science*, **337**, 453–457.
10. Yang, Y., Wang, C. (2009) Hierarchical construction of self-assembled low-dimensional molecular architectures observed by using scanning tunneling microscopy. *Chem. Soc. Rev.*, **38**, 2576–2589.
11. Wang, A., Huang, J., Yan, Y. (2014) Hierarchical molecular self-assemblies: construction and advantages. *Soft Matter*, **10**, 3362–3373.
12. Whitelam, S., Jack, R. L. (2015) The statistical mechanics of dynamic pathways to self-assembly. *Annu. Rev. Phys. Chem.*, **66**, 143–163.
13. Moughton, A. O., Hillmyer, M. A., Lodge, T. P. (2012) Multicompartment block polymer micelles. *Macromolecules*, **45**, 2–19.
14. Gröschel, A. H., Müller, A. H. E. (2015) Self-assembly concepts for multicompartment nanostructures. *Nanoscale*, **7**, 11841–11876.
15. Kao, J., Thorkelsson, K., Bai, P., Rancatore, B. J., Xu, T. (2013) Toward functional nanocomposites: taking the best of nanoparticles, polymers, and small molecules. *Chem. Soc. Rev.*, **42**, 2654–2678.
16. Thorkelsson, K., Bai, P., Xu, T. (2015) Self-assembly and applications of anisotropic materials: a review. *Nano Today*, **10**, 48–66.
17. Frenkel, D., Smit, B. (2007) *Understanding Molecular Simulation: From Algorithms to Applications*, 2nd edn, Academic Press.
18. Kremer, K., Müller-Plathe, F. (2001) Multiscale problems in polymer science: simulation approaches. *MRS Bull.*, **26**, 205–210.
19. Peter, C., Kremer, K. (2010) Multiscale simulation of soft matter systems. *Faraday Discuss.*, **144**, 9–24.
20. Fredrickson, G. H., Ganesan, V., Drolet, F. (2002) Field-theoretic computer simulation methods for polymers and complex fluids. *Macromolecules*, **35**, 16–39.
21. Müller-Plathe, F. (2002) Coarse-graining in polymer simulation: from the atomistic to the mesoscopic scale and back. *ChemPhysChem*, **3**, 754–769.

22. Landau, D. P., Binder, K. (2009) *A Guide to Monte Carlo Simulations in Statistical Physics*, 3rd edn. Cambridge University Press, New York.
23. Binder, K., Paul, W. (2008) Recent developments in Monte Carlo simulations of lattice models for polymer systems. *Macromolecules*, **41**, 4537–4550.
24. Öttinger, H. C. (1996) *Stochastic Processes in Polymeric Fluids*, Springer-Verlag, New York.
25. Hoogerbrugge, P. J., Koelman, J. M. V. A. (1992) Simulating microscopic hydrodynamic phenomena with dissipative particle dynamics. *Europhys. Lett.*, **19**, 155–160.
26. Español, P., Warren, P. B. (1995) Statistical-mechanics of dissipative particle dynamics. *Europhys. Lett.*, **30**, 191–196.
27. Fredrickson, G. H. (2006) *The Equilibrium Theory of Inhomogeneous Polymers*, Oxford University Press, New York.
28. Edwards, S. F. (1965) The statistical mechanics of polymers with excluded volume. *Proc. Phys. Soc. (London)*, **85**, 613–624.
29. Matsen, M. W., Schick, M. (1994) Stable and unstable phases of a diblock copolymer melt. *Phys. Rev. Lett.*, **72**, 2660–2663.
30. Drolet, F., Fredrickson, G. H. (1999) Combinatorial screening of complex block copolymer assembly with self-consistent field theory. *Phys. Rev. Lett.*, **83**, 4317–4320.
31. Guo, Z., Zhang, G., Qiu, F., Zhang, H., Yang, Y., Shi, A.-C. (2008) Discovering ordered phases of block copolymers: new results from a generic Fourier-space approach. *Phys. Rev. Lett.*, **101**, 028301.
32. Fraaije, J. G. E. M., van Vlimmeren, B. A. C., Maurits, N. M., Postma, M., Evers, O. A., Hoffmann, C., Altevogt, P., Goldbeck-Wood, G. (1997) The dynamic mean-field density functional method and its application to the mesoscopic dynamics of quenched block copolymer melts. *J. Chem. Phys.*, **106**, 4260–4269.
33. Maurits, N. M., Fraaije, J. G. E. M. (1997) Mesoscopic dynamics of copolymer melts: from density dynamics to external potential dynamics using nonlocal kinetic coupling. *J. Chem. Phys.*, **107**, 5879–5889.
34. Hasegawa, R., Doi, M. (1997) Adsorption dynamics. Extension of self-consistent field theory to dynamical problems. *Macromolecules*, **30**, 3086–3089.
35. Yeung, C., Shi, A.-C. (1999) Formation of interfaces in incompatible polymer blends: a dynamical mean field study. *Macromolecules*, **32**, 3637–3642.
36. Ruokolainen, J., Mäkinen, R., Torkkeli, M., Mäkelä, T., Serimaa, R., ten Brinke, G., Ikkala, O. (1998) Switching supramolecular polymeric materials with multiple length scales. *Science*, **280**, 557–560.
37. Ruokolainen, J., ten Brinke, G., Ikkala, O. (1999) Supramolecular polymeric materials with hierarchical structure-within-structure morphologies. *Adv. Mater.*, **11**, 777–780.
38. Ikkala, O., ten Brinke, G. (2002) Functional materials based on self-assembly of polymeric supramolecules. *Science*, **295**, 2407–2409.
39. ten Brinke, G., Loos, K., Vukovic, I., du Sart, G. G. (2011) Hierarchical self-assembly of two-length-scale multiblock copolymers. *J. Phys. Condens. Matter*, **23**, 284110.
40. Zheng, W., Wang, Z.-G. (1995) Morphology of ABC triblock copolymers. *Macromolecules*, **28**, 7215–7223.

41. Chatterjee, J., Jain, S., Bates, F. S. (2007) Comprehensive phase behavior of poly (isoprene-*b*-styrene-*b*-ethylene oxide) triblock copolymers. *Macromolecules*, **40**, 2882–2896.

42. Auschra, C., Stadler, R. (1993) New ordered morphologies in ABC triblock copolymers. *Macromolecules*, **26**, 2171–2174.

43. Bates, F. S., Hillmyer, M. A., Lodge, T. P., Bates, C. M., Delaney, K. T., Fredrickson, G. H. (2012) Multiblock polymers: panacea or pandora's box? *Science*, **336**, 434–440.

44. Matsushita, Y. (2007) Creation of hierarchically ordered nanophase structures in block polymers having various competing interactions. *Macromolecules*, **40**, 771–776.

45. Tyler, C. A., Qin, J., Bates, F. S., Morse, D. C. (2007) SCFT study of non-frustrated ABC triblock copolymer melts. *Macromolecules*, **40**, 4654–4668.

46. Sun, M., Wang, P., Qiu, F., Tang, P., Zhang, H., Yang, Y. (2008) Morphology and phase diagram of ABC linear triblock copolymers: parallel real-space self-consistent-field-theory simulation. *Phys. Rev. E*, **77**, 016701.

47. Guo, Z., Zhang, G., Qiu, F., Zhang, H., Yang, Y., Shi, A.-C. (2008) Discovering ordered phases of block copolymers: new results from a generic Fourier-space approach. *Phys. Rev. Lett.*, **101**, 028301.

48. Li, W. H., Qiu, F., Shi, A.-C. (2012) Emergence and stability of helical superstructures in ABC triblock copolymers. *Macromolecules*, **45**, 503–509.

49. Liu, M., Li, W. H., Qiu, F., Shi, A.-C. (2012) Theoretical study of phase behavior of frustrated ABC linear triblock copolymers. *Macromolecules*, **45**, 9522–9530.

50. Krappe, U., Stadler, R., Voigt-Martin, I. (1995) Chiral assembly in amorphous ABC triblock copolymers. Formation of a helical morphology in polystyrene-*b*-polybutadiene-*b*-poly(methyl methacrylate) block copolymers. *Macromolecules*, **28**, 4558–4561.

51. Elbs, H., Abetz, V., Hadziioannou, G., Drummer, C., Krausch, G. (2001) Antiferromagnetic ordering in a helical triblock copolymer mesostructure. *Macromolecules*, **34**, 7917–7919.

52. Jinnai, H., Kaneko, T., Matsunaga, K., Abetz, C., Abetz, V. (2009) A double helical structure formed from an amorphous, achiral ABC triblock terpolymer. *Soft Matter*, **5**, 2042–2046.

53. Gemma, T., Hatano, A., Dotera, T. (2002) Monte Carlo simulations of the morphology of ABC star polymers using the diagonal bond method. *Macromolecules*, **35**, 3225–3237.

54. Takano, A., Kawashima, W., Noro, A., Isono, Y., Tanaka, N., Dotera, T., Matsushita, Y. (2005) A mesoscopic Archimedean tiling having a new complexity in an ABC star polymer. *J. Polym. Sci., Part B: Polym. Phys.*, **43**, 2427–2432.

55. Hayashida, K., Kawashima, W., Takano, A., Shinohara, Y., Amemiya, Y., Nozue, Y., Matsushita, Y. (2006) Archimedean tiling patterns of ABC star-shaped terpolymers studied by microbeam small-angle X-ray scattering. *Macromolecules*, **39**, 4869–4872.

56. Huang, C.-I, Fang, H.-K., Lin, C.-H. (2008) Morphological transition behavior of ABC star copolymers by varying the interaction parameters. *Phys. Rev. E*, **77**, 031804.

57. Tang, P., Qiu, F., Zhang, H. D., Yang, Y. L. (2004) Morphology and phase diagram of complex block copolymers: ABC star triblock copolymers. *J. Phys. Chem. B*, **108**, 8434–8438.
58. Li, W. H., Xu, Y. C., Zhang, G. J., Qiu, F., Yang, Y. L., Shi, A.-C. (2010) Real-space self-consistent mean-field theory study of ABC star triblock copolymers. *J. Chem. Phys.*, **133**, 064904.
59. Zhang, G. J., Qiu, F., Zhang, H. D., Yang, Y. L., Shi, A.-C. (2010) SCFT study of tiling patterns in ABC star terpolymers. *Macromolecules*, **43**, 2981–2989.
60. Xu, W., Jiang, K., Zhang, P., Shi, A.-C. (2013) A strategy to explore stable and metastable ordered phases of block copolymers. *J. Phys. Chem. B*, **117**, 5296–5305.
61. Cao, X., Zhang, L., Wang, L., Lin, J. (2014) Insights into ordered microstructures and ordering mechanisms of ABC star terpolymers by integrating dynamic self-consistent field theory and variable cell shape methods. *Soft Matter*, **10**, 5916–5927.
62. Xia, J., Sun, M., Qiu, F., Zhang, H., Yang, Y. (2005) Microphase ordering mechanisms in linear ABC triblock copolymers. A dynamic density functional study. *Macromolecules*, **38**, 9324–9332.
63. Aissou, K., Choi, H. K., Nunns, A., Manners, I., Ross, C. A. (2013) Ordered nanoscale Archimedean tilings of a templated 3-miktoarm star terpolymer. *Nano Lett.*, **13**, 835–839.
64. Masuda, J., Takano, A., Nagata, Y., Noro, A., Matsushita, Y. (2006) Nanophase-separated synchronizing structure with parallel double periodicity from an undecablock terpolymer. *Phys. Rev. Lett.*, **97**, 098301.
65. Nagata, Y., Masuda, J., Noro, A., Cho, D., Takano, A., Matsushita, Y. (2005) Preparation and characterization of a styrene-isoprene undecablock copolymer and its hierarchical microdomain structure in bulk. *Macromolecules*, **38**, 10220–10225.
66. Nap, R., Erukhimovich, I. Y., ten Brinke, G. (2004) Self-assembling block copolymer systems involving competing length scales: a route toward responsive materials. *Macromolecules*, **37**, 4296–4303.
67. Nap, R., Sushko, N., Erukhimovich, I., ten Brinke, G. (2006) Double periodic lamellar-in-lamellar structure in multiblock copolymer melts with competing length scales. *Macromolecules*, **39**, 6765–6770.
68. Subbotin, A., Klymko, T., ten Brinke, G. (2007) Lamellar-in-lamellar structure of A-*b*-(B-*b*-C)$_m$-*b*-B-*b*-A multiblock copolymers. *Macromolecules*, **40**, 2915–2918.
69. Klymko, T., Markov, V., Subbotin, A., ten Brinke, G. (2009) Lamellar-in-lamellar self-assembled C-*b*-(B-*b*-A)$_m$-*b*-B-*b*-C multiblock copolymers: Alexander-de Gennes approach and dissipative particle dynamics simulations. *Soft Matter*, **5**, 98–103.
70. Markov, V., Kriksin, Y., Erukhimovich, I., ten Brinke, G. (2013) Perpendicular lamellar-in-lamellar and other planar morphologies in A-*b*-(B-*b*-A)$_2$-*b*-C and (B-*b*-A)$_2$-*b*-C ternary multiblock copolymer melts. *J. Chem. Phys.*, **139**, 084906.
71. Subbotin, A., Markov, V., ten Brinke, G. (2010) Parallel versus perpendicular lamellar-within-lamellar self-assembly of A-*b*-(B-*b*-A)$_n$-*b*-C ternary multiblock copolymer melts. *J. Phys. Chem. B*, **114**, 5250–5256.
72. Markov, V., Kriksin, Y., Erukhimovich, I., ten Brinke, G. (2013) Perpendicular lamellar-in-lamellar and other planar morphologies in A-*b*-(B-*b*-A)$_2$-*b*-C and (B-*b*-A)$_2$-*b*-C ternary multiblock copolymer melts. *J. Chem. Phys.*, **139**, 084906.

73. Subbotin, A., Markov, V., ten Brinke, G. (2010) Parllel versus perpendicular lamellar-within-lamellar self-assembly of A-*b*-(B-*b*-A)$_n$-*b*-C ternary multiblock copolymer melts. *J. Phys. Chem. B*, **114**, 5250–5256.
74. Xu, Y., Li, W., Qiu, F., Yang, Y., Shi, A.-C. (2011) The influence of volume fractions on the phase behaviors of linear A(BC)$_n$BA′ multiblock terpolymers. *Phys. Chem. Chem. Phys.*, **13**, 12421–12428.
75. Wang, L., Zhang, L., Lin, J. (2010) Hierarchically ordered microstructures self-assembled from A(BC)$_n$ multiblock copolymers. *Macromolecules*, **43**, 1602–1609.
76. Pryamitsyn, V., Ganesan, V. (2004) Self-assembly of rod–coil block copolymers. *J. Chem. Phys.*, **120**, 5824–5838.
77. Zhu, X., Wang, L., Lin, J. (2013) Self-assembly of rod–coil multiblock copolymers: a strategy for creating hierarchical smectic structures. *J. Phys. Chem. B*, **117**, 5748–5756.
78. Zhang, X., Wang, L., Zhang, L., Lin, J., Jiang, T. (2015) Controllable hierarchical microstructures self-assembled from multiblock copolymers confined in thin films. *Langmuir*, **31**, 2533–2544.
79. Fleury, G., Bates, F. S. (2009) Structure and properties of hexa- and undecablock terpolymers with hierarchical molecular architectures. *Macromolecules*, **42**, 3598–3610.
80. Zhu, X., Wang, L., Lin, J. (2011) Distinct elastic response to hierarchical nanostructures. *Macromolecules*, **44**, 8314–8323.
81. Zhang, L., Lin, J., Lin, S. (2007) Morphologies and bridging properties of graft copolymers. *J. Phys. Chem. B*, **111**, 351–357.
82. Zhang, L., Lin, J., Lin, S. (2009) Elastic properties of graft copolymers in the lamellar phase studied by self-consistent field theory. *Soft Matter*, **5**, 173–181.
83. Zhang, L., Lin, J., Lin, S. (2008) Effect of molecular architecture on phase behavior of graft copolymers. *J. Phys. Chem. B*, **112**, 9720–9728.
84. Wang, L., Zhang, L., Lin, J. (2008) Microphase separation in multigraft copolymer melts studied by random-phase approximation and self-consistent field theory. *J. Chem. Phys.*, **129**, 114905.
85. Wang, L., Lin, J., Zhang, L. (2009) Hierarchically ordered microstructures self-assembled from comb–coil block copolymers. *Langmuir*, **25**, 4735–4742.
86. Huang, C.-I, Lin, Y.-C. (2007) Hierarchical structure-within-structure morphologies in A-*block*-(B-*graft*-C) molecules. *Macromol. Rapid Commun.*, **28**, 1634–1639.
87. Tang, C., Lennon, E. M., Fredrickson, G. H., Kramer, E. J., Hawker, C. J. (2008) Evolution of block copolymer lithography to highly ordered square arrays. *Science*, **322**, 429–432.
88. Park, W. I., Kim, Y. J., Jeong, J. W., Kim, K., Yoo, J.-K., Hur, Y. H., Kim, J. M., Thomas, E. L., Alexander-Katz, A., Jung, Y. S. (2013) Host–guest self-assembly in block copolymer blends. *Sci. Rep.*, **3**, 3190.
89. Feng, E. H., Lee, W. B., Fredrickson, G. H. (2007) Supramolecular diblock copolymers: a field-theoretic model and mean-field solution. *Macromolecules*, **40**, 693–702.
90. Mohan, A., Elliott, R., Fredrickson, G. H. (2010) Field-theoretic model of inhomogeneous supramolecular polymer networks and gels. *J. Chem. Phys.*, **133**, 174903.

91. Mester, Z., Mohan, A., Fredrickson, G. H. (2011) Macro- and microphase separation in multifunctional supramolecular polymer networks. *Macromolecules*, **44**, 9411–9423.
92. Elliott, R., Fredrickson, G. H. (2009) Supramolecular assembly in telechelic polymer blends. *J. Chem. Phys.*, **131**, 144906.
93. Zhuang, Y., Wang, L., Lin, J. (2011) Hierarchical nanostructures self-assembled from diblock copolymer/homopolymer blends with supramolecular interactions. *J. Phys. Chem. B*, **115**, 7550–7560.
94. Delaney, K. T., Fredrickson, G. H. (2013) Polymer field-theory simulations on graphics processing units. *Comput. Phys. Commun.*, **184**, 2102–2110.
95. Zhang, L., Wang, L., Lin, J. (2014) Harnessing anisotropic nanoposts to enhance long-range orientation order of directed self-assembly nanostructures via large cell simulations. *ACS Macro Lett.*, **3**, 712–716.
96. Nguyen, T. D., Phillips, C. L., Anderson, J. A., Glotzer, S. C. (2011) Rigid body constraints realized in massively-parallel molecular dynamics on graphics processing units. *Comput. Phys. Commun.*, **182**, 2307–2313.
97. Laschewsky, A. (2003) Polymerized micelles with compartments. *Curr. Opin. Colloid Interface Sci.*, **8**, 274–281.
98. Lutz, J.-F., Laschewsky, A. (2005) Multicompartment micelles: has the long-standing dream become a reality? *Macromol. Chem. Phys.*, **206**, 813–817.
99. Gröschel, A. H., Schacher, F. H., Schmalz, H., Borisov, O. V., Zhulina, E. B., Walther, A., Müller, A. H. E. (2012) Precise hierarchical self-assembly of multicompartment micelles. *Nat. Commun.*, **3**, 710.
100. Fraaije, J. G. E. M., Sevink, G. J. A. (2003) Model for pattern formation in polymer surfactant nanodroplets. *Macromolecules*, **36**, 7891–7893.
101. Sevink, G. J. A., Zvelindovsky, A. V. (2005) Self-assembly of complex vesicles. *Macromolecules*, **38**, 7502–7513.
102. Sevink, G. J. A., Zvelindovsky, A. V. (2007) Mesoscopic dynamics of complex vesicle formation: kinetic versus thermodynamic factors. *Mol. Simul.*, **33**, 405–415.
103. Chi, P., Wang, Z., Li, B. H., Shi, A.-C. (2011) Soft confinement-induced morphologies of diblock copolymers. *Langmuir*, **27**, 11683–11689.
104. He, X. H., Schmid, F. (2006) Dynamics of spontaneous vesicle formation in dilute solutions of amphiphilic diblock copolymers. *Macromolecules*, **39**, 2654–2662.
105. He, X. H., Schmid, F. (2006) Using prenucleation to control complex copolymeric vesicle formation in solution. *Macromolecules*, **39**, 8908–8910.
106. He, X. H., Schmid, F. (2008) Spontaneous formation of complex micelles from a homogeneous solution. *Phys. Rev. Lett.*, **100**, 137802.
107. Wang, L., Jiang, T., Lin, J. (2013) Self-assembly of graft copolymers in backbone-selective solvents: a route toward stable hierarchical vesicles. *RSC Adv.*, **3**, 19481–19491.
108. Capone, B., Coluzza, I., Lo Verso, F., Likos, C. N., Blaak, R. (2012) Telechelic star polymers as self-assembling units from the molecular to the macroscopic scale. *Phys. Rev. Lett.*, **109**, 238301.
109. Capone, B., Coluzza, I., Blaak, R., Lo Verso, F., Likos, C. N. (2013) Hierarchical self-assembly of telechelic star polymers: from soft patchy particles to gels and diamond crystals. *New J. Phys.*, **2013**, **15**, 095002.

110. Koch, C., Panagiotopoulos, A. Z., Lo Verso, F., Likos, C. N. (2015) Customizing wormlike mesoscale structures via self-assembly of amphiphilic star polymers. *Soft Matter*, **11**, 3530–3535.

111. Li, Z., Kesselman, E., Talmon, Y., Hillmyer, M. A., Lodge, T. P. (2004) Multicompartment micelles from ABC miktoarm stars in water. *Science*, **306**, 98–101.

112. Li, Z., Hillmyer, M. A., Lodge, T. P. (2006) Laterally nanostructured vesicles, polygonal bilayer sheets, and segmented wormlike micelles. *Nano Lett.*, **6**, 1245–1249.

113. Li, Z., Hillmyer, M. A., Lodge, T. P. (2006) Morphologies of multicompartment micelles formed by ABC miktoarm star terpolymers. *Langmuir*, **22**, 9409–9417.

114. Kubowicz, S., Baussard, J.-F., Lutz, J.-F., Thunemann, A. F., von Berlepsch, H., Laschewsky, A. (2005) Multicompartment micelles formed by self-assembly of linear ABC triblock copolymers in aqueous medium. *Angew. Chem. Int. Ed.*, **44**, 5262–5265.

115. Zhu, J., Jiang, W. (2005) Self-assembly of ABC triblock copolymer into giant segmented wormlike micelles in dilute solution. *Macromolecules*, **38**, 9315–9323.

116. Jiang, T., Wang, L., Lin, S., Lin, J., Li, Y. (2011) Structural evolution of multicompartment micelles self-assembled from linear ABC triblock copolymer in selective solvents. *Langmuir*, **27**, 6440–6448.

117. Wang, L., Lin, J. (2011) Discovering multicore micelles: insights into the self-assembly of linear ABC terpolymers in midblock-selective solvents. *Soft Matter*, **7**, 3383–3391.

118. Zhulina, E. B., Borisov, O. V. (2008) Scaling theory of 3-miktoarm ABC copolymer micelles in selective solvent. *Macromolecules*, **41**, 5934–5944.

119. Xia, J., Zhong, C. (2006) Dissipative particle dynamics study of the formation of multicompartment micelles from ABC star triblock copolymers in water. *Macromol. Rapid Commun.*, **27**, 1110–1114.

120. Zhong, C., Liu, D. (2007) Understanding multicompartment micelles using dissipative particle dynamics simulation. *Macromol. Theory Simul.*, **16**, 141–157.

121. Kong, W. X., Li, B. H., Jin, Q. H., Ding, D. T., Shi, A. C. (2009) Helical vesicles, segmented semivesicles, and noncircular bilayer sheets from solution-state self-assembly of ABC miktoarm star terpolymers. *J. Am. Chem. Soc.*, **131**, 8503–8512.

122. Wang, L., Xu, R., Wang, Z., He, X. (2012) Kinetics of multicompartment micelle formation by self-assembly of ABC miktoarm star terpolymer in dilute solution. *Soft Matter*, **8**, 11462–11470.

123. Srinivas, G., Pitera, J. W. (2008) Soft patchy nanoparticles from solution-phase self-assembly of binary diblock copolymers. *Nano Lett.*, **8**, 611–618.

124. Sheng, Y., Yang, X., Yan, N., Zhu, Y. (2013) Janus-like spheres, disks, rings, cylinders, and vesicles from the self-assembly of mixture of AB and BC diblock copolymers in A- and C-selective solvents, *Soft Matter*, **2013**, **9**, 6254–6262.

125. Ma, J. W., Li, X., Tang, P., Yang, Y. (2007) Self-assembly of amphiphilic ABC star triblock copolymers and their blends with AB diblock copolymers in solution: self-consistent field theory simulations. *J. Phys. Chem. B*, **111**, 1552–1558.

126. Zhu, Y., Yang, X., Kong, W., Sheng, Y., Yan, N. (2012) Segmented and double-helix multicompartment micelles from self-assembly of blends of ABC and AB block copolymers in C block-selective solvents. *Soft Matter*, **8**, 11156–11162.

127. Xin, J., Liu, D., Zhong, C. (2007) Multicompartment micelles from star and linear triblock copolymer blends. *J. Phys. Chem. B*, **111**, 13675–13682.
128. Xin, J., Liu, D., Zhong, C. (2009) Morphology and structure control of multicompartment micelles from triblock copolymer blends. *J. Phys. Chem. B*, **113**, 9364–9372.
129. Cai, C., Li, Y., Lin, J., Wang, L., Lin, S., Wang, X.-S., Jiang, T. (2013) Simulation-assisted self-assembly of multicomponent polymers into hierarchical assemblies with varied morphologies. *Angew. Chem., Int. Ed.*, **52**, 7732–7736.
130. Li, Y., Jiang, T., Lin, S., Lin, J., Cai, C., Zhu, X. (2015) Hierarchical nanostructures self-assembled from a mixture system containing rod–coil block copolymers and rigid homopolymers. *Sci. Rep.*, **5**, 10137.
131. Balazs, A. C., Emrick, T., Russell, T. P. (2006) Nanoparticle polymer composites: where two small worlds meet. *Science*, **314**, 1107–1110.
132. Lin, Y., Boeker, A., He, J., Sill, K., Xiang, H., Abetz, C., Li, X., Wang, J., Emrick, T., Long, S., Wang, Q., Balazs, A. C., Russell, T. P. (2005) Self-directed self-assembly of nanoparticle/copolymer mixtures. *Nature*, **434**, 55–59.
133. Thompson, R. B., Ginzburg, V. V., Matsen, M. W., Balazs, A. C. (2001) Predicting the mesophases of copolymer–nanoparticle composites. *Science*, **292**, 2469–2472.
134. Sides, S. W., Kim, B. J., Kramer, E. J., Fredrickson, G. H. (2006) Hybrid particle–field simulations of polymer nanocomposites. *Phys. Rev. Lett.*, **96**, 250601.
135. Langner, K. M., Sevink, G. J. A. (2012) Mesoscale modeling of block copolymer nanocomposites. *Soft Matter*, **8**, 5102–5118.
136. Yan, L.-T., Xie, X.-M. (2013) Computational modeling and simulation of nanoparticle self-assembly in polymeric systems: structures, properties and external field effects. *Prog. Polym. Sci.*, **38**, 369–405.
137. Lee, J.-Y., Thompson, R. B., Jasnow, D., Balazs, A. C. (2002) Entropically driven formation of hierarchically ordered nanocomposites. *Phys. Rev. Lett.*, **89**, 155503.
138. Li, Z., Hur, K., Sai, H., Higuchi, T., Takahara, A., Jinnai, H., Gruner, S. M., Wiesner, U. (2014) Linking experiment and theory for three-dimensional networked binary metal nanoparticles-triblock terpolymer superstructures. *Nat. Commun.*, **5**, 3247.
139. Hur, K., Hennig, R. G., Escobedo, F. A., Wiesner, U. (2010) Mesoscopic structure prediction of nanoparticle assembly and coassembly: theoretical foundation. *J. Chem. Phys.*, **133**, 194108–194119.
140. Hur, K., Hennig, R. G., Escobedo, F. A., Wiesner, U. (2012) Predicting chiral nanostructures, lattices and superlattices in complex multicomponent nanoparticle self-assembly. *Nano Lett.*, **12**, 3218–3223.
141. Kang, H., Detcheverry, F. A., Mangham, A. N., Stoykovich, M. P., Daoulas, K. Ch., Hamers, R. J., Müller, M., de Pablo, J. J., Nealey, P. F. (2008) Hierarchical assembly of nanoparticle superstructures from block copolymer–nanoparticle composites. *Phys. Rev. Lett.*, **100**, 148303.
142. Zhang, L., Lin, J. (2009) Hierarchically ordered nanocomposites self-assembled from linear-alternating block copolymer/nanoparticle mixture. *Macromolecules*, **42**, 1410–1414.
143. Zhu, X., Wang, L., Lin, J., Zhang, L. (2010) Ordered nanostructures self- assembled from block copolymer tethered nanoparticles. *ACS Nano*, **4**, 4979–4988.

144. Ploshnik, E., Langner, K. M., Halevi, A., Ben-Lulu, M., Müller, A. H. E., Fraaije, J. G. E. M., Sevink, G., Shenhar, R. (2013) Hierarchical structuring in block copolymer nanocomposites through two phase-separation processes operating on different time scales. *Adv. Funct. Mater.*, **23**, 4215–4226.

145. Yu, X. F., Zhong, S., Li, X. P., Tu, Y. F., Yang, S. G., Van Horn, R. M., Ni, C. Y., Pochan, D. J., Quirk, R. P., Wesdemiotis, C., Zhang, W.-B., Cheng, S. Z. D. (2010) A giant surfactant of polystyrene-(carboxylic acid-functionalized polyhedral oligomeric silsesquioxane) amphiphile with highly stretched polystyrene tails in micellar assemblies. *J. Am. Chem. Soc.*, **132**, 16741–16744.

146. Wang, Z., Li, Y. W., Dong, X.-H., Yu, X. F., Guo, K., Su, H., Yue, K., Wesdemiotis, C., Cheng, S. Z. D., Zhang, W.-B (2013) Giant gemini surfactants based on polystyrene-hydrophilic polyhedral oligomeric silsesquioxane shape amphiphiles: sequential "click" chemistry and solution self-assembly. *Chem. Sci.*, **4**, 1345–1352.

147. Ma, S., Hu, Y., Wang, R. (2015) Self-assembly of polymer tethered molecular nanoparticle shape amphiphiles in selective solvents. *Macromolecules*, **48**, 3112–3120.

148. Estridge, C. E., Jayaraman, A. (2014) Assembly of diblock copolymer functionalized spherical nanoparticles as a function of copolymer composition. *J. Chem. Phys.*, **140**, 144905.

149. Zhang, L., Lin, J., Lin, S. (2007) Self-assembly behavior of amphiphilic block copolymer/nanoparticle mixture in dilute solution studied by self-consistent-field theory/density functional theory. *Macromolecules*, **40**, 5582–5592.

150. Hall, D. M., Lookman, T., Banerjee, S. (2009) Non-equilibrium particle–field simulations of polymer–nanocomposite dynamics. *Chem. Eng. Sci.*, **64**, 4754–4757.

151. Tang, Q.-y., Ma, Y.-q. (2009) Self-assembly of rod-shaped particles in diblock-copolymer templates. *J. Phys. Chem. B*, **113**, 10117–10120.

152. Whitelam, S., Jack, R. L. (2015) The statistical mechanics of dynamic pathways to self-assembly. *Annu. Rev. Phys. Chem.*, **66**, 143–163.

153. Zhang, L., Sevink, A., Schmid, F. (2011) Hybrid lattice Boltzmann/dynamic self-consistent field simulations of microphase separation and vesicle formation in block copolymer systems. *Macromolecules*, **44**, 9434–9447.

154. Jain, A., Bollinger, J. A., Truskett, T. M. (2014) Inverse methods for material design. *AIChE J.*, **60**, 2732–2740.

155. Hannon, A. F., Gotrik, K. W., Ross, C. A., Alexander-Katz, A. (2013) Inverse design of topographical templates for directed self-assembly of block copolymers. *ACS Macro Lett.*, **2**, 251–255.

156. Hannon, A. F., Ding, Y., Bai, W., Ross, C. A., Alexander-Katz, A. (2014) Optimizing topographical templates for directed self-assembly of block copolymers via inverse design simulations. *Nano Lett.*, **14**, 318–325.

11

Nucleation in Colloidal Systems: Theory and Simulation

Ran Ni

School of Chemical and Biomedical Engineering, Nanyang Technological University, Singapore

11.1 Introduction

As we know from high school physics courses, matter exists in different thermodynamic "phases," e.g. gas, liquid or solid, which are different states of aggregation with qualitatively different properties. By heat transfer, a thermodynamic system can undergo a "phase transition" from one state to another one. Phase transition occurs almost everywhere in our daily life. For instance, the melting of an ice cube is the phase transition process of water from solid to liquid, of which the reverse process is the freezing of liquid water. These phase transitions are normally triggered by nucleation. For example, water vaporization is normally triggered by the nucleation of bubbles, of which the understanding is also of great interest to the fabrication of many materials, such as polymer foams [1]. Nucleation is a rare event that is normally induced by thermal fluctuations. The detection and study of nucleation on the molecular scale are very difficult, since molecules are very small and move very fast. Colloids are particles within the size range of a nanometer to several micrometers that move around in a solvent. More than a hundred years ago, Perrin pointed out that the motion of colloids is caused by collisions with the molecules of the solvent, which causes the colloids to move in an irregular fashion, i.e. Brownian motion [2]. Due to the Brownian motion, colloidal particles can explore the phase space, and self-assemble similarly to atomic and molecular systems. Moreover, colloidal particles are several orders of magnitude larger than atoms or molecules, and they move much slower. This makes it possible to track the motion of the particles in real time using optical microscopy and to study the dynamics of physical processes, such as nucleation, which are very difficult to investigate in atomic or molecular systems. Therefore, colloids are outstanding model systems for studying the physics of nucleation.

In past decades, colloidal systems have been employed extensively in studying the physics of nucleation, and remarkable progress and new insights in nucleation have been obtained. However, even for the arguably simplest particle model systems, i.e. colloidal hard-sphere systems, the theoretical prediction on the crystal nucleation rate differs by orders of magnitude from the experimental values [3], which is still a controversial issue in the community of soft matter physics [4–7]. Partially because of the controversies on

Self-Assembling Systems: Theory and Simulation, First Edition. Edited by Li-Tang Yan.

calculating the nucleation rates, a number of new simulation methods, e.g. forward flux sampling [8], transition path sampling, transition interface sampling [9], etc., have been developed and employed to improve our understanding of nucleation from the theoretical point of view. In this chapter, we will review the recent progress of nucleation study in colloidal systems by using computer simulations, which includes the classical description of nucleation, i.e. classic nucleation theory, the state-of-the-art simulation methods for studying nucleation, and the new nucleation phenomena observed that violate our classic understanding of nucleation as well as the not well explored yet important future directions of nucleation study.

11.2 Theory of Nucleation

Nucleation is the onset of a first-order phase transition, which occurs when the first-order derivative of the free energy with respect to a thermodynamic variable is discontinuous. For example, the gas–liquid phase transition is a typical first-order transition, which has a discontinuity in the density ρ. Let us take the van der Waals fluid as an example. In his thesis of 1873, van der Waals proposed two correction terms to the ideal gas law $p = Nk_BT/V$ [10]. First, he argued that the actual volume available to a molecule is smaller than the total volume V of the container due to the excluded volume effect between molecules. Second, he argued that the attractions between the molecules reduce the pressure p by an amount $-a\rho^2$, where $a > 0$ is a measure of the attraction between the molecules. Therefore, van der Waals wrote

$$p = \frac{Nk_BT}{V - Nb} - a\rho^2 = \frac{\rho k_BT}{1 - \rho b} - a\rho^2, \tag{11.1}$$

with the two phenomenological parameters a and b. He also found that there is a critical temperature $T_c = 8a/27bk_B$, below which the system undergoes a gas–liquid phase transition upon increasing the density. A typical equation of state for the system of a van der Waals fluid at temperature $T < T_c$ is shown in Figure 11.1.

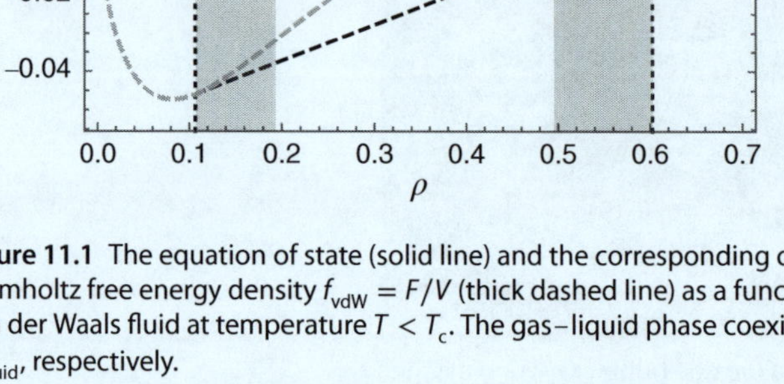

Figure 11.1 The equation of state (solid line) and the corresponding conveniently shifted and scaled Helmholtz free energy density $f_{vdW} = F/V$ (thick dashed line) as a function of density ρ for a system of van der Waals fluid at temperature $T < T_c$. The gas–liquid phase coexistence densities are ρ_{gas} and ρ_{liquid}, respectively.

By applying a common tangent construction on the free energy density curve as shown in Figure 11.1, one finds that the gas–liquid phase coexistence densities are ρ_{gas} and ρ_{liquid}, respectively, also called the *binodal* points. When the density of the system ρ lies well inside the region $\rho_{gas} < \rho < \rho_{liquid}$, the system is metastable (supersaturated or undersaturated), and in equilibrium it phase separates into a gas phase with density ρ_{gas} and a liquid phase with density ρ_{liquid}. The phase transition is triggered by fluctuations in the system. Moreover, for each isotherm $T < T_c$, there are two points where $\partial p/\partial \rho = 0$ equivalent to $\partial \rho/\partial p = \infty$, which means that an infinitely small fluctuation in the pressure can induce huge density fluctuations. These two points are called *spinodal* points, and are the stability limits of the metastable phases. Thus, in the shaded regions of Figure 11.1, the gas–liquid phase transition needs to be generated by finite and localized fluctuations, which is called *nucleation*. During nucleation, the metastable phase has to overcome a finite free energy barrier, after which the nuclei can grow spontaneously.

Moreover, when we talk about nucleation, it is important to distinguish between *heterogeneous* and *homogeneous* nucleation. Heterogeneous nucleation happens when it can be assisted by a heterogeneity in the system, such as walls or impurities. For instance, when we put a finger into a cup of Sprite as shown in Figure 11.2, there are more bubbles of carbon dioxide on the finger compared to the bulk phase, as nucleation of carbon dioxide bubbles is promoted by the interface created by the finger, i.e. heterogeneous nucleation. In contrast, homogeneous nucleation occurs due to spontaneous fluctuations in the bulk phase. Although heterogeneous nucleation is more likely to happen in real life, homogeneous nucleation is not just a theoretical simplification for studying the physics of phase transitions. There are still a lot of situations in the real world where homogeneous nucleation dominates, such as condensation in supersonic nozzles [11], explosions that occur when a cold liquid contacts a much hotter one [12], formation of heavily microcrystallized ceramics [13], (nano) particle synthesis, etc. Furthermore, experimental techniques have advanced significantly in recent years, which makes it possible to study homogeneous nucleation in experiments.

Figure 11.2 Nucleation of carbon dioxide bubbles around a finger.

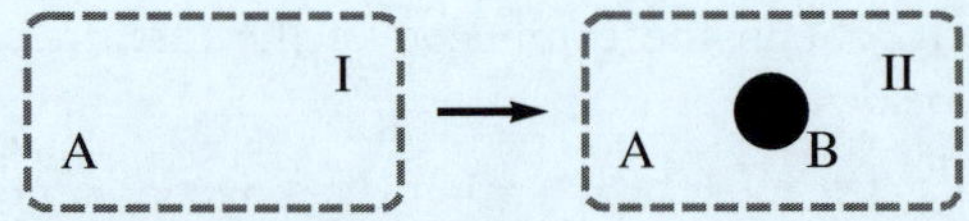

Figure 11.3 Illustration of the nucleation of phase B in a metastable phase A.

Nevertheless, the mechanism of homogeneous nucleation is still an open question, and, as mentioned before, even for a simple system of hard spheres, there is still an ongoing debate on the discrepancy in the measured nucleation rates between experimental and theoretical methods [3, 5–7]. This makes homogeneous nucleation an interesting and challenging topic. In the following, we briefly derive a commonly used theory to describe homogeneous nucleation, i.e. classical nucleation theory (CNT), which was first formulated by Volmer and Weber [14].

11.2.1 Free Energy Barrier

We consider a metastable phase A, e.g., a gas phase, at the thermodynamic condition where phase B, e.g., a liquid phase, is the stable phase. As shown in Figure 11.3, droplets of phase B exist in phase A, and they may grow and shrink due to thermal fluctuations.

In the following, we first consider a homogeneous system containing only the metastable phase A denoted as system I (left in Figure 11.3) and a system II containing the metastable phase A and a droplet (or a cluster) of phase B (right in Figure 11.3). The temperature of both systems is fixed and equal to T.

The internal energy of the homogeneous system I can be written as

$$U^{\mathrm{I}} = TS^{\mathrm{I}} - P^{\mathrm{I}}V^{\mathrm{I}} + \mu_A^{\mathrm{I}}N, \tag{11.2}$$

where T is the temperature of the system, S^{I} the total entropy, P^{I} the pressure, V^{I} the total volume, μ_A^{I} the chemical potential and N the total number of particles. The internal energy of system II, which contains the metastable phase A and a cluster of phase B, is

$$U^{\mathrm{II}} = TS^{\mathrm{II}} - P_A^{\mathrm{II}}V_A^{\mathrm{II}} - P_B^{\mathrm{II}}V_B^{\mathrm{II}} + \mathcal{A}\gamma + \mu_A^{\mathrm{II}}N_A + \mu_B^{\mathrm{II}}N_B, \tag{11.3}$$

where $P_{A,B}^{\mathrm{II}}$, $V_{A,B}^{\mathrm{II}}$ and $\mu_{A,B}^{\mathrm{II}}$ are the pressure, volume and chemical potential of phase A, B, respectively, with the superscripts indicating the values in system II, γ the interfacial tension between A and B with $\mathcal{A}$ the surface area of cluster B, and N_A and N_B are the number of particles of phase A and B in system II, respectively.

If we keep the pressure in the metastable phase A fixed, i.e. $P^{\mathrm{I}} = P_A^{\mathrm{II}} = P$, and the total number of particles constrained, i.e. $N = N_A + N_B$, then the Gibbs free energy in systems I and II are

$$G^{\mathrm{I}} = U^{\mathrm{I}} - TS^{\mathrm{I}} + P^{\mathrm{I}}V^{\mathrm{I}} = \mu_A^{\mathrm{I}}N, \tag{11.4}$$

$$\begin{aligned} G^{\mathrm{II}} &= U^{\mathrm{II}} - TS^{\mathrm{II}} + P_A^{\mathrm{II}}(V_A^{\mathrm{II}} + V_B^{\mathrm{II}}) \\ &= (P - P_B^{\mathrm{II}})V_B^{\mathrm{II}} + \gamma\mathcal{A} + \mu_A^{\mathrm{II}}N_A + \mu_B^{\mathrm{II}}N_B. \end{aligned} \tag{11.5}$$

Furthermore, as the pressure and temperature of the metastable phase A are fixed, the chemical potential is constant, i.e. $\mu_A^{\mathrm{I}} = \mu_A^{\mathrm{II}}$, and the Gibbs free energy difference can be written as

$$\begin{aligned} \Delta G &= G^{\mathrm{II}} - G^{\mathrm{I}} \\ &= (P - P_B^{\mathrm{II}})V_B^{\mathrm{II}} + \mathcal{A}\gamma + [\mu_B^{\mathrm{II}}(P_B^{\mathrm{II}}) - \mu_A^{\mathrm{I}}(P_A)]N_B. \end{aligned} \tag{11.6}$$

To obtain the expression of the free energy in CNT, we make a few assumptions below:

1) The interfacial tension γ is independent of the size of the cluster, or $\gamma = \gamma_\infty$, with γ_∞ the interfacial tension of an infinite cluster or the planar interface.
2) The cluster is incompressible, meaning that its density ρ does not change with pressure. This assumption is valid for the nucleation of a denser phase from a dilute phase, e.g. gas–liquid or fluid–crystal nucleation. Using the Gibbs–Duhem equation, the chemical potential of phase B can be written as

$$\mu_B^{\mathrm{II}}(P_B^{\mathrm{II}}) = \mu_B^{\mathrm{II}}(P) + \int_P^{P_B^{\mathrm{II}}} \frac{1}{\rho(P')} \mathrm{d}P' \\ = \mu_B^{\mathrm{II}}(P) + \frac{P_B^{\mathrm{II}} - P}{\rho_B}, \tag{11.7}$$

where ρ_B is the density of phase B at pressure P.

Therefore, we can rewrite Equation 11.6 into

$$\Delta G(N_B) = \mathcal{A}(N_B)\gamma_\infty + [\mu_B^{\mathrm{II}}(P) - \mu_A^{\mathrm{I}}(P)]N_B \\ = \mathcal{A}(N_B)\gamma_\infty - |\Delta\mu| N_B, \tag{11.8}$$

where $\Delta\mu = \mu_B^{\mathrm{II}}(P) - \mu_A^{\mathrm{II}}(P)$ is the chemical potential difference between the two phases. Given that phase B is more stable than phase A, we have $\Delta\mu < 0$. In Equation (11.8), the surface area of the cluster $\mathcal{A}$ depends on the cluster size N_B and its shape. CNT assumes that the shape of the cluster is roughly spherical, and hence $\mathcal{A} = 4\pi R^2$ where R is the radius of the spherical cluster. The Gibbs free energy of a spherical cluster with radius R is given by

$$\Delta G(R) = 4\pi R^2 \gamma_\infty - \frac{4}{3}\pi R^3 \rho_B |\Delta\mu|, \tag{11.9}$$

which contains two terms:

- a "surface" term $4\pi R^2 \gamma_\infty$, that takes into account the free energy cost of creating an interface between phase A and B;
- a "volume" term $-\frac{4}{3}\pi R^3 \rho_B |\Delta\mu|$, that indicates the fact that phase B is more stable than phase A, and can be interpreted as the driving force for the formation of phase B.

A typical representation of ΔG as a function of the cluster radius R is shown in Figure 11.4.

One finds that the free energy ΔG goes through a maximum at the critical cluster size

$$R^* = \frac{2\gamma_\infty}{\rho_B |\Delta\mu|}, \tag{11.10}$$

beyond which the nuclei can spontaneously grow. The height of the free energy barrier reads

$$\Delta G^* = \frac{16\pi}{3} \frac{\gamma_\infty^3}{(\rho_B |\Delta\mu|)^2}. \tag{11.11}$$

We note that CNT always predicts a finite free energy barrier, but the region where nucleation and growth can happen is bounded by the spinodal as shown in Figure 11.1, which suggests that CNT might not be valid in describing nucleation close to spinodals.

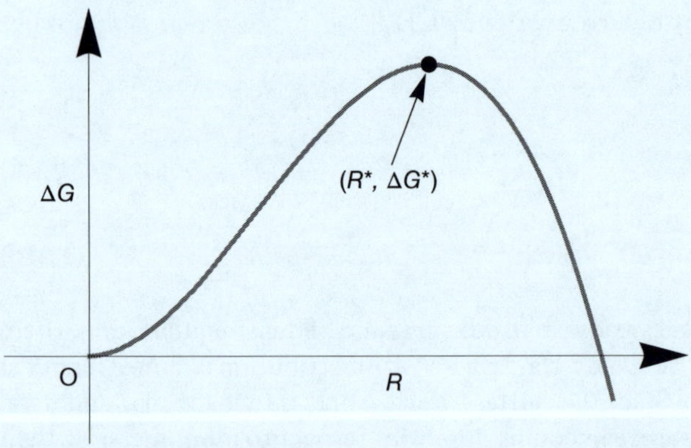

Figure 11.4 Gibbs free energy ΔG of a cluster as a function of the cluster radius R according to the classic nucleation theory where R^* and ΔG^* are the critical cluster and the height of the free energy barrier, respectively.

11.2.2 Kinetics of Nucleation

The kinetics of nucleation is generally interpreted via a phenomenological reaction rate theory, which was first formulated by Volmer and Weber [14].

We assume that clusters of phase B slowly grow or shrink via attachment or detachment of single particles:

$$B_{n-1} + B_1 \underset{b_{-,n}}{\overset{f_{+,n-1}}{\rightleftarrows}} B_n,$$

$$B_n + B_1 \underset{b_{-,n+1}}{\overset{f_{+,n}}{\rightleftarrows}} B_{n+1},$$

where B_{n-1} is a cluster of $n-1$ particles, B_1 a cluster of one particle (monomer), $f_{+,n-1}$ and $b_{-,n}$ are the attachment and detachment rates of a single monomer to a cluster of $n-1$ and from a cluster of n particles, respectively. This assumption is proposed on the basis that the concentration of monomers is much higher than that of dimers, trimers, etc., and the interaction between the growing clusters is negligible (ideal-gas-like) [15]. Hence, reactions of clusters with dimers, trimers, etc. are infrequent compared with single particle attachment and detachment. The time-dependent cluster distribution $N_n(t)$ obeys the following master equation:

$$\frac{dN_n(t)}{dt} = N_{n-1}(t)f_{+,n-1} + N_{n+1}(t)b_{-,n+1} - [N_n(t)f_{+,n} + N_n(t)b_{-,n}]. \tag{11.12}$$

The nucleation rate for a cluster size n is the time-dependent flux of clusters that reaches n:

$$I_{n,t} = N_n(t)f_{+,n} - N_{n+1}(t)b_{-,n+1}. \tag{11.13}$$

We now assume that the system is in a steady state, in which the cluster size distribution does not change in time, i.e. the nucleation rate is a constant:

$$I = N_n^s f_{+,n} - N_{n+1}^s b_{-,n+1}. \tag{11.14}$$

The equations can be solved by recurrence (see Ref. [16]):

$$I = N_1^s \left[\sum_{n=1}^{\infty} \frac{1}{f_{+,n} \xi_n} \right]^{-1}, \tag{11.15}$$

where

$$\xi_n = \prod_{i=1}^{n-1} \frac{f_{+,i}}{b_{-,i+1}} \quad \text{for} \quad n > 1. \tag{11.16}$$

The fluid is, due to the very small steady-state flux, in a kind of quasi-equilibrium, where for small cluster sizes $n \ll n^*$, the steady-state cluster size distribution is almost identical to the equilibrium distribution. Since the clusters of size n only interact with monomers, we assume that clusters of size n are in equilibrium with respect to monomers; we then find that

$$nN_1 \overset{K}{\rightleftarrows} N_n, \tag{11.17}$$

where N_1 and N_n are the equilibrium distributions of cluster sizes 1 and n, respectively, and K is the equilibrium constant. The ratio of products ξ_n in Equation 11.16 is just the equilibrium constant K [16]. The equilibrium constant K for this reaction is simply given by the free energy of forming a cluster with size n, and we find

$$\xi_n = K = \exp\ [-\beta \Delta G(n)]. \tag{11.18}$$

Thus, Equation 11.15 can be rewritten as

$$I = N_1 \left\{ \sum_{n=1}^{\infty} \frac{1}{f_{+,n} \exp\ [-\beta \Delta G(n)]} \right\}^{-1}. \tag{11.19}$$

In order to calculate the nucleation rate I, Becker and Döring made several approximations [17]:

1) The terms corresponding to the clusters near the top of the free energy barrier dominate the summation in Equation 11.19.
2) The shape of $\Delta G(n)$ around n^* can be approximated by a Taylor expansion at the top of the free energy barrier:

$$\Delta G(n) = \Delta G(n^*) + \frac{1}{2} \Delta G(n^*)''(n - n^*)^2.$$

3) $f_{+,n}$ is further replaced by f_{+,n^*}.
4) The sum gets replaced by an integral from $n - n^* = -\infty$ and $n - n^* = \infty$, by considering N_n to be a continuous function of n.

We then find the final expression of the steady-state nucleation rate,

$$I = N_1 f_{+,n^*} \left(\frac{|\Delta G''(n^*)|}{2\pi k_B T} \right)^{1/2} \exp\ [-\beta \Delta G(n^*)], \tag{11.20}$$

where $\Delta G''(n^*)$ is the second-order derivative of the Gibbs free energy with respect to the cluster size at the top of the free energy barrier n^*, and

$$Z = \left(\frac{|\Delta G''(n^*)|}{2\pi k_B T} \right)^{1/2} \tag{11.21}$$

is called the Zeldovitch factor. On the basis of this theory, Bennett [18] and Chandler [19] proposed a two-step scheme to calculate the nucleation rate in simulations: the free energy barrier $\Delta G(n)$ can be obtained by performing Monte Carlo simulations using the umbrella sampling technique, and molecular dynamics simulations starting from configurations on the top of the free energy barrier can be employed to compute the attachment rate f_{+,n^*}.

The most important assumption in this theory is that when the cluster size is smaller than the critical cluster size, the system is in quasi-equilibrium, which could be problematic in some cases. For instance, in systems of binary mixtures, dynamical heterogeneities may make the kinetic pathways of nucleation out of equilibrium [20], and it may be influenced by the order parameter used in the umbrella sampling simulations [21]. Later in this chapter, we will discuss the effect of the order parameter on the nucleation of binary hard-sphere mixtures.

11.2.3 Equilibrium Distribution of Cluster Sizes

In order to determine the nucleation rate, we need to calculate the probability of finding a critical cluster, see Equation 11.20. To this end, we consider a (meta)stable fluid in a cubic box consisting of N particles at pressure P and temperature T. The partition function is given by

$$Q(N,P,T) = \frac{\beta P}{\Lambda^{3N} N!} \int \mathrm{d}V V^N \exp\,(-\beta PV) \int \mathrm{d}\mathbf{s}^N \exp\,(-\beta \mathcal{U}), \tag{11.22}$$

where Λ is the de Broglie wavelength, and $\mathbf{s}^N$ are the factional coordinates of the particles. We define an order parameter of the configuration $(\mathbf{s}^N, V)$ as $f(\mathbf{s}^N, V) = n$, where n is the size of the largest cluster. This order parameter can be used to calculate the nucleation rate according to the transition state theory, when it can distinguish the two different states, i.e. fluid and crystal. The partition function of a system where the largest cluster size is n can be written as

$$Q_n(N,P,T) = \frac{\beta P}{\Lambda^{3N} N!} \int \mathrm{d}V V^N \exp\,(-\beta PV) \int \mathrm{d}\mathbf{s}^N \exp\,(-\beta \mathcal{U})\delta_{\mathrm{K}}[f(\mathbf{s}^N, V) - n], \tag{11.23}$$

where δ_{K} is the Kronecker delta function. Then the free energy difference between a system of N particles containing a largest cluster of size n and the (meta)stable fluid phase is

$$\begin{aligned}\Delta G(n)_{NPT} &= -k_{\mathrm{B}} T \ln\,\left[\frac{Q_n(N,P,T)}{Q(N,P,T)}\right] \\ &= -k_{\mathrm{B}} T \ln\,\langle \delta_{\mathrm{K}}[f(\mathbf{s}^N, V) - n]\rangle_{NPT}.\end{aligned} \tag{11.24}$$

The formation of a large cluster in the fluid phase is a rare event. When the cluster size is around the critical size, there is usually one large cluster present in the system. Thus

$$\Delta G(n)_{NPT} = -k_{\mathrm{B}} T \ln\,\langle \delta_{\mathrm{K}}[f(\mathbf{s}^N, V) - n]\rangle_{NPT} = -k_{\mathrm{B}} T \ln\,\langle N_n\rangle_{NPT}, \tag{11.25}$$

where N_n is the number of clusters of size n. This is the free energy of finding a cluster of size n in a system of N particles, and we scale this to a unit system by adding a $k_{\mathrm{B}} T \ln\, N$

term in the free energy, since the nucleation rate is system size independent. The free energy barrier in a unit system is

$$\Delta G(n) = \Delta G(n)_{NPT} + k_{\mathrm{B}}T \ln\ N = -k_{\mathrm{B}}T \ln \left\langle \frac{N_n}{N} \right\rangle_{NPT}. \tag{11.26}$$

We note that we can approximate $N \simeq \sum_{k=0}^{\infty} N_k$, as the number of fluid-like particles N_0 is dominant. We rewrite Equation 11.26 as

$$\begin{aligned}\Delta G(n) &= -k_{\mathrm{B}}T \ln \left\langle \frac{N_n}{\sum_{k=0}^{\infty} N_k} \right\rangle_{NPT} \\ &= -k_{\mathrm{B}}T \ln\ P(n),\end{aligned} \tag{11.27}$$

where $P(n)$ is the probability distribution function of finding a cluster of size n. This can be calculated by Monte Carlo simulations with the umbrella sampling technique. However, we should note that in the actual calculation, one can still use Equation 11.26 to compute the free energy barrier.

Moreover, when calculating the free energy barrier of nucleation, one should make sure that there is only one critical cluster in the simulation box. Thus the system size should be large but *not* too large that multiple critical clusters may exist.

11.3 Order Parameter

As shown in the previous section, nucleation is the onset of the formation process of a new phase from a metastable mother phase. Therefore, to study nucleation, we need to use an order parameter, e.g. the size of a large cluster of the new phase, to distinguish the new phase from the mother phase. In this chapter, we mainly discuss studies on crystal nucleation, for which the local bond-order parameter is employed to differentiate between liquid-like and solid-like particles. The local bond-order parameter was originally proposed by Steinhardt et al. [22] and employed by ten Wolde et al. [23, 24] in the study of crystal nucleation in a Lennard-Jones system, and it has been widely used in many crystal nucleation studies.

In the calculation of the local bond-order parameter, a list of "neighbors" is determined for each particle. The neighbors of particle i include all particles within a radial distance r_{c} of particle i, and the total number of neighbors is denoted $N_{\mathrm{b}}(i)$. A bond orientational order parameter $q_{l,m}(i)$ for each particle is then defined as

$$q_{l,m}(i) = \frac{1}{N_{\mathrm{b}}(i)} \sum_{j=1}^{N_{\mathrm{b}}(i)} Y_{l,m}(\theta_{i,j}, \phi_{i,j}), \tag{11.28}$$

where $Y_{l,m}(\theta, \phi)$ are the spherical harmonics, $m \in [-l, l]$ and $\theta_{i,j}$ and $\phi_{i,j}$ are the polar and azimuthal angles of the center-of-mass distance vector $\mathbf{r}_{ij} = \mathbf{r}_j - \mathbf{r}_i$ with $\mathbf{r}_i$ the position vector of particle i. Solid-like particles are identified as particles for which the number of connections per particle $\xi(i)$ is at least ξ_{c} and where

$$\xi(i) = \sum_{j=1}^{N_{\mathrm{b}}(i)} H[d_l(i,j) - d_{\mathrm{c}}], \tag{11.29}$$

H is the Heaviside step function, d_c is the dot-product cutoff, and

$$d_l(i,j) = \frac{\sum_{m=-l}^{l} q_{l,m}(i) q_{l,m}^*(j)}{\left(\sum_{m=-l}^{l} |q_{l,m}(i)|^2\right)^{1/2} \left(\sum_{m=-l}^{l} |q_{l,m}(j)|^2\right)^{1/2}}. \tag{11.30}$$

A cluster contains all solid-like particles which have a solid-like neighbor in the same cluster. Thus each particle can be a member of only a single cluster.

The parameters contained in this algorithm include the neighbor cutoff r_c, the dot-product cutoff d_c, the critical value for the number of solid-like neighbors ξ_c, and the symmetry index for the bond orientational order parameter l. A typical choice of r_c is the location of the first local minimum of the radial distribution function $g(r)$ in the metastable fluid phase, which only includes the first neighbors of the particles. The choice of d_c and ξ_c depends on the type of crystal nucleated, and for details, one can further read Refs. [25, 26]. For example, the solid nucleus of a hard-sphere crystal is expected to have randomly stacked hexagonal order, and one can choose $d_c = 0.7$ and $\xi_c = 6$ with $l = 6$. Note that this order parameter does not distinguish between particles with different crystalline symmetries, e.g. face-centered cubic (FCC), hexagonal close packed (HCP) and body-centered cubic (BCC).

To further distinguish between particles with different crystalline symmetries, one can use the averaged local bond order parameter $\overline{q}_l$ and $\overline{w}_l$ proposed by Lechner and Dellago [27], which can identify each particle as FCC-like, HCP-like or BCC-like, provided the number of neighboring particles $N_b(i) \geq 10$:

$$\overline{q}_l(i) = \sqrt{\frac{4\pi}{2l+1} \sum_{m=-l}^{l} |\overline{q}_{lm}(i)|^2}, \tag{11.31}$$

$$\overline{w}_l(i) = \frac{\sum_{m_1+m_2+m_3=0} \begin{pmatrix} l & l & l \\ m_1 & m_2 & m_3 \end{pmatrix} \overline{q}_{lm_1}(i)\overline{q}_{lm_2}(i)\overline{q}_{lm_3}(i)}{\left(\sum_{m=-l}^{l} |\overline{q}_{lm}(i)|^2\right)^{3/2}}, \tag{11.32}$$

where

$$\overline{q}_{lm}(i) = \frac{1}{N_b(i)+1} \sum_{k=0}^{N_b(i)} q_{lm}(i). \tag{11.33}$$

The sum from $k = 0$ to $N_b(i)$ runs over all neighbors of particle i plus the particle i itself. While $q_{lm}(i)$ takes into account the structure of the first shell around particle i, the averaged $\overline{q}_{lm}(i)$ also contains the information of the structure of the second shell, which increases the accuracy of the crystal structure determination. As shown in Figure 11.5, in the $\overline{q}_6$–$\overline{q}_4$ or $\overline{w}_4$–$\overline{q}_4$ planes, the distribution for FCC-, HCP- and BCC-like particles can be well separated for systems of Lennard-Jones particles [27].

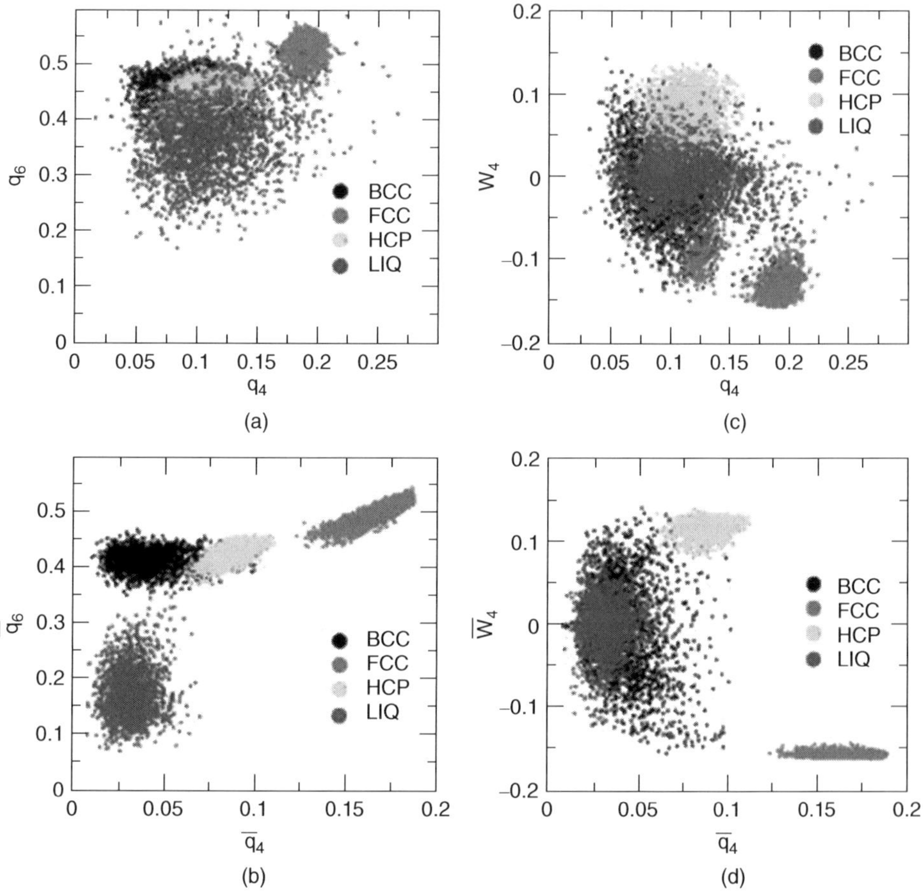

Figure 11.5 Top: Comparison between the q_4-q_6 plane (left) and the $\bar{q}_4-\bar{q}_6$ plane (right) for the Lennard-Jones system in three different crystalline structures and in the liquid phase. Each point corresponds to a particular particle, where 2000 points from each structure were chosen randomly. Bottom: The q_4-w_4 plane (left) and the $\bar{q}_4-\bar{w}_4$ plane (right). This figure is reproduced from Ref. [27] with permission of the American Institute of Physics.

11.4 Simulation Methods for Studying Nucleation

Nucleation is normally a rare event triggered by spontaneous fluctuations. The most straightforward way of studying nucleation would be brute force molecular dynamics (MD) simulations, but using brute force MD simulation would be very inefficient at low supersaturations. Because the nucleation rate drops exponentially with decreasing supersaturation, the waiting time for a nucleation event occurring in an MD simulation increases dramatically, and the statistics would be very bad. Therefore, besides brute force MD simulation, we also introduce two other rare event methods for studying nucleation, namely umbrella sampling and forward flux sampling methods.

11.4.1 Brute Force Molecular Dynamics Simulations

In theory, using an MD simulation to determine nucleation rates is quite simple. Starting with an equilibrated fluid configuration, an MD simulation is used to evolve the system until the largest cluster in the system exceeds the critical nucleus size. The MD time associated with such an event is then measured and averaged over many initial configurations. The nucleation rate is given by

$$I = \frac{1}{\langle t \rangle V}, \tag{11.34}$$

where V is the volume of the system and $\langle t \rangle$ is the average time to form a critical nucleus. Measuring this time is relatively easy for moderate or low supersaturations where the nucleation times are relatively long compared to the nucleation event itself, which corresponds with a steep increase in the crystalline fraction of the system. For high supersaturations, however, pinpointing the time of a nucleation event is more difficult. Additionally, the precise details of the initial configuration can play a role at high supersaturations since the equilibration time of the fluid is of the same order of magnitude as the nucleation time. Hence, to calculate the nucleation rate, one should perform a sufficient number of MD simulations starting from different initial configurations.

11.4.2 Umbrella Sampling

Umbrella sampling is a technique developed by Torrie and Valleau to study systems where Boltzmann-weighted sampling is inefficient [28]. This method has been applied frequently to study rare events, such as nucleation [29], and specifically has been applied in the past to study the nucleation of hard-sphere crystals [3]. In general, umbrella sampling is used to examine parts of a configurational space that are inaccessible by traditional schemes, e.g. Metropolis Monte Carlo simulations. Typically, a biasing potential is added to the true interaction potential causing the system to oversample a region of the configuration space. The biasing potential, however, is added in such a way that it is easy to "unbias" the measurables.

In the case of nucleation, while it is simple to sample the fluid, crystalline clusters of larger sizes will be rare, and as such, impossible to sample on reasonable time scales. The typical biasing potential for studying nucleation is given by [23, 30]

$$U_{\text{bias}}[n(\mathbf{r}^N)] = \frac{\lambda}{2}[n(\mathbf{r}^N) - n_C]^2, \tag{11.35}$$

where λ is a coupling parameter, $n(\mathbf{r}^N)$ is the size of the largest cluster associated with configuration $\mathbf{r}^N$, and n_C is the targeted cluster size. By choosing λ carefully, the simulation will fluctuate around the part of configurational space with $n(\mathbf{r}^N)$ in the vicinity of n_C. The expectation value of an observable A is then given by

$$\langle A \rangle = \frac{\langle A/W(n(\mathbf{r}^N)) \rangle_{\text{bias}}}{\langle 1/W(n(\mathbf{r}^N)) \rangle_{\text{bias}}}, \tag{11.36}$$

where

$$W(n) = e^{-\beta U_{\text{bias}}(n)}. \tag{11.37}$$

Using this scheme to measure the probability distribution $P(n)$ for clusters of size n, the Gibbs free energy barrier can be determined by [31]

$$\beta\Delta G(n) = \text{constant} - \ln\ [P(n)]. \tag{11.38}$$

Many more details on this method are given elsewhere [31, 32].

As long as we have obtained the free energy barrier, the nucleation rate I can be determined by [3]

$$I = \kappa e^{-\beta\Delta G(n^*)}, \tag{11.39}$$

where

$$\kappa \approx \rho f_{n^*} \sqrt{\frac{|\beta\Delta G''(n^*)|}{2\pi}}, \tag{11.40}$$

n^* is the number of particles in the critical nucleus, ρ is the number density of the supersaturated fluid, f_{n^*} is the rate particles are attached to the critical cluster, and G'' is the second derivative of the Gibbs free energy barrier. Auer and Frenkel [3] showed that the attachment rate f_{n^*} could be related to the mean square deviation of the cluster size at the top of the barrier by

$$f_{n^*} = \frac{1}{2}\frac{\langle\Delta n^2(t)\rangle}{t}. \tag{11.41}$$

The mean square deviation (MSD) of the cluster size $\Delta n^2(t) = \langle n(t) - n^*\rangle$ can then be calculated by employing either a kinetic MC or MD simulation at the top of the barrier.

The mean square deviation, or variance, in the cluster size appearing in Equation 11.41 has both a short- and long-time behavior. At short times, fluctuations are due to particles performing Brownian motion around their average positions, while the long-time behavior is caused by rearrangements of particles required for the barrier crossings. The slope of the variance is large at short times where only the fast rattling is sampled. However, the longer the time the further the system has diffused away from the critical cluster size at the top of the nucleation barrier. Auer [33] states that runs need to be selected that remain at the top of the barrier. However, when this is done the attachment rate is lower than when the average over all runs is taken since it excludes the runs that move off the barrier fast and have the largest attachment rate. This problem is analogous to determining the diffusion constant of a particle performing a random walk. By only including walks which remain in the vicinity of the origin, the measurement is biased and excludes trajectories which quickly move away from the origin. This result is an underestimation of the diffusion constant, and similarly, in this case, an underestimate of the attachment rate. Hence, in the calculation of attachment rate, one should include all trajectories starting from the top of the barrier. To illustrate this, we show the size of nucleus as a function of time for different trajectories starting from the top of the free energy barrier in the crystal nucleation of a hard-sphere system in Figure 11.6. One can see that the cluster can completely disappear or double in size within $0.3\tau_l$, where τ_l is the time that it takes a particle on average to diffuse over a distance equal to its diameter, i.e. $\tau_l = \sigma^2/(6D_l)$. From these trajectories, one can calculate the kinetic attachment rate from the mean square deviation as shown in Figure 11.7.

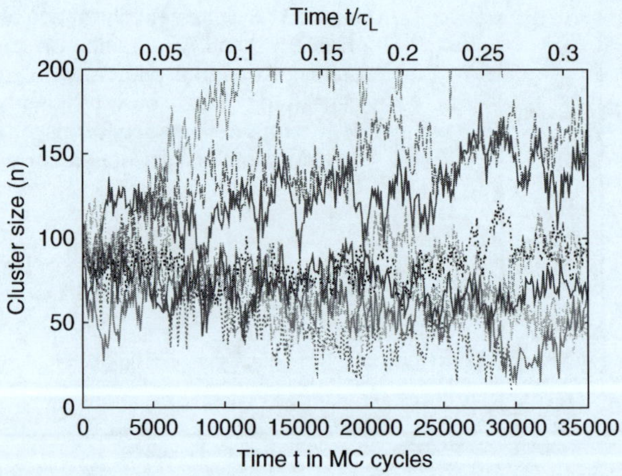

Figure 11.6 The cluster size ($n(t)$) as a function of time in MC cycles for a random selection of clusters that start at the top of the nucleation barrier.

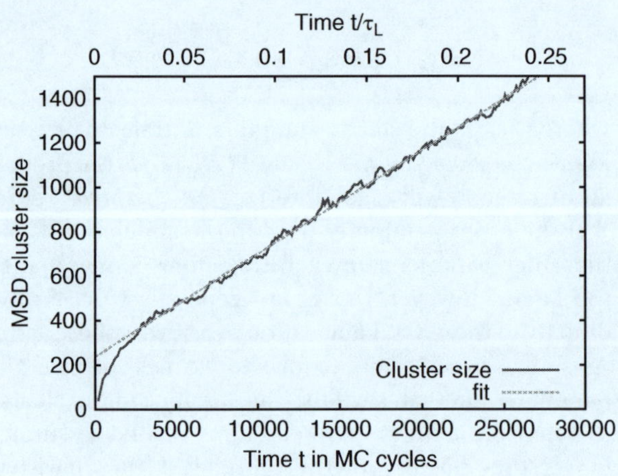

Figure 11.7 The mean squared deviation (MSD) of the cluster size $\langle \Delta n^2(t) \rangle$ as a function of time t in MC cycles. The cluster size has been measured every cycle and averaged over 100 cycles to reduce the short-time fluctuations. The slope of this graph is twice the attachment rate (Equation 11.41).

11.4.3 Forward Flux Sampling

In the classic nucleation theory and umbrella sampling method, one assumes that the process of nucleation is a quasi-equilibrium process, and the trajectory follows the minimum free energy pathway, which is not always true. Therefore, we introduce another rare event method, namely forward flux sampling (FFS), which can also be used to sample non-equilibrium phase transitions.

The forward flux sampling method was introduced by Allen et al. [8] in 2005 to study rare events and has since been applied to a wide variety of systems. Two review articles on the subject (Refs. [34, 35]) have appeared recently and provide a thorough overview of the method. As shown in Figure 11.8, the basic idea behind FFS is that the whole transition from state A (fluid) to state B (solid) can be divided into many steps, and there are a number of interfaces $\{\lambda_i, 0 \leq i \leq n\}$ in between A and B [34]. Then the nucleation

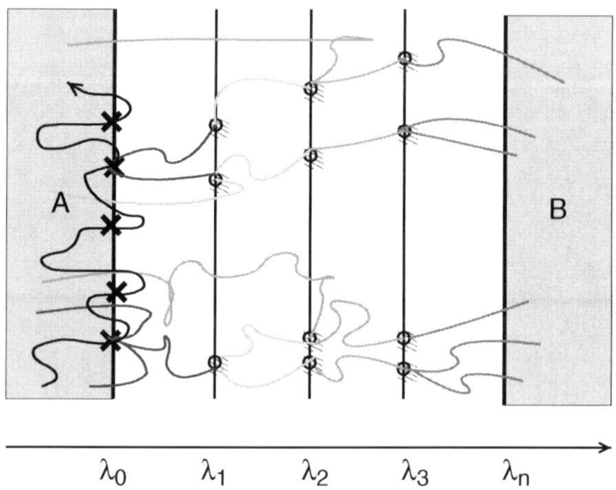

Figure 11.8 Schematic illustration of the FFS method. An ensemble of branched transition paths is generated simultaneously by firing M_i trial runs from randomly chosen configurations at each interface in turn.

rate from the fluid phase A to the solid phase B can be written as

$$k_{AB} = \Phi_{A\lambda_0} P(\lambda_N|\lambda_0) \tag{11.42}$$

$$= \Phi_{A\lambda_0} \prod_{i=0}^{N-1} P(\lambda_{i+1}|\lambda_i), \tag{11.43}$$

where $\Phi_{A\lambda_0}$ is the steady-state flux of trajectories leaving the A state and crossing the interface λ_0 in a volume V, and $P(\lambda_{i+1}|\lambda_i)$ is the probability that a configuration starting at interface λ_i will reach interface λ_{i+1} before it returns to the fluid (A). This assumes that the order parameter used in the simulation is perfect, and that the fluctuation of the order parameter in each trajectory connecting two neighboring interfaces is not too large. However, this is not generally true. For instance, if we apply this method directly to the crystal nucleation of a hard-sphere system, a number of problems appear. As shown in Figure 11.6, for short times the size of a cluster measured by the order parameter fluctuates wildly. The variance in the cluster size displays two different types of behavior: short-time fluctuations related to surface fluctuations of the cluster, and longer-time cluster growth (Figure 11.7). Thus, if we try to measure the flux $\Phi_{A\lambda_0}$ directly, we encounter difficulties due to these short-time surface fluctuations. In theory, FFS should be able to handle these types of fluctuations; however, they increase the amount of statistics necessary to properly measure the flux and the first probability window. In the second part of the FFS calculations, probabilities of the form $P(\lambda_{i+1}|\lambda_i)$ need to be determined. In calculating these probabilities it is important to be able to determine if a cluster has returned to the fluid (A). For pre-critical clusters we find large fluctuations of the order parameter, as shown in Figure 11.9, which can lead to a cluster being misidentified as the fluid (A). Specifically, in this figure the darkest trajectory (black) shows a cluster containing 43 particles that shrinks to around 5 particles before it returns to 40, and finally reaches a cluster size of 60 particles. Hence, if we had set $\lambda_0 = 5$, this trajectory would have been identified as melting back to the fluid phase (A). However, since the growth of a cluster from size 5 to 60 is a rare event in our system, we presume that this was simply a short-time fluctuation of the cluster and not a "real" melting of the instantaneously measured cluster. For pre-critical clusters, these fluctuations result in cluster sizes that are smaller than the cluster "really" is. We suggest that these fluctuations are

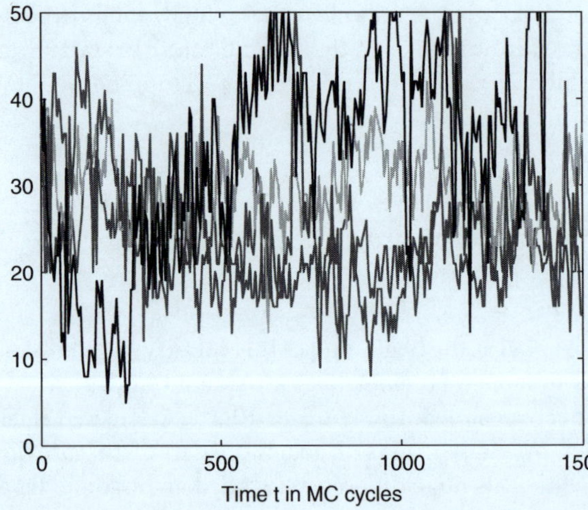

Figure 11.9 The cluster size as a function of time t in MC cycles for four random trajectories at pressure $\beta p\sigma^3 = 17$ starting with a cluster size of $n = 43$ using kinetic MC simulations with step size $\Delta_{KMC} = 0.1\sigma$ and measuring the order parameter every $\Delta t_{ord} = 5$ MC steps.

largely related to the difficulty that this order parameter has in distinguishing between solid- and fluid-like particles at the fluid–solid interface. For larger clusters, where the surface to volume ratio is small, this problem is minimal. However, for elongated or rough pre-critical clusters, where the surface to volume ratio is large, these surface fluctuations and rearrangements are important, and can cause problems in measuring the order parameter.

Thus, to try to address these problems, we suggest a modification to the original FFS method [5]. Essentially, we regroup the elements of the rate calculation such that

$$k_{AB} = \tilde{\Phi}_{A\lambda_1} \prod_{i=1}^{N-1} P(\lambda_{i+1}|\lambda_i), \tag{11.44}$$

where

$$\tilde{\Phi}_{A\lambda_1} = \Phi_{A\lambda_0} P(\lambda_1|\lambda_0). \tag{11.45}$$

We note that if λ_1 is chosen such that it is a relatively rare event for trajectories starting in A to reach λ_1, then

$$\tilde{\Phi}_{A\lambda_1} \approx \frac{1}{\langle t_{A\lambda_1} \rangle V}, \tag{11.46}$$

where $\langle t_{A\lambda_1} \rangle$ is the average time it takes a trajectory in A to reach λ_1. The approximation made here, in contrast to normal FFS simulations, is that the time the system spends with an order parameter greater than λ_1 is negligible. Since even reaching this interface is a rare event, this approximation should have a minimal effect on the resulting rate. Additionally, in this way we are relatively free to place the first interface (λ_0) anywhere under λ_1. One can choose to use $\lambda_0 = 1$ to minimize the effect of fluctuations, as seen in Figure 11.9, on the probability of reaching the following interface. Here we assume that any crystalline order in a system with an order parameter of 1 likely does not arise from fluctuation of a much larger cluster, but rather is very close to the fluid, and is expected to fully melt and not grow out to the next interface. In this manner one can start several parallel trajectories from the fluid phase in order to measure $\langle t_{A\lambda_1} \rangle$, stopping whenever the trajectory first hits interface λ_1.

11.5 Crystal Nucleation of Hard Spheres: Debates and Progress

In the seminal paper by Auer and Frenkel at the beginning of this century [3], the crystal nucleation rate for systems of hard spheres was calculated for the first time by using the umbrella sampling method. However, as shown in Figure 11.10, the calculated nucleation rates from simulations differ by many orders of magnitude from the previous experimental measurements [36–38]. This huge difference invoked an ongoing debate in the community of soft matter physics, and also motivated much further research using colloidal systems to study the physics of nucleation.

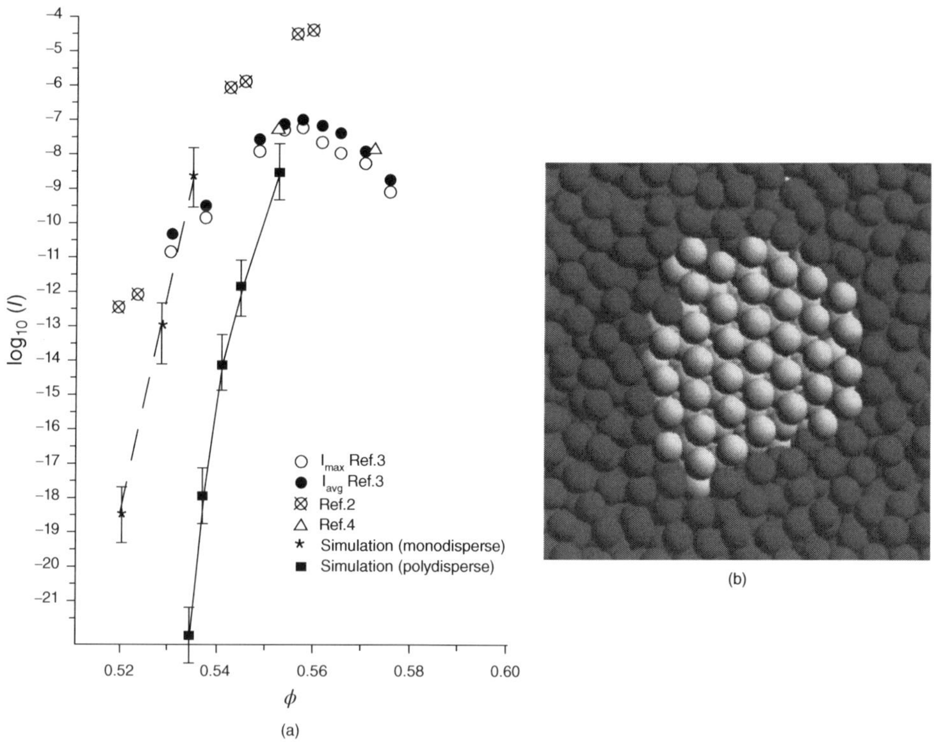

Figure 11.10 (a) Crystal nucleation rates I obtained from computer simulations for systems of monodisperse and polydisperse hard spheres as a function of packing fraction ϕ, and its comparison with experimental measurements. This figure is reproduced from Ref. [3], and the Ref.2, 3 and 4 in the figure correspond to Refs. [36], [37] and [38] in this chapter, respectively. (b) Snapshot of a cross-section of a critical nucleus of a hard-sphere crystal at a liquid volume fraction $\phi = 0.5207$. This figure is reproduced from Ref. [3] with permission of Nature Publishing Group.

Because of the large discrepancy between the nucleation rates obtained in simulations and experiments, a number of works were dedicated to understanding the physics of nucleation in hard-sphere systems. It was speculated that the crystal nucleation in hard-sphere systems may not be a quasi-equilibrium process, and one should use some simulation methods without the quasi-equilibrium assumption. Therefore, by using three different simulation methods, i.e. brute force MD simulation, umbrella sampling Monte Carlo simulation and forward flux sampling simulation, we systematically investigated the crystal nucleation in hard-sphere systems [5]. The brute force MD simulations and FFS simulations aim to follow the real physical trajectories of the system, which do not assume the nucleation to be a quasi-equilibrium process, and they are also weakly affected by the choice of order parameter, given that the order parameter is not too bad. Essentially, as shown in Figure 11.11, we found that within these estimated error bars (1–2 orders of magnitude), nucleation rates obtained from computer simulations are all in agreement, while the experimentally obtained rates show a markedly different behavior, particularly for low supersaturations where the difference between the simulations and experiments can be as large as 12 orders of magnitude. Note that here we plot the nucleation rates in units of long-time diffusion time τ_l to avoid the effect of diffusion time. This agreement was further confirmed by Schilling and co-workers, who simulated the crystal nucleation in a huge system of one million hard spheres to exclude the effect of system size [4, 39].

Moreover, to be even closer to experimental situations, in which the colloidal particles are charge stabilized and not perfect hard spheres, we also investigated the effect of softness in the potential [6]. We found that the softness of the potential has a minor effect on the crystal nucleation, if the potential is not extremely soft, i.e. penetrable, and with a proper normalization one can map the nucleation rates of soft repulsive spheres to systems of hard spheres [5, 7, 40]. Besides the softness of the potential, another issue

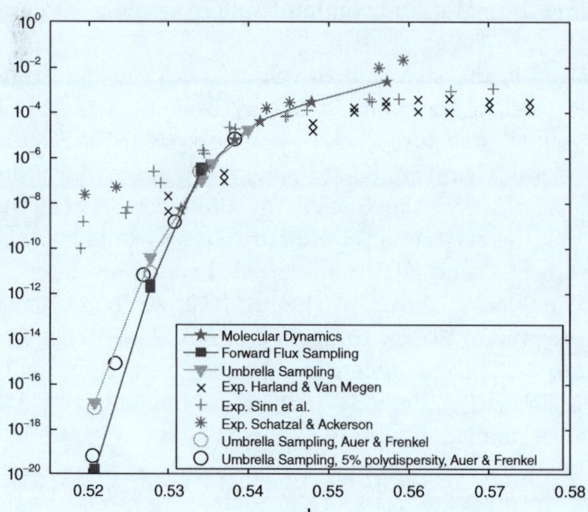

Figure 11.11 A comparison of the crystal nucleation rates of hard spheres as determined by the three methods described in this chapter, FFS, US and MD, with the experimental results from Refs. [36–38] and theoretical results from Ref. [3, 5].

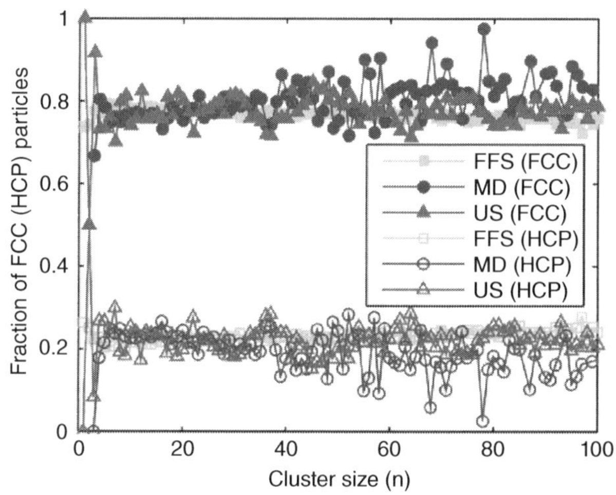

Figure 11.12 Fraction of particles identified as either FCC or HCP, respectively, in clusters produced via molecular dynamics (MD), forward flux sampling (FFS) and umbrella sampling (US) simulations as a function of cluster size *n*. All three methods agree and find the pre-critical clusters are predominately FCC.

is that in computer simulations, due to limitations of computing power, the hydrodynamic effect was mostly neglected, which may change the dynamics of nucleation and the resulting rate. In a recent work by Radu and Schilling, they simulated the crystal nucleation in a dense hard-sphere system with full hydrodynamic interactions, and they found that the viscosity of the solvent can, to some extent, speed up the nucleation rate [41]. However, due to the large uncertainty (1–2 orders of magnitude) in measuring nucleation rate in general, the effect of hydrodynamics on the nucleation rate remains unclear, and needs further more systematic study on hard-sphere systems with larger density ranges.

In addition to the nucleation rate, the structure of critical nuclei in hard-sphere systems is also not well understood. It was pointed out by Bolhuis et al. that the thermodynamically stable crystalline structure in hard-sphere systems is FCC, but the free energy difference between FCC and HCP hard-sphere crystals is only around $10^{-3}k_BT$ per particle [42]. Because of such a tiny free energy difference between these two crystals, you might guess that the crystal nuclei of hard spheres would likely be a uniform mixture of particles with FCC and HCP symmetries, i.e. random hexagonal close packed (RHCP) nuclei. However, as shown in Figure 11.12, we found that the critical nuclei of hard spheres are predominantly comprised of FCC-like particles, e.g. more than 80% of particles have FCC-like symmetry, which has been confirmed by using three completely different simulation methods: umbrella sampling, forward flux sampling and molecular dynamics simulations. Moreover, as shown in Figure 11.13a, most of the FCC-like particles are in the center of the nucleus, and the small number of HCP-like particles are found on the outside surface of the nucleus. Here we employed the order parameter proposed by Lechner and Dellago [27] to distinguish the FCC- and HCP-like particles, which may not be perfect. Therefore, it is fair to conclude that supersaturated hard-sphere fluids tend to nucleate FCC crystals instead of RHCP crystals. According to CNT, the free energy of forming a nucleus consists of both the volume and

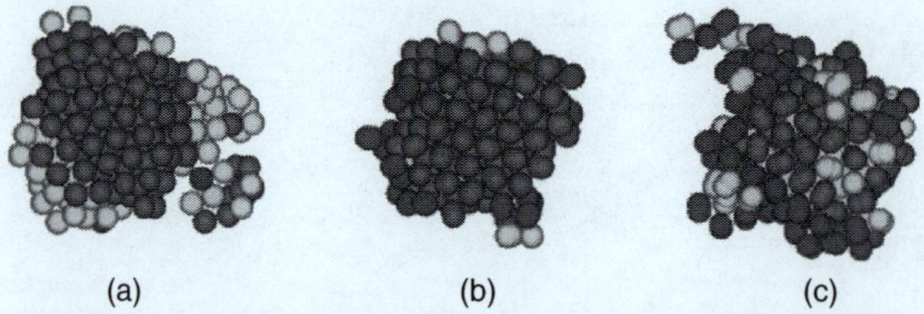

Figure 11.13 Typical configurations of critical nuclei for the plastic crystal nucleation of hard dumbbells with aspect ratios $L/\sigma = 0$ (a), 0.15 (b), and 0.3 (c) at supersaturation $|\Delta\mu| = 0.34k_BT$. The dark gray particles are FCC-like, the light black particles are HCP-like particles, while the light gray particles are undetermined. Here L is the center-to-center distance between two spheres of diameter σ in the dumbbell.

surface term, and this suggests that possibly the preference of nucleating FCC crystal is due to the low surface tension of FCC nuclei. By doing a free energy calculation, it was found that in systems of hard dumbbells consisting of two connected spheres with aspect ratio $L/\sigma < 0.3$, where L is the center-to-center distance between two spheres of diameter σ in the dumbbell, the system can form a stable plastic crystal phase, in which the positions of particles are on a crystal lattice while the orientation of each particle is random [43]. With increasing aspect ratio L/σ, the stability of the HCP plastic crystal increases, and when $L/\sigma \gtrsim 0.15$, the HCP plastic crystal becomes a thermodynamically stable crystal instead of the FCC plastic crystal. Interestingly, in the nucleation of plastic crystals in a system of hard dumbbells, we also found that with increasing L/σ, there are more HCP-like particles in the critical nucleus, as shown in Figure 11.13b,c. These suggest that the free energy difference between FCC and HCP crystals, although it is tiny, does play a subtle role in determining the structure of critical nuclei, and complete understanding of the physics behind this requires further investigation.

Following the existence of a critical nucleus in the supersaturated fluid, the system would crystallize quickly, for which the process is highly out of equilibrium. It was found by O'Malley and Snook [45] that during the growth of a post-critical hard-sphere nucleus, a number of morphologies were observed ranging from multiply twinned crystals to a crystal with a single stacking direction. As shown in Figure 11.14a, because of the stacking defects produced in the growth of the nucleus, the system can form a large crystalline cluster with RHCP symmetry. Interestingly, besides the RHCP crystal, the defects produced during growth can drive the formation of multiply twinned structures exhibiting in some cases structures with an intriguing five-fold symmetry [45, 47] as

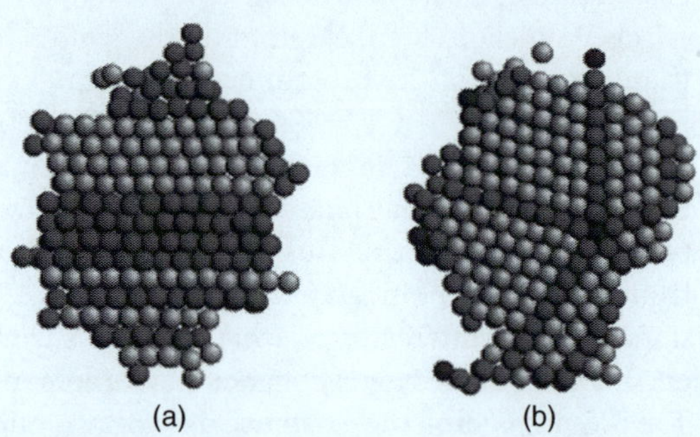

Figure 11.14 A cross-section through the center of several nuclei at the induction time. The light gray particles are face-centered cubic ordered while the dark gray particles are hexagonally close-packed ordered. (a) A nucleus possessing a single dominant stacking direction; (b) a multiply twinned nucleus orientated with its main five-fold axis into the page. This figure is reproduced from Ref. [45] with permission of the American Physical Society.

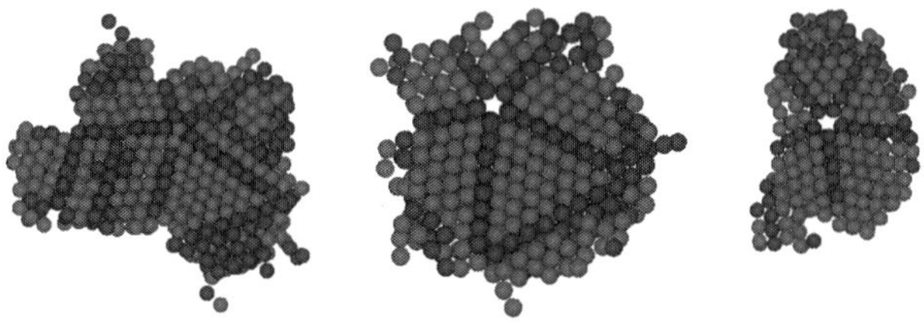

Figure 11.15 Typical configurations of the crystal structures for linear hard-sphere chains with chain length $N = 20$ (a) and for ring-like polymers with $N = 3$ (b) and $N = 5$ (c). Only crystalline spheres are shown here. The black and gray spheres are HCP-like and FCC-like particles, respectively.

shown in Figure 11.14b. These five-fold symmetric structures are not compatible with any close packing structure in hard-sphere crystals. Perhaps more surprisingly, even in the crystallization of hard-sphere polymers consisting of tangentially connected hard spheres, we found the existence of such five-fold symmetric patterns as shown in Figure 11.15. Additionally, it seems not dependent on the topology of the polymer, and we found it exists in all hard-sphere polymer systems that we studied, i.e. both linear and circular polymer systems, including even a system of circular hard-sphere polymers of length 3 [46].

All these discrepancies, debates and intriguing phenomena have made crystal nucleation in hard-sphere systems, i.e. arguably the simplest particle-based model system for studying phase transition, a big mystery, in which there are still many open questions. To fully understand crystal nucleation, we need to develop new simulation techniques as well as better experiments to perform more comprehensive investigations.

11.6 Two-Step Nucleation in Systems of Attractive Colloids

According to the theory of van der Waals, a relatively long-range attraction between particles can induce a gas–liquid transition [10], and there exists a critical temperature T_c above which the gas and liquid phases are indistinguishable. Below T_c, with increasing density the system undergoes a first-order gas–liquid transition, and the phase diagram of a typical example of such a system is shown in Figure 11.16a. If we decrease the range of attraction to the scale of 10% of the particle size (as shown in the inset of Figure 11.16b), which normally occurs in suspensions of protein particles, the gas–liquid critical point becomes metastable with respect to crystallization.

Ten Wolde and Frenkel found that when the temperature of the system is close to T_c, the nucleation of the protein crystal occurs via a two-step process [49]. As shown in Figure 11.17b, when $T = T_c$, the lowest free energy path to the critical nucleus (indicated by the dashed line) is strongly biased towards the axis of N_ρ, which suggests that the supersaturated gas phase first nucleates into a metastable liquid phase, in which stable crystal nuclei form. This is different from the conventional nucleation pathway far from the critical point, as shown in Figure 11.17a, where N_ρ and N_{crys} increase proportionally. More interestingly, when close to the metastable gas–liquid critical point, the crystal nucleation rate is enhanced by orders of magnitude by the critical fluctuations. The physics behind the existence of two-step nucleation is the competition of interfacial

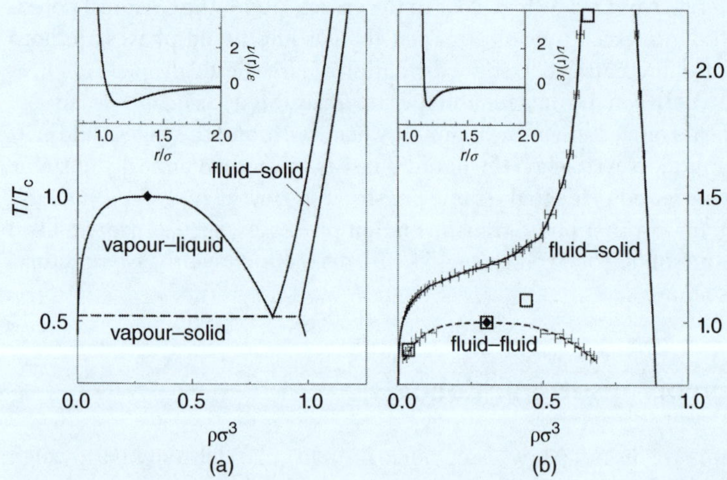

Figure 11.16 (a) Typical phase diagram of a molecular substance with a relatively long-range attractive interaction. This phase diagram corresponds to the Lennard-Jones 6-12 potential (solid curve in insert). This figure is reproduced from Ref. [48] with permission of the American Physical Society. (b) Typical phase diagram of colloids with short-range attraction, which is shown as the solid line in the inset [49]. This figure is reproduced from Ref. [49] with permission of The American Association for the Advancement of Science.

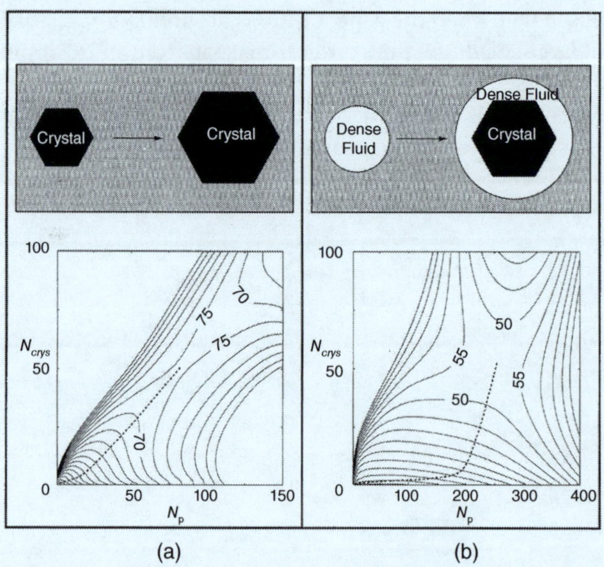

Figure 11.17 Contour plots of the free energy landscape along the path from the metastable fluid to the critical crystal nucleus for a system of spherical particles with short-range attraction. (a) The free energy landscape well below the critical temperature ($T/T_c = 0.89$). The lowest free energy path to the critical nucleus is indicated by a dashed curve. This curve corresponds to the formation and growth of a highly crystalline cluster. (b) As (a), but for $T = T_c$. Here, N_ρ and N_{crys} are the numbers of particles in the dense cluster and crystalline nucleus, respectively. This figure is reproduced from Ref. [49] with permission of The American Association for the Advancement of Science.

tension between difference phases. When close to the critical point, the chemical potential difference and the interfacial tension between the gas and liquid phases are both almost zero, and critical fluctuation drives the formation of large liquid droplets to lower the interfacial tension between the nucleus and the supersaturated gas phase. Because of this, the two-step crystal nucleation process not only occurs around the gas–liquid critical point, it is even found to exist when the liquid phase is thermodynamically unstable [50]. Besides crystal nucleation, due to the same physical reason, the two-step transition process also exists in many other phase transformation processes, such as protein fibril formation [51–53], the solid–solid transition [54, 55], nucleation near the isostructural transition [56] and so on.

11.7 Nucleation of Anisotropic Colloids

In the previous section, we discussed two-step nucleation in colloidal systems, in which there exists a metastable phase promoting the nucleation of the thermodynamically stable phase. In this section, we show you a few colloidal systems, in which there are also some metastable phases between the supersaturated phase and the stable phase, but the metastable phases strongly prevent the nucleation of the stable phase.

We consider a system of hard spherocylinders, and the phase diagram of the system in the $\rho^* = \rho/\rho_{cp}$ (reduced density) versus L/σ (aspect ratio) representation is shown in Figure 11.18, where ρ and ρ_{cp} are the density and densest close packing density of the system, respectively. We can see that when the aspect ratio is around $L/\sigma = 2$, with increasing density the system undergoes a direct first-order transition from an isotropic fluid phase to an ordered crystal phase [57]. It was found that crystal nucleation in a system of such short rods is extremely difficult [58]. Nuclei of single-layered clusters form and their further growth was prevented by surrounding particles, which is called *self-poisoning nucleation*. By using direct event-driven molecular dynamics simulations,

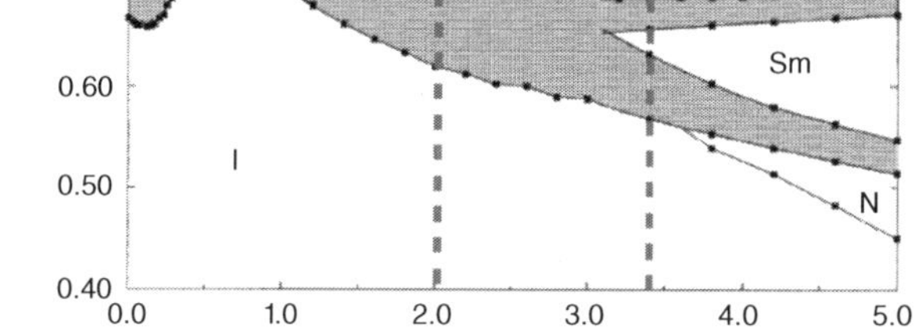

Figure 11.18 Phase diagram of hard spherocylinder in the ρ^* (density) versus L/σ (aspect ratio), representation from Ref. [57], where I, P, N, Sm, S denote isotropic, plastic crystal, nematic, smectic and crystal phases respectively. Inset: illustration of a hard spherocylinder.

we found that when the system is weakly supersaturated, it can nucleate multi-layered clusters to crystallize, and a typical dynamic trajectory is shown in Figure 11.19a. Moreover, by performing umbrella sampling Monte Carlo simulations with a newly developed order parameter [59], we indeed found that the pressure window where the nucleation growth can occur is very narrow (Figure 11.19b), and at high pressure, the system transforms into a metastable cubatic-like structure quickly, where the system gets stuck due to the glassy dynamics [60]. Similar behavior was also observed in the nucleation of the columnar phase in a system of hard platelets [61], which was suppressed by the metastable cubatic phase [62]. This suggests that glassy dynamics and the metastable phase play an important role in the nucleation of anisotropic particles.

Besides getting stuck in the metastable phase due to the glassy dynamics, for systems of hard rods we also found that the metastable spinodal instability can prevent nucleation of the stable phase. For instance, as shown in Figure 11.18, when $L/\sigma = 3.4$, with increasing density the system should undergo a direct phase transition from an isotropic

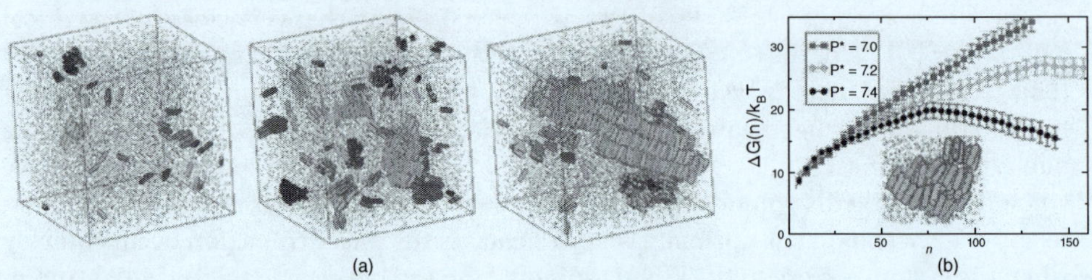

Figure 11.19 (a) Configurations for spontaneous crystal nucleation from a typical molecular dynamics trajectory at $P^* = 7.6$ and $t/\tau_{MD} = 0$, 1000 and 3000 (from left to right). Isotropic-like particles are drawn 10 times smaller than their actual size. (b) Gibbs free energy $\Delta G(n)$ as a function of the number of rods n in the crystalline cluster at pressure $P^* = 7.0$, 7.2 and 7.4. Inset: a typical configuration of a critical cluster ($n = 81$) at $P^* = 7.4$. *(See insert for color representation of the figure.).*

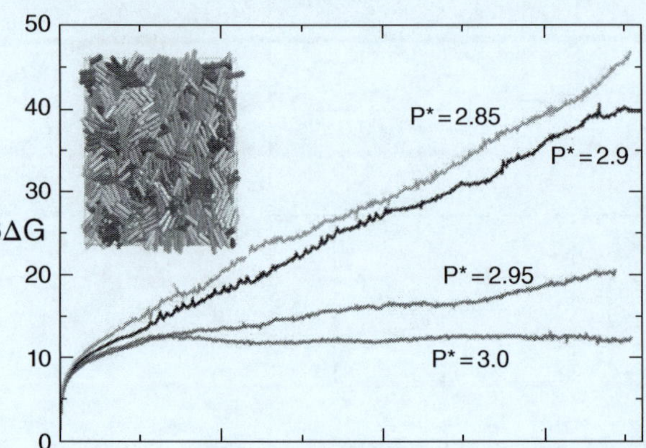

Figure 11.20 Gibbs free energy $\beta\Delta G$ of a hard spherocylinder fluid with $L/\sigma = 3.4$ as a function of the number of particles in the biggest cluster n, as calculated by umbrella sampling MC simulations at pressures $P^* = 2.85$, 2.90, 2.95 and 3.00. Inset: a typical snapshot of the system with large clusters. This figure is reproduced from Ref. [63] with permission of the Royal Society of Chemistry.

fluid to a smectic phase. Computer simulation shows that the calculated free energy barrier for a supersaturated fluid phase cannot reach a maximum. As shown in Figure 11.20, even at high pressure $P^* = P\sigma^3/k_B T = 3.0$, the free energy first increases with the cluster size n, and then reaches a plateau, and the size of the cluster fluctuates largely around several hundred particles without transforming into a bulk smectic phase. However, at a slightly higher pressure, i.e. $P^* = 3.05$, the system can quickly phase transform into a smectic phase via spinodal decomposition [63].

To resolve this mystery, we further characterize the system by calculating the positional and orientation structure factors, i.e. $S(k)$ and $S_{or}(k)$ of the system as shown in Figure 11.21. We can see that with increasing density, $S_{or}(k)$ shows a small-k divergence while $S(k)$ does not, which suggests that there is a huge orientational fluctuation in the system, while the density fluctuation remains tiny. This agrees with our observation that there exist large nematic-like clusters in the system without forming any denser smectic clusters. We also calculate the correlation length ξ by fitting the orientation correlation function $g_{or}(r) \sim \exp\ (-r/\xi)/r$ as shown in the inset of Figure 11.21, and we clearly see a power-law divergence of $\xi \sim |P^* - P_c^*|^{-\nu}$ with $\nu = 0.47$, which is very close to the critical exponent $\nu = 1/2$ for the isotropic–nematic (IN) spinodal obtained from Landau theory [65]. This implies that the system is approaching an IN spinodal before phase transforming into the stable smectic phase. The existence of this IN spinodal severely hinders the nucleation of smectic phase, and the system can only transform into a smectic phase when it is over the IN spinodal [60]. These results are surprising for two reasons. First, to approach a spinodal is normally very difficult, as the phase transition occurs quickly when the system is close to the kinetic spinodal instead of the true spinodal [66], and in this case the IN spinodal is metastable, which makes it possible to approach an IN spinodal in a realistic particle system for the first time. Second, that the existence of such a metastable IN spinodal hinders the nucleation of the smectic phase is counterintuitive.

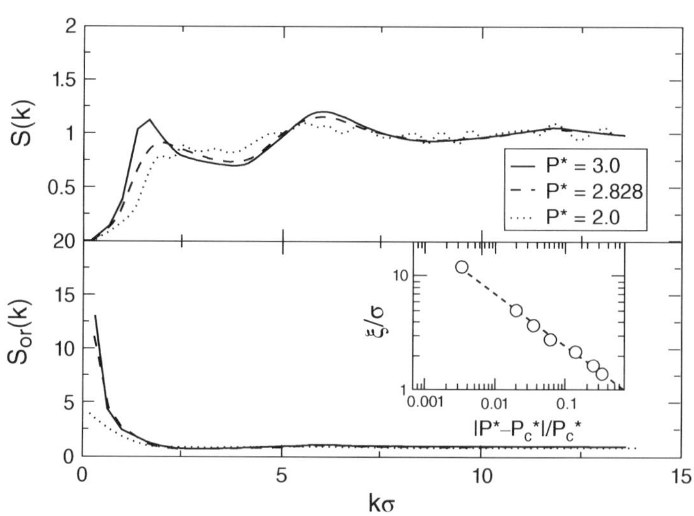

Figure 11.21 Positional (top) and orientational (bottom) structure factor of hard rods with $L/\sigma = 3.4$ at varying P^*. The inset shows the pressure dependence of the orientation correlation length ξ. The dashed line is the power-law fitting $\xi \sim |P^* - P_c^*|^{-\nu}$ with exponent $\nu = 0.47$.

One of the basic principles of statistical mechanics is that no matter what the starting configuration of the system is, after a long enough time of dynamical evolution, it can always reach the thermodynamically stable state of the system. However, we found here that in a supersaturated fluid of hard rods before reaching the metastable IN spinodal, although by definition the system is in a metastable state, the kinetic pathway of nucleation of the stable smectic phase does not exist, and the waiting time for forming a stable smectic phase in the thermodynamic limit is infinite. This finding challenges the basic principle of statistical mechanics and offers new insights in understanding the role of metastability in the process of phase transition.

11.8 Crystal Nucleation in Binary Mixtures

So far we have focused on nucleation in monodisperse colloidal systems, which have been employed to study the physics of nucleation. Furthermore, in systems of binary mixtures, there are a variety of different superlattice crystals found experimentally, which have shown great promise in photonic applications [64, 67–69]. For instance, in systems of oppositely charged colloidal particles, depending on the size ratio and charge ratio between the two types of colloids, a number of AB_n crystals have been observed in experiments [64].

As shown in Figure 11.22, when the size ratio $a_1/a_2 = 1$, depending on the charge ratio and the Debye screening length, the system can form stable NaCl or CsCl crystals. Moreover, when the sizes of the oppositely charged colloids are not the same, they can form many other superlattice crystals. For instance, at $a_1/a_2 = 0.31$, ReO_3, Li_3N, SiF_4, LS_6, CaF_2, Cr_3Si, A_4C_{60}, A_6C_{60}, LS_4 and LS_8 are predicted to be stable; A_6C_{60}, LS_6 and LS_8 have already been observed in experiments [67]. These predictions and findings open up new possibilities for fabricating photonic crystals by using binary colloidal mixtures. Therefore, the study on crystal nucleation of binary mixtures is not only of fundamental interest but also practical importance.

However, the nucleation of binary crystals is far more complicated compared to a monodisperse system. For instance, in a study of nucleation of oppositely charged colloids with size ratio $a_1/a_2 = 1$, where CsCl is the thermodynamically stable phase [70], it was found that the dynamics of the system can drive the formation of a metastable disordered FCC crystal, which suggests that the nucleation pathway is intrinsically out of equilibrium [20]. This motivated us to look at the nucleation of binary mixtures. We first formulated the classic nucleation theory for nucleation in binary mixtures, and we found that the choice of order parameter has a strong influence in the calculation of the free energy barrier [21]. To illustrate the effect of the order parameter, we investigated binary crystal nucleation in a toy model of hard spheres, in which we considered a system consisting of two species of hard spheres with identical sizes, but tagged with different colors, i.e. species 1 is gray and species 2 is black. Obviously, the stable solid phase to be nucleated is a substitutional disordered FCC crystal phase with the gray and black particles randomly distributed on the FCC lattice. We choose $\Phi = N_1 + \lambda N_2$ as the order parameter, in which N_1 and N_2 are the numbers of gray and black spheres in the largest crystalline cluster, respectively.

As shown in Figure 11.23 (top), we can see that by changing λ in the order parameter, one can obtain nuclei with different composition. Even in the steady-state distribution

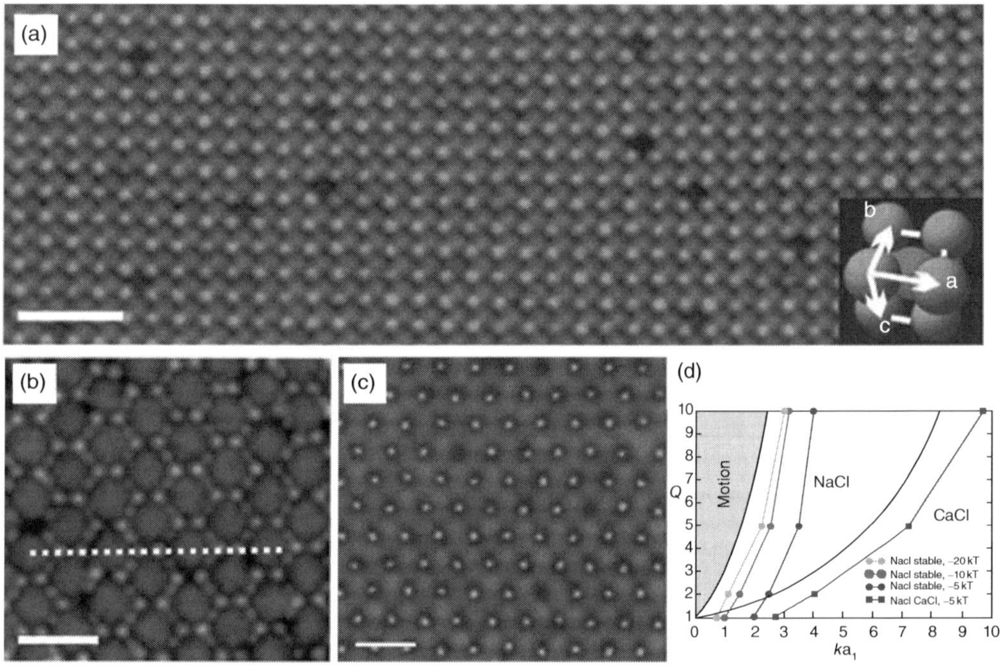

Figure 11.22 (a) A CsCl crystal formed by positive (radius 1.08 μm) and negative (0.99 μm) poly(methyl methacrylate) (PMMA) spheres. (b) LS_6-type crystal formed by positive (radius 0.36 μm) and negative (1.16 μm) PMMA particles. (c) NaCl-type crystal formed by charged (radius 1.16 μm) and uncharged (0.36 μm) PMMA particles. (d) Theoretical phase diagram of oppositely charged colloids with size ratio $a_1/a_2 = 1$ based on the screened Coulomb energy, where a_1 and a_2 are the radius of the oppositely charged colloids. Q is the charge ratio $Q = Z_1/Z_2$, and κ is inverse Debye screening length of the system. This figure is reproduced from Ref. [64] with permission of Nature Publishing Group.

calculated using FFS, which was believed much less dependent on the choice of order parameter, one can still find significant dependence of composition x on λ in the pre-critical nucleus. In the two-dimensional free energy landscape shown in Figure 11.23 (bottom), we can see that by using different order parameters, umbrella sampling simulations bias the nucleation of the system to follow different pathways. However, interestingly, all the pathways travel through the same saddle point, which is due to the special property of saddle points. This suggests that to calculate the nucleation free energy barrier of binary mixtures, the properties of the pre- and post-critical nucleus, e.g. structure, composition, etc., depend on the choice of order parameter, while the critical nucleus is independent of the order parameter.

Besides the influence of the order parameter, to study the nucleation of binary mixtures one should also be very careful with the finite size effect of the system. As the nucleated crystal may have a different stoichiometry from the supersaturated binary fluid phase, to calculate the nucleation barrier of binary mixtures, in principle, one should perform simulations either using a very large system or using the semi-grand canonical ensemble ($NPT - \Delta\mu$) to exclude the finite size effect, in which $\Delta\mu$ is the chemical potential difference between the two species in the metastable bulk phase [71].

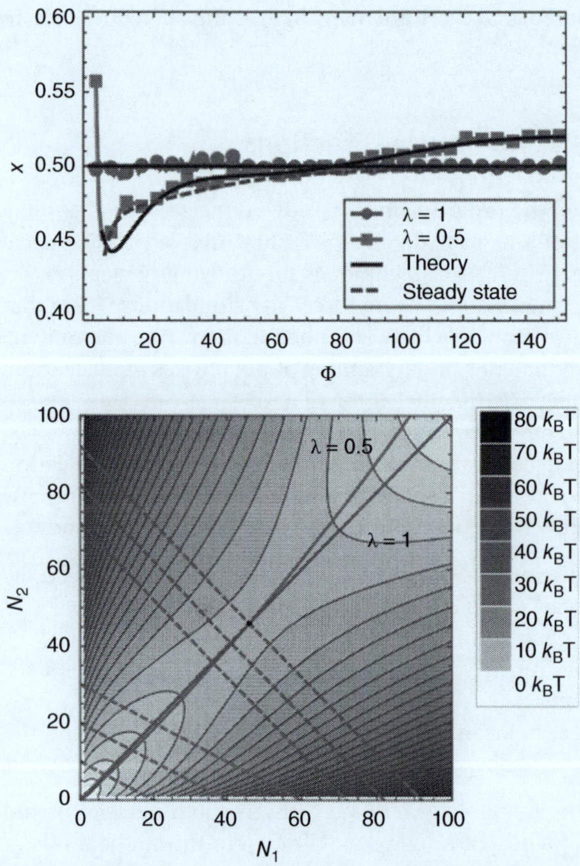

Figure 11.23 Top: Composition $x = N_2/(N_1 + N_2)$ of the largest crystalline cluster as a function of the order parameter $\Phi = N_1 + \lambda N_2$ for a binary mixture of gray (species 1) and black (species 2) hard spheres with equal diameter σ as obtained from umbrella sampling simulations at pressure $P^* = P\sigma^3/k_BT = 17$. The steady-state distribution was obtained by using FFS with $\lambda = 0.5$. Bottom: Contour plot of the two-dimensional free energy landscape $\Delta G(N_1, N_2)/k_BT$ as a function of N_1 and N_2.

However, to simulate the crystal nucleation of binary mixtures, semi-grand canonical ensemble Monte Carlo simulations are prohibitively slow. To solve this problem, we devised a successive umbrella sampling method; for details of the method, please refer to Ref. [21]. By using this novel method, we successfully obtained the free energy barrier for the nucleation of an interstitial solid solution, in which the composition of the crystal phase can be continuously tuned by changing the pressure of the system [72]. We found that instead of nucleating the crystal with the composition corresponding to the stable bulk phase, the system nucleates a crystalline cluster, in which the chemical potential of the small spheres equals that of the metastable fluid. This suggests that the counterintuitive Ostwald's step rule may generally appear in the nucleation of most binary mixtures, and it makes the study of the nucleation of binary mixtures highly challenging from both aspects of physics and simulation techniques. Therefore, nucleation in binary mixtures is still largely underexplored, and there are many interesting open questions to investigate.

11.9 Concluding Remarks and Future Directions

In this chapter, we briefly reviewed the progress on the study of nucleation in colloidal systems using computer simulations in past decades. We first derived the classical nucleation and the state-of-the-art simulation techniques for studying nucleation, i.e. umbrella sampling, forward flux sampling and brute force MD simulations. After that we discussed four hot topics of nucleation, in which a number of interesting phenomena were found in past decades, and refined our understanding of the physics of nucleation. Some of the debates are still ongoing, e.g. the nucleation of hard-sphere crystals, and to resolve these mysteries, we need to devise better theories, better simulation techniques, and better experiments. All these debates and new phenomena have made nucleation study one of hottest research directions in soft matter physics. In the following few years, I personally feel that the following topics could be interesting for future study:

- large-scale dynamic simulations incorporating hydrodynamic effects to study the influence of hydrodynamics on the crystal nucleation of colloidal systems;
- polymorph selection in crystal nucleation and theoretical explanations;
- crystal nucleation pathways for anisotropic crystals and many superlattice crystals, e.g. Laves phase, AB_{13} crystals, etc.;
- nucleation in non-equilibrium systems, e.g. active matter systems [73–90], and the theoretical treatment.

The study of these problems will definitely give us new insights into nucleation and help to build a complete physical picture of nucleation. Ultimately, the study of nucleation can hopefully help and guide the fabrication of functional materials in experiments.

Acknowledgments

This work is supported by the Start-Up Grant from Nanyang Technological University.

References

1. A. Nawaby, Y. Handa, X. Liao, Y. Yoshitaka, and M. Tomohiro, *Polym. Int.* **56**, 67 (2007).
2. J. Perrin, *Ann. Chim. Phys.* **18**, 5 (1909).
3. S. Auer and D. Frenkel, *Nature* **409**, 1020 (2001).
4. T. Schilling, H. J. Schöpe, M. Oettel, G. Opletal, and I. Snook, *Phys. Rev. Lett.* **105**, 025701 (2010).
5. L. Filion, M. Hermes, R. Ni, and M. Dijkstra, *J. Chem. Phys.* **133**, 244115 (2010).
6. L. Filion, R. Ni, D. Frenkel, and M. Dijkstra, *The Journal of Chemical Physics* **134**, 134901 (2011).

7. T. Kawasaki and H. Tanaka, *Proc. Nat. Ac. of Sci.* **107**, 14036 (2010).
8. R. J. Allen, P. B. Warren, and P. R. ten Wolde, *Phys. Rev. Lett.* **94**, 018104 (2005).
9. D. Moroni, P. R. ten Wolde, and P. G. Bolhuis, *Phys. Rev. Lett.* **94**, 235703 (2005).
10. J. D. van der Waals, PhD thesis, Leiden University (1873).
11. P. Wegener, *Acta Mechanica* **21**, 65 (1975).
12. R. Reid, *Amer. Sci.* **64**, 146 (1976).
13. A. I. Berezhnoi, *Glass-Ceramics and Photo-Sitalls* (Springer, New York, 1970).
14. M. Volmer and A. Weber, *Z. Phys. Chem.* **119**, 277 (1926).
15. J. L. Katz, *Pure and Applied Chemistry* **64**, 1661 (1992).
16. D. W. Oxtoby, *J. Phys: Condens. Matter* **4**, 7627 (1992).
17. R. Becker and W. Döring, *Ann. Phys.* **24**, 719 (1935).
18. C. H. Bennett, in *ACS Symposium Series No. 46*, ed. R. Christofferson (1977).
19. D. Chandler, *The Journal of Chemical Physics* **68**, 2959 (1978).
20. E. Sanz, C. Valeriani, D. Frenkel, and M. Dijkstra, *Phys. Rev. Lett.* **99**, 055501 (2007).
21. R. Ni, F. Smallenburg, L. Filion, and M. Dijkstra, *Mol. Phys.* **109**, 1213 (2011).
22. P. J. Steinhardt, D. R. Nelson, and M. Ronchetti, *Phys. Rev. B* **28**, 784 (1983).
23. P. R. ten Wolde, M. J. Ruiz-Montero, and D. Frenkel, *Faraday Discuss.* **104**, 93 (1996).
24. P. R. ten Wolde, PhD thesis, University of Amsterdam (1998).
25. C. Valeriani, PhD thesis, University of Amsterdam (2007).
26. R. Ni, PhD thesis, Utrecht University (2012).
27. W. Lechner and C. Dellago, *The Journal of Chemical Physics* **129**, 114707 (2008).
28. G. M. Torrie and J. P. Valleau, *Chem. Phys. Lett.* **28**, 578 (1974).
29. J. S. van Duijneveldt and D. Frenkel, *J. Chem. Phys.* **96**, 4655 (1992).
30. P. R. ten Wolde, M. J. Ruiz-Montero, and D. Frenkel, *J. Chem. Phys.* **104**, 9932 (1996).
31. S. Auer and D. Frenkel, *J. Chem. Phys.* **120**, 3015 (2004).
32. D. Frenkel and B. Smit, *Understanding Molecular Simulations: From Algorithms to Applications* (Academic Press, Waltham, MA, 2002).
33. S. Auer, PhD thesis, University of Amsterdam (2002).
34. R. J. Allen, C. Valeriani, and P.-R. ten Wolde, *J. Phys.: Cond. Matter* **21**, 463102 (2009).
35. F. A. Escobedo, E. E. Borrero, and J. C. Araque, *J. Phys.: Cond. Matter* **21**, 333101 (2009).
36. K. Schätzel and B. J. Ackerson, *Phys. Rev. E* **48**, 3766 (1994).
37. J. L. Harland and W. van Megen, *Phys. Rev. E* **55**, 3054 (1997).
38. C. Sinn, A. Heymann, A. Stipp, and T. Palberg, *Prog. Colloid Polym. Sci.* **118**, 266 (2001).
39. T. Schilling, S. Dorosz, H. Schöpe, and G. Opletal, *J. Phys.: Condens. Matter* **23**, 194120 (2011).
40. D. Richard and T. Speck, *Sci. Rep.* **5**, 14610 (2015).
41. M. Radu and T. Schilling, *Europhys. Lett.* **105**, 26001 (2014).
42. P. G. Bolhuis, D. Frenkel, S. C. Mau, and D. A. Huse, *Nature* **388**, 235 (1997).
43. M. Marechal and M. Dijkstra, *Phys. Rev. E* **77**, 061405 (2008).
44. R. Ni and M. Dijkstra, *The Journal of Chemical Physics* **134**, 034501 (2011).
45. B. O'Malley and I. Snook, *Phys. Rev. Lett.* **90**, 085702 (2003).
46. R. Ni and M. Dijkstra, *Soft Matter* **9**, 365 (2013).

47. N. C. Karayiannis, R. Malshe, M. Kroger, J. J. de Pablo, and M. Laso, *Soft Matter* **8**, 844 (2012).
48. J. Hansen and L. Verlet, *Phys. Rev. A* **184** (1969).
49. P. R. ten Wolde and D. Frenkel, *Science* **277**, 1975 (1997).
50. H. Liu, S. Kumar, and J. Douglas, *Phys. Rev. Lett.* **103**, 018101 (2009).
51. R. Ni, S. Abeln, M. Schor, M. Cohen-Stuart, and P. Bolhuis, *Phys. Rev. Lett.* **111**, 058101 (2013).
52. R. Ni, J. Kleijn, S. Abeln, M. Cohen-Stuart, and P. Bolhuis, *Phys. Rev. E* **91**, 022711 (2015).
53. A. Saric, Y. Chebaro, T. Knowles, and D. Frenkel, *Proc. Natl Acad. Sci. USA* **111**, 17869 (2014).
54. Y. Peng, F. Wang, Z. Wang, A. Alsayed, Z. Zhang, A. Yodh, and Y. Han, *Nature Materials* **14**, 101 (2015).
55. W. Qi, Y. Peng, Y. Han, R. Bowles, and M. Dijkstra, *Phys. Rev. Lett.* **115**, 185701 (2015).
56. A. Cacciuto, S. Auer, and D. Frenkel, *Phys. Rev. Lett.* **93**, 166105 (2004).
57. P. Bolhuis and D. Frenkel, *J. Chem. Phys* **106**, 666 (1997).
58. T. Schilling and D. Frenkel, *Phys. Rev. Lett.* **92**, 085505 (2004).
59. A. Cuetos and M. Dijkstra, *Phys. Rev. Lett.* **98**, 095701 (2007).
60. R. Ni, S. Belli, R. van Roij, and M. Dijkstra, *Phys. Rev. Lett.* **105**, 088302 (2010).
61. M. Marechal, A. Patti, M. Dennison, and M. Dijkstra, *Phys. Rev. Lett.* **108**, 206101 (2012).
62. M. Marechal, A. Cuetos, B. Martínez-Haya, and M. Dijkstra, *The Journal of Chemical Physics* **134**, 094501 (2011).
63. A. Cuetos, E. Sanz, and M. Dijkstra, *Faraday Discuss.* **144**, 253 (2010).
64. M. E. Leunissen, C. G. Christova, A. Hynninen, C. P. Royall, A. I. Campbell, A. Imhof, M. Dijkstra, R. van Roij, and A. van Blaaderen, *Nature* **437**, 235 (2005).
65. E. F. Gramsbergen, L. Longa, and W. H. de Jeu, *Physics Reports* **135**, 195 (1986).
66. J. Wedekind, G. Chkonia, J. Wölk, R. Strey, and D. Reguera, *The Journal of Chemical Physics* **131**, 114506 (2009).
67. A.-P. Hynninen, C. G. Christova, R. van Roij, A. van Blaaderen, and M. Dijkstra, *Phys. Rev. Lett.* **96**, 138308 (2006).
68. A.-P. Hynninen, J. H. J. Thijssen, E. C. M. Vermolen, M. Dijkstra, and A. Van Blaaderen, *Nature Materials* **6**, 202 (2007).
69. Y. Wang, Y. Wang, X. Zheng, E. Ducrot, J. Yodh, M. Weck, and D. Pine, *Nat. Commun.* **6**, 7253 (2015).
70. A.-P. Hynninen, M. E. Leunissen, A. van Blaaderen, and M. Dijkstra, *Phys. Rev. Lett.* **96**, 018303 (2006).
71. S. Punnathanam and P. A. Monson, *J. Chem. Phys.* **125**, 024508 (2006).
72. L. Filion, M. Hermes, R. Ni, E. C. M. Vermolen, A. Kuijk, C. G. Christova, J. C. P. Stiefelhagen, T. Vissers, A. van Blaaderen, and M. Dijkstra, *Phys. Rev. Lett* **107**, 168302 (2011).
73. J. Tailleur and M. E. Cates, *Phys. Rev. Lett.* **100**, 218103 (2008).
74. A. G. Thompson, J. Tailleur, M. E. Cates, and R. A. Blythe, *Journal of Statistical Mechanics: Theory and Experiment* **2011**, P02029 (2011).
75. J. Bialké, H. Löwen, and T. Speck, *Europhys. Lett.* **103**, 30008 (2013).

76. J. Stenhammar, A. Tiribocchi, R. J. Allen, D. Marenduzzo, and M. E. Cates, *Phys. Rev. Lett.* **111**, 145702 (2013).
77. R. Wittkowski, A. Tiribocchi, J. Stenhammar, R. Allen, D. Marenduzzo, and M. Cates, arXiv:1311.1256 (2014).
78. T. Speck, J. Bialké, A. M. Menzel, and H. Löwen, *Phys. Rev. Lett.* **112**, 218304 (2014).
79. S. C. Takatori, W. Yan, and J. F. Brady, *Phys. Rev. Lett.* **113**, 028103 (2014).
80. R. Ni, M. A. Cohen Stuart, and P. G. Bolhuis, *Phys. Rev. Lett.* **114**, 018302 (2015).
81. I. Buttinoni, J. Bialké, F. Kümmel, H. Löwen, C. Bechinger, and T. Speck, *Phys. Rev. Lett.* **110**, 238301 (2013).
82. G. Redner, M. Hagan, and A. Baskaran, *Phys. Rev. Lett.* **110**, 055701 (2013).
83. Y. Fily and M. C. Marchetti, *Phys. Rev. Lett.* **108**, 235702 (2012).
84. R. Ni, M. Cohen-Stuart, and M. Dijkstra, *Nat. Commun.* **4**, 2704 (2013).
85. R. Ni, M. Cohen-Stuart, M. Dijkstra, and P. Bolhuis, *Soft Matter* **10**, 6609 (2014).
86. M. E. Cates and J. Tailleur, *Europhys. Lett.* **101**, 20010 (2013).
87. Y. Fily, S. Henkes, and M. C. Marchetti, *Soft Matter* **10**, 2132 (2014).
88. J. Stenhammar, D. Marenduzzo, R. J. Allen, and M. E. Cates, *Soft Matter* **10**, 1489 (2014).
89. I. Theurkauff, C. Cottin-Bizonne, J. Palacci, C. Ybert, and L. Bocquet, *Phys. Rev. Lett.* **108**, 268303 (2012).
90. J. Palacci, S. Sacanna, A. P. Steinberg, D. J. Pine, and P. M. Chaikin, *Science* **339**, 936 (2013).

12

Atomistic and Coarse-Grained Simulation of Liquid Crystals

Saientan Bag[1], *Suman Saurabh*[1], *Yves Lansac*[2] *and Prabal K. Maiti*[1]

[1] *Centre for Condensed Matter Theory, Department of Physics, Indian Institute of Science, Bangalore, India*
[2] *GREMAN, Université François Rabelais, CNRS UMR 7347, France*

12.1 Introduction

The liquid crystalline phase is ubiquitous in nature and is exhibited by a variety of systems like surfactants, nucleic acids, lipid molecules, etc. Depending on the phase behaviour, liquid crystals can be classified into thermotropic and lyotropic liquid crystals. The two varieties of liquid crystals are different in the conditions under which they undergo phase transformations. Thermotropic liquid crystals undergo phase transformation mainly as a function of temperature, while in the case of lyotropic liquid crystals, phase change occurs primarily as a function of composition and concentration [1]. Thermotropic liquid crystals are traditionally single component melts and show a wide range of phases that are classified based on the positional and orientational order of the molecules as a function of temperature [2, 3]. These intermediate phases include nematic (N) and smectic (Sm), as well as more exotic phases such as twist grain boundary (TGB), blue phase, etc., and the formation of a particular phase depends on the chemical structure and architecture of the constituent molecules. Additionally, pressure can also induce phase transformations and replace the role of temperature. When more than one component is present, then the presence of a given phase is dependent on the concentration of the mixture. Materials forming liquid crystal phases upon addition of a solvent are called lyotropic. Commonly occurring examples of lyotropic liquid crystals are water-amphiphilic systems where the amphiphilic molecule is characterized by a hydrophilic head group and a hydrophobic tail. Lyotropic liquid crystals can arrange into hexagonal, micellar, reverse micellar and lamellar phases depending on the balance between the hydrophilic and hydrophobic character of the molecule and the concentration relative to the aqueous phase [4, 5]. Examples of amphiphilic molecules are surfactants, the common ingredients of soaps and phospholipids, which are integral constituents of cell membranes. Due to the wide range of microstructures possible with liquid crystals, these systems are commonly referred to as complex fluids.

Various phases formed in both thermotropic and lyotropic liquid crystals are very sensitive to the shape of the constituent molecules. The macroscopic properties of a

Self-Assembling Systems: Theory and Simulation, First Edition. Edited by Li-Tang Yan.

given liquid crystal phase are strongly influenced by the structure of the constituent molecule in a very complex way. A proper understanding of this structure–property relationship is very important both from a fundamental point of view and for developing potential technologies based on liquid crystals. Due to the complex interplay of molecular shape and interaction, it is difficult to develop a predictive theoretical framework for liquid crystals. As a result, computer simulations have emerged as a powerful tool to investigate the dependence of the phase behaviour of liquid crystals on the structure of the constituent molecules [6, 7]. Various levels of description have been used to develop models for liquid crystals. These models include simple spherocylinders or assemblies of spheres which interact through hard or soft excluded volume interaction [8–11], molecules with ellipsoidal shape interacting through the Gay–Berne potential [12, 13], and simple 'bead–spring' representation of molecules interacting through Lennard-Jones potentials [14]. And finally, there exists a large body of work involving atomic-level models of real molecules that show liquid crystalline ordering [15–17].

Simulation methods employed to study liquid crystal (LC) phases range from molecular dynamics (MD), Monte Carlo (MC), Brownian dynamics (BD) to dissipative particle dynamics (DPD). In this chapter we give an overview of the various methods used in simulating self-assembly phenomena in liquid crystals. From a microscopic point of view, atomistic MD is a realistic approach as it considers all atom dynamics, but suffers from the drawback that a huge computational effort is required to explore the full phase space and longer length and time scales involved at the mesoscopic level. An effective coarse-grained model reduces the computational effort considerably, but lacks the required chemical details to be useful as a predictive tool. Developing suitable multiscaling strategies that cover the range of length and time scales typically involved in these complex mesophases is an active area of research [18–20].

In the first part of this chapter, we give a comprehensive overview of the atomistic simulations carried out to date on various thermotropic liquid crystals. We focus on thermotropic liquid crystals and describe prominent atomistic MD studies of the LC phase behaviour of various rod-like molecules. Then we discuss how Monte Carlo simulations, along with coarse-grained models based on excluded volume interactions can be used to study the wide variety of phases that occur in systems consisting of a collection of bent-core or banana-shaped liquid crystal molecules. Then we move on to describe the thermotropic LC phases formed by various disk-shaped molecules which have many important technological applications. The second part of the chapter deals with chromonic liquid crystals, which are basically lyotropic in nature. In this section we describe important atomistic as well as coarse-grained studies carried out to date on the chromonic LC phase. We also refer the reader to some earlier reviews on this subject [21, 22].

12.2 Thermotropic Liquid Crystal

12.2.1 Fully Atomistic Simulation

Early atomistic simulation studies mostly dealt with the nematic phase and were limited to very small system size and short time scales due to the unavailability of reliable force fields as well as inadequate computer resources. For example, early all-atom MD simulation studies by Wilson and Allen [23, 24] demonstrated the formation of the nematic phase in the *trans*-4-(*trans*-4-*n*-pentylcyclohexyl)cyclohexylcarbonitrile (CCHS) compound. The isotropic phase was reported at 390 K. Cooling from the isotropic phase, the nematic phase was obtained at 350 K. The formation of the LC phase was monitored by calculating the orientational order parameter or the so-called nematic order parameter given by $S = \langle P_2 \cos\theta \rangle = \left\langle \frac{3}{2}\cos^2\theta - \frac{1}{2} \right\rangle$, where P_2 is the second Legendre polynomial, θ is the angle between the nematic director and the long axis of the molecule and $\langle\ \rangle$ refers to ensemble average.

This was followed by 'The first thousand-molecule simulation of a mesogen at the fully atomistic level' by Cook and Wilson [25], where the authors reported LC phases of the 4-(trans-4-*n* pentylcyclohexyl)benzonitrile (PCHS) compound. The authors used the OPLS-AA force field (FF) from Jorgenson and co-workers [26] and employed the DL_POLY software to perform the MD simulations. Simulation was started from an ordered phase and was heated to gradually get to the isotropic phase. The measured density and diffusion constant of the system in the isotropic phase matched quite well with the experimental data, making this study an early success for atomistic simulation in this field. The first simulation in which, starting from an isotropic phase, the LC phase was shown to self-assemble was also reported by Wilson and co-workers in a system of 4,4-din-pentyl-bibicyclo[2.2.2]octane (5,5-BBCO) [27]. In this study, the authors used the MM3 force field [28] to describe inter- and intra-molecular interactions.

Lansac et al. [29] simulated the formation of the partial bilayer SmA_d of 4-octyl-4′-cyanobiphenyl (8CB) (Figure 12.1) starting from an initial bilayer condition using a force field combining *ab initio* calculations (bonded interactions) and emirical data taken mostly from the OPLS FF (non-bonded interactions). The layer spacing was found to be close to the experimental value and a microscopic picture of the partial bilayer showing the importance of the interactions between permanent cyano-electric dipoles was provided. The same approach was used to study the

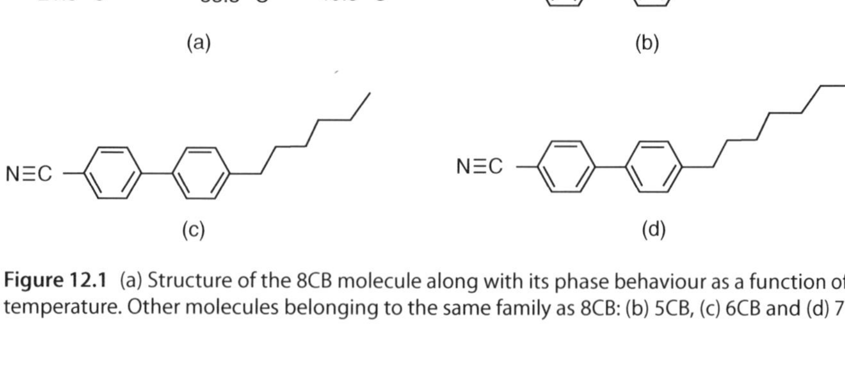

Figure 12.1 (a) Structure of the 8CB molecule along with its phase behaviour as a function of temperature. Other molecules belonging to the same family as 8CB: (b) 5CB, (c) 6CB and (d) 7CB.

reversible increase in layer spacing observed experimentally in an 8CB solvent doped with a small number of photosensitive azobenzene (7AB) molecules [30]. Atomistic MD simulations showed that the phenomenon was due to a nanophase segregation of the azobenzene molecules from within the 8CB layers to locations in between the layers during the trans-to-cis photo-isomerization (Figure 12.2). Considering the small fraction of azobenzene molecules undergoing trans-to-cis isomerization under UV illumination, the simulations were able to quantitatively reproduce the experimental increase in layer spacing. This study was one of the early successes of atomistic MD simulation in the field of LC simulation.

Continuing their effort to make a more accurate FF for LC simulation, Wilson and co-workers started developing what is now called the liquid crystal force field (LCFF) to simulate various LC phases at the all-atom level [31]. The force field employed the same energy expression as in AMBER [26] to treat the various non-bonded and bonded terms and used the AMBER atom type convention. To test the accuracy of this FF, Wilson and co-workers used LCFF to reproduce the densities, heat of vaporization and diffusion constant of fragments of several liquid crystal compounds as tabulated in Table 12.1.

Wilson et al. later [32] simulated the biaxial nematic phase in bent core compounds made of 2,5-bis-(*p*-hydroxyphenyl)-1,3,4-oxadiazole mesogen (ODBP-Ph-C(7)) using all-atom MD simulations. They used a combination of LCFF and OPLS-AA force fields to describe the intra- and inter-molecular interaction. The degree of biaxiality was found to be very small and disappears when the electrostatic interaction is turned off. In their simulation, they were also able to show the formation of ferroelectric domains in the nematic phase characterized by parallel orientation of the short axis of the molecule, which is aligned such that the oxidazole dipoles are parallel to each other. Later, the same group also reported simulation of the commercial liquid crystal mixture E7, which is a mixture of the following four components mixed in a certain weight percentage: 4-cyano-4′-*n*-pentyl-biphenyl (5CB), 4-cyano-4′-*n*-heptyl-biphenyl (7CB), 4-cyano-4′-*n*-octyloxy-biphenyl (8CB) and 4-cyano-4′-*n*-pentyl-*p*-terphenyl (5CT). They demonstrated the formation of the nematic phase from an isotropic solution of E7 mixture during a 100 ns-long atomistic simulation [33] using a combination of LCFF and OPLS-AA. Interestingly they observed a more ordered arrangement of the cyanoterphenyl component 5CT than any other components in the mixture. LCFF was also successfully used to study the liquid crystal phase of *n*-4-(trans-4-*n*-pentylcyclohexyl)benzonitrile (PCH5) using a fully atomistic model [34]. The authors also calculated flexoelectric coefficients in the nematic phase of PCH5, which were found to be in good agreement with the experimental results.

Later, the same group used GAFF FF [35] to simulate various LC phases of 8CB compounds [36]. 8CB compounds have attracted a lot of attention as they are a good testing ground for various LC force fields due to the wealth of experimental data available. The authors predicted a crystal–SmA transition in the temperature window of 300–320 K (corresponding to an experimental transition temperature of 295 K), a SmA–nematic (N) transition in the temperature window of 340–360 K (corresponding to an experimental transition temperature of 307 K) and a N–isotropic (I) transition at 375 K (corresponding to an experimental transition temperature of 314 K). The authors could simulate the low-temperature SmA phase of 8CB materials, which also reproduced the correct layer spacing as observed in experiments. In the same work, they also simulated

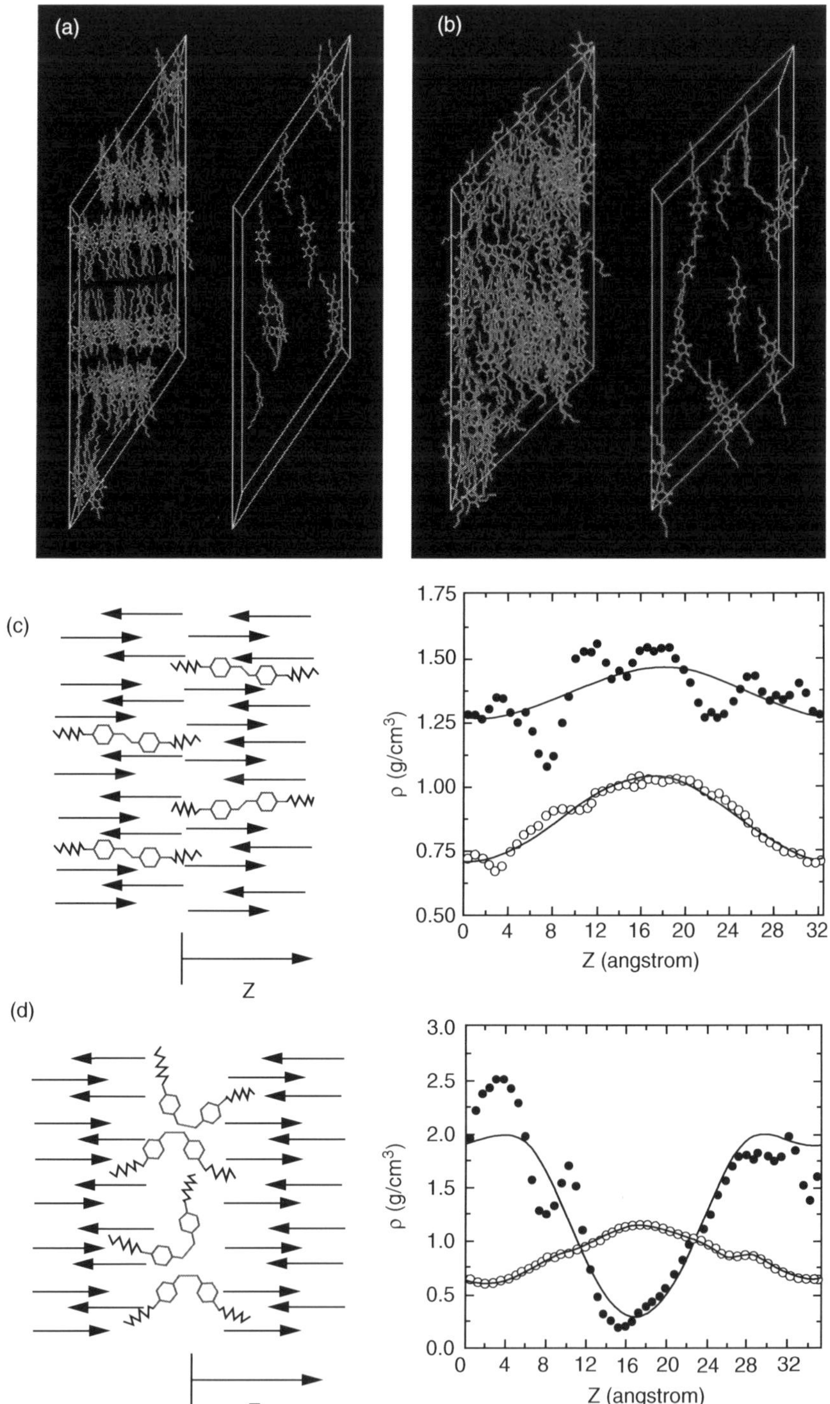

(a)
(b)
(c)
(d)
ρ (g/cm^3)
Z (angstrom)
Z

Figure 12.2 (a) Initial system for the mixture of 8CB and trans-7AB. (b) Final system for the same mixture. The 7AB molecules are shown separately for clarity. Mass density profiles along the layer normal for 8CB (open circles) and 7AB (filled circles) averaged over the last 2 ns of (c) trans-7AB and (d) cis-7AB simulations. Mass densities of 7AB have been multiplied by a factor of 10 for clarity. The solid curves are fits to a low-order Fourier expansion. One notes that trans-7AB is preferentially incorporated into the 8CB layers; cis-7AB exhibits a deep minimum near the centre of the 8CB layer, indicating that it is excluded quite effectively from the 8CB layers.

Table 12.1 Comparison of experimental densities and heat of vaporization with simulated numbers using LCFF for several standard liquid fragments [31].

Liquid	T (K)	$\langle \rho \rangle_{simu}$ (kg m^{-3})	$\langle \rho \rangle_{expt}$ (kg m^{-3})	$\Delta_{vap}H_{simu}$ (kJ mol^{-1})	$\Delta_{vap}H_{expt}$ (kJ mol^{-1})
Biphenyl	343	981.3±7.1	990	62.99	61.23
4-cyanobiphenyl	360	1028.7±7.1	—	77.119	—
Butane	272.5	605.3±10.6	602	21.67	22.44
n-propylbenzene	298	861.8±7.8	860	45.88	46.48
Phenylcyclohexane	298	954.9	939	66.41	64.14
2-fluorobiphenyl	343	1085.6	—	31.66	—
Hexane	298	664.36±8.0	656	33.13	31.98
Terphenyl	493	932.46 ±8.8	957.7	72.02	75.89

the electron paramagnetic resonance (EPR) spectra of the 8CB liquid crystal across the N–I phase transition and found good agreement with the experimental data.

McDonald and Hanna [37] used the united atom (UA) model to simulate the LC phases of 8CB and reported isotropic, nematic and smectic phases. In their simulation, they neglected long-range electrostatic interactions. Although the study predicted a smectic A phase, as observed in experiments, the predicted periodicity of the smectic phase (19.5 Å) was equal to the length of a fully extended 8CB molecule, rather than an 8CB dimer (31.4 Å) as seen in experiments. The neglect of electrostatic interactions precluded dimer formation. In addition to this, the model also predicted a higher disorder–order transition temperature. The 8CB molecules, with flexible tails, which lead to greater entropy, are expected to order at lower temperatures than the model predicted.

Zannoni and co-workers used atomistic simulation to probe the odd–even effect in the phenyl alkyl-4-(4-cyanobenzylidene)aminocinnamates mesogenic series [38], as well as to calculate the temperature corresponding to the isotropic to nematic (I–N) phase transition. Back then, this was one of the largest simulations covering almost 40–50 ns for each temperature in the range of 500–560 K. The equilibration was monitored by plotting the variation of nematic order parameter S as a function of time at each temperature, and the simulation was continued until a plateau was observed. A combination of AMBER [39] and OPLS force fields [26] was used. The I–N phase transition was monitored by fitting the temperature dependence of the order parameter to the Haller [40] expression given by

$$\langle S_2 \rangle_{\text{Haller}} = (1 - \langle S_2 \rangle_{\text{Iso}}) \left(1 - \frac{T}{T_{\text{NI}}} \right)^{\beta} + \langle S_2 \rangle_{\text{Iso}},$$

where β is the critical exponent corresponding to the N–I transition. The N–I transition temperature (T_{NI}) obtained through such a fit and shown in Table 12.2 is in good agreement with the available experimental results.

Later Zannoni and co-workers [41] used a modified AMBER FF to model various properties of the nCB series of liquid crystals; their results are summarized in Table 12.3. They modified the AMBER LJ parameter set to reproduce the experimental nematic to isotropic transition temperature for 5CB. The final set of LJ parameters was chosen in such a way that they reproduced the experimental densities of the nCB series over a wide range of temperatures. The temperature dependence of the density along with the experimental densities are shown in Figure 12.3. Although the agreement with the experiment in terms of density and transition temperature is very encouraging, one can see systematic deviation in the simulated density from the experimental one particularly

Table 12.2 Comparison of N–I transition temperatures from simulation and experiments. n = 0, 1 and 2 correspond to the first three homologues of the phenyl alkyl-4-(4′-cyanobenzylidene)aminocinnamate series. Table taken from [38]. Reproduced with permission of John Wiley and Sons Ltd.

n	β	T_{NI}(sim)(K)	$\langle S_2 \rangle_{\text{Iso}}$	T_{NI}(expt)(K)	ΔT(K)
0	0.384	540	0.13	553.4	−13.4
1	0.304	430	0.19	433.7	−3.7
2	0.228	465	0.09	460.9	+4.1

Table 12.3 Phase sequence and transition temperature for the nCB families of liquid crystals [41].

n	Alkyl chain	Cry		SmA		N	
		Expt	Simu	Expt	Simu	Expt	Simu
4	C_4H_9	321.2	—	—	—	289.6	—
5	C_5H_{11}	297.2	—	—	—	308.2	310
6	C_6H_{13}	287.6	—	—	—	302.2	300
7	C_7H_{15}	303.2	—	—	—	316.2	320
8	C_8H_{17}	294.6	—	306.6	—	313.6	317
9	C_9H_{19}	315.6	—	321.2	—	322.6	—
10	$C_{10}H_{21}$	317.2	—	323.6	—	—	—
11	$C_{11}H_{23}$	326.2	—	330.6	—	—	—
12	$C_{12}H_{25}$	321.2	—	332.0	—	—	—

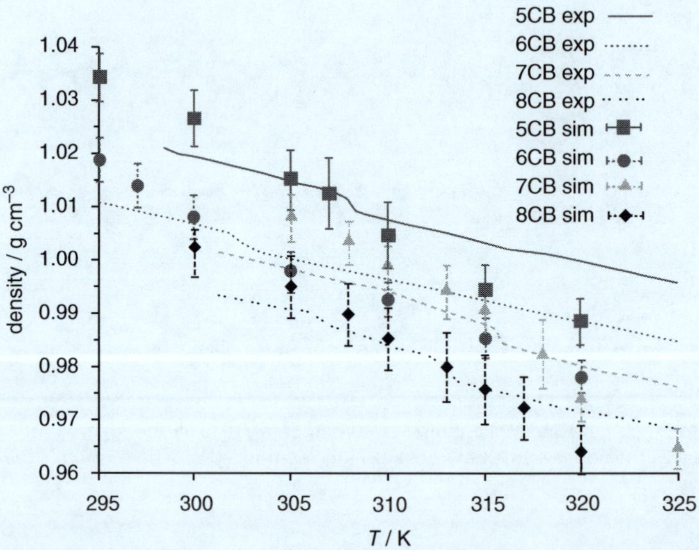

Figure 12.3 Comparison between simulated (symbols) and experimental (continuous lines) mass densities for 8CB homologues. Reproduced from [41] with permission of John Wiley and Sons Ltd.

in the low-temperature regime. Improvements in the force field parameters are needed to have better quantitative agreement with available experimental data.

Recently, Zhang et al. [42] have optimized some of the parameters in the well-known TraPPE-UA force field from the Siepmann group [43] to simulate various LC phases of 5CB molecules and reproduced various structural properties in good agreement with experimental results. They demonstrated the transferability of this FF to reproduce the phase behaviour of 8CB molecules as well. Although these new parameters are able to produce various LC phases in the correct temperature range, the dynamical quantities such as the diffusion constant of 8CB over a wide range of temperatures does not show good agreement with experiment.

Earl and co-workers have simulated various LC phases of di-, tri- and tetrafluoroterphenyls (Figure 12.4) using all-atom MD simulations [44]. They used the LCFF force field developed earlier by Wilson and co-workers. The same group have also simulated the binary mixture of alkoxy substituted phenylpyrimides 2-[4-(butyloxy)phenyl]-5- (octyloxy)pyrimidine (2PhP) and 2-[4-(tetradecyloxy)phenyl]-5-(tetradecyloxy)pyrimidine (PhP14) and presented a phase diagram of this mixture using fully atomistic simulation [45]. The molecular structures of these compounds are shown in Figure 12.5.

Earl et al. [45] were able to simulate very large system involving more than 100 000 atoms over 100 ns and produced liquid crystal phases of this mixture at various mixing ratios of 2PhP and PhP14. The binary phase diagram showing various LC phases is shown in Figure 12.6. To the best of our knowledge this is the only full phase diagram for such a complex thermotropic LC system using fully atomistic simulation. The pure 2PhP system exhibits isotropic (Iso), nematic (N), SmA, smectic C (SmC) and crystal (Cr) phases (Figure 12.6). Pure PhP14 exhibits similar phase behaviour except for the appearance of the SmA phase. The addition of PhP14 in the 2PhP system destabilizes

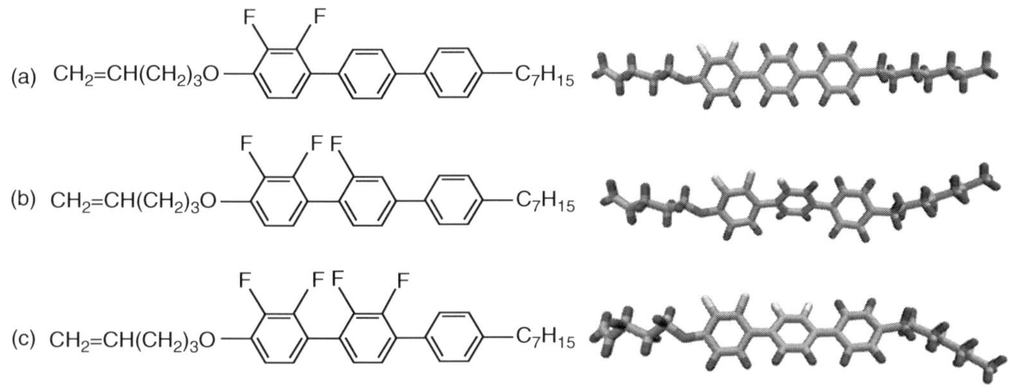

Figure 12.4 Molecular structures (left) and MM optimized geometries (right) of alkenic (a) di-, (b) tri- and (c) tetrafluoroterphenyls. The figure is adapted from Ref. [44]. Reproduced with permission of the Royal Society of Chemistry.

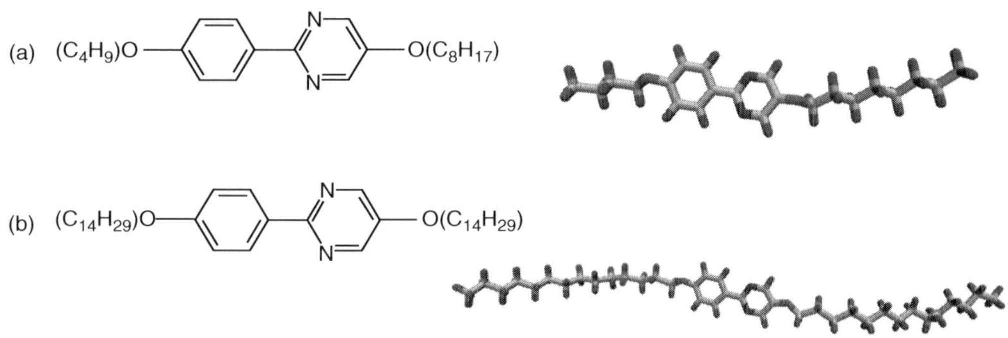

Figure 12.5 Molecular structures (left) and MM optimized geometries (right) of (a) 2-[4-(butyloxy) phenyl]-5-(octyloxy)pyrimidine (2PhP) and (b) 2-[4-(tetradecyloxy)phenyl]-5-(tetradecyloxy) pyrimidine (PhP14).

the SmC phase. The SmA layer spacing gradually increases with the addition of more and more PhP14 compound in the pure 2PhP system.

The authors also give a molecular level picture of the organization of PhP14 in the Sm host of 2PhP. The proposed organization is shown in Figure 12.7.

12.2.2 Coarse-Grained Model

So far we have given an overview of the simulation of various LC phases using fully atomistic models with a variety of atomistic FFs. Over the years both the accuracy of the employed FF and the size of the simulated systems have increased and helped us to understand the structure–property relationship at the molecular level. However, studying the self-assembly of various LC phases starting from an isotropic configuration is still computationally challenging and precludes simulation of large systems. Coarse-grained models are ideal for studying such self-assembly phenomena. Hard-core models are a class of coarse-grained models that have been extensively used in the context of self-assembly in thermotropic LC phases. They are relatively

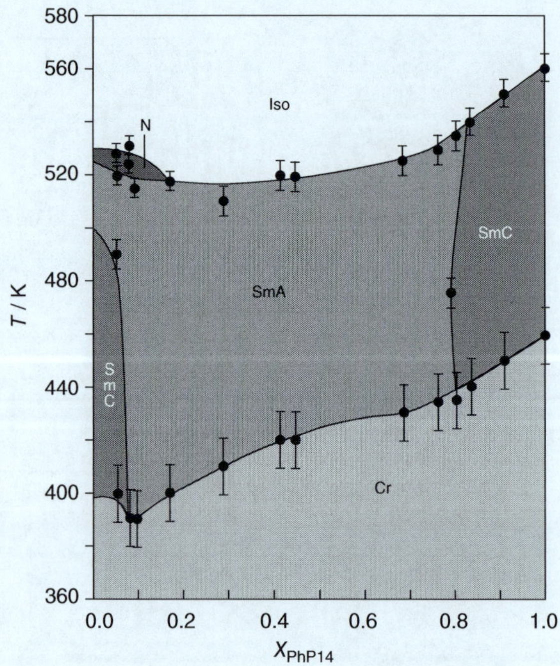

Figure 12.6 Phase diagram of the binary system 2PhP/PhP14 at a pressure of 1 atm. x_{PhP14} is the mole fraction of PhP14 in the mixture. Different phases exist for different ranges of x_{PhP14}. The nematic (N), phase exists at $0 \cdot 0 \leq x_{PhP14} \leq 0 \cdot 167$. The smectic A (SmA), phase exists at a very broad phase region. The smectic C (SmC), phase exists at lower mole fractions of PhP14, $0 \cdot 0 \leq x_{PhP14} < 0 \cdot 076$, and at higher mole fractions of PhP14, $0 \cdot 762 < x_{PhP14} \leq 1 \cdot 0$. Both isotropic (Iso), and crystal (Cr), phases cover the whole phase region at high and low temperatures respectively. Reproduced from [45] with permission of AIP.

simple to code, computationally very efficient, require modest computational resources and are also easy to treat theoretically. In particular, hard spherocylinders have been widely used to study various phases of liquid crystals [8, 9]. This model exhibits a rich phase behaviour which includes the formation of isotropic (I), nematic (N), smectic A (SmA), columnar (Col) and crystal (Cr) phases. The phase transitions in a system of hard spherocylinders are driven by the competition between two major entropic contributions: the orientational entropy that favours the isotropic phase and the positional entropy that favors the ordered phases, as shown by Onsager in the limit of infinitely thin rods [46]. In the literature, various simulation methodologies such as MC, MD and BD techniques have been used to simulate collections of thin rods with certain aspect ratios. Below we give a brief outline of the MC methodology for simulating hard spherocylinders. We refer the reader to the well-known and excellent review article on this subject by Allen et al. [47].

Two main types of models of thin rods can be envisaged: it can be viewed as a collection of connected spheres as shown in Figure 12.8a or a spherocylinder, as shown in Figure 12.8b.

Each MC step consists of translation and rotation of the rods. Euler-angle-based quaternions can be used to generate rotation moves during MC simulation. A move is accepted if there is no overlap between two rods after an MC step, otherwise the move is rejected. Hence an efficient overlap test is the key ingredient in this technique. In the case of rods consisting of spherical objects the overlap test is very simple. Two hard spheres of radius R_1 and R_2 (Figure 12.9) overlap if the distance between the centres of the two spheres r_{12} is less than $\sigma_{12} = R_1 + R_2$.

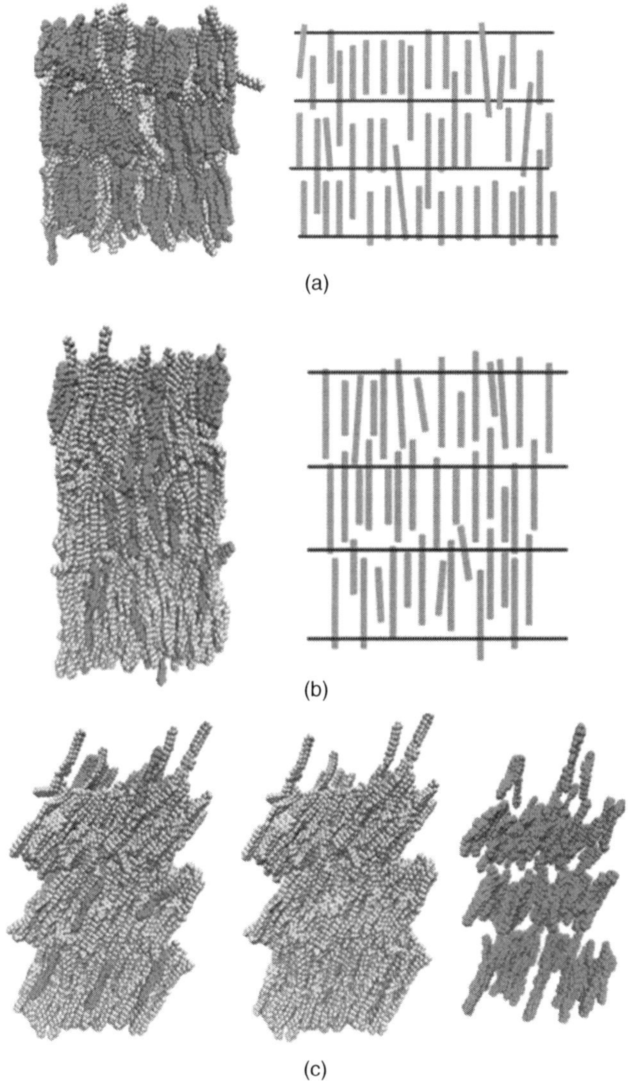

Figure 12.7 (a) Snapshot of an SmA phase for a mixture with xPhP14 = 0 · 167 at 485 K (left). Schematic sketch (right). (b) Snapshot of an SmA phase for a mixture with xPhP14 = 0 · 684 at 480 K (left). Schematic version (right). (c) Snapshot of an SmC phase for a mixture with xPhP14 = 0 · 833 at 460 K showing the whole mixture (left), only the 2PhP molecules (centre) and only the PhP14 molecules (right). The PhP14 molecules are represented in white and 2PhP molecules in gray in the snapshots. In the schematic PhP14 molecules are represented by tall lines while 2PhP molecules are represented by short lines. Reproduced from [45] with permission of AIP.

However, the computational expense is enormous as of the order of N^2 calculations are involved at each MC step, where N is the number of spherical beads that constitute one molecule of the system. To get rid of this issue cell/neighbour list are used (like in atomistic simulations) to reduce this cost to nN, where N is the number of beads constituting a molecule and n the average number of 'neighbours'. Another elegant way

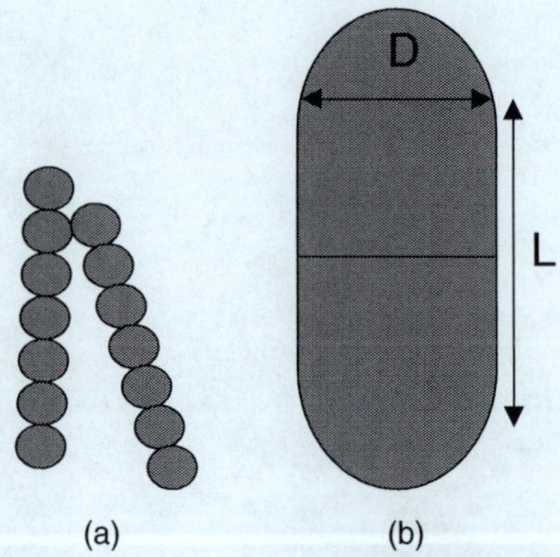

Figure 12.8 Models for thin rods: (a) connected spheres and (b) hard spherocylinder with aspect ratio L/D.

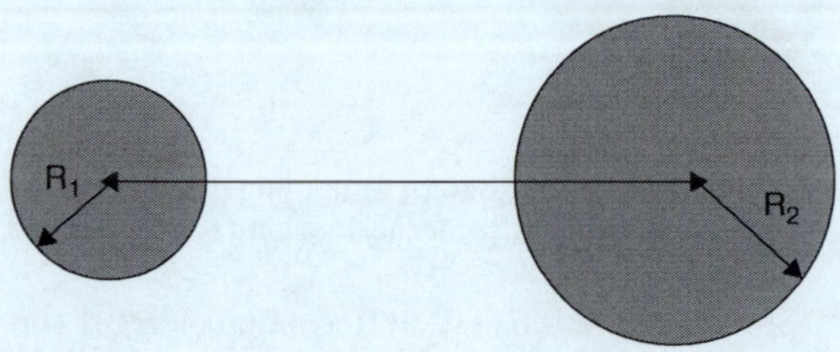

Figure 12.9 Overlap test for two rigid spheres of radius R_1 and R_2.

to treat the thin rod is to consider the molecule to be a spherocylinder. The overlap test for the spherocylinder shape is very efficient and in the case of very long molecules it is more efficient than the neighbour list method. The key steps are outlined below. More details can be found out in the brilliant article by Allen and co-workers [48]. The spherocylinder can be thought of as a set of points falling within a distance R from a line segment of length L. We can draw around every point on this line segment a sphere of radius R that contains all points that are within a distance R from that point.

The problem reduces to determining the minimum distance between two finite line segments i and j with orientations $\hat{u}_i$ and $\hat{u}_j$ and centres r_i and r_j. Any point on line i can be parametrically written as

$$\vec{r}_i(\lambda) = \vec{r}_i + \lambda\hat{u}_i \tag{12.1}$$

For line j, similarly, it is given by

$$\vec{r}_j(\mu) = \vec{r}_j + \mu\hat{u}_j \tag{12.2}$$

The vector distance between these two points is given by

$$\vec{r}_{ij}(\lambda, \mu) = (\vec{r}_i - \vec{r}_j) + \lambda\hat{u}_i - \mu\hat{u}_j \tag{12.3}$$

Now, we want to determine the values of λ and μ for which the distance r_{ij} is minimum. A simple method to find these values of λ and μ is the following. Construct the dot product of $\vec{r}_{ij}$ (Equation (12.3)) with $\hat{u}_i$ and $\hat{u}_j$. The shortest distance vector must be perpendicular to both $\hat{u}_i$ and $\hat{u}_j$, so we need to solve the following equations:

$$(\vec{r}_i - \vec{r}_j) \cdot \hat{u}_i = -\lambda\hat{u}_i \cdot \hat{u}_i + \mu\hat{u}_j \cdot \hat{u}_i \tag{12.4}$$

$$(\vec{r}_i - \vec{r}_j) \cdot \hat{u}_j = -\lambda\hat{u}_i \cdot \hat{u}_j + \mu\hat{u}_j \cdot \hat{u}_j \tag{12.5}$$

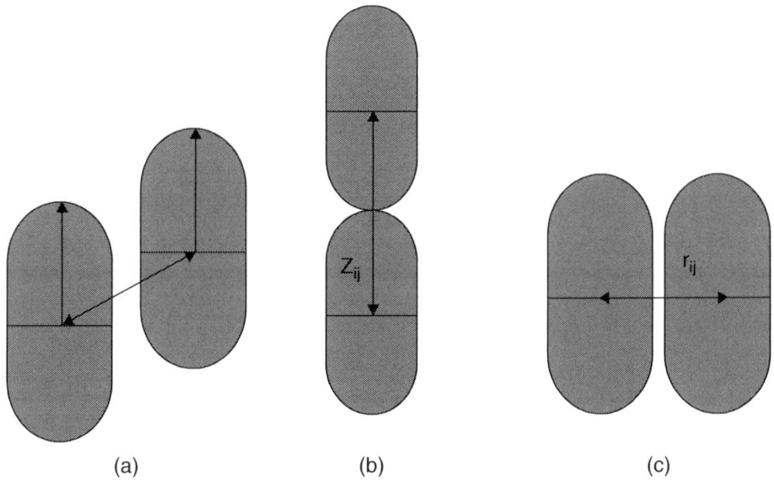

Figure 12.10 Various cases for computing overlap criteria for spherocylinders: (a) parallel to each other, (b) on top of each other, (c) touching each other in parallel fashion.

Solving Equations (12.4) and (12.5) we get the following values for λ and μ, which minimize the distance between the two line segments:

$$\begin{pmatrix}\lambda_0\\ \mu_0\end{pmatrix} = \frac{1}{1-(\hat{u}_i\cdot\hat{u}_j)^2}\begin{pmatrix}-\hat{u}_i\cdot\vec{\mathrm{r}}_{ij} + (\hat{u}_i\cdot\hat{u}_j)(\hat{u}_j\cdot\vec{\mathbf{r}}_{ij})\\ \hat{u}_j\cdot\vec{\mathrm{r}}_{ij} - (\hat{u}_i\cdot\hat{u}_j)(\hat{u}_i\cdot\vec{\mathbf{r}}_{ij})\end{pmatrix} \tag{12.6}$$

Equation (12.6) is a general expression which helps to check the overlap criteria for any general orientation of the spherocylinders. Below we discuss several special cases for which Equation (12.6) reduces to a much simpler form.

Case I: For an arrangement of spherocylinders as shown in Figure 12.10a, the overlap test reduces to a much simpler case as follows. For the parallel case we have

$$\vec{r}_{ij}\cdot\hat{u}_i = \vec{r}_{ij}\cdot\hat{u}_j \tag{12.7}$$

Using the above criteria in Equation (12.6), we get the following values for λ and μ:

$$\begin{aligned}\lambda_0 &= \frac{1}{1-(\hat{u}_i\cdot\hat{u}_j)^2}(-\hat{u}_i\cdot\vec{r}_{ij} + (\hat{u}_i\cdot\hat{u}_j)(\hat{u}_i\cdot\vec{r}_{ij}))\\ &= \frac{-\hat{u}_i\cdot\vec{r}_{ij}(1-\hat{u}_i\cdot\hat{u}_j)}{(1+\hat{u}_i\cdot\hat{u}_j)(1-\hat{u}_i\cdot\hat{u}_j)} = \frac{-\hat{u}_i\cdot\vec{r}_{ij}}{2}\end{aligned} \tag{12.8}$$

Similarly,

$$\mu_0 = \frac{\hat{u}_j\cdot\vec{r}_{ij}}{2} \tag{12.9}$$

Case II: For spherocylinders with a relative arrangement as shown in Figure 12.10b, the overlap test can be performed by calculating the distance between the centres of the two spherocylinders given by

$$z_{ij} = \frac{L_i}{2} + \frac{L_j}{2} + D \tag{12.10}$$

If $z_{ij} < \frac{L_i}{2} + \frac{L_j}{2} + D$ then there is overlap between the spherocylinders.

Case III: For an arrangement of spherocylinders similar to that in Figure 12.10c, the overlap can be detected if $r_{ij} < D$.

For all other general orientations of the spherocylinder we have to make a series of approximations to find the condition for overlap. In one case we set $\lambda = |L/2|$ and find out μ, which will minimize the distance between two spherocylinders. From Equation (12.6) we have

$$\vec{r}_{ij}^{\,2}(\lambda, \mu) = \vec{r}_{ij}^{\,2} + \lambda^2 + \mu^2 - 2\lambda\vec{r}_{ij} \cdot \hat{u}_i + 2\mu\vec{r}_{ij} \cdot \hat{u}_j - 2\lambda\mu\hat{u}_i \cdot \hat{u}_j \tag{12.11}$$

Minimizing the above expression with respect to μ, $\frac{d\vec{r}_{ij}^{\,2}(\lambda,\mu)}{d\mu} = 0$, we have

$$\mu = \lambda\hat{u}_i \cdot \hat{u}_j - \vec{r}_{ij}^{\,2} \cdot \hat{u}_j \tag{12.12}$$

Similarly, if we set $\mu = |L/2|$ and find out λ, which will minimize the separation between the spherocylinders, we get

$$\lambda = \mu\hat{u}_i \cdot \hat{u}_j - \vec{r}_{ij} \cdot \hat{u}_i \tag{12.13}$$

Frenkel and co-workers pioneered the use of the spherocylinder model to study LC phases [49–51] and for the first time demonstrated the existence of LC phases in such systems. At the same time, McGrother et al. [52] used the spherocylinder model and presented various LC phases for aspect ratios up to $L/D = 5$. For $L/D = 5$ they reported isotropic (I), nematic (N), smectic A (SmA) and crystal (Cr) phases. All the phase transitions were accompanied by a discontinuous jump in the density and so were concluded to be first order in nature. Later, Bolhuis and Frenkel [53] presented the complete phase diagram of the hard spherocylinder model from $L/D = 0$ to $L/D = \infty$. The phase diagram is shown in Figure 12.11.

The spherocylinder model has also been used to study the LC phase behaviour of bent-core molecules. Bent-core or banana-shaped molecules give rise to a new class of smectic LC phase (SmCP phase) [54, 55] which has attracted tremendous attention in the recent past due to its potential application in display and non-linear optical (NLO) devices. These molecules also present a very rare example of spontaneous formation of macroscopic chiral layers from achiral molecules. The molecules comprising these phases have 'bow' or 'banana' shaped cores (shown in the inset of Figure 12.12). We earlier reported MC simulation studies for a system of 400 banana molecules and mapped out the phase diagram as a function of the opening angle of the banana molecule [56]. The detailed phase diagram of the LC phases formed by banana molecules is shown in Figure 12.12. Our MC simulation gives rise to a variety of LC and crystal phases such as: isotropic fluid (I), nematic (N), polar smectic A (SmAP), smectic A (SmA), columnar (Col), polar crystal (XP) and crystal (X). In Figures 12.13 and 12.14, equilibrium snapshots for few LC phases for two opening angles are shown. The SmAP phase exists up to a certain value of the opening angle, disappears beyond that and changes to non-polar SmA. The polar and non-polar phases were differentiated from each other by monitoring in-plane polarization of the smectic phase. The in-plane polarization is calculated as $m = \sum_i m_i/N$, where m_i is the bow vector of the banana molecules defined in Figure 12.14. We also show, in Figure 12.14, the equilibrium configuration of the polar and non-polar smectic phases corresponding to opening angles of 165° and 172.5°

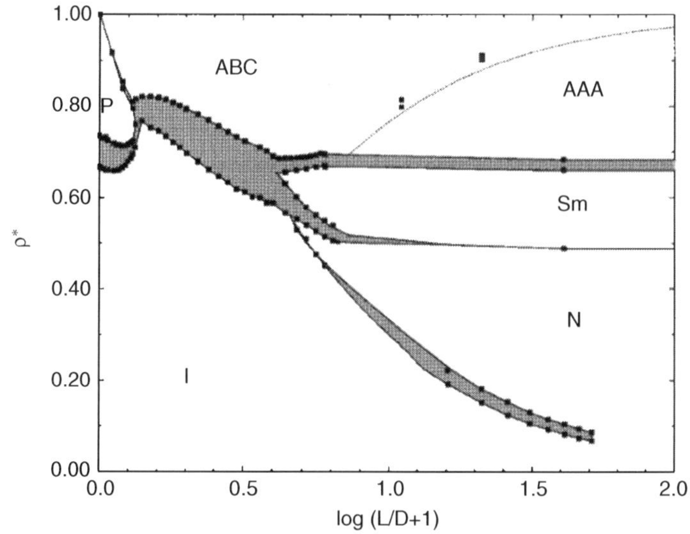

Figure 12.11 Phase diagram of the hard spherocylinder of aspect ratio L/D ranging from 0 to 100. AAA and ABC correspond to different arrangements of hexagonal crystal lattices. Reproduced from [53] with permission of AIP.

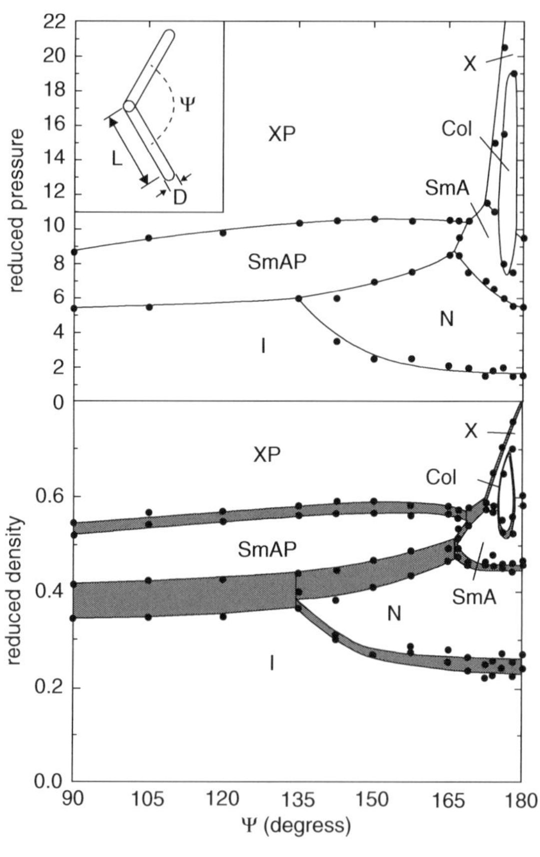

Figure 12.12 Phase diagram of bent-core liquid crystal as a function of opening angle obtained through MC simulation. The various phases shown include: isotropic fluid (I), nematic (N), polar smectic A (SmAP), smectic A (SmA), columnar (Col), polar crystal (XP) and crystal (X).

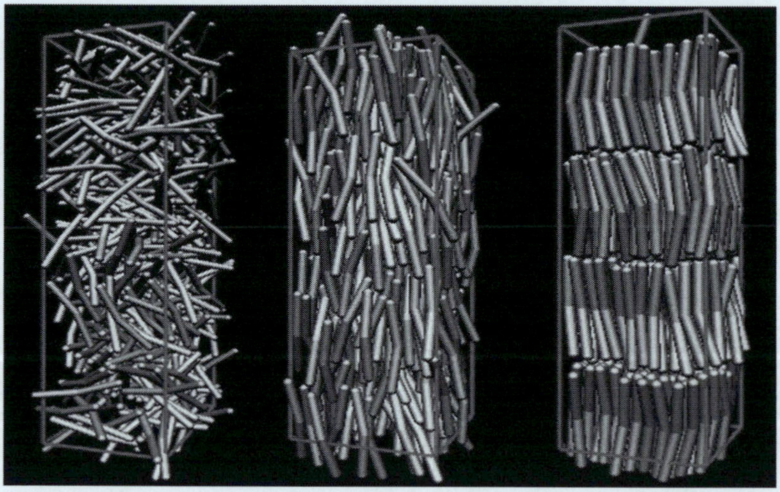

Figure 12.13 Final configurations from Monte Carlo simulations of $N = 400$ bent-core molecules with opening angle $\psi = 165°$ as a function of reduced pressure P^* ($P^* = \beta P v_0$, where v_0 is the volume of a zigzag molecule and P is the actual pressure). From left to right: isotropic phase ($P^* = 1$), nematic ($P^* = 5$) and polar smectic A ($P^* = 10$). *(See insert for color representation of the figure.)*.

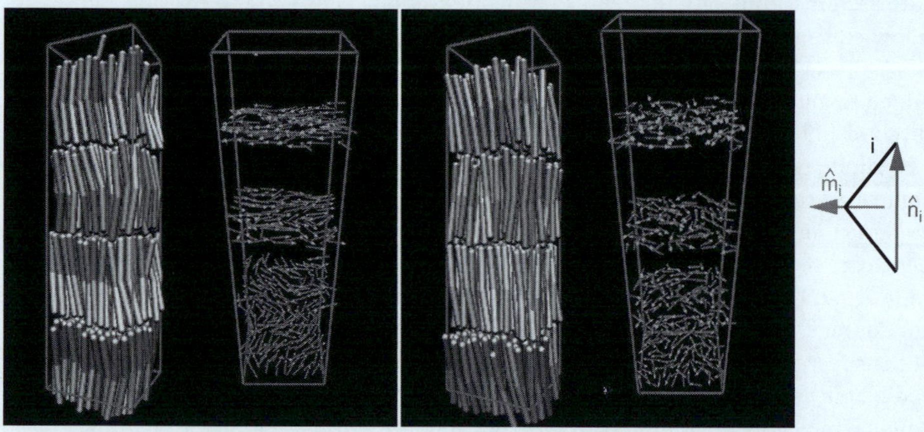

Figure 12.14 Configuration showing the polar smectic (SmAP) phase for an opening angle $\psi = 165°$ and $P^* = 10$. This phase disappears and becomes a non-polar smectic (SmA) at an opening angle of $\psi = 172.5°$ for the same value of P^*. We have also shown the in-plane arrangement of the bow vector for both cases.

respectively. Our MC simulation results also demonstrate the existence of the columnar phase for a narrow range of opening angle. Figure 12.15 shows the equilibrium configuration of a columnar phase for an opening angle of 178°.

We were also able to show, through umbrella sampling calculations, that the SmA antipolar phase was more stable than the SmAP phase, in good agreement with experiments.

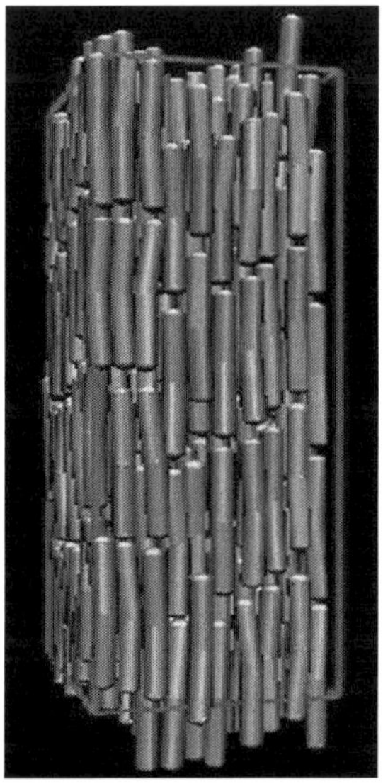

Figure 12.15 Columnar phase formed at an opening angle of $\psi = 178°$.

Using the spherocylinder model, Maiti et al. [57] have studied the effect of the addition of a small concentration of bent-core molecules in a smectic host. This study was motivated by the experiment performed by the Bangalore group at the Raman Research Institute [58]. In their experiment, Pratibha et al. showed that the addition of a low concentration of banana molecules to a smectic host gives rise to a new biaxial SmA phase. It is known in the literature that a small concentration of chiral solute added to a nematic LC phase formed of achiral molecules induces a helical twist of the director [59]. Maiti et al. simulated a mixture of rod-like and banana molecules using MC simulation in an NPT ensemble. They observed that the effect of adding bent-core solutes on the SmA phase depends on the shape (characterized by the banana opening angle $\boldsymbol{\psi}$) and size (characterized by the aspect ratio of the spherocylinders) of the bent-core molecule. Initially, a SmA phase was prepared using spherocylinders of aspect ratio $L_{rod}/D = 5$. Simulations were performed for two different mixtures. In one set of simulations, a 3% concentration of bent-core molecules made of two spherocylinders with aspect ratios of $L_{ban}/D = 5$ (the overall length of the banana molecules is twice the rod length) was added to the SmA host. In another case, a 3% concentration of shorter banana molecules made of two spherocylinders with aspect ratios of $L_{ban}/D = 2.5$ was added to the SmA host. It was found that in the case of bent-core molecules with aspect ratio $L_{ban}/D = 5$, the initial SmA phase changes to an anticlinic SmC_A ordering for large values of the opening angle ψ. The anticlinic tilt angle θ as a function of ψ is shown in Figure 12.16a. For $100° < \psi < 150°$, the rods in each layer attain a uniform tilt angle $\theta = (\pi - \psi)/2$ (Figure 12.16). For $\psi < 100°$, half of each banana molecule is nanophase segregated between the untilted smectic layers leading to an increase in smectic layer spacing (Figures 12.16 and 12.17). The tilt angle at different values of ψ is shown in Figure 12.16. It clearly depicts the presence of the SmC phase for a certain range of values of ψ for $L_{ban}/D = 5$. For $\psi > 150°$, molecules within each layer are uniformly tilted, but there is no anticlinic ordering on average (Figure 12.16). For smaller banana molecules with $L_{ban}/D = 2.5$, no anticlinic ordering is induced

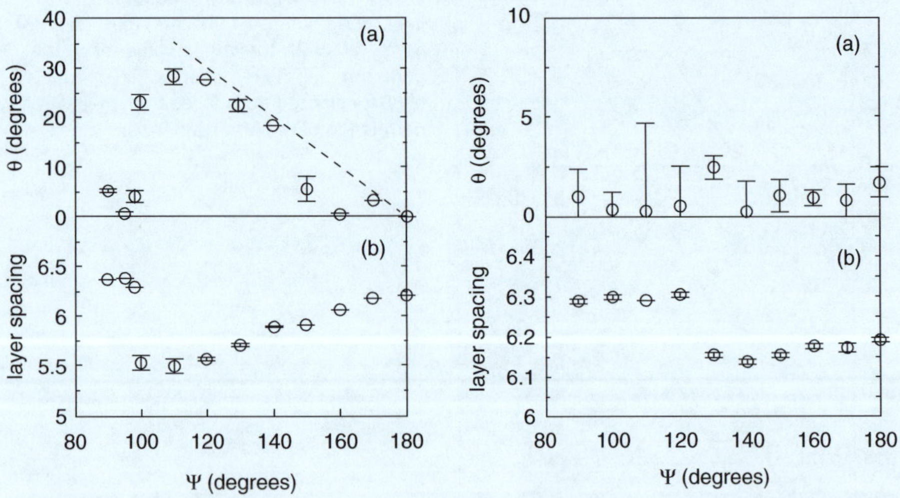

Figure 12.16 (a) SmC tilt angle θ and (b) layer spacing as a function of ψ for $L_{\text{ban}}/D = 5$ (left) and $L_{\text{ban}}/D = 2.5$ (right).

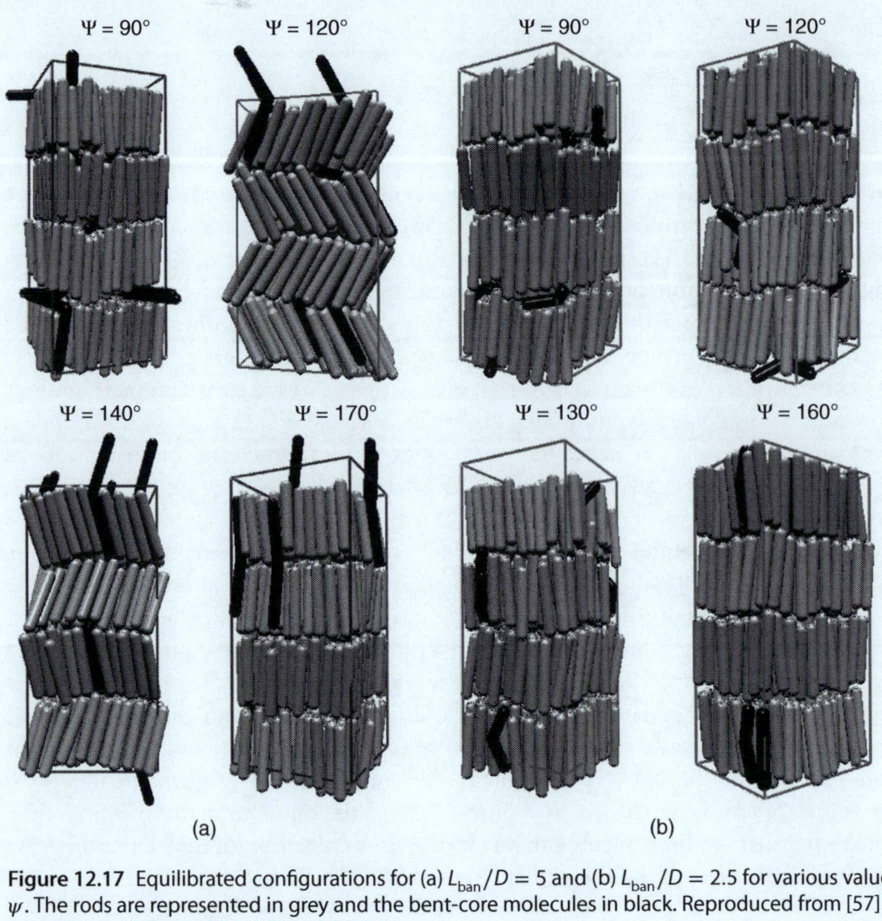

Figure 12.17 Equilibrated configurations for (a) $L_{\text{ban}}/D = 5$ and (b) $L_{\text{ban}}/D = 2.5$ for various values of ψ. The rods are represented in grey and the bent-core molecules in black. Reproduced from [57] with permission of the American Physical Society.

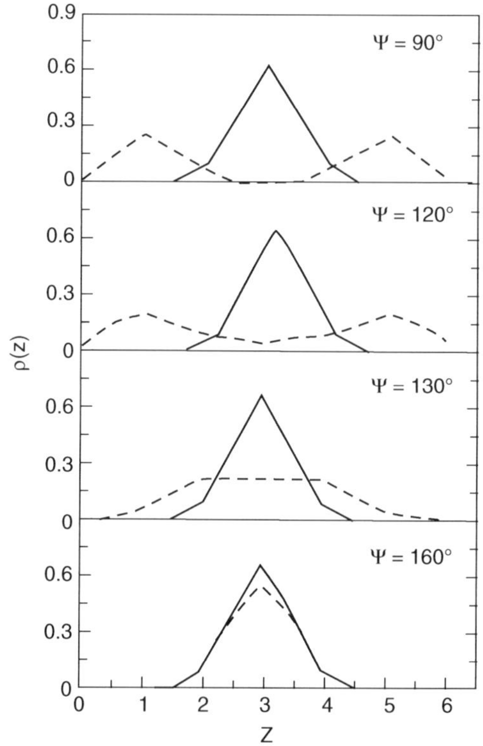

Figure 12.18 Density profiles for rods (solid lines) and bent-core molecules (dashed lines) for $L_{ban}/D = 2.5$ for various values of ψ. The bent-core density has been scaled by a factor of 30 for clarity. Reproduced from [57] with permission of the American Physical Society.

(Figures 12.16 and 12.17), but a sharp increase in smectic layer spacing is observed at $\psi = 130°$ resulting from an entropy-driven transition from intralamellar to interlamellar nanophase segregation of the bent-core molecules (Figure 12.16). The density profiles of the rods and bent-core molecules along the smectic layer at different values of ψ demonstrate this effect (Figure 12.18). This study with a coarse-grained spherocylinder model provides an understanding of the relation between shape and size of low-concentration solute on a solvent in the LC phase. The results and methodology can be used in molecular engineering to select a suitable solute shape to induce a desired tilt. The results on nanophase segregation due to smectic ordering and its dependence on the molecular shape can also be used to tune polymerization rates in monomer–LC mixtures.

In yet another study at the coarse-grained level, using spherocylinders as the building blocks, Maiti et al. [60] simulated a collection of zigzag-shaped molecules using the Monte Carlo technique in the NPT ensemble and demonstrated that simple excluded volume interactions are sufficient to produce a tilted lamellar (SmC) phase. The zigzag molecules are constructed using three rigidly fixed spherocylinders connected in a zigzag fashion (Figure 12.19 inset). For $35° < \psi < 80°$, the system exhibits the SmC phase. The phase diagram of the system with varying zigzag angle ψ and reduced pressure defined as $P^* = \beta P v_0$, where v_0 is the volume of a zigzag molecule and P is the actual pressure, is shown in Figure 12.19. This simulation for the first time demonstrated that the SmC phase can be induced in a collection of molecules by steric interactions arising out of the molecular shape alone.

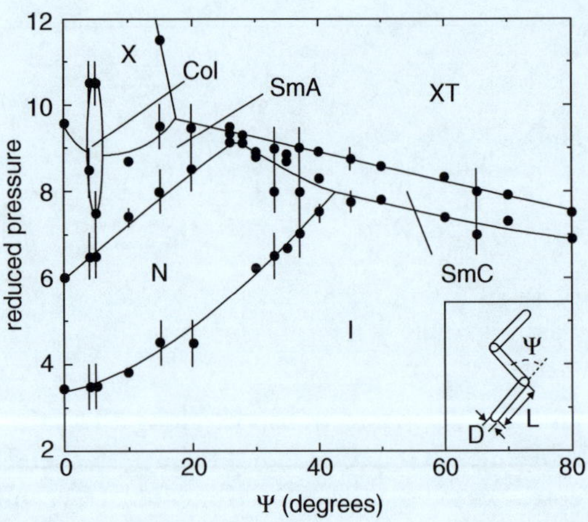

Figure 12.19 Phase diagram of zigzag-shaped molecules with aspect ratio $L/D = 2$ as a function of opening angle (ψ) and reduced pressure (P^*). The phase diagram shows the existence of various phases: isotropic liquid (I), nematic (N), smectic A (SmA), smectic C (SmC), columnar (Col), tilted crystal (XT) and crystal (X). The zigzag molecule is shown in the inset and the zigzag angle (ψ) defined. Reproduced from [60] with permission of the American Physical Society.

12.3 Discotic Liquid Crystals

The thermotropic liquid crystal phases described so far are mostly formed by rod-like molecules. In this section we are going to discuss discotic columnar liquid crystalline (DCLC) phases, which are typically formed by the self-assembly of disk-like molecules, consisting of a disk-shaped aromatic core with several aliphatic chains attached along the edge of the core in a symmetric manner. The DCLC phase was first observed by Chandrasekhar, Sadashiva and Suresh in 1977 [61]. The compounds they investigated were benzene-hexa-*n*-alkanoates (BHn-alkanoates), whose molecular structure is shown in Figure 12.20.

In DCLC phases the discotics stack on top of each other to form a column and the columns are further arranged in a regular lattice. Depending on the arrangement of the columns, various DCLC phases can form: hexagonal columnar discotic LC phase, triangular LC phase, etc. The core–core and tail–tail van der Waals interactions are mainly responsible for the stacking of the disks in a column [62]. Like other classes of liquid crystal phases, the DCLC phase shares some crystal-like properties like optical, electrical and magnetic anisotropy, and fluid-like properties like the inability to support shear stress (in one or more direction).

DCLC materials are also promising candidates for molecular electronics [63–65] because they can provide anisotropic charge transport. Charge transport in the DCLC phases mainly happens through the stacked aromatic cores along the column. To get

Figure 12.20 Molecular structure of BHn-alkanoates.

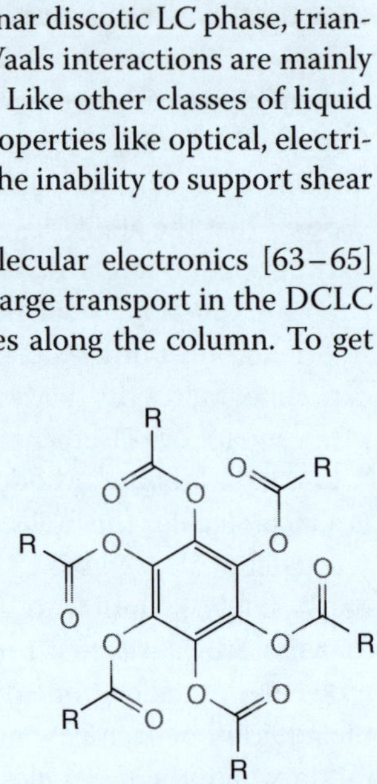

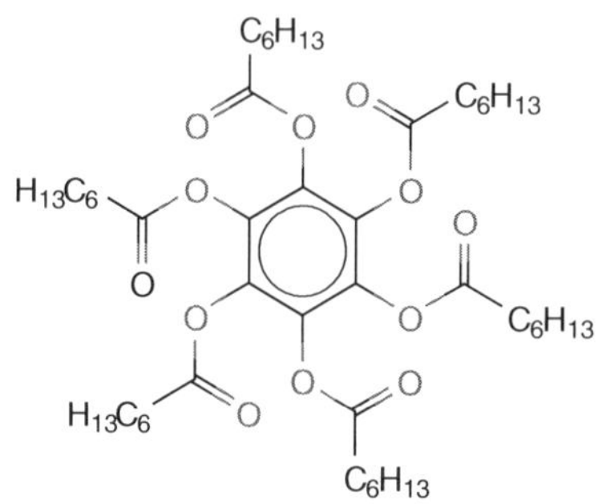

Figure 12.21 Molecular structure of BHA7.

an efficient charge transport along the columnar direction the arrangement of the molecules in the column has to be perfect and the $\pi - \pi$ overlap between the aromatic cores has to be sufficiently strong [66]. Developing a DCLC material with a well-ordered column and high $\pi - \pi$ overlap between the aromatic cores by tuning the structure of the core and tails of the discotic molecules has been an active area of research for the past several years. The technique of confining the DCLC in a nanochannel to get a high-quality columnar order has recently been studied by few groups [67, 68].

Benzene-hexa-*n*-alkanoates (BHn-alkanoates) [69], discotic hexakis triphenylene (HAT) [62, 70] and discotic hexabenzocorone (HBC) [71] and its derivatives are some of the widely studied discotic materials in the literature. In most of these cases the tails are aliphatic. Recently, a disotic liquid crystal with aromatic oligothiophene connected with a long alkyl chain as the tail was studied [72]. Beltran et al. [73] have studied self-assembly of star-shaped discotic liquid crystal.

One of the early simulation studies on the DCLC phase was performed by Maliniak et al. [69]. They simulated the isotropic phase of BHA7 molecules (Figure 12.21), which have the following phase sequence [74]: $solid \xrightarrow{81.2^{\circ}\mathrm{C}} D \xrightarrow{87.0^{\circ}\mathrm{C}} isotropic.$

Back in 1992, these authors managed to simulate a system with 32 molecules in an isotropic phase using the united atom force field, where the bonded terms were taken from the MM2(87) force field [75] and non-bonded terms were treated using the OPLS force field [26]. The final structure of the molecules at the end of a 100 ps MD simulation showed that the aliphatic chains of the molecules are alternately up and down relative to the aromatic core. An association of molecules in pairs was revealed by the radial distribution function calculation. The order parameter calculation pointed out a decrease in orientational order along the chains of the discotic, which was consistent with the experimental results that were performed as part of the same work. Calculation of the mean square displacement showed that the chains are more mobile than the cores.

In a later attempt, Mulder et al. [62] simulated the discotic columnar phase of the HAT6 molecule. The molecular structure of HAT6 is shown in Figure 12.22.

Mulder et al. used the COMPASS force field [76] in their simulation. They simulated only a single column with four molecules. The initial configuration of the column with four molecules was first prepared using the following two different methods: (i) Four molecules are added one by one to the column. Energy minimization was done after the addition of each new molecule. This gives rise to what is known as the S configuration where the molecules are rotated 30° or 60° with respect to each other. (ii) The

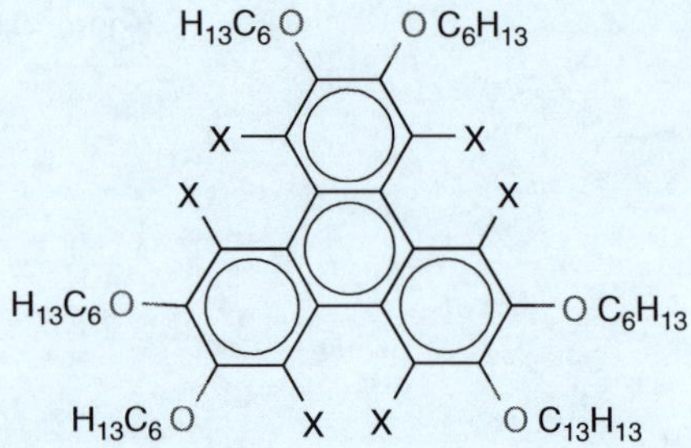

Figure 12.22 Molecular structure of discotic HAT6.

four-molecule column was heated and then annealed. This gives rise to an L configuration where the molecules are on top of each other. MD simulations were performed at 300 K starting from the above mentioned initial configurations.

At the end of the MD simulation the starting S configuration turned out to be the one with the lowest energy and it matched quite well with the crystallographic report. In their article the authors hypothesized that core/core and tail/tail van der Waals interactions are responsible for the self-assembly of the columns. To substantiate their claim, they also calculated the force constant related to the in-plane and out-of-plane motion of the molecules by numerically evaluating the derivative of the energy. They further divided the contribution of the force constant into core and tail parts. The results are presented in Table 12.4.

From the data in Table 12.4 it is clear that the van der Waals interactions dominate and that the in-plane force constants are considerably higher than their out-of-plane counterparts. Interestingly, the in-plane force constant for the tails is much higher than that for the cores. These results prove that the tail/tail van der Waals interaction is responsible for the rigidity of the columns.

The first simulation study where multiple columns were taken into account was reported by Giorgio et al. [70] in 2004. They managed to simulate 80 HAT5 molecules arranged in 8 columns with 10 molecules in each column. The chemical structure of a HAT5 molecule is shown in Figure 12.23.

Columns were further arranged in hexagonal lattices. Hydrogen atoms belonging to the core were explicitly considered while the united atom approach was used for the alkyl chains. After an initial energy minimization they performed NPT simulation at $T = 375$ K corresponding to the middle of the temperature range of the discotic columnar phase. A production run of almost 2 ns was performed. Berendsen temperature and

Table 12.4 Force constants resulting from MD simulations. Reproduced from [62] with permission of the American Chemical Society.

	Force constant of the core (kcal mol^{-1} Å^{-2})	**Force constant of the tail (kcal mol^{-1} Å^{-2})**
Δ van der Waals' out of plane	3.3	60.1
Δ Coulomb out of plane	1.1	3.1
Δ Coulomb in plane	5.7	4.4
Δ van der Waals' in plane	165.2	455.3

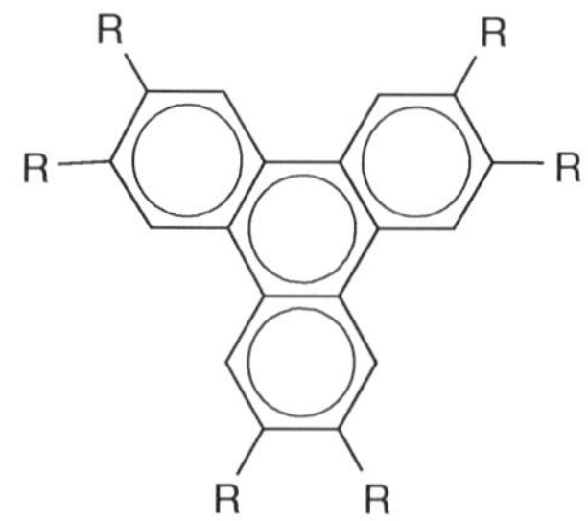

Figure 12.23 Molecular structure of HAT5 molecule.

pressure coupling [77] was used for temperature and pressure regulation. During the simulation the system maintained the hexagonal structure, which was further confirmed by measuring the value of the 6-fold bond orientational order parameter Ψ_6. Various correlation functions were measured to examine the ordering of the molecules in a column. They also measured the mean squared displacement in the NVE ensemble to compute the diffusion constant, but the time scale was too short for a reliable estimate of dynamical quantities. The mean square displacement measurement showed a solid-like behaviour along the direction perpendicular to the column, whereas a fluid-like sliding motion of the column was detected along the columnar direction.

The discotics of hexabenzocoronene (HBC) derivatives were first simulated by Andrienko et al. [71]. The different systems simulated by Andrienko et al. are shown in Figure 12.24. They also followed the united atom approach for the side chains, and the hydrogen attached to the cores was explicitly considered. A system of 160 molecules within 16 columns and 10 molecules in each column was simulated. The columns were arranged in a hexagonal lattice. In this study also, the same kinds of order parameters and correlation functions were measured as mentioned in the previous paragraph.

HBC derivatives in which the side chains are also aromatic were first simulated by Saientan et al. [78]. They simulated the discotic hexabenzocoronene/oligothiophene hybrid (W732) which has a molecular structure as shown in Figure 12.25a. Initially, a system of 160 molecules arranged in 16 columns with 10 molecules in each column was

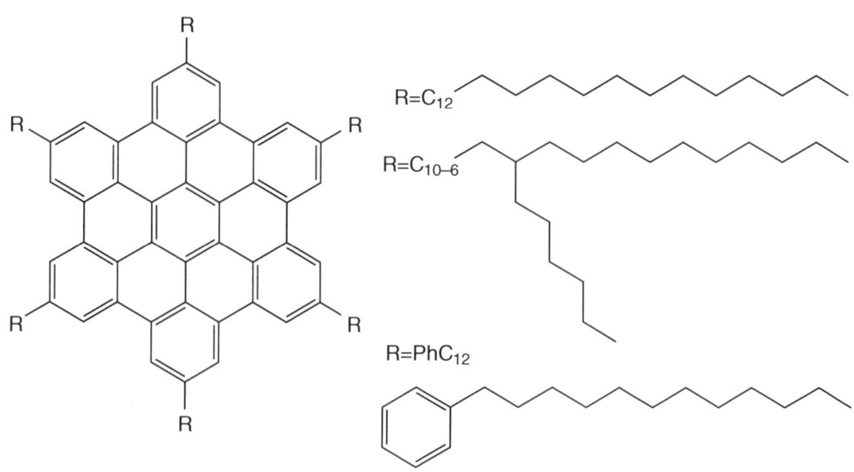

Figure 12.24 Molecular structure of HBC derivatives.

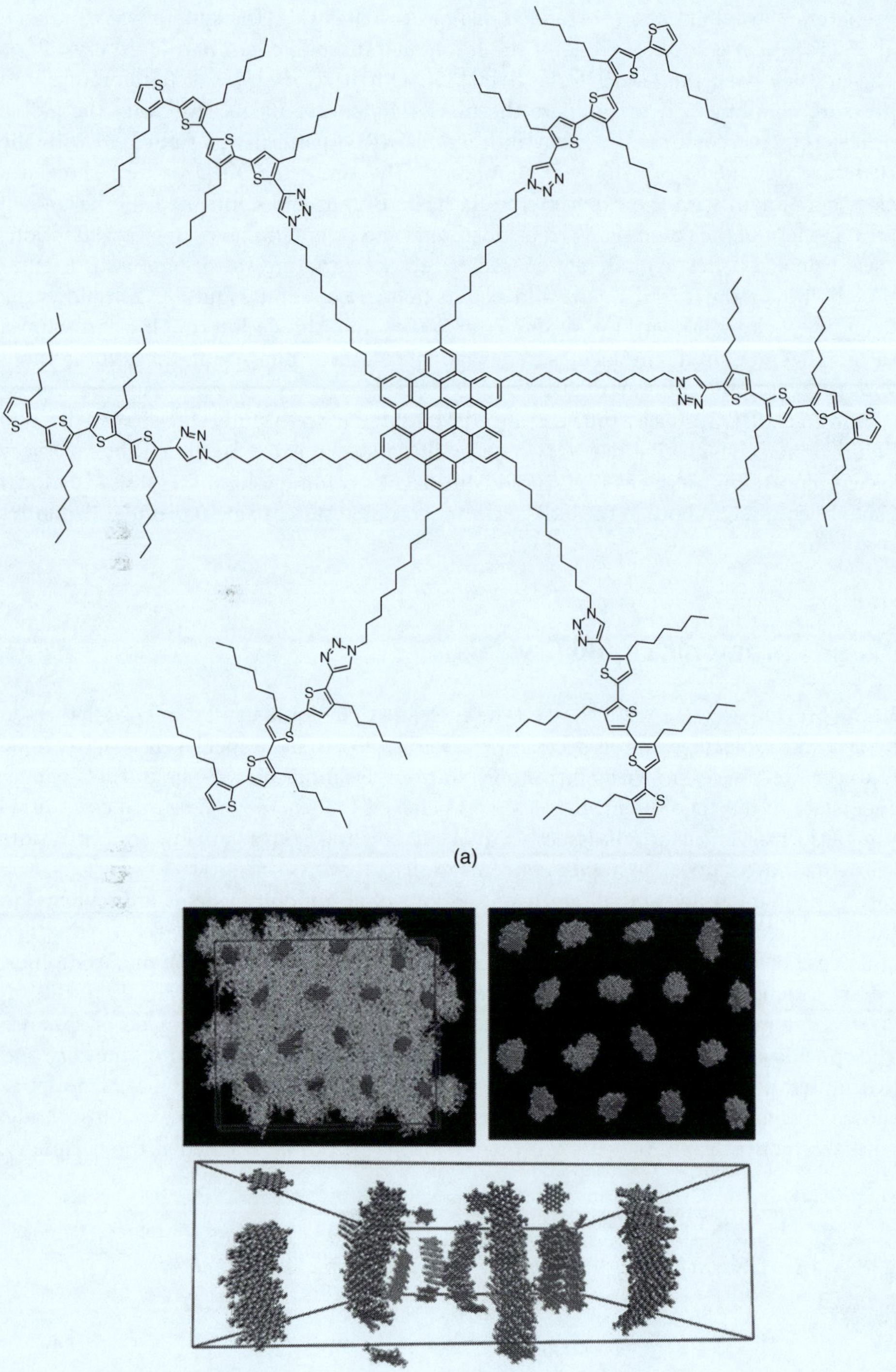

Figure 12.25 (a) Molecular structure of the HBC/oligothiophene hybrid. (b) Molecular arrangement in the hexagonal columnar phase after 200 ns of MD simulation.

prepared. The columns were arranged in a hexagonal lattice. The system was simulated at 357 K, which is in the middle of the hexagonal discotic columnar liquid crystalline phase. They used the GAFF force field [35] with Berendsen weak temperature and pressure coupling [77] to simulate the phase for almost 500 ns. To verify the global ordering of the columns, an X-ray diffraction (XRD) pattern was simulated with the structure obtained from the MD simulation. The simulated XRD pattern showed a clear agreement with the experimental lattice constant and confirmed the hexagonal arrangement of the columns. Various positional and orientational correlation functions were calculated that completely characterized the ordering (positional and orientational) of the aromatic HBC core and also its tails in a column. Figure 12.25b shows the instantaneous snapshot of the system after 200 ns long simulation run. It clearly shows the arrangement of the molecules in hexagonal columnar phase with the column tilted with respect to the columnar axis.

Beltrán et al. [73] successfully employed MD simulation to study star-shaped discotic liquid crystals which give rise to two different nanosegregated architectures. In a very recent report, Busselez et al. [79] studied the discotic columnar liquid crystals of a model Gay – Berne system both in the bulk and nanoconfined states using molecular dynamics simulation.

12.4 Chromonic Liquid Crystals

In the previous section we discussed the thermotropic liquid crystalline phase formed by disk-like molecules. In this section we are going to discuss a special class of lyotropic liquid crystal phase, known as chromonic. Various dye molecules, drugs and anti-cancer agents are in the class of chromonics. It is believed that there is no critical concentration of the molecules for self-assembly to occur: starting from a very low concentration, self-assembly occurs at all higher concentrations.

Both coarse-grained and all-atom simulations of chromonic systems are reported in the literature [80, 81], but the number of simulations attempted in this area is rather low and begs for more attention to be given to the technological application of chromonics.

First attempt at simulating the chromonic phase was made by Maiti et al. [80] in 2001. They modelled the chromonic molecules as disc-like objects made of spheres. The spheres are made hydrophobic or hydrophilic to mimic the hydrophobic core and hydrophilic periphery of the chromonic. They developed two different types of model, as shown in Figure 12.26, with hydrophobic and hydrophilic units shown explicitly. Model 1 has a larger hydrophobic core with two hydrophilic beads attached at the periphery.

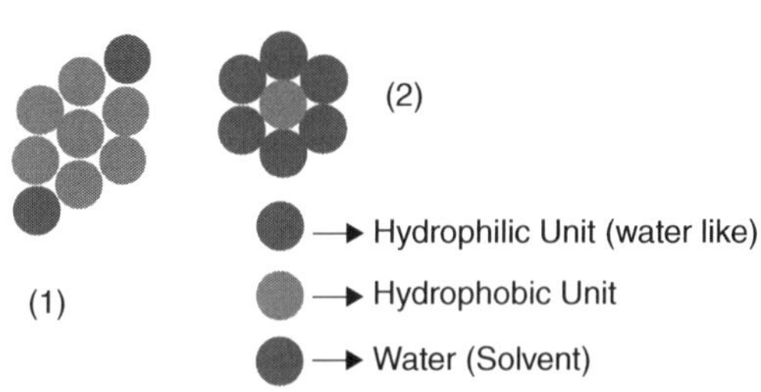

Figure 12.26 A schematic representation of model chromonic and water molecules.

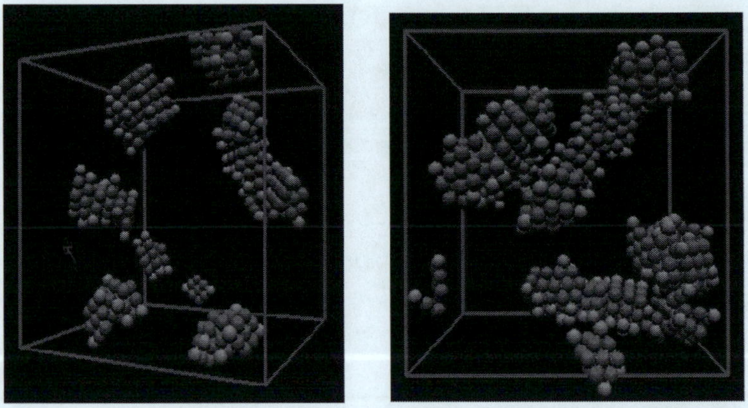

(a) (b)

Figure 12.27 Simulation snapshot of the system after a simulation run starting from isotropic conditions for low (a) and high (b) concentration.

Model 2 has a single hydrophobic bead at the core surrounded by a larger number of hydrophilic beads at the periphery. Water was explicitly treated and also modelled as a hydrophilic sphere in their work.

Interactions between the same type of particles and between the water and hydrophilic particles were modelled as Lennard-Jones (LJ) potentials. To model the water–hydrophobic and hydrophobic–hydrophilic interactions, a short-range repulsive WCA potential (i.e. truncated and shifted LJ potential) was used.

The model chromonic and water molecules were dispersed randomly in a box and Monte Carlo simulation was performed in an NPT ensemble. Simulations were performed at various concentrations of the chromonic. Chromonic molecules were found to form short columns (Figure 12.27) at low concentration while increases in columnar length and the number of aggregated structures were observed (Figure 12.27) with an increase in concentration. To understand the molecular order present in the columnar structure, various positional and orientational correlation functions were calculated. The calculations confirmed the arrangement of the molecules in a columnar fashion with the molecular planes parallel to each other. They also studied the variation of the average cluster size (L) with increasing concentration (ϕ). At low concentration, L was found to vary with ϕ in a power-law fashion with an exponent equal to 0.6, but at higher concentrations the dependence of L on ϕ turned out to be stronger than the power law.

Starting from an isotropic initial organization, the formation of well-arranged columnar aggregates is very time consuming and was not possible with the computational resources available in 2001. So, they simulated a set of preassembled columns of chromonics for both models 1 and 2. The columnar arrangement for model 1 was stable throughout the simulation (Figure 12.28). For model 2 the columnar structure was unstable and the system ended up in an isotropic phase. This proved the importance of the ratio of hydrophobic and hydrophilic units in a chromonic for spontaneous self-assembly to occur.

Recently, Glaser et. al [81] reported a Monte Carlo simulation of a collection of hard cylinders with aspect ratios (L/D) 0.5, 0.9 and 2. The cylinders interact through hard

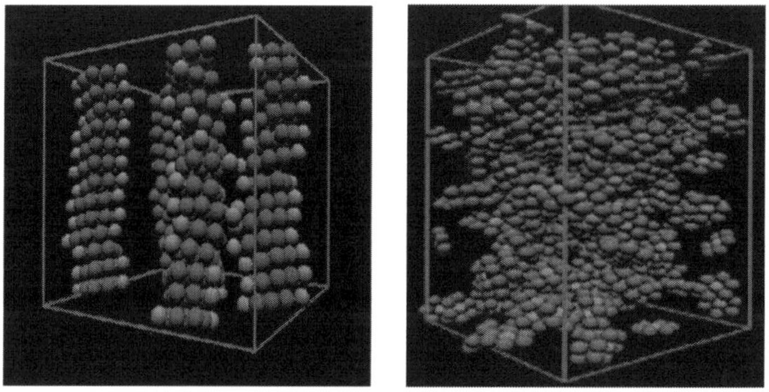

(a) (b)

Figure 12.28 Simulation snapshot for the system with (a) model 1 and (b) model 2, after the simulation run starting from a pre-assembled configuration.

core excluded volume interaction but can stack and bind end-to-end. To maintain the sticky end condition the following form of potential was incorporated:

$$U(r) = -E_{\mathrm{bond}}[1 - (r/r_c)^{\gamma}] \quad \text{for } r < r_c$$
$$= 0 \quad \text{otherwise}$$

where r is the distance between the centres of the circular faces of two neighbouring cylinders, $r_c = D/2$ and γ and E_{bond} are two adjustable parameters. This model applied originally to the stacking of short DNA strands but it is rather similar to chromonic assembly. In the context of chromonics, only the simulation with L/D ratio 0.5 is relevant because the cylinder can mimic the flat disk like chromonic molecules only for very low aspect ratios.

The system was seen to order spontaneously only above a particular concentration. At lower concentrations a nematic phase of the columns was formed while higher concentrations gave rise to hexagonally ordered columns as typically found in the chromonic phase. The model failed to predict a very important property of the chromonic: the occurrence of self-assembly at all concentrations, for the obvious reason that no effect of the underlying solvent was taken into account. But this very simplified model with a simple interaction could predict the qualitative behaviour of the chromonics satisfactorily.

Very recently, F. Chami et al. [82] simulated the chromonic phase of disk-like anionic azo dye Sunset Yellow (SSY) in an aqueous solution using fully atomistic simulation. The molecular structure of SSY is shown in Figure 12.29.

The MD simulation was perfomed using GAFF [35] parameters in the AMBER force field [39]. The TIP3P model was used to describe water. They used the CHELPG [83] method to calculate the atomic charges. Na^+ counter ions were added to neutralize the system. They simulated the chromonic phase at two different concentrations of 5 wt% and 33 wt%. They studied systems consisting of both pre-assembled columns and isotropic dispersed molecules. The pre-assembled stack of eight molecules system (5 wt%) turned out to be stable over the course of a long simulation run (200 ns). The molecules in a column preferred anti-parallel orientation with infrequent but rapid rotational transitions between parallel and anti-parallel configurations. For the

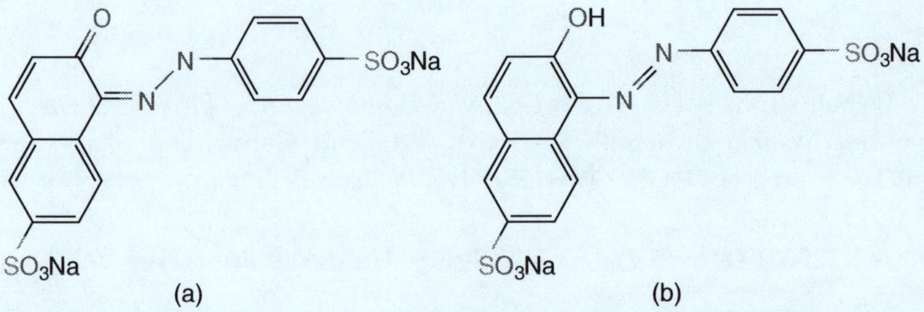

Figure 12.29 Molecular structure of SSY in NH hydrazone form (a) and hydroxy azo form (b).

dispersed system, full aggregation of eight molecules in a stack was achieved within 200 ns. At the concentration of 33 wt% (matches with the nematic phase of SSY) the columns showed a pseudohexagonal packing with a distance of 2.35 nm between the columns. They also calculated the free energy of binding of a molecule in a stack. A value of $7\,k_B T$ was measured for stacks containing three and eight molecules while an increase in binding energy by $\%2\,k_B T$ was obtained for the dimer. This led them to confirm the isodesmic (free energy change due to adding a new molecule to a stack is independent of the stack size after a particular stack size is reached) nature of self-assembly in chromonic systems.

12.5 Conclusion and Outlook

In this chapter, we have given a comprehensive summary of the status of the existing computational studies of various liquid crystal phases. The simulation methodologies covered in this article range from the idealized spherocylinder model to the fully atomistic model. Liquid crystals have wide range of applications in various technological applications ranging from display, non-linear optics, charge transport to drug delivery. These applications require a proper understanding of the structure–property relationship and computer simulation has provided many key results in this aspect. In particular, atomistic simulations have provided deeper understanding of the effect of molecular shape on the phase behaviour for a variety of complex LC systems. With the increase in computational power and development of advanced and efficient simulation methodologies and accurate force fields, future computer simulations will help in the engineering of various self-assembly phenomena with their predictive power. This will also enable the study of the liquid crystalline phase behaviour of aggregates of much more complex systems like large nucleic acids and biomolecules, hence providing an insight into the complex self-assembly phenomena occurring inside living cells.

Acknowledgment

We thank DST, India for financial assistance. PKM thanks USIEF for a Fulbright-Nehru Senior Research Fellowship that enabled his stay at the University of Colorado, Boulder where part of this work was carried out.

References

1. Evans DF & Wennerström H (1999) *The Colloidal Domain: Where Physics, Chemistry, Biology, and Technology Meet* (Wiley-VCH, Weinheim, Germany) p. 672.
2. de Gennes PG & Prost J (1995) *The Physics of Liquid Crystals* (Oxford University Press, New York) p 616.
3. Chandrasekhar S (2008) *Liquid Crystals* (Cambridge University Press, New York) p. 480.
4. Taylor MP & Herzfeld J (1993) Liquid-crystal phases of self-assembled molecular aggregates. *Journal of Physics-Condensed Matter* **5**(17):2651–2678.
5. Lydon J (2004) Chromonic mesophases. *Curr. Opin. Colloid Interface Sci.* **8**(6):480–490.
6. Wilson MR (2007) Molecular simulation of liquid crystals: Progress towards a better understanding of bulk structure and the prediction of material properties. *Chemical Society Reviews* **36**(12):1881–1888.
7. Wilson MR (2005) Progress in computer simulations of liquid crystals. *International Reviews in Physical Chemistry* **24**(3–4):421–455.
8. McGrother SC, Williamson DC, & Jackson G (1996) *J. Chem. Phys.* **104**:6755.
9. Bolhuis P & Frenkel D (1997) *J. Chem. Phys.* **106**:666.
10. Williamson DC & Jackson G (1998) *J. Chem. Phys.* **108**:10294.
11. Maiti PK, Kumar AV, & Ayappa K (2009) Computer simulation of liquid crystals. *Journal of the Indian Institute of Science* **89**(2):229–241.
12. Gay JG & Berne BJ (1981) *J. Chem. Phys.* **74**:3316.
13. Bates MA & Luckhurst GR (1999) *J. Chem. Phys.* **110**:7087.
14. Affouard F, Kroeger M, & Hess S (1996) *Phys. Rev. E.* **54**:5178.
15. Wilson MR & Allen MP (1991) *Mol. Cryst. Liq. Cryst.* **198**:465.
16. Glaser MA, Malzbender R, Clark NA, & Walba DM (1994) *J. Phys.: Condens. Matter* **6**:A261.
17. Glaser M (2000) Atomistic Simulation and Modeling of Smectic Liquid Crystals. *Advances in the Computer Simulatons of Liquid Crystals, NATO Science Series*, eds Pasini P & Zannoni C (Springer Netherlands), Vol **545**, pp. 263–331.
18. Praprotnik M, Delle Site L, & Kremer K (2008) Multiscale simulation of soft matter: From scale bridging to adaptive resolution. *Annual Review of Physical Chemistry* **59**:545–571.
19. Murtola T, Bunker A, Vattulainen I, Deserno M, & Karttunen M (2009) Multiscale modeling of emergent materials: Biological and soft matter. *Physical Chemistry Chemical Physics* **11**(12):1869–1892.
20. Muller-Plathe F (2002) Coarse-graining in polymer simulation: From the atomistic to the mesoscopic scale and back. *Chemphyschem* **3**(9):754–769.
21. Care CM & Cleaver DJ (2005) Computer simulation of liquid crystals. *Rep. Prog. Phys.* **68**(11):2665–2700.
22. Shelley JC & Shelley MY (2000) Computer simulation of surfactant solutions. *Curr. Opin. Colloid Interface Sci.* **5**(1–2):101–110.
23. Wilson MR & Allen MP (1991) Computer simulations of mesogenic molecules using realistic atom–atom potentials. *Molecular Crystals and Liquid Crystals* **198**(1):465–477.

24. Wilson MR & Allen MP (1992) Structure of trans-4-(trans-4-*n*-pentylcyclohexyl) cyclohexylcarbonitrile (CCH5) in the isotropic and nematic phases: A computer simulation study. *Liquid Crystals* **12**(1):157–176.
25. Cook MJ & Wilson MR (2001) The first thousand-molecule simulation of a mesogen at the fully atomistic level. *Molecular Crystals and Liquid Crystals Science and Technology. Section A. Molecular Crystals and Liquid Crystals* **363**(1):181–193.
26. Jorgensen WL, Maxwell DS, & Tirado-Rives J (1996) Development and testing of the OPLS all-atom force field on conformational energetics and properties of organic liquids. *Journal of the American Chemical Society* **118**(45):11225–11236.
27. McBride C, Wilson MR, & Howard JAK (1998) Molecular dynamics simulations of liquid crystal phases using atomistic potentials. *Molecular Physics* **93**(6):955–964.
28. Allinger NL, Yuh YH, & Lii JH (1989) Molecular mechanics. The MM3 force field for hydrocarbons. 1. *Journal of the American Chemical Society* **111**(23):8551–8566.
29. Lansac Y, Glaser MA, & Clark NA (2001) Microscopic structure and dynamics of a partial bilayer smectic liquid crystal. *Phys Rev E Stat Nonlin Soft Matter Phys* **64**(5 Pt 1):16.
30. Lansac Y, Glaser MA, Clark NA, & Lavrentovich OD (1999) Photocontrolled nanophase segregation in a liquid-crystal solvent. *Nature* **398**(6722):54–57.
31. Cheung DL, Clark SJ, & Wilson MR (2002) Parametrization and validation of a force field for liquid-crystal forming molecules. *Phys Rev E Stat Nonlin Soft Matter Phys* **65**(5 Pt 1):21.
32. Pelaez J & Wilson MR (2006) Atomistic simulations of a thermotropic biaxial liquid crystal. *Phys Rev Lett* **97**(26):26.
33. Pelaez J & Wilson M (2007) Molecular orientational and dipolar correlation in the liquid crystal mixture E7: A molecular dynamics simulation study at a fully atomistic level. *Physical Chemistry Chemical Physics* **9**(23):2968–2975.
34. Cheung DL, Clark SJ, & Wilson MR (2004) Calculation of flexoelectric coefficients for a nematic liquid crystal by atomistic simulation. *The Journal of Chemical Physics* **121**(18):9131–9139.
35. Wang J, Wolf RM, Caldwell JW, Kollman PA, & Case DA (2004) Development and testing of a general AMBER force field. *J Comput Chem* **25**(9):1157–1174.
36. Chami F, Wilson MR, & Oganesyan VS (2012) Molecular dynamics and EPR spectroscopic studies of 8CB liquid crystal. *Soft Matter* **8**(25):6823–6833.
37. McDonald AJ & Hanna S (2006) Atomistic simulation of a model liquid crystal. *J Chem Phys* **124**(16):164906.
38. Berardi R, Muccioli L, & Zannoni C (2004) Can nematic transitions be predicted by atomistic simulations? A computational study of the odd–even effect. *ChemPhysChem* **5**(1):104–111.
39. Cornell WD, et al. (1995) A second-generation force field for the simulation of proteins, nucleic acids, and organic molecules. *Journal of the American Chemical Society* **117**(19):5179–5197.
40. Haller I (1975) Thermodynamic and static properties of liquid crystals. *Progress in Solid State Chemistry 10*, Part **2**:103–118.
41. Tiberio G, Muccioli L, Berardi R, & Zannoni C (2009) Towards in silico liquid crystals. Realistic transition temperatures and physical properties for *n*-cyanobiphenyls via molecular dynamics simulations. *ChemPhysChem* **10**(1):125–136.

42. Zhang J, Su J, & Guo H (2011) An atomistic simulation for 4-cyano-4′-pentylbiphenyl and its homologue with a reoptimized force field. *The Journal of Physical Chemistry B* **115**(10):2214–2227.
43. Maerzke KA, Schultz NE, Ross RB, & Siepmann JI (2009) TraPPE-UA force field for acrylates and Monte Carlo simulations for their mixtures with alkanes and alcohols. *J Phys Chem B* **113**(18):6415–6425.
44. Yan F & Earl DJ (2011) An all-atom computer simulation study of the liquid crystalline phase behaviour of alkenic fluoroterphenyls. *Soft Matter* 7(21):10266–10273.
45. Yan F & Earl DJ (2012) Atomistic simulations of liquid crystal mixtures of alkoxy substituted phenylpyrimidines 2PhP and PhP14. *The Journal of Chemical Physics* **136**(12):124506.
46. Onsager L (1949) The effects of shape on the interaction of colloidal particles. *Annals of the New York Academy of Sciences* **51**(4):627–659.
47. Allen MP (1993) Simulations using hard particles. *Philosophical Transactions of the Royal Society of London A: Mathematical, Physical and Engineering Sciences* **344**(1672):323–337.
48. Allen MP, Evans GT, Frenkel D, & Mulder BM (1993) Hard convex body fluids. *Advances in Chemical Physics* **86**:1.
49. Stroobants A, Lekkerkerker HN, & Frenkel D (1986) Evidence for smectic order in a fluid of hard parallel spherocylinders. *Phys Rev Lett* **57**(12):1452–1455.
50. Stroobants A, Lekkerkerker HN, & Frenkel D (1987) Evidence for one-, two-, and three-dimensional order in a system of hard parallel spherocylinders. *Phys Rev A* **36**(6):2929–2945.
51. Veerman JA & Frenkel D (1991) Relative stability of columnar and crystalline phases in a system of parallel hard spherocylinders. *Phys Rev A* **43**(8):4334–4343.
52. McGrother SC, Williamson DC, & Jackson G (1996) A re-examination of the phase diagram of hard spherocylinders. *The Journal of Chemical Physics* **104**(17):6755–6771.
53. Bolhuis P & Frenkel D (1997) Tracing the phase boundaries of hard spherocylinders. *The Journal of Chemical Physics* **106**(2):666–687.
54. Niori T, Sekine T, Watanabe J, Furukawa T, & Takezoe H (1996) Distinct ferroelectric smectic liquid crystals consisting of banana-shaped achiral molecules. *J. Mater. Chem.* **6**(7):1231–1233.
55. Link DR, et al. (1997) Spontaneous formation of macroscopic chiral domains in a fluid smectic phase of achiral molecules. *Science* **278**(5345):1924–1927.
56. Lansac Y, Maiti PK, Clark NA, & Glaser MA (2003) Phase behavior of bent-core molecules. *Phys. Rev. E* **67**(1):011703.
57. Maiti PK, Lansac Y, Glaser MA, & Clark NA (2002) Induced anticlinic ordering and nanophase segregation of bow-shaped molecules in a smectic solvent. *Phys Rev Lett* **88**(6):25.
58. Pratibha R, Madhusudana NV, & Sadashiva BK (2000) An orientational transition of bent-core molecules in an anisotropic matrix. *Science* **288**(5474):2184–2187.
59. Lavrentovich OD (1994) The Physics of Liquid Crystals, Second Edition, by PG de Gennes and J Prost, International Series of Monographs on Physics No 83, published OUP (1993) ISBN 0 19852024 7. *Liquid Crystals Today* **4**(3):7–7.
60. Maiti PK, Lansac Y, Glaser MA, & Clark NA (2004) Entropy-stabilized smectic C phase in a system of zigzag-shaped molecules. *Phys Rev Lett* **92**(2):12.

61. Chandrasekhar S, Sadashiva BK, & Suresh KA (1977) Liquid crystals of disc-like molecules. *(Translated from English) Pramana - J Phys* **9**(5):471–480 (in English).
62. Mulder FM, et al. (2003) Dynamics of a triphenylene discotic molecule, HAT6, in the columnar and isotropic liquid phases. *Journal of the American Chemical Society* **125**(13):3860–3866.
63. Wu J, Pisula W, & Mullen K (2007) Graphenes as potential material for electronics. *Chem Rev* **107**(3):718–747.
64. Laschat S, et al. (2007) Diskotische Flüssigkristalle: Von der maßgeschneiderten Synthese zur Kunststoffelektronik. *Angewandte Chemie* **119**(26):4916–4973.
65. Sergeyev S, Pisula W, & Geerts YH (2007) Discotic liquid crystals: A new generation of organic semiconductors. *Chem Soc Rev* **36**(12):1902–1929.
66. Feng X, et al. (2009) Towards high charge-carrier mobilities by rational design of the shape and periphery of discotics. *Nat Mater* **8**(5):421–426.
67. Zhang R, et al. (2015) Columnar liquid crystals in cylindrical nanoconfinement. *ACS Nano* **9**(2):1759–1766.
68. Busselez R, et al. (2014) Discotic columnar liquid crystal studied in the bulk and nanoconfined states by molecular dynamics simulation. *The Journal of Chemical Physics* **141**(13):134902.
69. Maliniak A (1992) Molecular dynamics of discotics in the isotropic phase. *The Journal of Chemical Physics* **96**(3):2306–2317.
70. Cinacchi G, Colle R, & Tani A (2004) Atomistic molecular dynamics simulation of hexakis(pentyloxy)triphenylene: structure and translational dynamics of its columnar state. *The Journal of Physical Chemistry B* **108**(23):7969–7977.
71. Andrienko D, Marcon V, & Kremer K (2006) Atomistic simulation of structure and dynamics of columnar phases of hexabenzocoronene derivatives. *The Journal of Chemical Physics* **125**(12):124902.
72. Hu N, et al. (2014) An electric-field-responsive discotic liquid-crystalline hexa-peri-hexabenzocoronene/oligothiophene hybrid. *Adv Mater* **26**(13):2066–2071.
73. Beltran E, et al. (2015) Self-organization of star-shaped columnar liquid crystals with a coaxial nanophase segregation revealed by a combined experimental and simulation approach. *Chemical Communications* **51**(10):1811–1814.
74. Frank FC & Chandrasekhar S (1980) Evidence of a tilted columnar structure for mesomorphic phases of benzene-hexa-*n*-alkanoates. (Translated from English) *Journal de Physique* **41**(11):1285–1288 (in English).
75. Allinger NL (1977) Conformational analysis. 130. MM2. A hydrocarbon force field utilizing V1 and V2 torsional terms. *Journal of the American Chemical Society* **99**(25):8127–8134.
76. Sun H (1998) COMPASS: An ab initio force-field optimized for condensed-phase applications. Overview with details on alkane and benzene compounds. *The Journal of Physical Chemistry B* **102**(38):7338–7364.
77. Berendsen HJC, Postma JPM, van Gunsteren WF, DiNola A, & Haak JR (1984) Molecular dynamics with coupling to an external bath. *The Journal of Chemical Physics* **81**(8):3684–3690.
78. Bag S, et al. (2015) Molecular structure of the discotic liquid crystalline phase of hexa-peri-hexabenzocoronene/oligothiophene hybrid and their charge transport properties. *The Journal of Chemical Physics* **143**(14):144505.

79. Busselez R, et al. (2014) Discotic columnar liquid crystal studied in the bulk and nanoconfined states by molecular dynamics simulation. *J Chem Phys* **141**(13):4896052.

80. Maiti PK, Lansac Y, Glaser MA, & Clark NA (2002) Isodesmic self-assembly in lyotropic chromonic systems. *Liquid Crystals* **29**(5):619–626.

81. Kuriabova T, Betterton MD, & Glaser MA (2010) Linear aggregation and liquid-crystalline order: Comparison of Monte Carlo simulation and analytic theory. *Journal of Materials Chemistry* **20**(46):10366–10383.

82. Chami F & Wilson MR (2010) Molecular order in a chromonic liquid crystal: A molecular simulation study of the anionic azo dye Sunset Yellow. *Journal of the American Chemical Society* **132**(22):7794–7802.

83. Breneman CM & Wiberg KB (1990) Determining atom-centered monopoles from molecular electrostatic potentials. The need for high sampling density in formamide conformational analysis. *Journal of Computational Chemistry* **11**(3):361–373.

Index

l

m

n

o

p

q

r

s

t

u

v

w

X